Studies in Computational Intelligence

Volume 904

The series "Studies in Computational Intelligence" (SCI) publishes new developments and advances in the various areas of computational intelligence—quickly and with a high quality. The intent is to cover the theory, applications, and design methods of computational intelligence, as embedded in the fields of engineering, computer science, physics and life sciences, as well as the methodologies behind them. The series contains monographs, lecture notes and edited volumes in computational intelligence spanning the areas of neural networks, connectionist systems, genetic algorithms, evolutionary computation, artificial intelligence, cellular automata, self-organizing systems, soft computing, fuzzy systems, and hybrid intelligent systems. Of particular value to both the contributors and the readership are the short publication timeframe and the world-wide distribution, which enable both wide and rapid dissemination of research output.

The books of this series are submitted to indexing to Web of Science, EI-Compendex, DBLP, SCOPUS, Google Scholar and Springerlink.

More information about this series at http://www.springer.com/series/7092

Michael Zgurovsky · Victor Sineglazov ·
Elena Chumachenko

Artificial Intelligence Systems Based on Hybrid Neural Networks

Theory and Applications

Springer

Michael Zgurovsky
Kyiv, Ukraine

Victor Sineglazov
Kyiv, Ukraine

Elena Chumachenko
Kyiv, Ukraine

ISSN 1860-949X ISSN 1860-9503 (electronic)
Studies in Computational Intelligence
ISBN 978-3-030-48452-1 ISBN 978-3-030-48453-8 (eBook)
https://doi.org/10.1007/978-3-030-48453-8

This Springer imprint is published by the registered company Springer Nature Switzerland AG
The registered company address is: Gewerbestrasse 11, 6330 Cham, Switzerland

Preface

According to the analysis of many think tanks of the world, over the next five years, the total volume of artificial intelligence (AI) technologies market will increase by at least 4 times. Compound Annual Growth Rate (CAGR) in the forecast period will exceed 30%. It can be concluded that AI in the near future will become an integral part of the personal and professional people activities.

In particular, in the field of healthcare, AI is increasingly being used to identify regularities in the medical data of patients, which permits significantly to improve the process of establishing their diagnoses and to increase the effectiveness of treatment. In the field of cybersecurity, AI provides the protection against information threats that cannot be implemented with traditional network security tools.

In the current decade, steady growth of AI technologies is also observed in the aerospace and defense industries, energy, unmanned vehicles, robotics, ICT, banking and finance, the video game industry, retail, cognitive, neuromorphic, quantum and large-scale computing and in many other areas of human activities.

At the same time, the main limitations of the known methods and technologies of AI are due to the lack of their training effectiveness, the difficulty of setting up and adapting to the problem area in the context of incomplete and inaccurate initial information, the difficulty of accumulating expert knowledge and other features.

Thus, one of the actual problems in the development of modern AI systems is the development of integrated, hybrid systems based on deep learning. Unfortunately, today there is no methodology for hybrid neural networks (HNN) topologies design and hybrid technologies of their structural-parametric synthesis using deep learning.

The main factor that contributes to the development of such systems is the expansion of neural networks (NN) use for solving tasks of recognition, classification, optimization and nother problems. Using the other technologies for solving such class of problems leads to difficult symbolic calculations or to the big computing problems.

The monograph is devoted to an important direction in the development of artificial intelligence systems, based on the creation of a unified methodology for constructing hybrid neural networks (HNN) with the possibility of choosing models of artificial neurons. In order to increase the efficiency of solving the tasks, it is proposed to gradually increase the complexity of the structure of these models and use hybrid learning algorithms, including deep learning algorithms.

Unfortunately, today there is no methodology for constructing topologies of hybrid neural networks and hybrid technologies for their structural and parametric synthesis using deep learning. In recent years, there has been an active growth in the successful use of hybrid intelligent systems in various fields such as robotics, medical diagnostics, speech recognition, fault diagnosis of industrial equipment, monitoring (control of production processes) and applications in the field of finance. The main factor contributing to the development of intelligent hybrid systems is the increased use of neural networks for recognition, classification and optimization. The ability of neural networks to perform tasks that, if applied by other technologies, would be difficult to solve or those that are difficult to symbolic calculations, is now recognized, and they are often used as modules in intelligent hybrid systems.

The monograph is designed for specialists in the field of artificial intelligence, information technology, students and graduate students in these areas and can be useful to a wide range of readers involved in solving applied problems and interested in expanding the functionality of existing systems through the use of elements of artificial intelligence.

Kyiv, Ukraine

Michael Zgurovsky
Victor Sineglazov
Elena Chumachenko

Introduction

The development and implementation of artificial neural networks (ANNs) based on advanced technologies is one of the priority areas for the development of the branches of science and technology in all industrialized countries.

When solving applied problems in order to increase accuracy and reduce complexity of calculations, problems arise of finding the optimal network topology and, accordingly, structural (determining the number of hidden layers and neurons in them, interneuronal connections of individual neural networks) and parametric (setting weight coefficients) optimization.

The main limitations of the known methods and technologies currently used are due to the insufficient efficiency of solving the problem of training ANNs, tuning and adapting to a problem area, processing incomplete and inaccurate source information, interpreting data and accumulating expert knowledge, presenting information from various sources, and like that.

Thus, one of the leading trends in modern computer science has been the development of integrated, hybrid systems based on deep learning. Such systems consist of various elements (components), combined in the interests of achieving the goals. Integration and hybridization of various methods and technologies allows to solve complex problems that cannot be solved on the basis of individual methods or technologies.

Integration, as a fundamental property of a complex system, which is closely related to its integrity, involves mutual adaptation and joint evolution of its components and provides the emergence of new qualities that are not peculiar to its components separately.

The construction of hybrid neural networks (HNN), consisting of various types, each of which is trained according to a certain algorithm in layers, in many cases can significantly increase the efficiency of the ANN.

The study of the principles of hybridization of ANNs, fuzzy logic and genetic algorithms allows you to create new types of models that have a higher quality of recognition, prediction, decision support while reducing computational costs for training.

A significant contribution to the development of hybrid neural networks was made by such scientists as Jeffrey Hinton, Ian Le Coon, Ian Benggio, E. V. Bodyansky, V. V. Kruglov and V. V. Borisov.

Despite the certain achievements of these and other authors in the development of HNN, there is currently no single approach to their creation. The basis of existing methods is mainly based on the choice of a base NN (fuzzy NN of various topologies) proposed without sufficient justification, with the addition of other networks (the network is considered to be a HNN due to this), which ensures improved accuracy in solving the problem. For each HNN, built on this principle, an individual learning algorithm is developed.

An additional element that improves the quality of training is the use of the argument grouping method algorithm. In addition, the proposed approaches do not use such a powerful mechanism as deep learning.

Thus, the main problems of synthesis of HNN are currently

- lack of formal methods for choosing the type of NN, adequate class of tasks that need to be addressed;
- insufficient study of issues of automatic formation of the topology of the NN, which does not allow the creation of NN of high accuracy and minimal complexity (minimum structural and computational costs);
- insufficient validity of the choice of optimization methods in the learning process of the National Assembly, which leads to significant errors.

Given the above, we can conclude that it is necessary to create a unified methodology for constructing HNN with the possibility of choosing models of artificial neurons that make up HNN, gradually increasing the complexity of their structure and using hybrid learning algorithms, including deep learning, in order to improve the solution of the tasks.

Contents

Chapter 1
Classification and Analysis Topologies Known Artificial Neurons and Neural Networks

1.1 The Concept of Artificial Neural Network

Development and application of artificial neural networks (ANN) based on advanced technologies is one of the priority areas of science and technology in all industrialized countries.

A neural network (NN)—a distributed parallel processor consisting of elementary units of information processing, accumulating experimental knowledge and provide them for subsequent processing [1].

Although there are differences, some types of NN have several common features: NN basis of every make relatively simple in most cases—the same type of components (cells) that simulate the work of neurons in the brain. Then when neurons will have in mind the artificial neuron, neural cell is [1–6].

Artificial neurons—a unit of information processing in neural networks. The model neuron that underlies neural network shown in Fig. 1.1. In this model, there are four basic elements.

1. A set of synapses, or connections, each of which is characterized by its weight. In particular, the signal at the input of the synapse x_j j, associated with neuron k, multiplied by the weight w_{kj}.
2. The adder adds the input signals weighted by the respective neuron synapses.
3. Activation function $f(s)$ limits the amplitude of the output neuron. Usually the normalized amplitude range of the output neuron is in the range [0, 1] or [−1, 1]. Different types of activation functions are shown in Figs. 1.2, 1.3, 1.4 and 1.5.
4. The threshold element, labeled b_0. This value reflects an increase or decrease in the signal induced local field served the function activation.

The current state of the neuron is defined as the weighted sum of its inputs

$$s = \sum_{i=1}^{n} X_i w_i \tag{1.1}$$

M. Zgurovsky et al., *Artificial Intelligence Systems Based on Hybrid Neural Networks*, Studies in Computational Intelligence 904,
https://doi.org/10.1007/978-3-030-48453-8_1

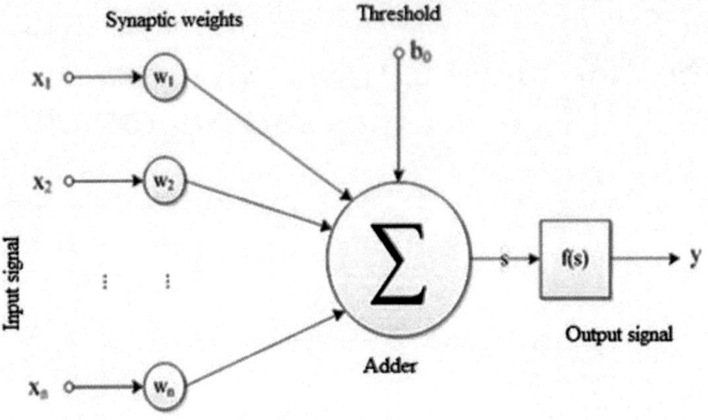

Fig. 1.1 Artificial neurons

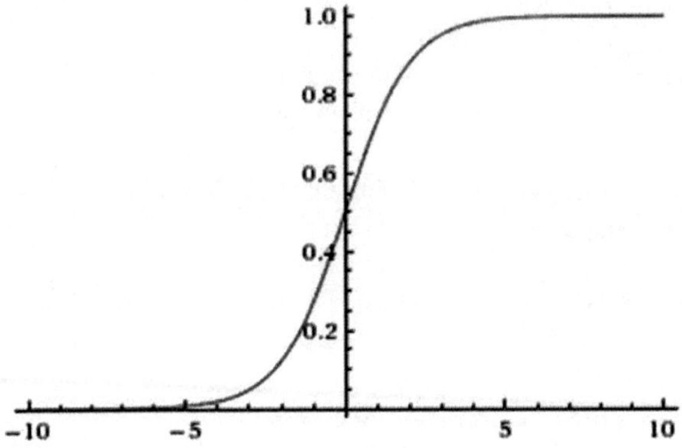

Fig. 1.2 Sigmoid function

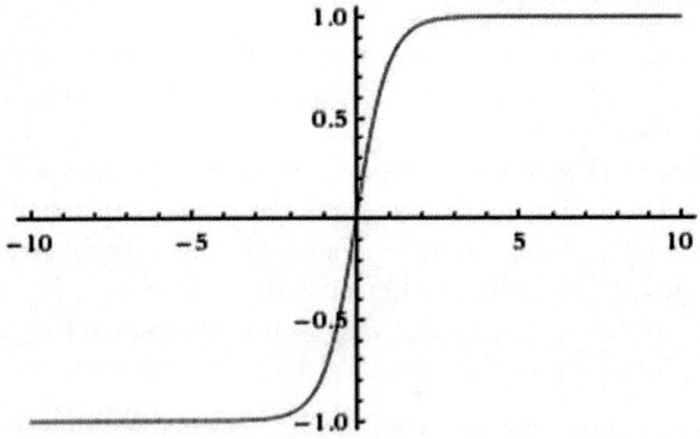

Fig. 1.3 Hyperbolic tangent

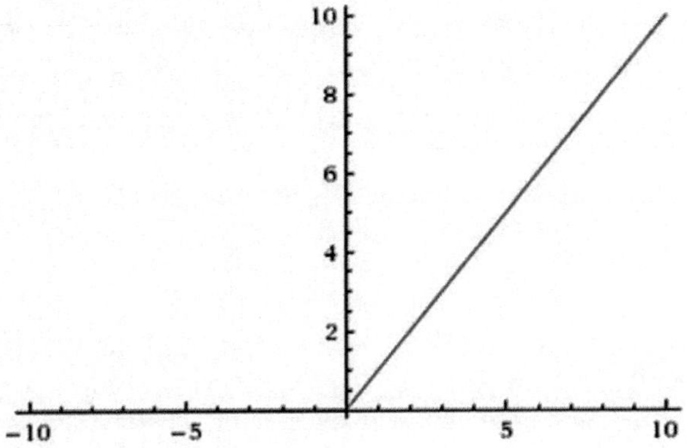

Fig. 1.4 Activation function ReLU

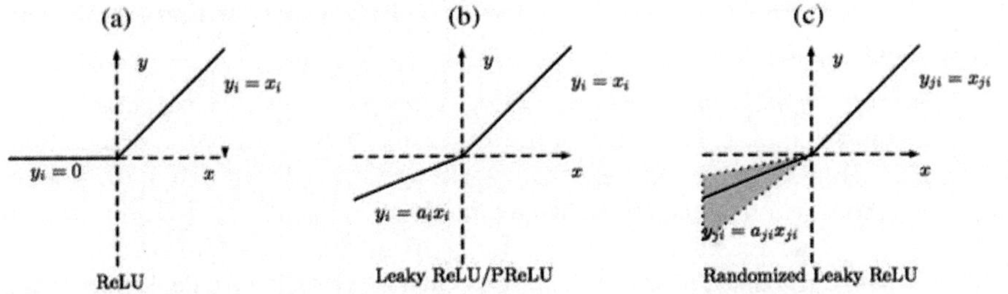

Fig. 1.5 Family modifications ReLU **a** a_i is fixed; **b** a_i based on data; **c** a_{ji} randomly generated from a given time interval under study and remains constant during testing

The output neuron is a function of its state

$$y = f(s). \tag{1.2}$$

Neurons network in some way arranged in layers. The input layer is used to input values of input variables. Each of the hidden and output neuron is connected to all elements of the previous layer. When the (use of) incoming network elements submitted values of input variables that consistently treated neurons intermediate and output layers. Each activation calculates its value using a weighted sum of outputs of the previous layer and subtracting from it the threshold. Then the value of the activation function becomes activated, and the result is the output neuron. After a run for the entire network, the output value output layer elements are taken out for the entire network as a whole.

Currently, there are many different topologies NN, each of which has different qualities. The next point will be given a general classification of existing ANN.

1.2 Classification Activation Function

As activation function can use the following functions: sigmoid function (Fig. 1.2), hyperbolic tangent (Fig. 1.3) and ReLU (Figs. 1.4 and 1.5).

Sigmoid function (sigmoid) expressed by the following formula:

$$\sigma(x) = 1/\left(1 + e^{x}\right).$$

This function takes inlet arbitrary real number, and the output provides real number between 0 and 1. In particular, large (in absolute value) negative numbers are converted to zero and positive large—in the unit. Historically sigmoid function found wide application because of its good yield interpreted as the activation level of the neuron from the absence of activation (0) to completely saturated activation (1).

Currently sigmoid function lost its former popularity and is used very rarely. This feature has two major drawbacks:

1. *Saturation sigmoid function leads to attenuation gradients.* It is undesirable characteristic sigmoid function is that the saturation of the functions of a hand (0 or 1), the gradient in these areas becomes close to zero. Recall that in the present back propagation (local) multiplied by the total gradient. So if the local gradient is very small, it actually clears generic gradient. As a result, almost no signal will pass through the neuron to its weights and recursively its data. Furthermore, it should be very careful when initializing weights sigmoid neurons to prevent saturation. For example, if the initial weight values are too high, most neurons pass into a state of saturation, resulting in poor network will learn.
2. *Exit sigmoid function is not centered relative to zero.* This property is undesirable because the neurons in these layers receive a value that is not centered relative to the ground, affecting the dynamics of gradient descent (gradient descent). If the value received by the neuron is always positive ($x > 0, f = \omega^T x + b$), then the process back propagation all gradient scales ω will be either positive or negative (depending on the gradient of the entire expression f). This can lead to unwanted zigzag dynamics. Note, however, that when these gradients are summed for package updates the final weights may have different signs, which partly eliminates the described disadvantage. Thus, no alignment is an inconvenience, but has less serious consequences, compared with the problem of saturation.

Hyperbolic tangent (hyperbolic tangent, tanh) accepts arbitrary entrance real number, and the output provides real number in the range from minus 1 to 1. Like the sigmoid function, hyperbolic tangent can be satisfied. However, unlike the sigmoid function, the output of this function centered on zero. Thus, in practice, always better to use hyperbolic tangent, not sigmoid function.

ReLU

In recent years gained much popularity activation function called "rectifier" (rectifier, similar to the half-wave rectifier in electrical engineering). Neurons this activation function called ReLU (rectified linear unit). ReLU has the following formula $f(x) = \max(0, x)$ and implements a simple threshold shift at zero (Fig. 1.4).

Consider the positive and negative sides ReLU.

Positive aspects

1. Calculating the hyperbolic tangent sigmoid function and performance requires intensive operations, such as exponentiation, while ReLU can be implemented using a simple matrix transformation activation threshold at zero. In addition, ReLU not prone to saturation.
2. Application ReLU significantly increases the rate of convergence of stochastic gradient descent (in some cases up to six times) compared to the sigmoid function and hyperbolic tangent. It is believed that this is due to the linear nature of saturation and lack of function.

Disadvantages

Unfortunately, ReLU not always sufficiently reliable and the learning process may go down ("dead"). For example, a large gradient, through ReLU, can lead to the upgrade of weights that this neuron is never activated. If this happens, then, starting from this moment gradient that passes through the neuron is always zero. Accordingly, this neuron is irreversibly incapacitated. For example, if a high speed training (learning rate), it may be that 40% ReLU "dead" (i.e., never activated). This problem is solved by selecting the proper speed training.

Currently, there is a whole family of different modifications ReLU. Next, consider their features (Fig. 1.5).

Leaky ReLU

ReLU the "source" (leaky ReLU, LReLU) is one of the attempts to solve the problem described above failure of conventional ReLU. Straight ReLU the interval $x < 0$ gives zero output, while LReLU this range is small negative (angular coefficient of about 0.01). That feature for LReLU has the form $f(x) = ax$ if $x < 0$ and $f(x) = x$ if $x \geq 0$, where a is the was constant.

Parametric ReLU

For parametric ReLU (parametric ReLU, PReLU) on a negative slope coefficient range is not set in advance, and based on the data. Back propagation process and updates to PReLU is simple and similar to the corresponding process for traditional ReLU.

Randomized ReLU

For randomized ReLU (randomized ReLU, RReLU) slope coefficient on the negative range during the study randomly generated from a given interval, and during testing remained constant. Within Kaggle-race National Data Science Bowl (NDSB) RReLU allowed to reduce the conversion due to the inherent element of randomness [7–9]. The above analytical functions have the following form:

$$f(x) = \frac{1}{1+e^{-x}}, f(x) = \tanh(x) = \frac{2}{1+e^{-2x}} - 1\, f(x) = arctg^{-1}(x)$$
$$f(x) = \begin{cases} 0\ for\ x < 0 \\ x\ for\ x \geq 0 \end{cases}, f(x) = \begin{cases} \alpha x\ for\ x < 0 \\ x\ \ for\ x \geq 0 \end{cases}, f(x) = \begin{cases} \alpha(e^x - 1)\ for\ x < 0 \\ x \qquad for\ x \geq 0 \end{cases}.$$

Artificial neurons constitute the neural network and that of their properties and connectivity options depend on the properties of neural networks from which they are formed.

1.3 Classification Neural Networks

Classification scheme neural networks [8, 9] Author modification shown in Fig. 1.6.
Neural networks can be divided into several grounds.

(1) The number of layers:

   (a) layered network (*layer perceptron*)
   (b) *Multilayer network.*

(2) By way of setting weights:

   (a) *fixed communications;*
   (b) *dynamic links.*

(3) The structure of relationships:

   (a) *fully connected (Hopfield network)*
   (b) *layered;*
   (c) *weakly (Kohonen map).*

(4) For a neural network model:

   (a) *direct distribution network—are used for solving problems of approximation, prediction and classification (based network perceptron).*
   (b) *counter distribution networks—associative memory used for data compression (Kohonen network-Hrosberha);*
   (c) recurrent network—used in associative *memory to solve problems of classification and filtering:*

      - cross-connections (*Boltzmann machine, the machine Cauchy*);
      - feedback (*Hopfield network, Hamming network, network Elmana network Jordan, RLMP, RTRN, Maxnet*);

   (d) network with time delay—used in pattern recognition tasks (*TDNN, TLFN*)

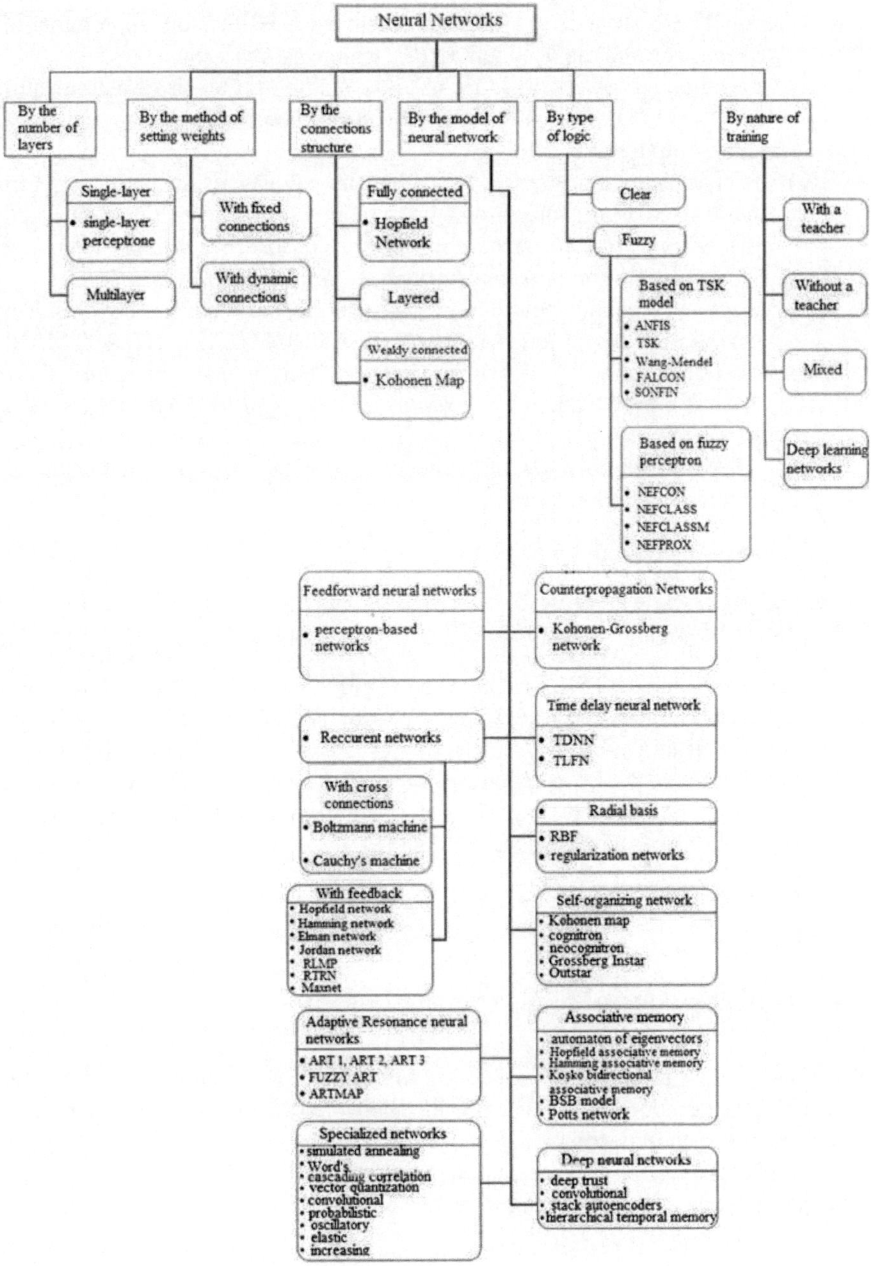

Fig. 1.6 Classification of neural networks

(e) radial basis—are used to solve problems of classification, approximation, forecasting and management (*RBN, regularization network*)
(f) self-organizing network—are used for solving problems of classification and clustering (Kohonen map, Kognitron, neocognitron, star Hrosberha input, output star);
(g) associative memory serves as a filter that adjusts results NN, comparing them to existing images and use compression information (*Automatic own vectors Hopfield associative memory, associative memory Hamming, bidirectional associative memory Koski, BSB-model network Potts*).
(h) adaptive resonance neural network—used for problem solving clustering and pattern recognition (ART1, ART2, ART3, FUZZY ART, ARTMAP);
(i) specialized networks (*neural networks that mimic annealing, Ward, cascade-correlation, vector quantization, probability, oscillating, flexible, growing*).
(j) a network of deep learning (deep trust, rolls, stackable avtoenkodery, hierarchical temporal memory).

5. The type of logic used:

   (a) Clear network;
   (b) fuzzy network:

      - based on the TSK model—used to solving problems of approximation, management and forecasting (ANFIS, TSK, Wang-Mendel, FALCON, SONFIN);
      - based on fuzzy perceptron—used to solving problems of classification, management, and forecasting approximation (NEFCLASS, NEFSLASSM, NEFCON, NEFPROX) [6].

6. The nature of learning.

1.4 Training Neural Networks

Neural network training is a process, during which the network settings are configured by modeling environment in which embedded network [8]. Product training is defined way to configure these settings.

General classification scheme training methods shown in Fig. 1.7.

Training includes the following sequence of events.

1. In the neural network receives stimuli (inputs) to the environment.
2. As a result of receipt of incentive changing values of free parameters of the neural network, such as weights.
3. After changing the internal structure of the network responds to stimuli otherwise.
4. The above sequence is called learning algorithm. Universal learning algorithm networks exist because different neural networks in terms of architecture and

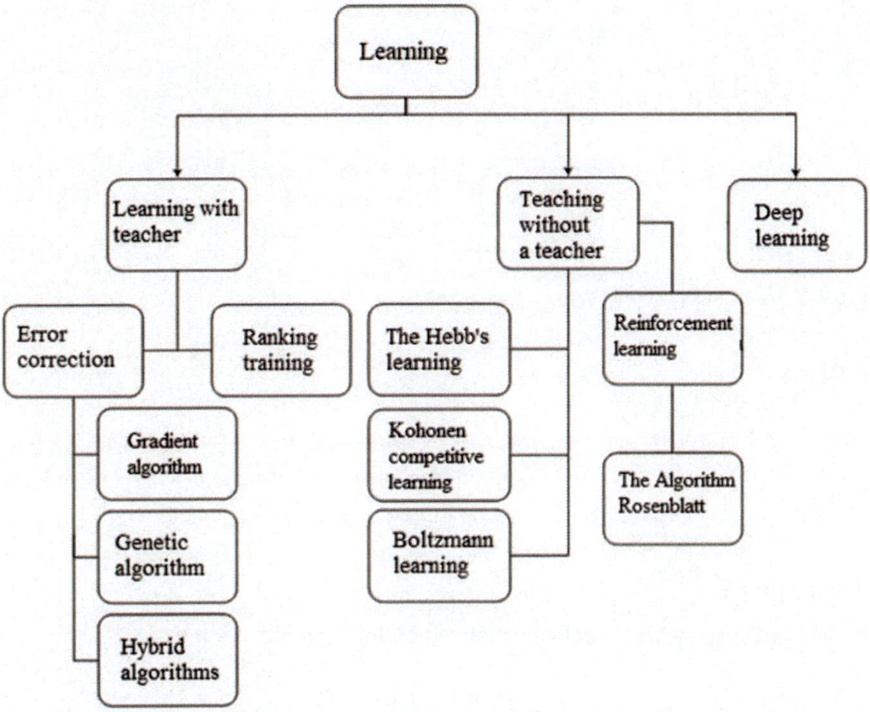

Fig. 1.7 Training neural networks

the tasks for which they are used. The same reason has generated set of learning algorithms that are different way to configure the synaptic weights, each of which has both advantages and disadvantages.

There are two paradigms of education: teacher training and learning without a teacher.

1.5 Synthesis Converting Unit

For further work is necessary to bring all data types into a single type, which will operate the network, i.e. to clear data. In this paper we propose conversion unit that converts fuzzy binary and linguistic variables in distinct variables [10]. Conversion unit consists of two elements and has the form shown in Fig. 1.8.

Fuzzy variables

The second element of the conversion unit is the defuzer, that is, the fuzzy number converter, to clear numbers.

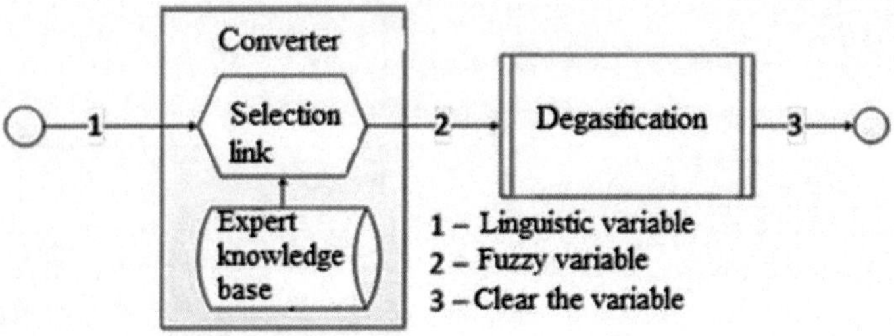

Fig. 1.8 Unit conversion scheme

Example Suppose we have a fuzzy variable x given as. Then the converted value: $X = \{1/0.5, 2/0.9, 3/0.4\}$.

$$x' = \frac{\sum_{j=1}^{k} a_j X_{ij}}{\sum_{j=1}^{k} a_j} = \frac{1 \times 0.6 + 2 \times 0.9 + 3 \times 0.5}{0.6 + 0.9 + 0.5} = \frac{3.9}{2} = 1.95$$

Binary data

To handle binary variable data normalized by the formula

$$X_{norm} = \frac{(x - x_{\min})(d_2 - d_1)}{x_{\max} - x_{\min}} + d_1$$

where x a binary value to be normalization; $x_{\max}$ is the maximum input; $x_{\min}$ is the minimum value of the input data.

After this procedure the data are reduced to the desired interval $[d_1, d_2]$.

Example Suppose we have a binary variable $X = 1$ (i.e., “true”), $x_{\max} = 1, x_{\min} = 0$. Lead X to the interval [11, 12]:

$$X_{norm} = \frac{(1 - 0)(50 - 25)}{1 - 0} + 25 = 50$$

Linguistic variables. The first element unit conversions (“transducer”) with the knowledge base in correspondence Terme linguistic variable fuzzy variable.

$$X_i = \begin{cases} A_1(x) \ \ if \ \ X_i = T_1 \\ \cdots \\ A_n(x) \ \ if \ \ X_i = T_n \end{cases}$$

where T_1 is the term linguistic variable; $A_1(x)$ is the fuzzy subset that matches the input terms (given by an expert).

The second block ("Defrag") converts the fuzzy variables into clear ones. This work uses the centroid conversion method.
Centroid Method—a clear value is found using one of the following formulas:

- for a continuously defined function

$$X = \frac{\int_b^a x\mu(x)dx}{\int_b^a \mu(x)dx} \tag{1.3}$$

where X is the obtained a clear value; x is the fuzzy variable; $\mu(x)$ is the function of belonging to a given fuzzy variable; $[a, b]$ is the scope for defining a fuzzy variable;

- for a discrete variable

$$X = \frac{\sum_{j=1}^{k} a_j X_{ij}}{\sum_{j=1}^{k} a_j}, \tag{1.4}$$

where X is the obtained a clear value; X_{ij} is the fuzzy plural element; a_j is the value of the membership function of the corresponding element; k is the number of discretes.

Example Let linguistic variables are given the following set of terms. $X =$ "*water temperature*" $X = \{$"Cold", "Warm", "Hot"$\}$. Each term in the base set on fuzzy variable

1. *Let the input unit conversion comes to "hot". Based on the available knowledge base of information experts put this value into line fuzzy variable A*

   ("*Hot*") $= \{60/0.5, 70/0.6, 80/0.9, 90/1, 100/1\}$.

2. *Use the defuzzifier to convert the resulting fuzzy variable to a clear one:*

$$X' = \frac{60 \times 0.5 + 70 \times 0.6 + 80 \times 0.9 + 90 \times 1 + 100 \times 1}{0.5 + 0.6 + 0.9 + 1.1} = \frac{334}{4} = 83.5.$$

1.6 Classification of Artificial Neurons and Their Properties

The type of activation function largely determines the properties of the artificial neuron and often the name of the neuron, the mathematical model of which is shown in Fig. 1.1, is determined by the type of activation function used, for example, a ReLU neuron, etc.

Function activation is not the only thing that can be changed in order to learn neuron approximating functions or properties. New discoveries in biology quickly showed that even the human brain is far from being a structure of neurons, and thus it makes sense in further studies. Not all topologies artificial neurons that were later invented based on the ideas inherited in nature. Much of it is based on mathematical and logical reasoning or experiments.

For further consideration of hybrid neural systems, their synthesis, consider constructing first and try to classify types of neurons of interest in this work.

Neurons can be roughly classified according to different criteria, we list only the most important ones. The type of computing: those that perform calculations and those that carry the signal intensifying or weakening. The position activation function, those containing activation function of synapses and after adder. The type of logic: clear and unclear. Consider the main ones.

1.6.1 N-Neuron

Artificial neurons called nonlinear Adalina presented in Fig. 1.9. N-adalina or N- a neuron has two inputs, consisting of a nonlinear preprocessor layer and a linear layer. The input signal enters the multiplication units that perform the conversion form

$$\hat{y}_1 = w_{l0}^{ij} + w_{l1}^{ij} x_i + w_{l2}^{ij} x_i^2 + w_{l3}^{ij} x_i x_j + w_{l4}^{ij} x_j^2 + w_{l5}^{ij} x_j .$$

This structure can influence the output neuron field is not only the absolute value of the inputs, but their combinations $(x_i x_j)$ or square (x_i^2). Using the multiplication between input parameters $(x_i x_j)$ actually realizes logical multiplication (conjunction) as large in magnitude to the positive input signal synapse, coming after the element will only multiply if both parameters have positive values for large absolute size. Synapses receiving input box input parameters effectively (x_i^2), responsive to the absolute value of the parameter regardless of the sign of the signal.

Nonlinear Adalina can be trained using any of teacher learning algorithms. For example, the method of least squares to form.

$$w_l^{ij}(N) = \left(\sum_{k=1}^{N} \varphi^{ij}(x(k)) (\varphi^{ij}(x)(k)) \right)^{-1} \sum_{k=1}^{N} \varphi^{ij}(x(k)) \, y_1(k)$$

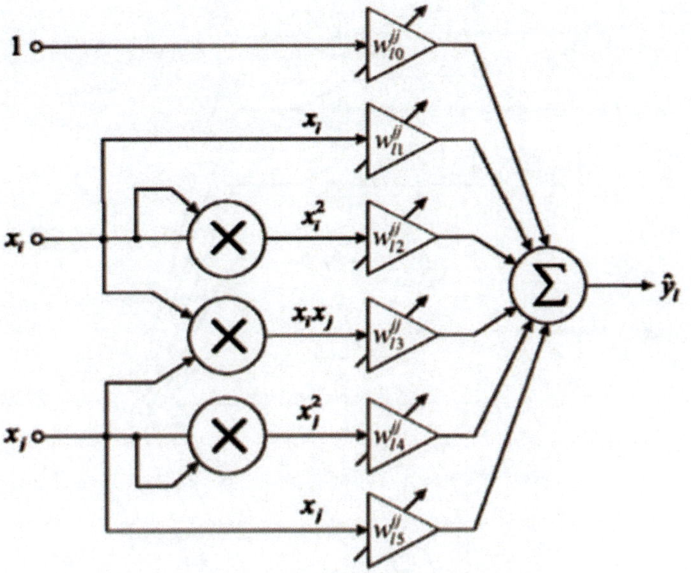

Fig. 1.9 Structure nonlinear adaliny

where $w_l^{ij}(N) = w_{l0}^{ij}(N), w_{l2}^{ij}(N), w_{l3}^{ij}(N), w_{l4}^{ij}(N), w_{l5}^{ij}(N)^T$

$$\varphi^{ij}(x(k)) = \left(1, x_i(k), x_i^2(k), x_i(k)\,x_j(k), x_j^2(k), x_j(k)\right)^T, \; k = 1, 2, \ldots, N$$

k is the number of observations in the study sample or discrete time index.

In the sequential processing can be used each line learning algorithms, such as popular for optimal performance algorithm Kachmazha Uidrou-Hoff [13].

$$w_l^{ij}(k) = w_l^{ij}(k-1) + \frac{y_l(k) - \left(w_l^{ij}(k-1)\right)^T \varphi^{ij}(x(k))}{\varphi^{ij}\,(x(k))^2} = \varphi^{ij}(x(k))$$

N-neuron performs a quadratic approximation of the two input signals. It is necessary to determine during six training synaptic weights.

1.6.2 *Q*-Neuron

Q-neuron performs a quadratic approximation in general, when the number of inputs $N > 2$. Where $n = 2$ N-adalina does not structurally different from Q-neuron [14]. Q-neuron structure shown in Fig. 1.10.

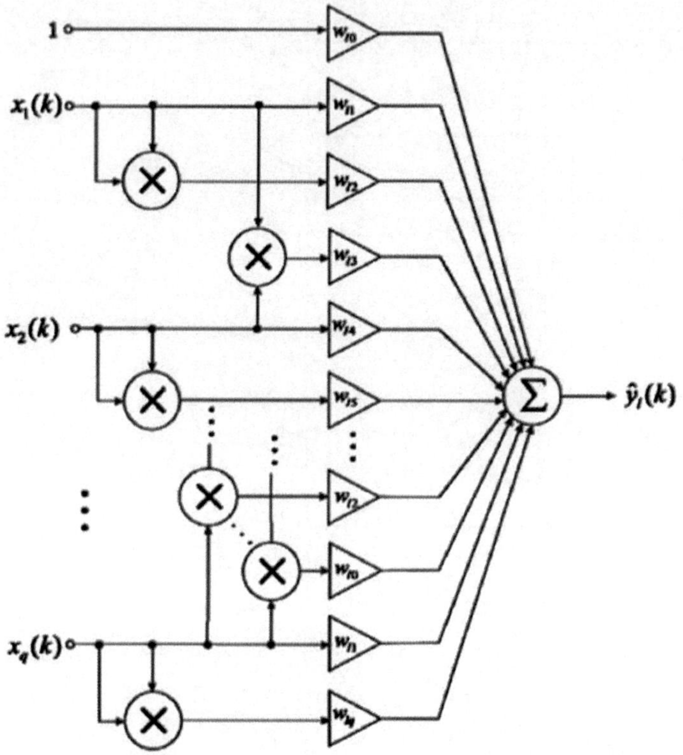

Fig. 1.10 Structure of *Q*-neuron

Defining the matrix of synaptic weights can be accomplished by minimizing the learning criterion gradient procedure form $W_l(k)$

$$W_l(k) = W_l(k-1) + \eta(k)\, e_l(k)\vec{x}(k)\vec{x}^T(k)$$

where $\eta(k)$ speed setting learning, which can be determined by impirychnym or procedure proposed in [14], and *e*-difference between the reference signal and actual output neuron

$$E_l(k) = \frac{1}{2}(y_1(k)) - \hat{y}_1(k)^2 = \frac{1}{2}(e)^2$$

The output neuron is determined by the transformation of the form

$$\hat{y}_l = w_{l0}(k-1) + \sum_{i=1}^{q} w_{li}(k-1)\, x_i(k) + \sum_{i_1=1}^{n}\sum_{i_2=1}^{n} w_{li_1i_2}(k-1)\, x_{i_1}(k)\, x_{i_2}(k)$$

or in matrix form

$$\hat{y}_1(k) = \vec{x}^T(k)\, W_l(k-1)\vec{x}(k),$$

where the matrix looks like scales

$$W_l(k-1) \;=\; \begin{Bmatrix} w_{l0}(k-1) & \cdots & 0.5b_l^T(k-1) \\ \vdots & & \vdots \\ 0.5b_l(k-1) & \cdots & C_1(k-1) \end{Bmatrix} \text{—block } (q+1)\times(q+1)$$

matrix

$$b_l(k-1) = \left(w_{l1}(k-1), w_{l2}(k-1), \ldots, w_{lq}(k-1)\right)^T - (q\times 1)\text{—vector}$$
$$C_l(k-1) = \left\{w_{l_{i_1 i_2}}(k-1)\right\} - (q\times q)matrix$$
$$x(k) = (x1(k), x2(k), \ldots xq(k))^T - (q\times 1)\text{—}vector,\ \vec{x}^T(k) = \left(1, x^T(k)\right).$$

1.6.3 R-Neuron

Radial basis function network with two inputs is the basis for *R*-neuron [14] its representation in Fig. 1.11. *R*-neuron contains nine ringing activation functions $\varphi(x, c, \sigma)$, whose arguments are the center c and the width parameter σ. As activation function can be used, for example, Gaussians

$$\varphi_h^{ij}(x, c, \sigma) = \exp\left(\frac{\left(x_i - c_h^i\right)^2 + \left(x_j - c_h^j\right)^2}{2\sigma^2}\right) = \exp\left(\frac{\left|x^{ij} - c_h^{ij}\right|^2}{2\sigma^2}\right),$$

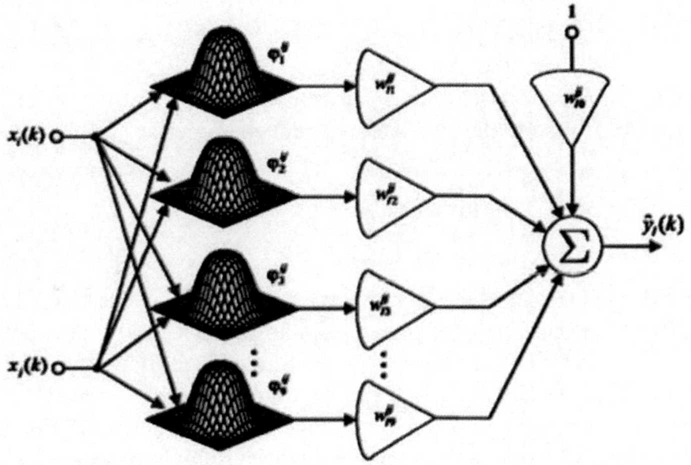

Fig. 1.11 Structure *R*-neuron

Parameters σ and c in this case, they can be selected so that the activation functions form a grid covering all possible values.

After passing through the activation functions, a weighting is applied to the signal. The peculiarity of this neuron is a function that is activated before the adder, unlike a typical neuron. Since the radial basis function network is a universal aproximator, neuron, repeating its struturu also has this feature. There are many approaches to initialization parameters of activation functions. In one case, the parameters must be selected so that the membership function created a grid in parameter space, with covers all possible values of input parameters without "holes".

The learning process of the neuron then consist in determining the vector of weights w_l^{ij} for the training set that contains a reference output. The criterion of training can be used

$$E_l^N = \sum_{k=1}^{N}\left(y(k) - \hat{y}_l(k)\right)^2 = \sum_{k=1}^{N} e_l^2(k) = \sum_{k=1}^{N}(y(k) - \left(w_l^{ij}\right)^T \varphi^{ij}(x(k)))^2,$$

using a standard method of least squares is easy to get the desired assessment in a

$$w_l^{ij} = \left(\sum_{k=1}^{N} \varphi^{ij}(x(k))\,\varphi^{ij}(x(k))^T\right)^{+} \sum_{k=1}^{N} \varphi^{ij}(x(k))(y(k)).$$

For finding scales can be used adaptive algorithm having filtering pяroperties and making sure [15]

$$\begin{cases} w_l^{ij}(k) = w_l^{ij}(k-1) + \eta_w(k)(y(k) - \left(w_l^{ij}(k-1)\right)^T \varphi^{ij}(x(k)))\,\varphi^{ij}(x(k)) \\ \quad = w_l^{ij}(k-1) + \eta_w(k)\,e_l(k)\,\varphi^{ij}(x(k)) \\ \eta_w^{-1}(k) = r_w(k) = ar_w(k-1) + \left\|\varphi^{ij}(x(k))\right\|^2, 0 \le a \le 1 \end{cases}$$

As an activation function φ_h^{ij} of the R-neuron, it was suggested to use the Epaknikov kernels because of their additional learning-related properties [14]

$$\varphi_h^{ij}\left(x^{ij}, c_h^{ij}, \sum_h^{ij}\right) = 1 - \left\|x^{ij} - c_h^{ij}\right\|^2_{(\sum_h^{ij})^{-1}}.$$

When the receptor matrix is positively defined, the nucleus has a bell-shaped shape. The advantage of using nuclear Epaknikova as activation function is that the first derivative contains all parameters in linear form, which allows to establish centers as activation function and its width during training. Study parameters activation function lets not use any heuristics to initialize the center functions and parameters width and generally increases the accuracy of the system. Introducing consideration $(p+1) \times 1$ vector activation functions

$$\varphi^{ij}(x(k)) = \left(1, \varphi_1^{ij}\left(x^{ij}(k), c_1^{ij} \sum\nolimits_1^{ij}\right), \ldots, \varphi_p^{ij}\left(x^{ij}(k), c_p^{ij} \sum\nolimits_p^{ij}\right)\right)^T,$$

criteria and training

$$E_l(k) = e_l^2(k) = \left(y(k) - \hat{y}_l(k)\right)^2 = \left(y(k) - w_{l0}^{ij} - \sum_{h=1}^{9} w_{lh}^{ij} \varphi_h^{ij}\left(x^{ij}(k)\right)\right)^2$$

The procedure to minimize not only to adjust synaptic weights but also centers and width function is as follows

$$\begin{cases} w_l^{ij}(k) = w_l^{ij}(k-1) + \eta_w(k) e_l(k) \varphi^{ij}(x(k)) \\ \eta_w^{-1}(k) = r_w(k) = a r_w(k-1) + \left\|\varphi^{ij}(x(k))\right\|^2 \\ c_h^{ij}(k) = c_h^{ij}(k-1) + \eta_c(k) e_l(k) w_l^{ij}(k) \left(\sum\nolimits_h^{ij}(k-1)\right)^{-1} \left(x^{ij}(k) - c_h^{ij}(k-1)\right) \\ \quad = c_h^{ij}(k-1) + n_c(k) e_l(k) g_h(k) \\ \eta_c^{-1}(k) = r_c(k) = a r_c(k-1) + \|g_h(k)\|^2 \\ \left(\sum\nolimits_h^{ij}(k)\right)^{-1} = \sum\nolimits_h^{ij}(k-1)^{-1} - \eta_\Sigma(k) e_l(k) w_l^{ij}(k) \left(x^{ij}(k) - c_h^{ij}(k-1)\right)^T \\ \quad = \left(\sum\nolimits_h^{ij}(k-1)\right)^{-1} - \eta_\Sigma(k) e_l(k) G_h(k) \\ \eta_\Sigma^{-1}(k) = \Gamma_\Sigma(k) = a\Gamma_\Sigma(k-1) + Tr G_h(k) G_h^T(k) \end{cases}$$

1.6.4 W-Neuron

Widespread theory of wavelet transformation [16, 17]. Wavelet features allow you to analyze the local features of the input signal. A neuron containing a wavelet function as an activation function is called an adaptive wavelength (W-neuron) [18]. Its structure is somewhat similar to the structure of the R-neuron discussed earlier and shown in Fig. 1.12.

Neuron has two inputs and receives input signal in the form of two-dimensional vector

$$x^{ij}(k) = \left(x_i(k), x_j(k)\right)^T.$$

The first layer as the activation function is used wavelet view

$$\varphi_{lh}^{ij}(\vec{x}(k)) = \varphi_{lh}^{ij}\left(\vec{x}(k) - c_{lh}^{ij}(k)\right)^T \left(Q_{lh}^{ij}(k)\right)^{-1} \left(\vec{x}(k) - c_{lh}^{ij}(k)\right), h = 1, 2, \ldots, p,$$

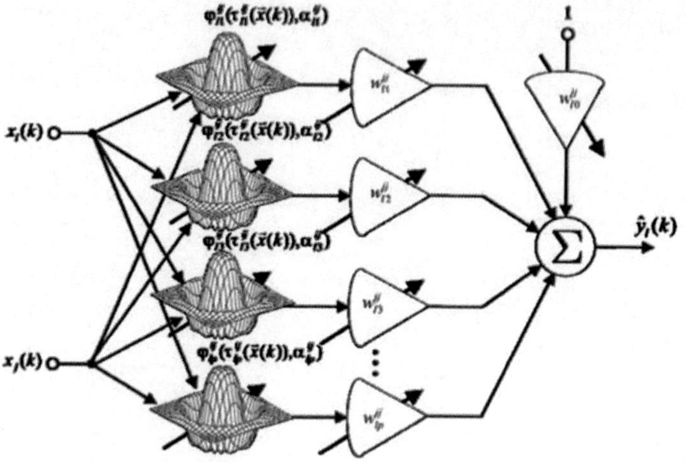

Fig. 1.12 Structure W-neuron

whose width instead of scalar parameters used matrix $\left(Q_{lh}^{ij}(k)\right)^{-1}$ which determines the shape and orientation relative to the axes of the activation function space.

As activation function can use the popular feature "Mexican Hat" [19], that the network will look like

$$\varphi_{lh}^{ij} = (\vec{x}(k)) = \left(1 - \alpha_{lh}^{ij}(k)\left(\tau_{lh}^{ij}(x^{ij}(k))\right)^{2}\right)\exp\left(-(\tau_{lh}^{ij}(x^{ij}(k)))^{2}/2\right),$$

where

$$\tau_{lh}^{ij}(x^{ij}(k)) = \left(\left(x^{ij}(k) - c_{lh}^{ij}(k)\right)^{T}\left(Q_{lh}^{ij}(k)\right)^{-1}\left(x^{ij}(k) - c_{lh}^{ij}(k)\right)\right),$$

a_{h}^{ij}—an option to adjust the shape of activation functions $0 < a_{lh}^{ij} < 1$ With the value parmetra $a_{lh}^{ij} = 0$ get Gaussian activation function, while $a_{lh}^{ij} = 1$ function "Mexican Hat".

After passing the activation functions, linear weights are applied to the input signal. The final output of the adaptive wavelon is written as

$$\hat{y}_l(k) = w_{l0}^{ij} + \sum_{h=1}^{p} w_{lh}^{ij}\, \varphi_l^{ij}\left(\left(x^{ij}(k) - c_{lh}^{ij}(k)\right)\left(Q_{lh}^{ij}(k)\right)^{-1}\left(x^{ij}(k) - c_{lh}^{ij}(k)\right)\right)$$
$$= \left(w_l^{ij}\right)^{T}\varphi_l^{ij}(x^{ij}(k)).$$

1.6.5 Neo-Fuzzy-Neuron

Multi-layer system consisting of neurons clearly has known significant drawback: it works for the observer as a "black box." After training the system can not establish what rules she learned and interpret the results in a form understandable to people. In view of this interesting and promising direction is the development of fuzzy systems, including distinct networks that are used as nodes fuzzy neurons or their combination. In this direction, working Japanese researcher T. Yamakava. He developed neuron structure that implements fuzzy logic system. Networks formed from these neurons is called neo-fuzzy networks.

Figure 1.13 is the framework of neo-fuzzy-neuron [12, 20, 21] with n inputs at the synaptic weights. Neuron consists of synapses in n h wavelet functions each. When applying for admission neo-fuzzy-neuron vector signal $x(k) = (x_1(k), x_2(k), \ldots, x_n(k))^T$ (where $k = 0, 1, 2, \ldots$ is the number of observations in the study sample or discrete time) appears at the output signal form

$$y = \sum_{i=1}^{n} (f_i(x(k)) = \sum_{i=1}^{n} \sum_{j=1}^{h_i} w_{ji}(k)\, \varphi_{ji}(x_i(k))$$

The resulting signal is determined by the values of synaptic weight coefficients and values used membership functions.

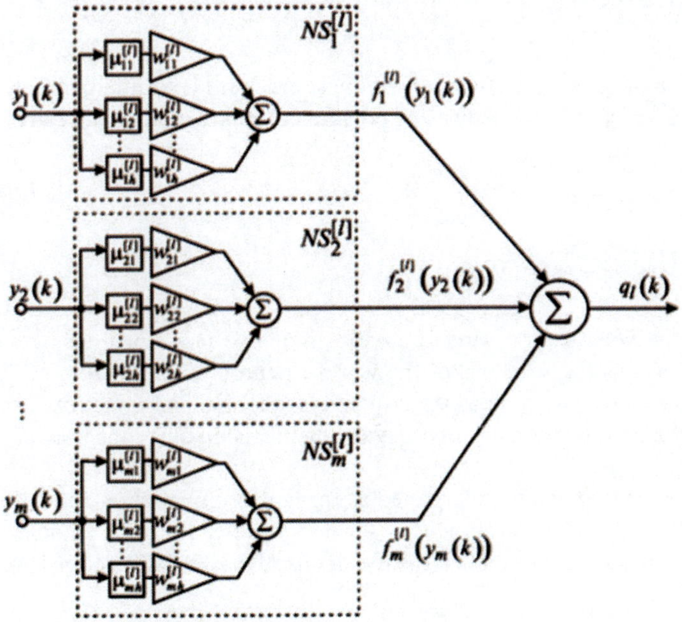

Fig. 1.13 Structure of neo-fuzzy-neuron

Given system is comparable with a typical rule fuzzy inference system Takagi-Sugeno first order. If the first input signal is x and the second input signal is equal y, then $z = ax + by + c$. In the zero-order model $a = 0$ and $b = 0$ therefore conclude determined constant c. Nonlinear neo-phase neuron synapses are similar to the structure of the W-neuron and actually realize fuzzy zero-order Takagi-Sugeno IF x_i are equal w_{ji} then EXIT w_{ji} where x_{ji} fuzzy set membership function which μ_{ji}, w_{ji} is synaptic weight, which serves as a conclusion (constant). Total exits nonlinear synapses—total output neuron. Compared neo-fuzzy neuron with a typical neuron, it is easy Compared neo-fuzzy neuron with a typical neuron, it is easy to see that the main vidminnisttyu was a significant complication of the structure synapses, allowing not only better interpret the results, but approximated more advanced features. The output of a typical neuron synapse depends linearly on the input signal, the lower the signal to get the input of those less.

As coefficient synapse weight affects the overall output neuron and vice versa. Neo-fuzzy neuron through fuzzy inference system may respond differently to the input signal synapse, depending on the range to which the value of the input signal.

In [22] developed the idea of a fuzzy neuron and instead of the traditional to neo-fuzzy neuron, triangular membership functions used cubic membership function

$$\begin{cases} 0.25\left(2 + 3\dfrac{2x - x_i - x_{i-1}}{x_i - x_{i-1}} - \left(\dfrac{2x - x_i - x_{i-1}}{x_i - x_{i-1}}\right)^3\right), x \notin \left[x_{i-1}, x_{i-1}\right] \\ 0.25\left(2 + 3\dfrac{2x - x_{i+1} - x_i}{x_{i+1} - x_i} - \left(\dfrac{2x - x_{i+1} - x_i}{x_{i+1} - x_i}\right)^3\right), x \notin \left[x_i, x_{i+1}\right] \end{cases}$$

The use of such cubic functions in most cases leads to improved accuracy and simplification of the system by using polynomial approximation instead of piecewise linear [22].

1.6.6 Wavelet Neuron

In [23], a wavelet neuron, structurally based on neo-phase neurons, was proposed. Fig. 1.14 shows the structure of the wavelet neuron. The structure of the wavelet neuron is exactly the same as the structure of the neo-phase neuron, except in the first layer that instead of cubic or triangular membership functions, wavelet activation functions are used.

Wavelet neuron can be interpreted as fuzzy neuron-variation neo-fuzzy neuron. It should be noted that each neuron synapse wavelet structure coincides with the W-neuron. It was suggested that the use of adaptive wavelet function form

$$\varphi_{ji}(x(k)) = \left(1 - a_{ji}(k)\,\tau_{ji}^2\right)\exp\left(-\tau_{ji}^2(k)\big/2\right),$$

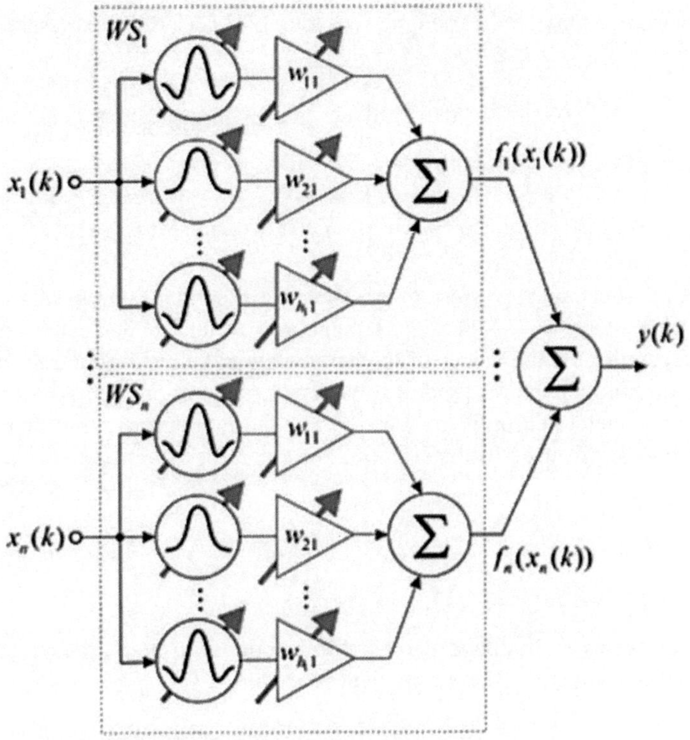

Fig. 1.14 Structure wavelet neuron

entitled "Mexican hat". Here $\tau_{ji}(k) = \left(x_i(k) - c_{ji}(k)\right)\sigma_{ji}^{-1}(k), c_{ji}(k), \sigma_{ji}(k)$ are parameters determining the center (offset) and wide functions, and a_{ji} shape parameter function. Instead of setting the width σ_{ji}^{-1} can also be used $(n \times n)$ transformation matrix Q_{ji}^{-1}.

Wavelets and families POLYWOG RASP can be obtained by using the tool based on analytic wavelet generator proposed in [24]. Most wavelets can be divided into even and odd. Depending on the type of signal being processed, often need to choose even or odd wavelet generator. Doubles generator can be described by the expression

$$y_{even}(x(k)) = \sum_{i=1}^{n} a_l \cos\cos(ix(k)) = a^T \varphi_{even}(x(k))$$

$$y_{odd}(x(k)) = \sum_{i=1}^{n} b_i \sin\sin(ix(k)) = b^T \varphi_{odd}(x(k))$$

where $\varphi_{even}(x(k)) = \cos x(k), \cos 2x(k), \ldots \cos nx(k)$; a_i are spectral coefficients wavelet decomposition odd.

An alternative may be the triangular wavelet [24] Described expression

$$\varphi(x,[a,b,c,d,e,h_1]) = \begin{cases} 0,\ if x < a \,\&\, x > e \\ -h_1 \frac{(x-a)}{(b-1)},\ if a \le x \le b \\ (h_1+1)\frac{x-b}{c-b} - h_1,\ if b \le x \le c \\ -\frac{(h_2+1)(x-c)}{d-c} + 1,\ if c \le x \le d \\ h_2 \frac{(x-d)}{(e-d)} - h_2,\ if d \le x \le e \end{cases}$$

In order to meet the conditions of existence wavelet must comply with conditions $\int_{-\infty}^{\infty} \varphi(x)dx = 0,\ h_2 = \frac{(b-d)+h_2(c-a)}{(c-e)}$. Triangular wavelet function can be either even or odd, depending on the values of the parameters to be set during training, so this membership function can be used to solve a wide class of problems.

Training wavelet neuron is the teacher. The output neuron and reference output $y(k)$ training sample used for determining the error, for example quadratic

$$E(k) = \frac{1}{2}\left(y(k) - \hat{y}(k)^2\right) = \frac{1}{2}e(k)^2 = \frac{1}{2}\left(y(k) - \sum_{i=1}^{n}\sum_{j=1}^{h} w_{ji}(k)\,\mu_{ji}(x_i(k))\right)^2$$

where μ can be any of the above membership functions. To clarify the weights matrix can be used traditional first order gradient procedure

$$\begin{aligned} w_{ji}(k+1) &= w_{ji}(k) + \eta e(k+1)\,\mu_{ji}(x_i(k+1)) \\ &= w_{ji}(k) + \eta(y(k+1) - \sum\nolimits_{i=1}^{n}\sum\nolimits_{j=1}^{h} w_{ji}(k)\,\mu_{ji}(x_i(k+1))\,\mu_{ji}(x_i(k+1)) \end{aligned}$$

where η parameter that determines the speed of convergence of the algorithm. To increase the rate of convergence [25] and training, second-order procedures, such as Widrow-Hoff or Levenberg–Marquardt, can be used. For this case, the Wydrow–Hoff algorithm [26–28] can be written in the form

$$w(k+1) = w(k) + \frac{y(k+1) - w^T(k)\mu(x(k+1))}{\|\mu(x(k+1))\|^2}\mu(x(k+1))$$

Its exponentially weighted modification

$$\begin{cases} w(k+1) = w(k) + r^{-1}(k+1)\big(y(k+1) - w^T(k)\mu(x(k+1)\mu(x(k+1) \\ r(k+1) = ar(k) + \|\mu(x(k+1)\|^2, 0 \le a \le 1 \end{cases}$$

1.6.7 *Wavelet Fuzzy-Neuron Type-2*

In [10], a complicated topology based on a wavelet neuron was proposed, the type-2 phase-wavelet wavelet (denoted by WN, the structure of which is shown in Fig. 1.15

Here *WN* has parameters corresponding to lower limits membership function, and—the top. Block reductions in some way unites otrimani of *WN* signals output neuron $WN\hat{y}(k)$.

Bodyanskiy was asked the following rule

$$\hat{y}(k) = c(k)\overline{y}(k) + (1 - c(k))\overline{y}(k)$$

here $c(k)$ is a configurable parameter. Criterion learning unit reduction model can be written as

$$E(k) = \sum_{k=0}^{N} \frac{1}{2}\hat{e}^2(k) = \sum_{k=0}^{N} \frac{1}{2}\left(d(k) - c(k)\overline{y}(k) - (1 - c(k)\overline{y}(k)\right)^2$$

In [10] was proposed optimal algorithm parameter setting unit reduction.

$$\begin{cases} y(k+1) = y(k) + (\overline{y}(k+1) - \overline{y}\ ((k+1))^2 \\ c(k+1) = c(k)\frac{y(k)}{y(k+1)} + \frac{(d(k+1) - \overline{y}(k+1)\overline{y}(k+1) - \overline{y}(k+1)}{y(k+1)} \end{cases}$$

1.6.8 *Multivariate Neo-Fuzzy-Neuron*

The wavelet-fuzzy-neuron often fails to provide the required accuracy in the prediction problems of non-stationary series with high uncertainty. The solution may be to complicate the model, namely the use of *n* phasezines in parallel [29]. Such a system

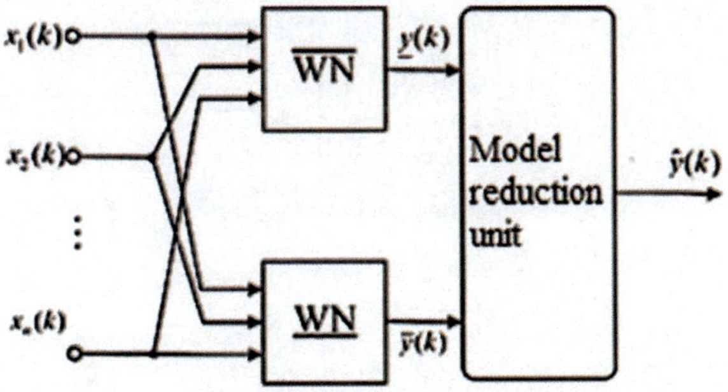

Fig. 1.15 Wavelet fuzzy-neuron type-2

can define n times more rules, however, it has a significant drawback due to the presence of linear synapses with a similar membership function but different weight. To avoid this drawback, the x_n architecture of the multidimensional neo-phase-neuron presented in Fig. 1.16.

The nodes of the neuron is enhanced nonlinear synapses containing h functions and hn synaptic weights that can be set during training. Introducing the consideration of a new matrix of synaptic weights we come to the learning algorithm equivalent to the neo-learning algorithm fuzzy neuron.

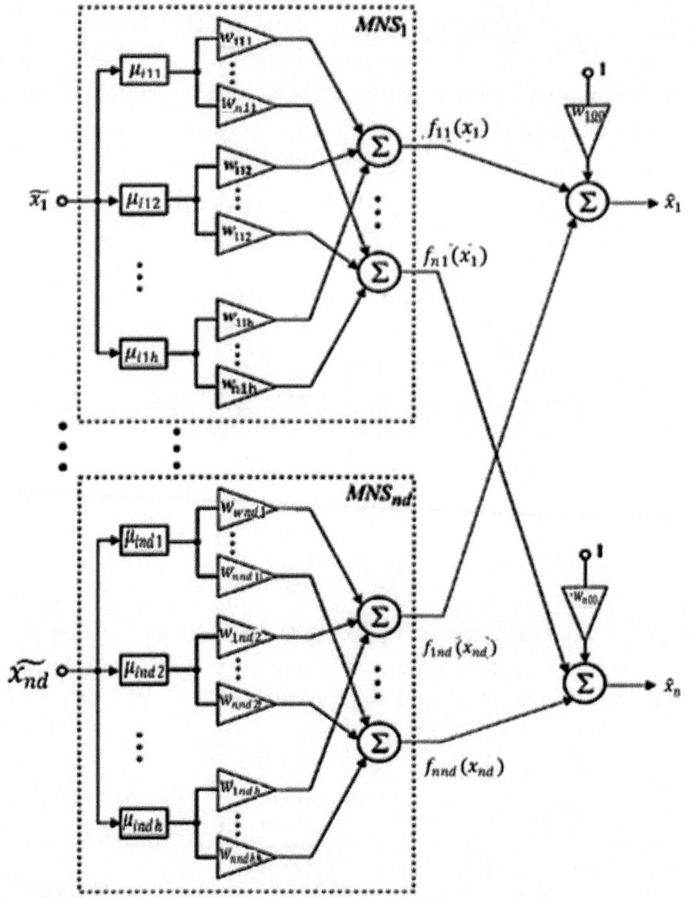

Fig. 1.16 Multidimensional neo-phase-neuron

1.6.9 Advanced Neo-Fuzzy-Neuron

Cynaps Neo-phase output neuron implements fuzzy Takagi-Sugeno first order and only the simplest fuzzy model VanhaMendelya. In [30] it was proposed to expand the properties approximating phase of neo-neurons and develop advanced neo-phase neuron (ANPN) for which advanced nonlinear synapse described by

$$\bar{y}(k) = w^T(k-1)\mu(x(k))$$

To adjust the neo-fuzzy neuron parameters, the authors used a gradient procedure that minimizes the learning criterion

$$\varphi_{ji}(x_i) = \mu_{ji}(x_i)\left(w_{ji}^0 + w_{ji}^1 x_i + w_{ji}^2 x_i^2 + \cdots + w_{ji}^p x_i^p\right)$$

where is the ith component of the n-dimensional input signal vector x_i, w_{ji} is the jth synaptic weight of the i th nonlinear synapse, $\mu_{ji}(x_1)$ is the l th membership function in the i th nonlinear synapse performing fuzzification of a crisp component x_i.

And he realizes the system outputs ANPN. If $x_i \in X_{ji}$ then the output is equal

$$w_{ij}^0 + w_{ij}^1 x_i + \cdots + w_{ij}^p x_i^p,\ i = 1, 2, \ldots, h,$$

where h is the number of synaptic weights.

Architecture enlarged neo-fuzzy neuron extensions and neo-fuzzy synapse shown in Figs. 1.17 and 1.18.

Learning advanced neo-fuzzy neuron can be accomplished by minimizing criterion

$$E_J^{[m]}(k) = \frac{1}{2}\left(e_J^{[m]}(k)\right)^2 = \frac{1}{2}\left(y(k) - w_J^{[m]}(k-1)\,\mu^{[m]}\left(x^{[m]}(k)\right)\right)^2$$

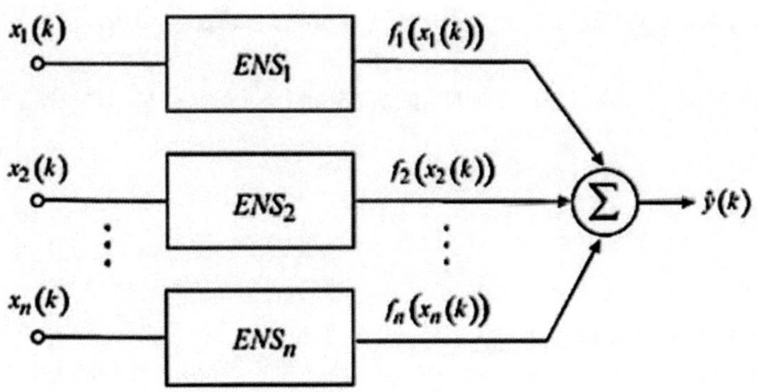

Fig. 1.17 Architecture enlarged neo-fuzzy neuron

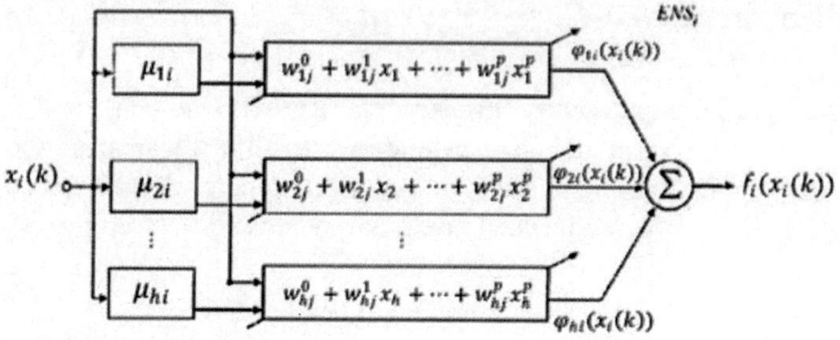

Fig. 1.18 Architecture enlarged neo-fuzzy synapse

Or, in more general terms

$$E_J^{[m]}(k) = \sum_{\tau=1}^{k} a^{k-t} \left(e_J^{[m]}(k)\right)^2$$

where $0 < a < 1$ criterion forgetting.

Minimize given criterion can be a variety of ways, such as recurrent least squares method. In this case, the procedure can be written as

$$\begin{cases} w_j^{[m]}(k) = w_j^{[m]}(k-1) + \frac{p_j^{[m]}(k-1)\, e_j^{[m]}(k)\, \mu^{[m]}\left(x^{[m]}(k)\right)}{a_j + \mu^{[m]}\left(x^{[m]}(k)\right) p_j^{[m]}(k-1)\, \mu^{[m]}\left(x^{[m]}(k)\right)} \\ p_j^{[m]}(k) = \frac{1}{a_j} p_j^{[m]}(k-1) - \frac{p_j^{[m]}(k-1)\, \mu^{[m]}\left(x^{[m]}(k)\right) \mu^{[m]}\left(x^{[m]}(k)\right) p_j^{[m]}(k-1)}{a_j + \mu^{[m]}\left(x^{[m]}(k)\right) p_j^{[m]}(k-1)\, \mu^{[m]}\left(x^{[m]}(k)\right)} \end{cases}$$

Increasing convergence can also be achieved by more complex procedures, for example, using the Hurwitz and Poison-Goff algorithm [10].

1.6.10 Developing a New Topology of the New Neuron

1.6.10.1 Description of a New Neuron Model

1. Mathematical model neuron is represented as:

$$y = F(S(x, W)),$$

where $S(x, W)$ is the induced local field, x is the vector of inputs, W is the vector weighting coefficients, F is the vector functions activation, y is the output signal.

The properties of the neuron, and thus the possibility of ANN, which consists of these neurons, depending on the type of activation functions.

Work network is reduced to classification (generalization) inputs that are n-dimensional hyperspace for a number of classes and is by partitioning the hyper hyperspace.

This is equivalent to the following: continuous function of many variables $f(\vec{x}), \rightarrow \in R^n$ which has some additional features can be with any given accuracy approximated piecewise linear function that breaks the space R^n on convex sets $S_1, \ldots, S_N \subseteq R^n$ which do not overlap, and each set of corresponding linear function set

$$f_i(\vec{x}) = \vec{w}_i^T \vec{x}, i = 1, \ldots, N, \vec{w}_i \in R^n.$$

Each region is an area obtained a class. For example, a single-layer perceptron with Heaviside function as activation function, number of classes does not exceed 2m, where m is the number of outputs. However, not all classes can be divided by this neural network. Single-layer perceptron, which consists of a single neuron with two inputs, can realize the logic function "excluding OR", that can be divided plane (two-dimensional hyperspace) into two half-plane so as to carry out a classification of input signals in classes A and B [10, 29]. Functions not implemented a single layer perceptron called linearly inseparable.

An alternative to this solution is the use of piecewise linear neuron as a basic element for the creation of difficult piecewise linear functions in the optimization problem [21]:

$$y = piecewice\_linear(x, w+, w\_, h) = \begin{cases} w_+^T x, h^T x \geq 0 \\ w_-^T x, h^T x < 0 \end{cases} w_+, w_-, h \in R^n$$

The structure of the neuron shown in Fig. 1.19.

Vectors parameters $\vec{w}_+, \vec{w}_-, \vec{h}$ have the following meaning:

- vector $\vec{h}$ asks hyperplane separating space R^n two half-space;
- vector $\vec{w}_+$ asks weight piecewise linear function piecewise_linear in half-space wher $\vec{h}^T x \geq 0$;

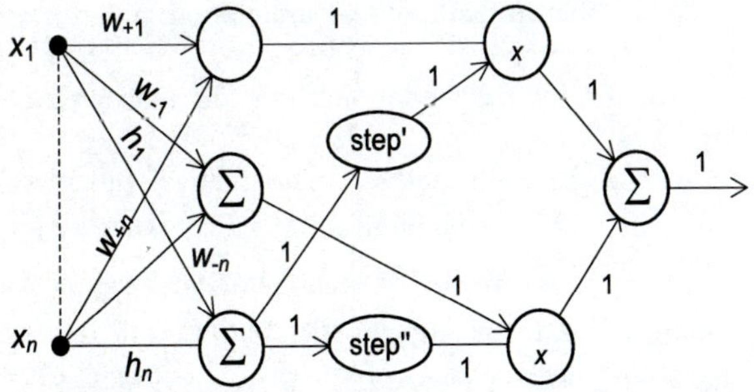

Fig. 1.19 The structure of the neuron

- vector $\overrightarrow{w}_-$ asks weight piecewise linear function piecewise_linear in half-space where $\overrightarrow{h}^T x < 0$.

Obviously, the function piecewise_linear not differentiated in points $x : \overrightarrow{h}^T \overrightarrow{x} = 0$—so if you use such a model neurons, the neural network learning using gradient methods for finding the optimal values of weighting coefficients is somewhat complicated.

To solve this problem, we propose the following model neuron:

$$sigm\_piecewise\left(\overrightarrow{x}, \overrightarrow{w_+}, \overrightarrow{w_-, h}\right) = \frac{w_+^T \overrightarrow{x}}{1 + e^{-\overrightarrow{khT x}}} + \frac{w_+^T \overrightarrow{x}}{1 + e^{-\overrightarrow{khT x}}}; k \succ 0,$$

where all the vectors $\overrightarrow{x}$ Such that $\overrightarrow{h}^T \overrightarrow{x} \neq 0$ We have:$k \rightarrow \infty$: $sigm\_piecewise\left(\overrightarrow{x}, \overrightarrow{w_+}, \overrightarrow{w_-, h}\right) \rightarrow piecewise\_linear\left(\overrightarrow{x}, \overrightarrow{w_+}, \overrightarrow{w_-, h}\right)$.

Although the function sigm_piecewise "Becomes" function piecewise_linear (to all $\overrightarrow{x} : \overrightarrow{h}^T \overrightarrow{x} \neq 0$). Only on the border, is even with small values k it is very well approximates the function. This can ensure recording function piecewise_linear as follows:

$$piecewise\_linear\left(\overrightarrow{x}, \overrightarrow{w_+}, \overrightarrow{w_-, h}\right) = \overrightarrow{w}_+^T \overrightarrow{x} * step\prime\left(\overrightarrow{h}^T \overrightarrow{x}\right) + \overrightarrow{w}_+^T \overrightarrow{x} * step\prime\prime\left(\overrightarrow{h}^T \overrightarrow{x}\right),$$

where $step\prime(x) = \begin{cases} 1, x \geq 0 \\ 0, x \prec 0 \end{cases}$, $step\prime\prime(x) = \begin{cases} 1, x \succ 0 \\ 0, x \leq 0 \end{cases}$.

Actually, the function sigm_piecewise different from the piecewise_linear that instead $step\prime\left(\overrightarrow{h}^T \overrightarrow{x}\right)$ and $step\prime\prime\left(\overrightarrow{h}^T \overrightarrow{x}\right)$ using multipliers $\frac{1}{1+e^{-k \overrightarrow{h}^T \overrightarrow{x}}}$ and $\frac{1}{1+e^{-k \overrightarrow{h}^T \overrightarrow{x}}}$, respectively—i.e. instead of using functions $step\prime(x)$ and $step\prime\prime(x)$ using logistic sigmoid function $sigm(x, k) = \frac{1}{1+e^{-kx}}$, $k \succ 0$. Compare the graphics functions $step\prime(x)$ (function $step\prime\prime(x)$ different from the $step\prime(x)$ Only in value $x = 0$) and $sigm(x, k)$ for $k = 100$ (Fig. 1.20)

As we see, in fact quite accurate approximation even for relatively small values of the parameter k.

Structurally, the model neuron sigm_piecewise can be represented as shown in Fig. 1.21.

As you know, usually a mathematical model of artificial neuron consists of two parts: block-weighted adder $s(\overrightarrow{x}; \overrightarrow{w}) = \overrightarrow{w}^T \overrightarrow{x}$ and activation functions $TF\left(x; \overrightarrow{\theta}\right)$ (where $\overrightarrow{\theta}$—a parameter vector function activation may be absent), which is fed to the input adder output, and together these two parts describing the model neuron view $f\left(\overrightarrow{x}; \overrightarrow{w}; \overrightarrow{\theta}\right) = TF\left(s(\overrightarrow{x}; \overrightarrow{w})\overrightarrow{\theta}\right)$. In the case of the

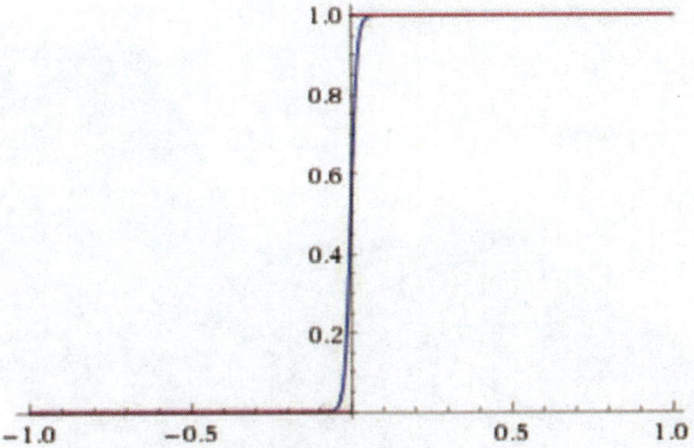

Fig. 1.20 Graphics features *step*′(*x*) and *sigm*(*x*, *k*) for $k = 100$

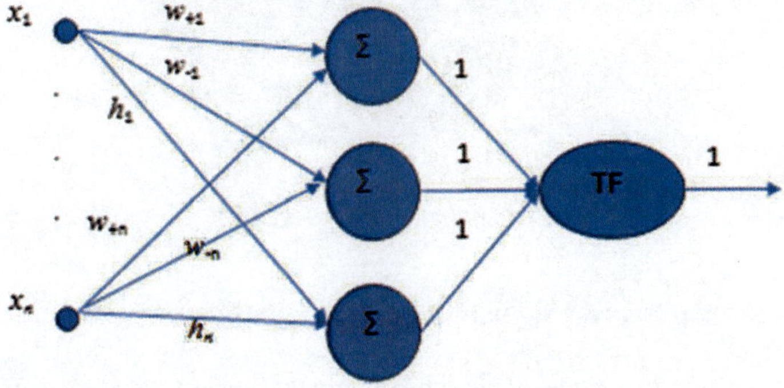

Fig. 1.21 The structure of the neuron model sigm_piecewise

neuron mathematical model sigm_piecewise complex structure, have three adders—$S_+(\vec{x}, \vec{w}_+)$, $S_-(\vec{x}, \vec{w}_-)$, $S_h\left(\vec{x}, \vec{h}\right)$ and hence activation function, which depends on three variables—$TF(S_+, S_-, S_h, k) = \frac{S_+}{1+e^{-kS_h}} + \frac{S_-}{1+e^{-kS_h}}$ and using these pieces full neuron model can be written as:

$$sigm\_piecewice\left(\vec{x}, \vec{w}_+, \vec{w}_-, \vec{h}\right) = TF\left(S_+(\vec{x}, \vec{w}_+), S_-(\vec{x}, \vec{w}_-), S_h\left(\vec{x}, \vec{h}\right); k\right), k \succ 0,$$

Unfortunately, because the activation function $TF(S_+, S_-, S_h, k)$ is a function of three variables (if some fixed k) can not generally portray it as a two-dimensional or three-dimensional graphics. However, you can do it for some particular cases, for example, if we assume that $S_- = CS_+$ where C—some constant.

Graph activation function in the $C = 5$ and $k = 2$ look as shown in Fig. 1.22.

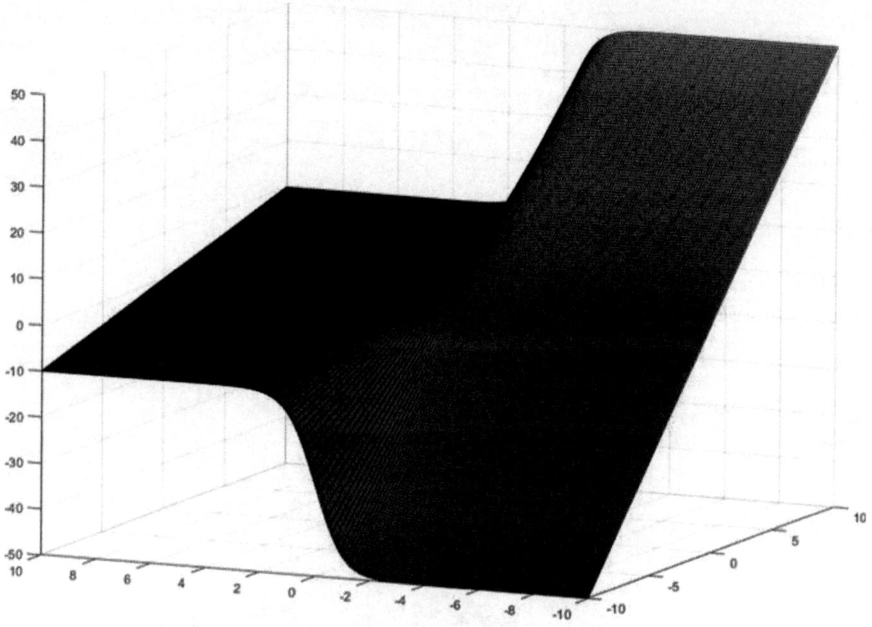

Fig. 1.22 Graph activation function in the $C = 5$ and $k = 2$

1.6.10.2 Setting Neuron Sigmoid Piecewise

Since the mathematical model of neuron Sigmoid Piecewise function is differentiated from their settings, their settings to minimize a function that depends on the output neuron, can be performed using a modification of the gradient descent algorithm. The formulas for calculating the first derivatives are as follows:

$$\frac{\partial f}{\partial w_{+q}} = \frac{x_q}{1 + e^{-k\overrightarrow{h^T \vec{x}}}},\ \frac{\partial f}{\partial w_{-q}} = \frac{x_q}{1 + e^{-k\overrightarrow{h^T \vec{x}}}},\ \frac{\partial f}{\partial h_q} = kx_q\left(\frac{\vec{w}_+^T\vec{x} - \vec{w}_-^T\vec{x}}{2 + e^{-k\overrightarrow{h^T \vec{x}}} + e^{-k\overrightarrow{h^T \vec{x}}}}\right).$$

Thus, we have the following meanings graph of derivatives $\partial f / \partial w_{+q}$ and $\partial f / \partial w_{-q}$ the values x_q and $S_h = \vec{h}^T \vec{x}$ $(k = 1)$ (Fig. 1.23).

And what a plot of the original value $\partial f / \partial h_q$ the values $t = kx_q\left(\vec{x}_+^T\vec{x} - \vec{x}_-^T\vec{x}\right)$ and $S_h = \vec{h}^T \vec{x}$ $(k = 1)$ (Fig. 1.24).

For further analysis of the "behavior" of partial derivatives neuron sigm_piecewise also worth noting that the amount $\frac{\partial f}{\partial w_{+q}} + \frac{\partial f}{\partial w_{-q}}$ can be simplified as follows:

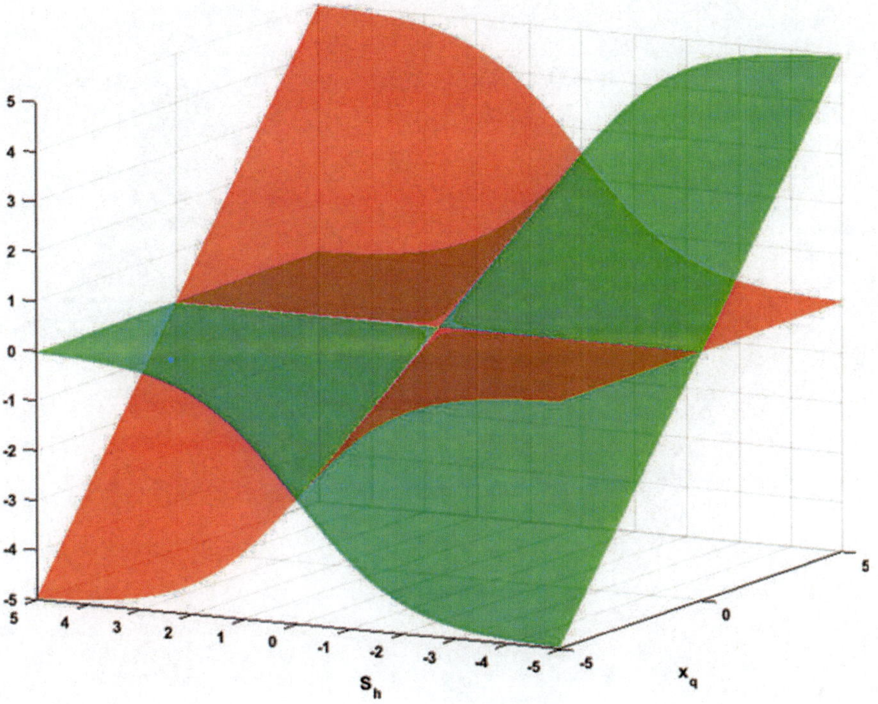

Fig. 1.23 Graph of derivatives values $\partial f / \partial w_{+q}$ and $\partial f / \partial w_{-q}$ of values x_q and $S_h = \vec{h}^T \vec{x}$ $(k = 1)$

$$\frac{\partial f}{\partial w_{+q}} + \frac{\partial f}{\partial w_{-q}} = \frac{x_q}{1 + e^{-k \vec{h}^T \vec{x}}} + \frac{x_q}{1 + e^{-k \vec{h}^T \vec{x}}}$$

$$= \frac{x_q\left(1 + e^{-k \vec{h}^T \vec{x}}\right) + x_q\left(1 + e^{-k \vec{h}^T \vec{x}}\right)}{\left(1 + e^{-k \vec{h}^T \vec{x}}\right)\left(1 + e^{-k \vec{h}^T \vec{x}}\right)}$$

$$= \frac{x_q\left(2 + e^{-k \vec{h}^T \vec{x}} + e^{-k \vec{h}^T \vec{x}}\right)}{2 + e^{-k \vec{h}^T \vec{x}} + e^{-k \vec{h}^T \vec{x}}} = x_q$$

Analyzing graphs and equations to the following conclusions:

1. An important influence on the value of all derivatives matters $S_h = \vec{h}^T \vec{x}$ Ie how far and in which direction from the dividing plane defined by the vector $\vec{h}$, Is an example $\vec{x}$.
2. The total value of derivatives $\frac{\partial f}{\partial w_{+q}} + \frac{\partial f}{\partial w_{-q}}$ is always x_q. However, depending on the sign and magnitude in absolute value S_h greater contribution to this amount

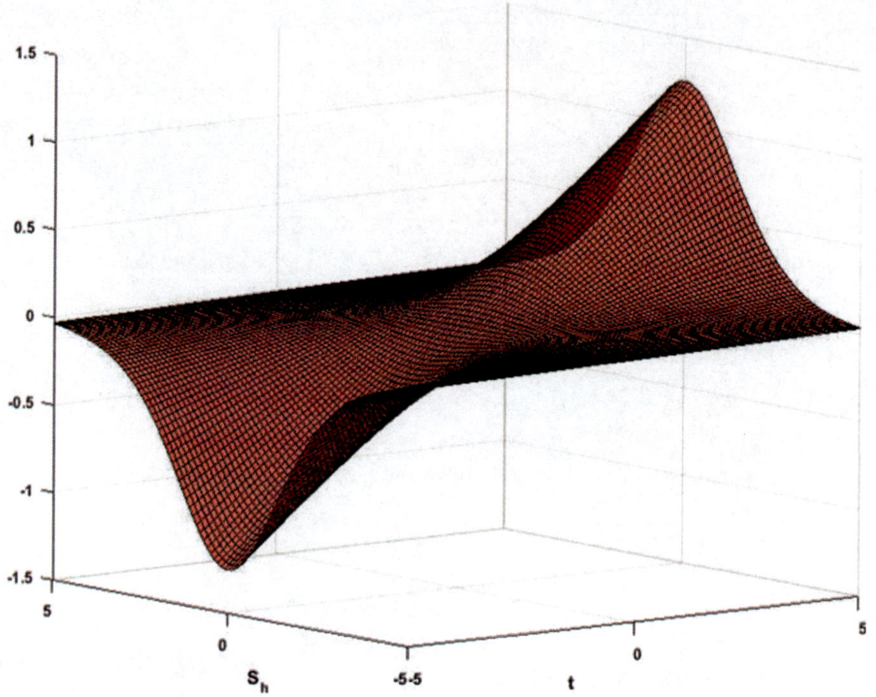

Fig. 1.24 Graph of the derivative values $\partial f / \partial h_q$ the values $t = kx_q\left(\vec{x}_+^T\vec{x} - \vec{x}_-^T\vec{x}\right)$ and $S_h = \vec{h}^T\vec{x}\,(k = 1)$

will make one of the original: if $S_h \ll 0$—the value of the derivative $\frac{\partial f}{\partial w_{-q}}$ will be very close to x_q And the value of the derivative $\frac{\partial f}{\partial w_{+q}}$—0; and vice versa when $S_h \gg 0$. If the $S_h \approx 0$—the contribution of both derivatives is approximately equal.

3. derivative $\frac{\partial f}{\partial h_q} \approx 0$ at $|S_h| \gg 0$ and $\approx kx_q\left(\vec{w}_+^T\vec{x} - \vec{w}_-^T\vec{x}\right)$ at $S_h \approx 0$.

Thus, if the vector $\vec{x}$ is far from dividing the hyper—it almost will not affect the configuration parameter vector $\vec{h}$; If the vector $\vec{x}$ close to the dividing hyperplane—the value of the derivative for this vector is proportional to the difference value $\vec{w}_+^T\vec{x} - \vec{w}_-^T\vec{x}$—that if $\vec{w}_+^T\vec{x} \approx \vec{w}_-^T\vec{x}$—then again, the impact of such vector the final correction vector to vector $\vec{h}$ is less than the impact of a hypothetical vector $\vec{x}\prime$ For which $\vec{h}^T\vec{x}' \approx \vec{h}^T\vec{x}$ and which has the same rate as the vector $\vec{x}$ But for which values $\vec{w}_+^T\vec{x}$ and $\vec{w}_-^T\vec{x}$ very different. Thus, the greatest impact on the correction parameter vector $\vec{h}$ will have such examples $\vec{x}$ For which $\vec{h}^T\vec{x} \approx 0$ and $\left|\vec{w}_+^T\vec{x} - \vec{w}_-^T\vec{x}\right| \gg 0$. From this we can conclude that if at some time settings neuron all examples of the training set will be of great value $\vec{h}^T\vec{x} \gg 0$—further setting vector $\vec{h}$ virtually stopped, because all derivatives $\frac{\partial f}{\partial h_q}$ will be very close to zero, with two possible situations:

1. All examples are on one side of the respective dividing the hyper—"bad" situation, because in such a situation, the neuron is actually converted into ordinary linear neuron and configured to be only one vector—$\vec{w}_+$ or $\vec{w}_-$;
2. Of examples is on one side of the respective dividing hyperplane, and the rest—on the other side—that is, in a study sample there are two or more clusters, examples of linearly separate a certain "gap" as a result find a hyperplane that separates these clusters that most of the "good"—for such clusters usually necessary to use different forecasting models.

If after correction vectors $\vec{w}_+$ and $\vec{w}_-$ will be the same—the next iteration of the correction vector $\vec{h}$ will be zero. But given the results of analyzing the behavior of derivatives ${}^{\partial f}/_{dw_{+q}}$ and ${}^{\partial f}/_{dw_{-q}}$—such a situation should occur very rarely, and do not take more than one iteration—as relevant derivatives will be very different for these iterations.

To use the neuron Sigmoid Piecewise in multilayer neural networks other than first derivatives neuron model parameters for its first derivatives are also required for the model input variables x_q:

$$\frac{\partial f}{\partial h_q} = \frac{k h_q\left(\vec{w}_+^T\vec{x} - \vec{w}_-^T\vec{x}\right)}{2 + e^{-k\vec{h}^T\vec{x}} + e^{-k\vec{h}^T\vec{x}}} + \frac{w_{+q}}{1 + e^{-k\vec{h}^T\vec{x}}} + \frac{w_{-q}}{1 + e^{-k\vec{h}^T\vec{x}}}$$
$$= \frac{\partial f}{\partial h_q}\frac{h_q}{x_q} + \frac{w_{+q}}{1 + e^{-k\vec{h}^T\vec{x}}} + \frac{w_{-q}}{1 + e^{-k\vec{h}^T\vec{x}}} = \frac{w_{+q}}{1 + e^{-k\vec{h}^T\vec{x}}} + \frac{w_{-q}}{1 + e^{-k\vec{h}^T\vec{x}}}$$

This derivative is nonzero whenever the conditions $w_{+q} \neq 0$ and $w_{-q} \neq 0$.

1.6.10.3 Comparison of the "Family" ReLU Neurons

In [22] proposed a neuron activation function of $\mathrm{Re}LU(\vec{x}, \vec{w}) = \max\left(0, \vec{w}^T, \vec{x}\right)$, The Graph of which is shown in Fig. 1.25.

Obviously, the model neuron And modifications—Parametric ReLU (PReLU) neuron, the model form:

$$\mathrm{Re}LU(\vec{x}, \vec{w}, a) = \begin{cases} \vec{w}^T\vec{x}, \vec{w}^T\vec{x} \succ 0 \\ a * \vec{w}^T\vec{x}, \vec{w}^T\vec{x} \leq 0 \end{cases}$$

is a particular case of a more general model piecewise_linear:

1. $\mathrm{Re}LU(\vec{x}, \vec{w}) = piecewise\_linear\left(\vec{x}, \vec{w}_+ = \vec{w}, \vec{w}_- = \vec{0}, \vec{h} = \vec{w}\right),$

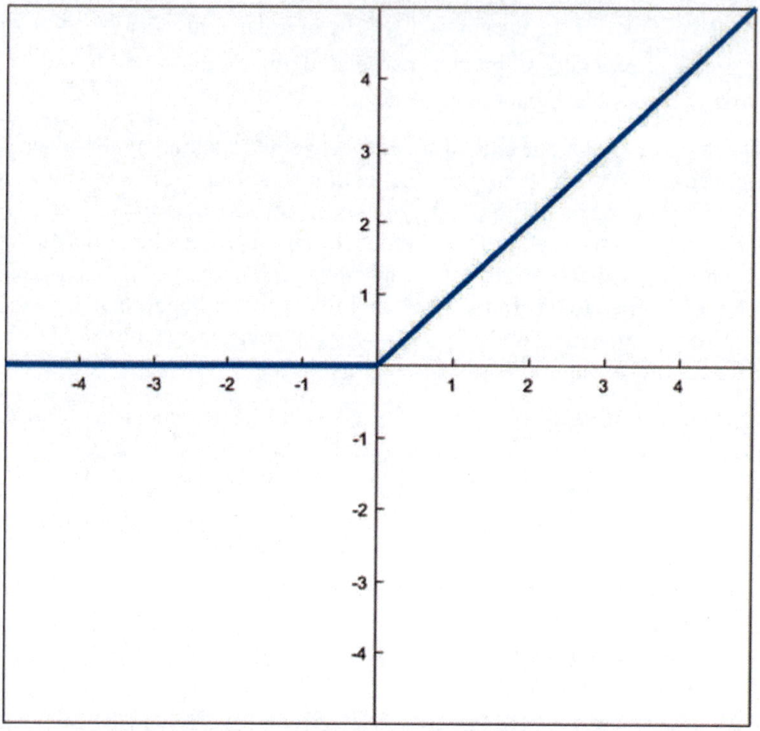

Fig. 1.25 Graph of neuron activation function $\mathrm{Re}LU(\vec{x}, \vec{w}) = \max\left(0, \vec{w}^T \vec{w}\right)$

2. $\mathrm{Re}LU(\vec{x}, \vec{w}, a) = piecewise\_linear\left(\vec{x}, \vec{w}_+ = \vec{w}, \vec{w}_- = a * \vec{w}, \vec{h} = \vec{w}\right)$.

In turn, function $\mathrm{Re}LU(\vec{x}, \vec{w}) = sigm\_piecewise\left(\vec{x}, \vec{w}_+, \vec{w}_-, \vec{h}\right)$ more common than $piecewise\_linear\left(\vec{x}, \vec{w}_+, \vec{w}_-, \vec{h}\right)$—hence the model sigm_piecewise more common than ReLU and PReLU But while it has three times more options.

Known issues in learning neural networks, which include activation function of neurons PReLU Is the so-called problem of "dying ReLU-neuron": if the result of some modification parameter vector $\vec{w}$ ReLU-neyrona for all examples of the training set $\vec{x} \in X_t$ the condition $\forall \vec{x} \in X_t : \vec{w}^T \vec{x} \leq 0$ then continue in learning gradient function of the neuron model parameters by vector $\vec{w}$ will always be equal $\vec{0}$ (Obviously, the output neuron for any example from the training set to 0). Thus, the neuron "die"—he learns and always have at its output 0.

Unlike ReLU-neuron, with neuron activation function $sigm\_piecewise\left(\vec{x}, \vec{w}_+, \vec{w}_-, \vec{h}\right)$ devoid of this problem—gradient vector parameters to functions $\vec{h}$ nonzero vector is almost always among the two gradients

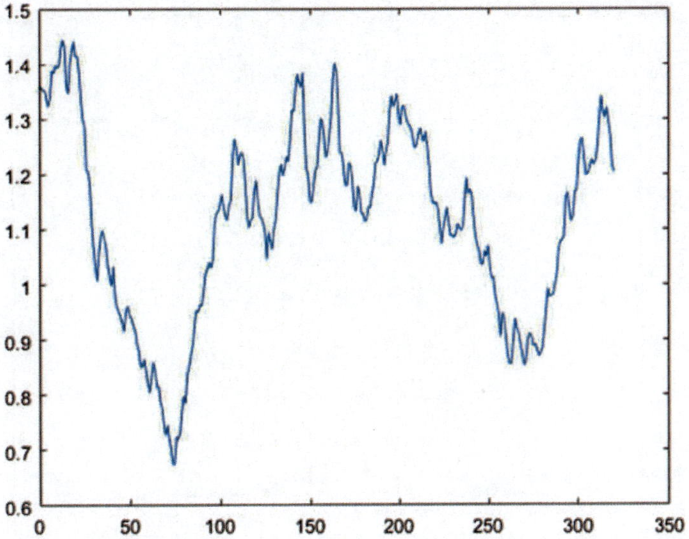

Fig. 1.26 Time series of daily value rate USD to EUR

to vector parameters $\vec{w}_+$ and vector parameters $\vec{w}_-$ at least one is nonzero vector for some $\vec{x}$, (Depending on the value $\vec{w}^T \vec{x}$).

1.6.10.4 Comparison on Real Samples

The first comparison was made on public domain some time daily value rate USD to EUR (Fig. 1.26) [31].

Training set $\prec X_t, y : X_t \rightarrow \Re \succ$ was built from the original time series by investing time series with the dimension of investments $m = 5$ and forecasting horizon $k = 4$—that for the prediction value g_{i+4} used values $g_{i-4}, g_{i-3}, g_{i-1}, g_i$.

Mean square error model naive view $g_{i+4} = g_i$ is $MSE_{naive} = 7.16 \times 10^{-2}$.

Define standard error of mean, you want to achieve $MSE_{target} = 5 \times 10^{-2}$ And be "greedy" way to teach both the network until a specified level of errors, and then construct a graph of the number of neurons from the mean network error, which consists of the number of "greedy" manner trained neurons in both networks (Fig. 1.27).

As you can see, starting with some error value $MSE_{treshold} \approx 76.5 \times 10^{-2}$ To achieve this error value needed three times more "weighted" according to the type of neurons ReLU than neuronal type sigm_piecewise Then the gap in right quantity of neurons only increased—and so, given that the neuron type sigm_piecewise is about three times more options—to achieve errors of less than $MSE_{treshold}$ using neuronal type sigm_piecewise needed fewer options than using "weighted" neuron type ReLU.

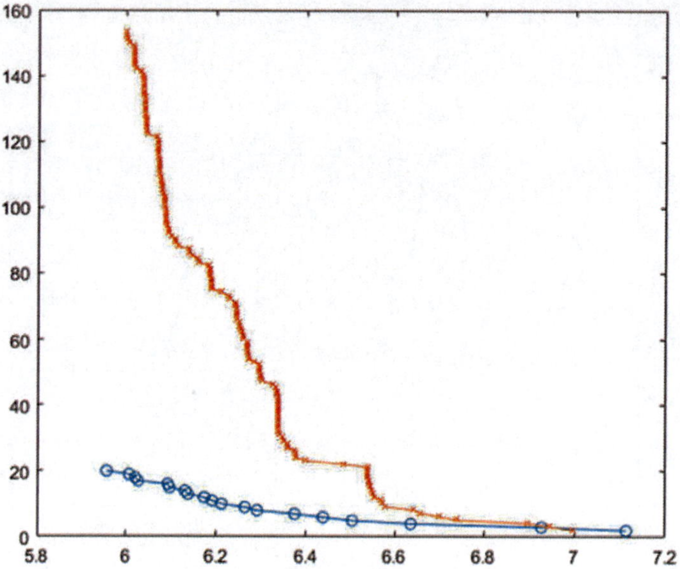

Fig. 1.27 The plot of the number of neurons on the mean square network errors

The second test was performed on a sample average interest rates on government bonds in Australia, 1969–1994 years (Figs. 1.28 and 1.29).

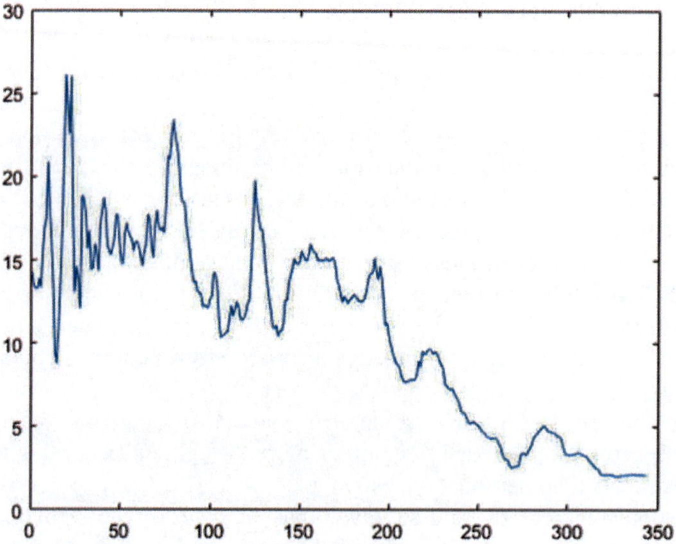

Fig. 1.28 Chart data used for the second comparative test

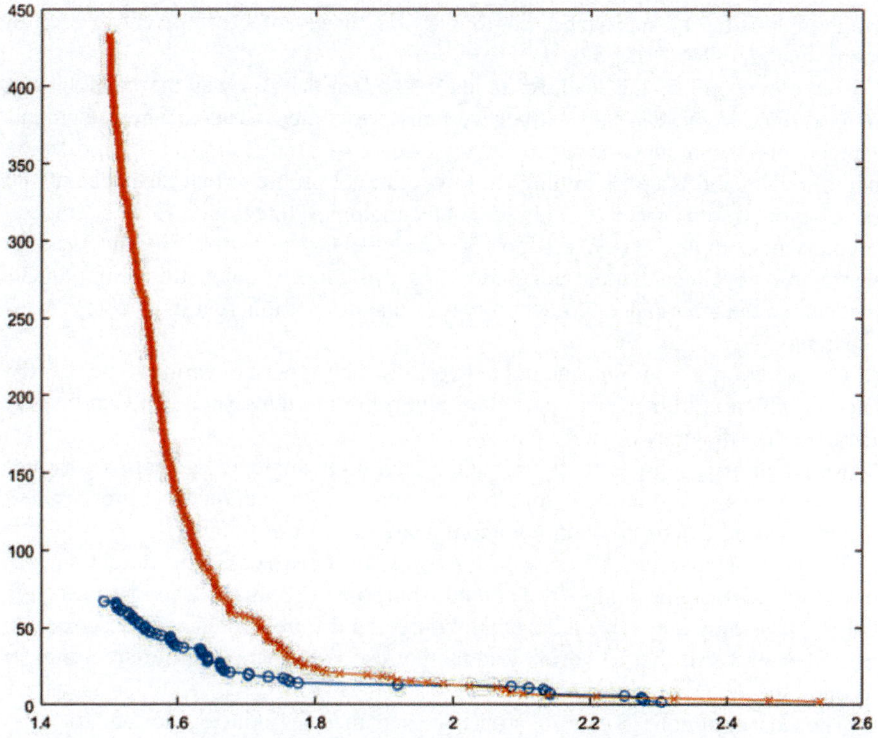

Fig. 1.29 Graph of the number of neurons of the achieved level of approximation errors in the second comparative test

Orange Graph—for ReLU network of neurons, blue—for a network of neurons SP. After performing similar steps for the training set $\prec X, y : X \to R \succ$ error was naive model $MSE_{naive} = 2.9157$. Define the error you want to achieve $MSE_{target} = 1.5$. After the test build similar Graphs:

Starting with about error value $MSE_{treshold} \approx 1.6$. To achieve this error value required 3 times more neurons than ReLU type of neuron type SP, then the gap in right quantity of neurons again only increased.

1.7 Rationalc for the Creation of Hybrid Artificial Neural Networks

The experience of recent years has shown that application in computer science techniques that meet one scientific paradigm to solve challenges and problems do not always lead to success. In the hybrid architecture that combines several paradigms effectiveness of one approach can compensate for the weakness of the other [9]. By combining different approaches can bypass the disadvantages inherent in each of them separately [4, 7, 10, 32].

Therefore, one of the leading trends in modern science was the development of integrated, hybrid systems. Such systems are composed of different elements (components) combined in order to achieve their goals. Integration and hybridization of different methods and technology to solve complex problems that can not be solved on the basis of some specific techniques or technologies. Integration as a fundamental property of complex systems, which is closely related to its integrity involves not just unions, but also mutual adaptation and evolution of compatible components that enable the emergence of new properties that are not inherent in its components separately.

Construction of combined neural networks, which consist of various types of NN, each of which is trained on a particular algorithm in many cases can significantly increase the efficiency of NN.

Research of principles of hybridization NN, fuzzy logic and genetic algorithms can create new neural network topology, which have a higher quality of recognition in case of simultaneous reduction in computational cost of training.

Define hybrid system (HS) as a system consisting of two or more different types of integrated subsystems, united by a common purpose or compatible actions (although these subsystems may have a different nature and different languages describe). In computer will call these hybrid systems that use two or more different computer technology.

The most common HS classification at present in terms of integration and transformation including autonomous, weakly, sylnozv'yazani and fully integrated system. Consider each of HS levels of integration under this classification.

Autonomous systems. Such systems contain independent software components that implement information processing on heterogeneous models. Despite the apparent degeneracy of the autonomous system, the development of autonomous models is relevant for hybridization, and may have several goals.

(1) These models represent a way of comparing features of solving the problem of two or more different methods.
(2) Consistent implementation of two or more independent models can confirm or deny the accuracy previously developed information processing.
(3) Standalone models can be used to quickly create an initial prototype system, then created applications that require more time to develop.

However, stand-alone models have a significant drawback—neither of them can "help" the other in a situation of updating information, all models must be modified at the same time.

Transformation systems are similar to standalone ones, since the end result of the processing is an independent, non-interacting model. The main difference is that such a model starts to work as a system that uses one standalone method, and ends—as a system that uses another standalone method (converting methods is implemented).

Transformational models have several advantages. They are quick to create and cost less. Because only one model is used and the final method adapts the results as best as possible to the environment. But there are also problems with creating a tool to automatically convert one model to another and the problem of significant modification of the model.

Weak systems. This is essentially the first real form of integration whereby an application breaks down into separate components that are linked through data files. Consider the basic types of such systems:

- Chain HS consists of two functionally completed components, one of which is the master processor and the other is the processor or postprocessor;
- hierarchical HS (subordinate) also uses components as completed components, but in this case one of them, being subordinate, is included in the other, which is the main problem solver;
- the target processor on the HS consists of the target processor and several functional components;
- co-processor HS uses equal components to solve problems. Each component can pass information to others, interact, processing subtasks of the same task.

The poorly coupled systems under consideration are easier to develop than other more integrated models, and allow commercially available applications to reduce development time.

Strongly coupled models. Increasing the degree of integration will lead to highly interconnected models. Highly coupled models can function in the same formats as weakly coupled ones, but their pre-, post- and coprocessor variants are inherently faster.

1.8 Approaches to the Creation of Hybrid Neural Networks

Actually, at the moment there is no single methodology or approach commonly used to create a hybrid neural networks, which would give the result in most cases at least. Structural and parametric synthesis currently being empirical, experimental or sometimes through the use of certain practices that have proven their worth on a certain class of problems.

The problem is structural and parametric synthesis can be divided into the following subtasks:

- selection of network types and structures;
- order networks and the way they mix;
- choice of topology neurons quantity order of their use;
- selection and activation of functions belonging;
- Internal recruitment system parameters.

Each of these stages has certain generally accepted practices and approaches.

First, you must determine which artificial neurons and networks should be used to solve the problem. For now you can create three basic approaches to creating hybrid neural networks.

The first is to use the known topologies of neural networks, such perceptron network Hopfield, radial-basis and others. networks and saturation of each neurons of different topologies (see. Sect. 1.3).

In the second approach, a hybrid neural network consists of different neural networks topologies using the same type of neurons. Structural and parametric synthesis of such hybrid networks discussed in Chap. 3.

The third approach is a combination of the first two approaches. That is combined of several known topologies and saturation of different types of neurons.

1.9 Overview of Topologies of Hybrid Neural Networks

Hybrid Multi Neural Network or systems are often used to improve the results of solving problems of classification and prediction [11, 14, 26, 27, 33–47]. When building such networks is necessary to solve the following tasks: designing network architecture; networks are selected, the results of which should be combined to give the best end result; use a limited set of data.

In recent years there has been strong growth in successful hybrid intelligent systems in various fields such as robotics [48, 49] medical diagnostics [50], speech recognition, diagnosis of faults industrial equipment [51], monitoring (control of production processes) [13] and applications in the field of finance. The main factor that contributes to the development of hybrid systems is the increased use of neural networks for problems of recognition, classification and optimization [52–54]. The ability of neural networks to perform tasks that otherwise would have been difficult to be solved or those that are difficult to symbolic computation now recognized, and they are often used as modules in intelligent hybrid systems.

Hybrid neuro-fuzzy network combined into a universal approximating characteristics of traditional neural networks transparency interpretovanist fuzzy inference systems. Further development of neuro-fuzzy networks became their hybridization methods of the theory of wavelet transform. Such systems are called wavelet neuro-fuzzy systems. Given the possibility of hybrid wavelet neuro-fuzzy systems computational intelligence about their approximate properties of linguistic interpretovanosti, opportunities to identify local features of the processed signals and the ability to implement soft computing based on fuzzy sets in [32, 55–57] was implemented creating hybrid evolutionary adaptive (fuzzy system parameters and structure that can be set in a sequential (online) mode) wavelet neuro-fuzzy systems.

In [16, 17, 19, 22, 58–65] developed a mathematical model of adaptive univariate and multivariate wavelet function activated, belonging to which synthesized methods for setting parameters width, center and shape features based on generalized metric Itakura-Saito, thereby improving the properties approximating and extrapolating hybrid evolutionary adaptive wavelet neuro-fuzzy systems. Based introduced wavelet function, activation of belonging developed univariate and multivariate wavelet function activation-origin type 2, which are different from the standard features of type 2 types of uncertainty, namely the width, in the center and form membership functions.

The proposed fuzzy wavelet function-activating type-2 membership allow reasonably select membership functions for solving specific problems.

In [15, 16, 20, 21, 32, 55, 56, 66, 67] proposed adaptive composite structure W-neuron (wavelet) with multidimensional adaptive wavelet-activation functions belonging. The method of training based on modified procedure to filter and Smoothing properties. The method of training based on robust criteria optimization process allowing transient signals with abnormal emissions from non gaussian distribution. Architecture W-neuron can be used as a standalone hybrid wavelet neuro-fuzzy network or as part of a hybrid adaptive evolutionary wavelet neuro-fuzzy systems, which has improved approximation and extrapolating the properties through the introduction of multi-dimensional wavelet function activated, accessories and configuration of all their parameters in a sequential mode.

In [19, 60, 65, 68, 69] developed a hybrid architecture adaptive evolutionary wavelet neuro-fuzzy system based on W-responsive neurons in the consequent nalashtovnymy and with one-dimensional or multidimensional adaptive wavelet-activation functions belonging to antecedents. This system has significantly improved properties approximating and extrapolating that allows transient nonlinear signal processing arbitrary nature at its current and prior uncertainty.

In [12, 16, 22, 58, 63, 70] developed a hybrid evolutionary adaptive wavelet neuro-fuzzy system type-2 defazyfikatsiyi-reduction procedure in a sequential mode. A neuro-fuzzy veyvlon type-2 fuzzy multidimensional wavelet membership functions type-2 fuzzy adaptive wavelet-neuron type-2 fuzzy one-dimensional wavelet membership functions type-2. Synthesized adaptive wavelet neuro-fuzzy system type-2 based bank neural networks, each of which is characterized by an individual set of parameters fuzzy wavelet functions of type-2 form in perenalashtovnoyu antecedents. The proposed hybrid evolutionary adaptive wavelet neuro-fuzzy system type-2 have flexibility, increased performance, enhanced approximation and extrapolating properties.

In [24, 71, 72] developed a hybrid architecture evolutionary cascade GMDH neural network. A use as nodes hybrid neurons, namely: W-neuron, Q-neuron wavelet neuron, to improve and extrapolating the approximating properties and expand the number of entrances to the site. The proposed hybrid evolutionary cascade GMDH neural network combines the advantages of GMDH neural (selection of input signals with the most informative) and cascaded networks and allows adaptation architecture and parameters in a sequential mode.

In [16] proposed adaptive fuzzy wavelet neural network with consequent line that can be used as a standalone network or as part of a more complex wavelet neuro-fuzzy systems type-2.

In [16, 23, 60] developed a modified training methods hybrid adaptive wavelet neuro-fuzzy-systems for adaptive W-neurons from nalashtovnymy wavelet functions activate-affiliation in the hidden layer based on quadratic and robust criteria providing increased speed of convergence learning process.

In [24, 73] proposed a modification of the architecture multi Hybrid GMDH neural network by entering into the structure node hybrid Q-neurons and W-neurons, thus improving aproksymuvalni properties GMDH neural networks to expand the number of inputs to the node network to optimize the structure hybrid networks in the learning process.

In [12, 19, 74] proposed modification of architectural evolution wavelet neural network by introducing wavelet neuron with adaptive wavelet function activated, belonging to the structure node cascade network, enabling to increase network architecture in sequential mode processing non-stationary time series, conditions of uncertainty.

In [75] developed a hybrid system of computational intelligence (which combines the idea of deep learning and evolution/ development cascade of neuro-fuzzy systems) for processing online. Each node of the system solves the problem of fuzzy clustering independently. The process of quality assessment is determined by finding the optimal values of the validity of the used cluster.

In [76] proposed a hybrid neuro-fuzzy system that combines various concepts such as deep learning, group method of data processing and evolutionary system. It is also proposed to adjust all settings online. As evolutionary unit multilayer system uses advanced fuzzy neo-neurons that have high approximating properties. At the stage of training offered by evolving a system that develops deep, calculates the parameters and adjusts its architecture. The system architecture can evolve over time as synaptic weight and width parameters centers neuro-fuzzy nodes configured to improve the properties approximations system. To train the system do not need to study a large sample.

In [25] the problem of optimizing the structure of neuro-fuzzy neural network. For its solution proposed and described GMDH algorithm structure optimization. Experimental research and provides a comparative analysis of the accuracy of prediction obtained optimal neuro-fuzzy network and multilayer architecture of direct distribution. It is shown that the proposed neuro-fuzzy network solves the problem of classification is as good as the problem of forecasting.

In [77] the cascade GMDH-wavelet neuro-fuzzy network. As the nodes of the network allocated R-activation function of neurons and Yepanechnykova block adaptive fuzzy veyvlony with adaptive wavelet membership function. Algorithms studies that have properties of tracking and filtering to allow the on-line not only adjust synaptic weights, but activation function-parameters belonging. Computational experiments confirm the efficiency of the approach.

In [28] the evolution of neuro-fuzzy multi-system nodes which are neuro-fuzzy systems, all options are subject to adjustment by using the algorithms at work.

In [31] the TreNet, continuous new hybrid neural network to study the local and global context functions for forecasting trends in time series, which are connected sequence trends. TreNet consists of recurrent neural network LSTM to capture long time series dependence, convolutional neural network to extract the characteristics of the local raw data time series and a layer of generalization features.

Next ANN construction approach based on the use of fuzzy polynomial neural networks (FPNN). In [78, 79] introduced the concept FPNN, hybrid architecture which is a union of polynomial neural networks (PNN) and Fuzzy Neural Networks (FNN). Networking technologies FPNN based on Computational Intelligence (CI), namely fuzzy sets, neural networks and genetic algorithms. The network consists of fuzzy polynomial neurons (FPN), forming components first (input) layer FPNN and polynomial neurons (PN) located in the layers of the network, coming one after another. Polynomial neural network generates fuzzy inference system (as part of the original rules considered as triangular and haussovski membership function, and in the final part of the rules—polynomial (constant, linear, quadratic and quadratic modified). Each network PN polynomial type implements a partial description of the mapping between input and output.

To build a dynamic network topology using GMDH. At the preliminary stage of training FPNN and FNN uses fuzzy logic method of training, implemented through a standard reverse circulation.

Parameters membership functions, learning rate adjusted using genetic optimization. The experimental results show that the proposed network demonstrates high accuracy and summarizing capabilities compared with other similar fuzzy model.

In [80–82] developed a new topology FPNN based on genetically optimized multilayer perceptron with fuzzy polynomial neurons (FPN).

The concept of neuro-fuzzy networks (SNFM), self-organizing, with a hybrid architecture and combining neuro-fuzzy network (NFN) and polynomial neural network (PNN) is to use the technology of computational intelligence, namely fuzzy sets, neural networks genetic algorithms.

CNFN architecture is the result of the synergistic use of neuro-fuzzy networks (NFN) and polynomial neural networks (PNN). Neuro-fuzzy network contributes to the formation of the original, rule-based structure CNFN. A consistent part CNFN developed using PNN. Polynomial neural network that is flexible and versatile structure built using GMDH method.

In [83, 84] studied the new neuro-fuzzy topology—genetically optimized hybrid polynomial neural network based on fuzzy sets (HPNNFS).

Each node of the first level HPNNFS, i.e. fuzzy polynomial neuron (FPN) is a compact system fuzzy inference.

In the other networks and higher levels HPNNFS as nodes used polynomial neurons (PN). Determining the optimal parameter values for each layer (the number of input variables, order polynomial best set of nodes and the number of membership functions) is performed based on the use of genetic algorithm (GA). Further, structural optimization by using HA, while the following detailed parametric optimization is performed with MNCs.

In [85] the campaign based on the use of parallel fuzzy polynomial neural networks (PFPNN) in combination with the method of fuzzy weighted clustering is used to form a heterogeneous inbox space, which leads to some significant subspaces which are associated respective local models implemented fuzzy polynomial neural networks. Roy particle method is used to optimize the tuning parameters PFPNN.

In [29] the architecture of the hybrid system consisting of surface agent training, and agent-depth study of the cognitive agent. The architecture of the hybrid system agent surface learning and deep learning agent working in parallel in response to changes in the environment. In particular, surface agent training structure captures short information signal and performs a quick analysis to respond to the stimulus. Agent-depth study seeks long-term structure information and identifies correlations longer present in the system. Cognitive agent following the high-level environment, as well as the decisions taken by the agents of superficial and deep learning and automatically generates the most desired output of the system as appropriate.

In [86] the architecture teaching methods multivariate hybrid cascade neural network optimization pool of neurons in each stage other than the famous cascade of computational intelligence possibility of processing multidimensional time series online that can handle non-stationary stochastic and chaotic signals Nonlinear about the objects with the required accuracy.

In [87] the problem of classification of breast tumors in medical images. To solve it proposed a new class of convolutional networks—fuzzy hybrid convolutional neural network in which convolutional network VGG-16 is used as an extractor features of the image and fuzzy neural network NEFClass—a classifier. Developed and researched learning algorithms hybrid convolutional network.

In [88] proposed hybrid neural network with advanced knowledge, which is a fundamental approach to the use of prior knowledge and multi-level semantics of the text in comparison. A hybrid neural network with improved knowledge (HNNIK) which uses prior knowledge to identify useful information and filter out noise in a long text and performs a comparison of different points of view.

The model combines previous knowledge of verbal representation through the gates of knowledge and establish appropriate channels to three words, sequential structure of the text and presentation, improved knowledge. The three channels are processed by convolutional neural network to generate a high level of features for the purpose of comparison, the signs are synthesized as conformity assessment multilayer perceptron.

In [89] the theoretical basis for the construction of ensemble methods significantly improved regression estimates. Given a set of regression estimates based hybrid rating that is as good or best in terms of mean square error (MSE) than any estimate in the sample. It is alleged that the ensemble presented method has several properties:

1. Effectively uses all network sampling—none of the networks should not be rejected.
2. Effectively using all available data to study without retraining.
3. Performed by smoothing regularization in functional space that avoids conversion.
4. Use local minima for the construction of improved ratings, while other neural-network algorithms prevent local minima.
5. Ideal for parallel computing.
6. Results in a very natural and useful evaluation of different estimates in the sample.

In [30] the evolutionary fuzzy hybrid neural network (EFHNN) to improve the accuracy of conceptual cost estimates. The approach involves the inclusion of hybrid neural network (HNN) neural networks (NN) and high-order neural networks (HNN).

Fuzzy logic (FL) is used for the treatment of uncertainties EFHNN used to optimize genetic algorithm for neural and for HNN therefore called EFHNN approach.

In [38] The problem of recursive filtering images using hybrid neural network comprising multiple spatially-varying recurrent neural networks (RNN) as equivalents of the group of recursive filters for each pixel and deep convolutional neural network (CNN), which teaches weight RNN.

In [90] the deep hybrid system with adaptation computational intelligence architecture for fuzzy medical diagnostics. This system can improve processing quality medical information in terms of classes, overlapping by using special adaptive architectures and learning algorithms.

This system can adjust its architecture in a situation where the number of signs and diagnoses can be variable. Developed special algorithms to study the situation of different architectures synapatychnyh system without retraining measures, which have been set in previous steps.

In [91] presented an approach that enhances the quality of solving the problem of classification for short samples, elements of which are noisy images through the use of hybrid neural network dzhhutovoyi architecture, which includes convolutional neural network and multilayer perceptrons. Improved productivity is achieved by obtaining and use in training additional features of the image.

Despite some progress in developing HNN currently there is no single approach to create them. At the heart of the main existing methods are proposed without sufficient justification for the choice of basic nm (fuzzy neural different topology) with the addition thereto of other networks (through a network that is considered HNN) that provides improved accuracy solve the problem. For each HNN built on this principle, developed individual learning algorithm. An additional element that improves the quality of education is the additional use GMDH algorithm. In addition, the proposed approach does not use such a powerful mechanism as deep learning.

Thus, the main problems HNN synthesis, currently are:

- lack of formal methods for selecting the type NN adequate class of problems that need to be addressed;
- insufficient processing for automatic creation of topology NN, which does not create NN precision and minimal complexity (minimum computational cost);
- insufficient justification of choice of optimization methods in teaching procedure nm, which leads to significant errors.

In view of the foregoing, we conclude the need for a unified methodology for constructing HNN with the possibility of scaling up the complexity of their structures and hybrid learning algorithms, including deep learning, to improve the resolution of tasks.

1.10 Criteria for Evaluating the Effectiveness of Neural Networks

The presence of obvious mutual dependence between complexity model, the size of the training set and the resulting ability of the umbrella model to independent data provides the ability to define this relationship in some way.

In other words, the method of building effective neural networks necessary to determine the possible ways of generalizing evaluation of the network.

Since the study is based on minimizing the value of some function, which shows the deviation of the results issued by the network in a given training set from the ideal required, you should choose an appropriate evaluation. Usually this assessment is taken mean square error—error or MSE (Mean Squared Error), defined as the average sum of squared differences between the desired output size E_{gen} and d_i real values obtained on the web L for each example:

$$E_{gen} = \frac{1}{P}\sum_{i=1}^{P} K_i\left(d_i - y_i\right)^2, \tag{1.5}$$

where P is the number of examples in the test sample, which is 20% of the training set.

Rating E_{gen} used in cases where the outputs must network with a given and equal for all signals match up with known vectors, where ε defined as a level of reliability.

For consideration of safety is commonly used modification of the formula

$$E = \frac{1}{P}\sum_{i=1}^{P} K_i\left(\frac{d_i - y_i}{\varepsilon}\right)^2 \tag{1.6}$$

where ε has a different range of variation for used methods of interpretation: $0 < \varepsilon < 1$—to sign interpretation; $0 < \varepsilon < 2$—for the rule of "winner takes all"; $0 < \varepsilon < 2/(n-1)$—for progressive interpretation, where n-dimension vector input signals.

The level of reliability is introduced to ensure the stable operation of the network. Stability criterion is formulated as follows: the work of the network is considered stable if output signals when changing network by an amount less ε Interpretation replies network changes. This can be used to provide accelerated training network: useful when calculating the score for formula (1.6) only use these outputs (set of correct answers), the interpretation of which does not change with changes in their values by an amount less than ε.

Evaluation E_{gen} can be generalized if used summing the squares of differences $(d_i - y_i)^2$ with corresponding weights:

$$E_{gen} = \frac{1}{P}\sum_{i=1}^{P} K_i\left(d_i - y_i\right)^2, \tag{1.7}$$

where K_i is the weight of i-example in the test sample. Using the evaluation reveals the most important examples of the training set, setting appropriate for this weight. In addition, this assessment should be used to balance the different groups of examples in problems of classification. To this end, K_i assign weights so that the total weight training examples in each class did not depend on the class (for example, you can assign any instance $K_i = 1/m$ where i is the number of class; t is the number of examples in class). In the case of fuzzy expert assessment "teacher" some examples of options to form a training set is also advisable to increase the weight of these options so that they can influence the learning network.

In cases where the comparison as forecasting models of heterogeneous objects, estimation errors in absolute terms unacceptable or causes difficulties with the interpretation of the results obtained. Therefore, the preferred transition to the estimation error percentage. For this purpose, using the average absolute error in percentage (Mean Absolute Percentage Error, MAPE):

$$MAPE = \frac{1}{P}\sum_{i=1}^{P}\frac{100||d_i| - |y_i||}{|y_i|}.$$

As the complexity of the criteria used in the construction of NM computing the total number of operations required to calculate output vector or number of interneuron ties.

1.11 Structural-Parametric Synthesis of Hybrid Neural Networks Based on the Use of Neurons of Different Topologies

As an example of solving the problem of structural-parametric synthesis of HNN based on the use of neurons of different topologies, let us consider without losing the commonality of the use of the following neural networks: multilayer perceptron, cascade radial basis and GMDH.

To build hybrid neural networks, we will use the following types of neurons: classic, Q-neural, wavelet neuron (a description of the topologies of these neurons is given in Sect. 1.6). As an optimization criterion, the generalized error criterion is used (Eq. 1.7).

The problem of structural-parametric synthesis of HNN is posed, which consists in the optimal choice of the number of layers, the number of neurons in the layers, the order of alternation of layers with different neurons.

A hybrid genetic algorithm (Sect. 2.4) was used to solve this problem. The structure of the chromosome consists of the following parameters: the type of neuron (Q-neuron, wavelet neuron, classic neuron), the number of neurons of one species in a particular layer and depending on the type of neuron the type and parameters of the activation function (membership).

The algorithm for solving the problem of structural-parametric synthesis of HNN based on the use of genetic algorithm has the following form:

1. Set the initial iteration number $i = 1$.
2. Form $x >$ $[x1 - x8]$-layer perceptron and solve the optimization problem of the optimal choice neuron type (Q-neuron, wavelet neuron, classic neuron), the number of neurons of one species in a particular layer and depending on the type of neuron the type and parameters of the activation function (membership) and the values of the weight coefficients.
3. If the generalized error for the NN found on the test data satisfies the set threshold then the learning process ends otherwise we proceed to point 4.
4. Form $i + 1$—layer perceptron and optimally using GA choice neuron type (Q-neuron, wavelet neuron, classic neuron), the number of neurons of one species in a particular layer and depending on the type of neuron the type and parameters of the activation function (membership) $i + 1$ layer and the values of the weight coefficients of the first and second layers (recalculation of the values of the weight coefficients of the first ith layers are repeated)
5. Repeat steps 3 and 4 until the optimal topology is found.

The solution algorithm is represented by the pseudocode below.

```
# create a single-layer network and write down its parameters

best_network = SPEA2.get_one_layer()
best_network.train()
best_loss = best_network.get_loss()
best_params = [best_network.get_params()]

for it in range(number_iterations):
  # form a network of layers that were found in previous iterations
  for i in best_params:
          cur_network +=
          get_layer_from_params(i)
  cur_network += SPEA2.get_one_layer()

  # train it
  cur_network.train()

  # check if the error is less than the best
  if cur_network.get_loss() < best_loss:
   # if the error is smaller, then we update the best network, the best error and add the last layer - the
configuration of this iteration

    best_params.append(
    cur_network.get_params())
    best_loss = cur_network.get_loss()
  else:
    # otherwise, if the error is not smaller, then the best network was found in the previous iteration, we
complete the search
    break
```

Table 1.1 Results of the operation of the optimal HNN

Number of layers	Type of neuron of a particular layer			Generalized error
	1st layer	2nd layer	3rd layer	
3	*Q*-neuron (64 neurons)	Classic neuron (32 neurons)	*Q*-neuron (16 neurons)	0.120621

Table 1.2 Results of the experiments on the forecasting problem

Number of layers	Type of neuron of a particular layer			Generalized error
	1st layer	2nd layer	3rd layer	
3	Classic neuron (64 neurons)	Wavelet neuron (32 neurons)	Wavelet neuron (16 neurons)	0.008060

The results of the solution of the problem of structural-parametric synthesis of HNN are considered in the example of the problem of classification and forecasting. A sampling of handwritten numbers from [18] was used for the classification task.

The results of the operation of the optimal HNN according to the criterion of generalized error in solving the classification problem are presented in Table 1.1.

The sample with the lowest temperatures was used for the forecasting task [19]. This sample has a minimum temperature for each day for a specified period. Sampling takes 64 consecutive days to enter the neural network and the next 65 days to exit the neural network.

The results of the experiments on the forecasting problem are presented in Table 1.2.

The results suggest that different architectures are required for different tasks. For the classification task, the best neuron is the Q-neuron, but for the prediction task, the wavelet neuron was the best.

Consider a radial basis network. It is a neural network that has an intermediate layer of radially basic neurons. Such a neuron converts the distance from the input vector to the center by a nonlinear law. Traditionally, this is a single-layer network, which is presented in Fig. 1.30.

Thus the output of the *RBN* network is a linear combination of some set of basic functions:

$$f(\overline{x}) = \sum_{j=1}^{m} w_j h_j(\overline{x})$$

where h_j means the function of a neuron, and w_j is the weighting for a neuron j.

Different radially basic functions can be considered as a neuron. In this paper, studies were conducted with the following:

- Gaussian function $h(\overline{x}) = \exp\left(-\frac{r^2}{2*s^2}\right)$.

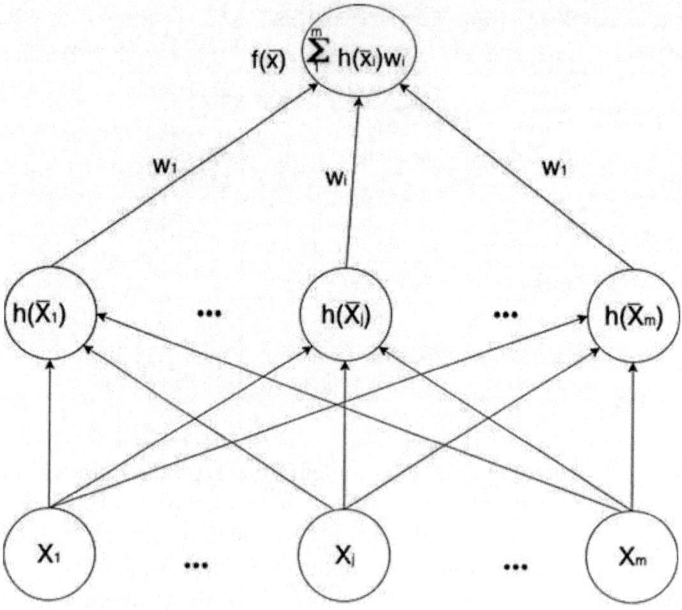

Fig. 1.30 Single-layer network

- Multiquadric $h(\overline{x}) = \sqrt{1 + \left(\frac{r^2}{2*s^2}\right)}$.
- Inverse quadratic $h(\overline{x}) = \frac{1}{1+\left(\frac{r^2}{2*s^2}\right)}$.
- Inverse multiquadratic $h(\overline{x}) = \frac{1}{\sqrt{1+\left(\frac{r^2}{2*s^2}\right)}}$.
- Thin plate spline $h(\overline{x}) = r^2 \ln(r)$.
- Linear $h(\overline{x}) = r$.
- Cubic $h(\overline{x}) = r^3$.

Where $r = \|x - c\|^2$, x is the vector input signal; c is the center vector, it is given at the beginning of training; s is the function width values.

A wavelet function can also be used as a radially basis function:

$$\varphi_{ji}(x_i)\left(1 - a_{ji}\, t_{ji}^2 \exp\left(-\frac{t_{ji}^2}{2}\right)\right)$$

where $t_{ji} = \left(x_i - c_{ji}\right)\sigma_{ji}^{-1}$, c_{ji} this is the center parameter, σ_{ji} width parameter and a_{ji} form parameter.

Cascade topology was used as the topology. Its principle is that initially the weights of a single layer, when its weights are adjusted, this layer freezes. The following layer is added. It accepts the input vector as well as the outputs of the previous layer. After

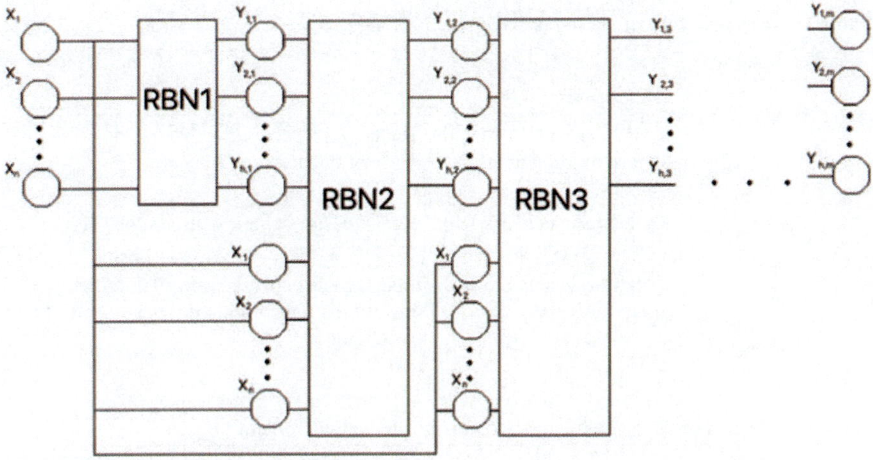

Fig. 1.31 Topology of the cascade network

adjusting the weights of the second layer, the following layer is similarly joined, etc. The topology of the cascade network is presented in Fig. 1.31.

The problem of structural-parametric synthesis of HNN is posed, which consists in the optimal choice of the number of layers, the number of neurons in the layers, the order of alternation of layers with different neurons.

Generalized error is selected as the optimization criterion:$E(n) = \frac{1}{2}\sum_{j \in C} e^2(n)$.

where $e_j(n)$ is the error signal of the original a_i neuron at the $a_i^n = \frac{a_i}{\sum_i a_i}$. iteration and is determined by the ratio:$e_j(n) = d_j(n) - y_j(n)$.

To solve this problem, we use a hybrid genetic algorithm (Sect. 2.4). The structure of the chromosome consists of the following parameters: number of network cascades, type of radial basis function in each cascade.

The algorithm for solving the problem of structural-parametric synthesis of HNN based on the use of genetic algorithm has the following form:

1. Set the initial iteration number $i = 1$.
2. Form i—cascading network and solve the problem of optimizing the optimal choice of the radial basis function and determining the appropriate weights.
3. If the generalized error for the NN found on the test data satisfies the set threshold then the learning process ends otherwise we proceed to point 4.
4. Form $(i + 1)$-cascading network and optimally using GA choice radial basis function and the corresponding weights of the $(i + 1)$ th cascade (the weights for the previous stages are unchanged).
5. Repeat steps 3 and 4 until the optimal topology is found.

The results of the solution of the problem of structural-parametric synthesis of HNN are considered in the example of the problem of classification and forecasting.

A sampling of handwritten numbers from [92] was used for the classification task.

Table 1.3 Results of the operation of the optimal HNN

Number of cascade	Type of radial basis network		Generalized error
	1st cascade	2nd cascade	
2	Radial basis network with inverse multiquadratic function	Radial basis network with wavelet function	0.2246878
2	Radial basis network with gaussian function	Radial basis network with wavelet function	0.2427712
2	Radial basis network with linear function	Radial basis network with inverse multiquadratic function	0.2506796

Table 1.4 Results of the synthesis of HNN on the prediction problem

Number of cascade	Type of radial basis network		Generalized error
	1st cascade	2nd cascade	
2	Radial basis network with gaussian function	Radial basis network with wavelet function	0.008361
1	Radial basis network with gaussian function		0.009199
2	Radial basis network with wavelet function	Radial basis network with gaussian function	0.009833

The results of the operation of the optimal HNN according to the criterion of generalized error in solving the classification problem are presented in Table 1.3.

The sample with the lowest temperatures was used for the forecasting task [93]. This sample has a minimum temperature for each day for a specified period. Sampling takes 64 consecutive days to enter the neural network and the next 65 days to exit the neural network.

The results of the synthesis of HNN on the prediction problem are presented in Table 1.4.

1.12 The Main Results Presented in the Book

Based on the analysis of scientific papers on the development HNN (Sect. 1.5) in order to overcome the existing difficulties in this paper the new structures, methods and algorithms shown in Fig. 1.32.

The book, based on newly developed methods and algorithms are resolved applications.

1. Building automated system traffic management (task approximation).
2. Intellectual Development of diagnostic systems in medicine (task classification).

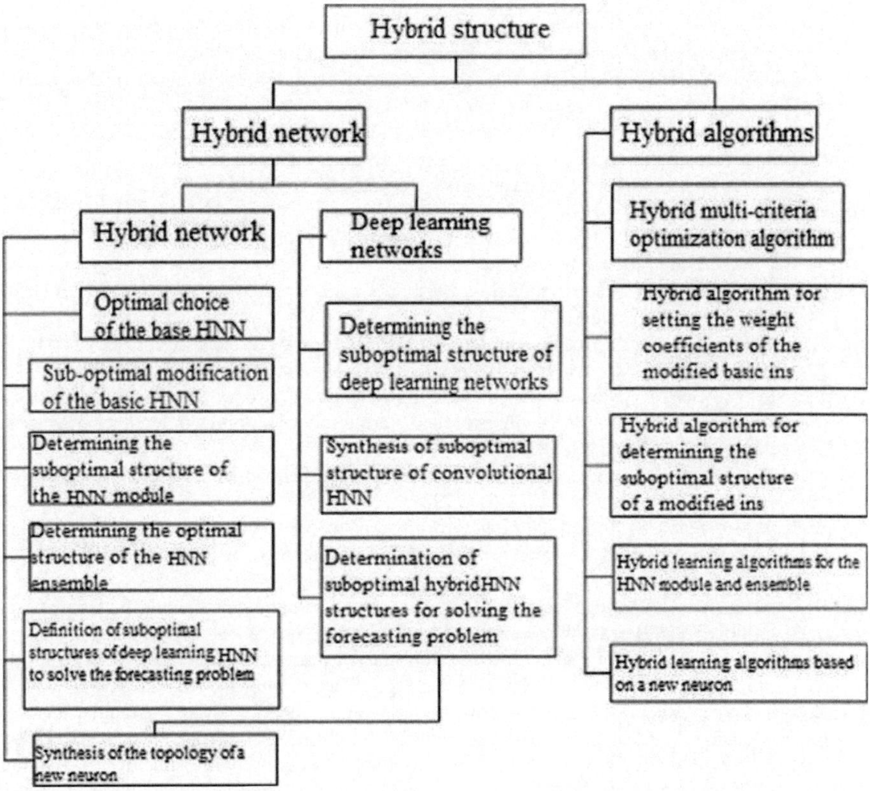

Fig. 1.32 The main results of the monograph

The development of the information system fire monitoring (forecasting, decision making).

References

1. Kruglov, V., Borisov, V.: Artificial Neural Networks. Theory and Practice, 2nd edn, Stereotype, 382 pp. M.: Hotline-Telecom (2002)
2. Zade, L.: The concept of a linguistic variable and its application to making approximate decisions. M.: Mir (1976)
3. Borisov, V., Kruglov, V., Fedulov, A.: Fuzzy models and networks. M.: Hotline-Telecom, 284 pp. (2007)
4. Chumachenko, H.: A hybrid evolutionary algorithm for the formation of a deep neural network topology. In: Chumachenko, H., Koval, D.: Proceedings of the International Scientific and Practical Conference on Information Technology and Computer Modeling. Ivano-Frankivsk – Yaremche, Ukraine, pp. 20–22, 23–28 May 2016
5. Khaykin, S.: Neural Networks. Full Course, 2nd edn., M.: Williams, 1104 pp. (2006)

6. Zaichenko, Y.: Fuzzy models and methods in intelligent systems. In: Textbook for Students of Higher Educational Institutions. Slovo Publishing House, 344 pp. (2008)
7. Bodyanskiy, Y.: Artificial neural networks: architectures, training, applications. In: Bodyanskiy, Y., Rudenko, O., Kharkiv: Teletech, 369 pp. (2004)
8. Golovko, V.: Neural networks: training, organization and application. In: M.: IPRZhR, 256 pp. (Series "Neurocomputers and their application". Book 4) (2001)
9. Gorban, A.: Neuroinformatics. Krasnoyarsk: SBRAS. 564 pp. (1999)
10. Chumachenko, H.: Features of hybrid neural networks use with input data of different types. In: Chumachenko, H., Koval, D., Sipakov, G., Shevchuk, D. (eds.) Electron. Control Syst. **4**(42), 91–97. NAU, Kyiv (2014)
11. Neal, R.M.: Bayesian mixture modeling by monte carlo simulation. Technical report crg-tr-91–2, Univeristy of Toronto (1992b)
12. Bodyanskiy, Y.: Hybrid cascade neural network based on wavelet-neuron. In: Bodyanskiy, Y., Kharchenko, O., Vynokurova, H. (eds.) Int. J. Inf. Theor. Appl. **18**(4), 335–343 (2008)
13. Becraft, W., Lee, P., Newell, R.: Integration of neural networks and expert systems for process fault diagnosis. In: Proceedings of the 12th International Joint Conference on Artificial Intelligence, vol. 1–2, pp. 832–837 (1991)
14. Baxt, W.G.: Improving the accuracy of an artificial neural network using multiple differently trained networks. Neural Comput. **4**(5) (1992)
15. Bodyanskiy, Y (2008) Robust learning algorithm for wavelet-neural-fuzzy network based on Polywog wavelet. In: Bodyanskiy, Y., Vynokurova, H. (eds.) Syst. Technol. **2**, **3**(56) 2, 129–134 (2008)
16. Bodyanskiy, Y.: An adaptive learning algorithm for a wavelet neural network. In: Bodyanskiy, Y., Lamonova, N., Pliss, I., Vynokurova, H. (eds.) Expert Syst. **22**(5), 235–240 (2005)
17. Bodyanskiy, Y.: Duble-wavelet neuron based on analytical activation functions. In: Bodyanskiy, Y., Lamonova, N., Vynokurova, H. (eds.) Int. J. Inf. Theor. Appl. **14**, 281–288 (2007)
18. Bodyanskiy, Y.: Adaptive wavelet-neuro-fuzzy network in the forecasting and emulation tasks. In: Bodyanskiy, Y., Pliss, I., Vynokurova, H. (eds.) Int. J. Inf. Theor. Appl. **15**(1), 47–55 (2008)
19. Bodyanskiy, Y.: Radial-basis-fuzzy-wavelet-neural network with adaptive activation-membership function. In: Bodyanskiy, Y., Vynokurova, H., Yegorova , E. (eds.) Int. J. Inf. Theor. Appl. **8**(II), 9–15 (2008)
20. Bodyanskiy, Y.: Outliers resistant learning algorithm for radial-basis-fuzzy-wavelet-neural network in stomach acute injury diagnosis tasks. In: Bodyanskiy, Y., Pavlov, O., Vynokurova, H. (eds.) Inf. Sci. Comput. Institute of Information Theories and Application, Sofia **2**, 55–62 (2008)
21. Bodyanskiy, Y.: Adaptive compartmental wavelon with robust learning algorithm. In: Bodyanskiy, Y., Pavlov, O., Vynokurova, H. (eds.) Int. J. Inf. Technol. Knowl. **3**, 24–36 (2009)
22. Bodyanskiy, Y.: Hybrid type-2 wavelet-neuro-fuzzy network for businesses process prediction. In: Bodyanskiy, Y., Vynokurova, H. (eds.) Bus. Inf. **21**, 9–21 (2011)
23. Bodyanskiy, Y.: Hybrid adaptive wavelet-neuro-fuzzy system for chaotic time series identification. In: Bodyanskiy, Y., Vynokurova, H. (eds.) Inf. Sci.[Electronic resource] Access mode: http://dx.doi.org/10.1016/j.ins.2012.07.044 (2012)
24. Bodyanskiy, Y.: Hybrid GMDH-neural network of computational intelligence. In: Bodyanskiy, Y., Pliss, I., Vynokurova, H. Proceedings of 3rd International Workshop on Inductive Modelling, pp. 100–107, Poland, Krynica (2009)
25. Bodyanskiy, Y., Zaychenko, Y., Pavlikovskaya, H., Samarina, M., Viktorov, Y.: The neofuzzy neural network structure optimization using the GMDH for the solving forecasting and classification problems. In: Proceedings of International Workshop on Inductive Modeling, Krynica, Poland, pp. 77–89 (2009)
26. Bridle, J.S., Cox, S.J.: RecNorm: simultaneous normalization and classification applied to speech recognition. In: Advances in Neural Information Processing Systems, vol. 3 (1991)
27. Buntine, W.L., Weigend, A.S.: Bayesian back-propagation. Complex Syst. **5**, 603–643 (1992)
28. Bodyanskiy, Y., Boiko, H.: Evolving multilayer system neuro-fuzzy system and its learning. Syst. Technol. **5**(100), 161–169 (2015)

29. Chen, Y., Kak, S., Wang, L.: Hybrid Neural Network Architecture for On-Line Learning
30. Cheng, M.Y., Tsai, H.C., Sudjono, E.: Evolutionary fuzzy hybrid neural network for conceptual cost estimates in construction projects. Inf. Comput. Technol, 512–519
31. Tao, L., Guo, T., Aberer, K.: Hybrid neural networks for learning the trend in time series. In: Proceedings of the Twenty-Sixth International Joint Conference on Artificial Intelligence (IJCAI-17), pp. 2273–2279
32. Sineglazov, V.: Improvement of hybrid genetic algorithm for synthesis of deep neural networks. In: Sineglazov, V., Chumachenko, H., Koval, D. (eds.) IV International Scientific-Practical Conference "Computational Intelligence", pp. 142–143, Kyiv, Ukraine (2017)
33. Cooper, L.N.: Hybrid neural network architectures: equilibrium systems that pay attention. In: Mammone, R.J., Zeevi, Y. (eds.) Neural Networks: Theory and Applications. Academic Press, vol. 1, pp. 81–96 (1991)
34. Hansen, L.K., Salamon, P.: Neural network ensembles. IEEE Trans. Pattern Anal. Mach. Intell. **12**(10), 993–1000 (1990)
35. Intrator, N., Reisfeld, D., Yeshurun, Y.: Face recognition using a hybrid supervised/unsupervised neural network. Preprint (1992)
36. Jacobs, R.A., Jordan, M.I., Nowlan, S.J., Hinton, G.E.: Adaptive mixtures of local experts. Neural Comput. **3**(2) (1991)
37. Kerber, R., Livezey, B., Simoudis, E.: A hybrid system for data mining. In: Goonatilake, S., Khebbal, S. (eds.) Intelligent Hybrid Systems, pp. 121–421. John Wiley and Sons, Chichester (1995)
38. Lincoln, W.P., Skrzypek, J.: Synergy of clustering multiple back propagation networks. In: Advances in Neural Information Processing Systems, vol. 2 (1990)
39. Neal, R.M.: Bayesian learning via stochastic dynamics. In: Moody, J.E., Hanson, S.J., Lippmann, R.P. (eds) Advances in Neural Information Processing Systems, vol. 5., Morgan Kaufmann, San Mateo, CA (1992a)
40. Pearlmutter, B.A., Rosenfeld, R.: Chaitin-kolmogorov complexity and generalization in neural networks. In: Advances in Neural Information Processing Systems, vol 3 (1991)
41. Reilly, D.L., Scofield, C.L., Cooper, L.N., Elbaum, C.: Gensep: A multiple neural network learning system with modifiable network topology. In: Abstracts of the First Annual International Neural Network Society Meeting (1988)
42. Reilly, R.L., Scofield, C.L., Elbaum, C., Cooper, L.N.: Learning system architectures composed of multiple learning modules. In: Proceedings of IEEE First International Conference on Neural Networks, vol. 2 (1987)
43. Scofield, C., Kenton, L., Chang, J.: Multiple neural net architectures for character recognition. In: Proceedings of Compcon, San Francisco, CA. IEEE Computer. Society Press, pp. 487–491, February, 1991
44. Wermter, S.: Hybrid approaches to neural network-based language processing, Technical Report TR-97-030. International Computer Science Institute, Berkeley, California (1997)
45. Wolpert, D. H.: Stacked generalization. Technical report LA-UR-90–3460, Complex Systems Group, Los Alamos, NM (1990)
46. Xu, L., Krzyzak, A., Suen, C.Y. (1990). Associative switch for combining classifiers. Technical report x9011, Department of Computer Science, Concordia University, Montreal, Canada
47. Xu, L., Krzyzak, A., Suen, C.Y.: (1992). Methods of combining multiple classifiers and their applications to handwriting recognition. IEEE Trans. Syst. Man Cybern. **22**(3)
48. Handelman, D., Lane, S., Gelfand, J.: Robotic skill acquisition based on biological principles. In: Kandel, A. (ed.) Hybrid Architectures for Intelligent Systems, ppp. 301–327. CRC Press (1992)
49. Khosla, R., Dillon, T.: Fusion of knowledge-based systems and neural networks and applications. In: 1st International Conference on Knowledge-Based Intelligent Electronic Systems, pp. 27–44, Adelaide, Australia, 21–23 May 1997
50. Hudson, D., Banda, P.W., Cohen, M.E., Blois, M.S.: Medical diagnosis and treatment plans derived from a hybrid expert system. In: Kandel, A. (ed.) Hybrid Architectures for Intelligent Systems, ppp. 330–244. CRC Press (1992)

51. Adgar, A., Emmanouilidis, C., MacIntyre, J., Mattison, P., McGarry, K., Oatley, G., Taylor, O.: The application of adaptive systems in condition monitoring. In: Rao, R. (ed.) Int. J. Cond. Monitor. Diagn. Eng. Manage. **1**(1), 13–17 (1998)
52. Hinton, G.E.: Connectionist learning procedures. Artifi. Intell. **40**(26), 185–234 (1989)
53. Hinton, G.E.: How neural networks learn from experience. Sci. Am. 105–109 (1992)
54. Lippmann, R.P.: An introduction to computing with neural nets. IEEE ASSP Bull. 4–22 (1987)
55. Bodyanskiy, Y.: Hybrid wavelet-neuro-fuzzy system using adaptive W-neurons. In: Bodyanskiy, Y., Pliss, I., Vynokurova, H. (eds.) Wissenschaftliche Berichte, FH Zittau/Goerlitz. **106**(№2454–2490), 301–308 (2010)
56. Vynokurova, H.: Functional coupled multidimensional wavelet-neuro phase system for processing chaotic time series. Sci. Works: Sci. Methodol. J. Comput. Technol. **130**(143), 71–76 (2010)
57. Vynokurova, H.: Adaptive algorithm for fuzzy wavelet neural network training. In: Vynokurova, H., Bednarskaya, G., Pliss, I. (eds.) Inf. Process. syst. **59**(1), 15–18 (2007)
58. Bodyanskiy, Y.: A wavelet neuron learning algorithm based on a combined criterion. In: Bodyanskiy, Y., Vynokurova, H., Lamonova, N., Pliss, I. (eds.) Radio Electron. Comput. Sci. Control **14**(2), 83–89 (2005)
59. Bodyanskiy, Y.: Double wavelet neuron and its learning algorithm. In: Bodyanskiy, Y., Vynokurova, H., Lamonova, N. (eds.) Radio Electroni. Comput. Sci. Control **16**(2), 85–91 (2006)
60. Bodyanskiy, Y.: Compound adaptive waveon and its learning algorithm. In: Bodyanskiy, Y., Vynokurova, H. (eds.) Control Syst Mach **1**(219), 47–53 (2009)
61. Bodyanskiy, Y.: Double wavelet neuron: triangular activation functions, architecture, training. In: Bodyanskiy, Y., Vynokurova, H. (eds.) Adapt. Autom. Control Syst. **29**(9), 16–22 (2005)
62. Bodyanskiy, Y.: Robust learning algorithm for radial-base fuzzy wavelet neural network. In: Bodyanskiy, Y., Vynokurova, H. (eds.) Adapt. Autom. Control Syst. **11**(31), 3–15 (2007)
63. Bodyanskiy, Y.: Wavelet-neural fuzzy type-2 system and algorithm of its training in problems of intellectual information processing. In: Bodyanskiy, Y., Vynokurova, H. (eds.) Adapt. Autom. Control Syst. **17**(37), 139–148 (2010)
64. Bodyanskiy, Y.: Wavelet-fuzzy neuron type-2. In: Bodyanskiy, Y., Vynokurova, H., Kharchenko, H. (eds.) Bulletin of Lviv Polytechnic National University. Computer science and Information Technology **170**, 175–181 (2011)
65. Vynokurova, H.: On a learning algorithm for an adaptive wavelet-neural network on a sliding window. In: Vynokurova, H., Lamonova, N. (eds.) Autom. Prod. Process. **21**(2), 71–75 (2005)
66. Bodyanskiy, Y.: Intelligent dara processing based on hybrid wavelet neural fuzzy system on adaptive W-neurons. In: Bodyanskiy, Y., Vynokurova, H. (eds.) Sci. Works: Sci. Methodol. J. Comput. Technol. **104**(117), pp. 88–98 (2009)
67. Vynokurova, H.: The algorithm of training of the radial-basis phase-wavelet-neural network. In: Vynokurova, H., Bednarskaya, G., Pliss, I. (eds.) Appl. Radio Electron. **6**(3), 427–431 (2007)
68. Bodyanskiy, Y.: Generalized multidimensional wavelet-neuro-fuzzy system in computational intelligence problems. In: Bodyanskiy, Y., Vynokurova, H., Kharchenko, H. (eds.) The Intellectual System Accepts the Problem of Calculating the Intellectual: Science and Technology Library for the Materials of the International Science Conference, 1st edn, pp. 215–220. Kherson(2011)
69. Loskutov, A.: Neural network algorithms for predicting and optimizing systems. In: Loskutov, A., Nazarov, A. (eds.) Sci. Technol. **384** (2003)
70. Bodyanskiy, Y.: Hybrid evolving cascade GMDH neural network on fuzzy type-2 wavelet neurons. Inductive Model Fold. Syst. **3**, 17–26 (2011)
71. Bodyanskiy, Y. Adaptive wavelet as a node of artificial GMDH-neural networks. In: Bodyanskiy, Y., Vynokurova, H. (eds.) Modeling and management of the state of ecological and economic systems of regions **4**, 19–29 (2008)
72. Bodyanskiy, Y.: Functionally coupled GMDH-wavelet-neural-fuzzy system and robust algorithm for its learning. In: Bodyanskiy, Y., Vynokurova, H. (eds.) Inductive Model Fold. Syst. **2**, 15–24 (2010)

73. Vynokurova, H.: Generalized multidimensional wavelet-neuro-fuzzy system in computational intelligence problems. In: Intelligent Decision-Making Systems and Problems of Computational Intelligence: A Collection of Scientific Papers Based on the Materials of an International Scientific Conference, 2nd edn, pp. 329–333. Yevpatoria, Kherson (2010)
74. Vynokurova, H.: Hybrid adaptive neural fuzzy and wavelet-neural fuzzy systems of computational intelligence in the problems of signal processing in the presence of interference. Adaptive Autom. Control Syst. **15**(35), 113–120 (2009)
75. Hu, Z., Bodyanskiy, Y., Tyshchenko, O.: A cascade deep neuro-fuzzy system for high-dimensional online possibilistic fuzzy clustering. In: Proceedings of the XI-th International Scientific and Technical Conference on "Computer Science and Information Technologies" (CSIT 2016), pp. 119–122. https://doi.org/10.1109/stc-csit.2016.7589884
76. Zaychenko, Y., Bodyanskiy, Y., Tyshchenko, O., Boiko, H., Hamidov, G.: Hybrid GMDH-neuro-fuzzy system and its training scheme. Int. J. Inf. Theor. Appl. **24**(2), 156–172 (2018)
77. Bodyanskiy, Y., Vynokurova, H., Teslenko, N.: Cascade GMDH-Wavelet-Neuro-Fuzzy network. In: The 4th International Workshop on Inductive Modelling IWIM 2011, pp. 22–30 (2011)
78. Huang,W., Oh, S.K., Pedrycz,W.: Fuzzy polynomial neural networks: hybrid architectures of fuzzy modeling. IEEE Trans. Fuzzy Syst. **10**(5), 607–621 (2002)
79. Oh, S.K., Kim D.W., Pedrycz, W.: Hybrid fuzzy polynomial neural networks. Int. J. Uncertain. Fuzziness Knowl. Based Syst. **10**(3). https://doi.org/10.1142/S0218488502001478
80. Oh, S.K., Pedrycz, W., Park, B.J.: Self-organizing multi-layer fuzzy polynomial neural networks based on genetic optimization. Fuzzy Sets Syst. **145**(1), 165–181, July 2004
81. Oh, S.K., Park, B.J.: Self-organizing neuro-fuzzy networks in modeling software data. Neurocomputing. **64**, 397–431 (2005)
82. Oh, S.K., Pedrycz, W., Kim, H.K., Lee, J.B.: Self-organizing multi-layer fuzzy polynomial neural networks based on genetic optimization. In: 4-th International Conference on Computational Science-ICCS 2004, Proceedings, Part II, pp. 179–187, Krakov, Poland, June 2004
83. Oh, S.K., Pedrycz, W., Roh, S.B.: Genetically optimized hybrid fuzzy set-based polynomial neural networks. J. Franklin Inst. **348**(2), 415–425 (2011)
84. Park, B.J., Oh, S.K., Pedrycz, W., Ahn, T.C.: Information granulation-based multi-layer hybrid fuzzy neural networks: analysis and design. In: 4-th International Conference on Computational Science—ICCS 2004, Proceedings, Part II, pp. 188–195, Krakov, Poland, June 2004
85. Huang, W., Oh, S.K., Pedrycz, W.: Hybrid fuzzy polynomial neural networks with the aid of weighted fuzzy clustering method and fuzzy polynomial neurons. Article. First Online: 15 September 2016
86. Bodyanskiy, Y., Tishchenko, A., Kopaliyani, D.: Multidimensional cascade neuro-fuzzy system with neuron pool optimization. In: Bulletin of NTU "KhPI". Series: Mathematical Modeling in Engineering and Technology. NTU "KhPI", Kharkiv **18**(1061), 17–26 (2014)
87. Zaichenko, Y., Gamidov, G, Varga, I.: Diagnosis of medical images of tumors using hybrid fuzzy convolutional neural networks. System. Inf. Technol. **4**, 37–47 (2018)
88. Wu, Y, Wu, W, Xu, C, Li Z Knowledge enhanced hybrid neural network for text matching. In: The Thirty-Second AAAI Conference on Artificial Intelligence (AAAI-18)
89. Perrone, M.P., Cooper, L.N.: When networks disagree: ensemble methods for hybrid neural networks. In: Mammone, R.J. (eds.) Neural Networks for Speech and Image processing, 15 pp. Chapman-Hall (1992)
90. Perova, I., Pliss, I.: Deep hybrid system of computational intelligence with architecture adaptation for medical fuzzy diagnostics. Int. J. Intell. Syst. Appl. **7**, 12–21 (2017)
91. Janning, R., Schatten, C. Schmid t-Thieme, L.: HNNP—a hybrid neural network plait for improving image classification with additional side information. In: Information Systems and Machine Learning Lab (ISMLL), pp. 1–7, University of Hildesheim, Hildesheim, Germany

92. https://archive.ics.uci.edu/ml/datasets/Optical+Recognition+of+Handwritten+Digits
93. https://www.kaggle.com/paulbrabban/daily-minimum-temperatures-in-melbourne

Chapter 2
Classification and Analysis of Multicriteria Optimization Methods

2.1 Classification of Optimization Methods

The problem of optimization is becoming one of the leading in the world of artificial intelligence. It is often representable in the form of an objective function that needs to be optimized (moreover, it is not always defined analytically, and sometimes defined as a "black box"), and a certain set of initial data and constraints on the solution.

For most of these problems, deterministic methods of solution are unacceptable or do not provide the necessary degree of accuracy. Therefore, an alternative approach is needed—the use of evolutionary methods of global optimization and the intentional introduction of an element of randomness in the search algorithm.

The main advantages of such methods are [1, 2]:

- increased performance;
- high reliability and noise immunity;
- high robustness, i.e. insensitivity to irregularities in the behavior of the objective
- function, the presence of random errors in the calculation of the function;
- a relatively simple internal implementation;
- low sensitivity to the growth of the dimension of the optimization set;
- the possibility of a natural input in the process of searching for the operation of learning self-training;
- within the framework of well-known random search schemes, new algorithms that implement various heuristic adaptation procedures are easily constructed. The classification of global optimization methods is presented in Fig. 2.1.

M. Zgurovsky et al., *Artificial Intelligence Systems Based on Hybrid Neural Networks*, Studies in Computational Intelligence 904,
https://doi.org/10.1007/978-3-030-48453-8_2

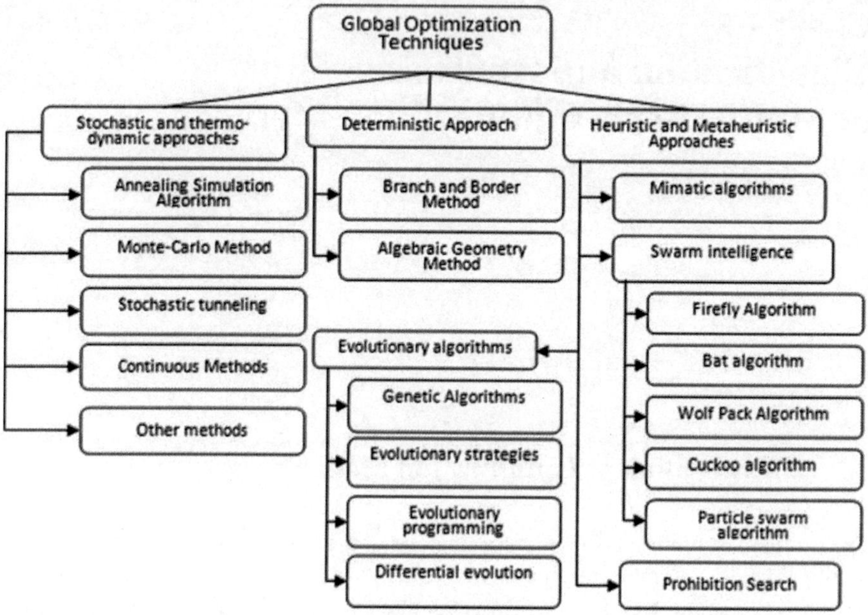

Fig. 2.1 Global optimization techniques

2.2 Overview of Optimization Techniques Used in Machine Learning

Modern neural network methods are among the most popular and continuously developing machine learning algorithms used in various fields of practical activity.

Due to the wide scope of their application, various tasks are formed that differ in the setting and types of input data: image recognition, parsing of texts, diagnosis of diseases [3, 4] etc.

In connection with the continuous improvement of existing neural network algorithms that differ in their properties and implementation features, the problem often arises of determining the most effective method of minimizing the error function, which guarantees the best results when solving a particular problem.

The main optimization methods used in training neural networks are evolutionary methods (genetic algorithms), bionic (swarm algorithms), gradient algorithms.

It is known that the most used method for training neural networks is the backpropagation algorithm. However, this algorithm has disadvantages [3]:

1. The possibility of a premature stop due to falling into the local minimum;
2. The need for multiple presentation of the entire training set to obtain a given recognition quality.
3. Lack of any acceptable estimates of study time.

The problems associated with the backpropagation algorithm have led to the development of alternative methods for calculating the weighting coefficients of neural networks. For the first time in 1989, David Montana and Lawrence Davis used genetic algorithms as a means of adjusting the weights of hidden and output layers for a fixed set of relationships.

The genetic algorithm is the most famous representative of evolutionary algorithms now and is essentially an algorithm for finding the global extremum of a multi-extreme function. It consists in parallel processing of many alternative solutions. At the same time, the search concentrates on the most promising of them. This suggests the possibility of using genetic algorithms in solving any problems of artificial intelligence, optimization, decision making.

The genetic algorithm has conceptual properties that give it a number of advantages in solving practical problems. One of them is high adaptability to environmental conditions. Everyday tasks of practical activity can have high dynamics in the process of their solution. When using traditional methods, the solution of such actively changing tasks leads to large expenditures of time and resources, which is not optimal. The evolutionary approach allows flexible analysis and population changes in relation to dynamic conditions. In this case, complete exhaustive search may not always be used. Another advantage of the genetic algorithm for solving problems is the ability to quickly generate sufficiently optimal solutions.

Genetic algorithms have the following features [5]:

- these algorithms operate with a variety of parameters in encoded form;
- these algorithms search for sets-populations of points, rather than a single point;
- these algorithms are not a gradient descent method. This means that it does not use the value of derivatives, hessians, wavelets or other auxiliary parameters of the function;
- in these algorithms, decision rules are probabilistic rather than deterministic.

The given detailing of the organization of the genetic algorithm significantly distinguishes it from traditional approaches characteristic of classical optimization.

In terms of use, the following advantages can be distinguished:

- wide scope—due to the possibility of problem-oriented, functional coding of solutions;
- genetic algorithms can be combined with other types of algorithms—a combination of both evolutionary and non-evolutionary algorithms is possible;
- genetic algorithms are easy to implement when searching in a space of large-dimensional solutions and are independent of its structure;
- genetic algorithms are invariant with respect to the type of objective function (fitness function).
- low dependence of efficiency on optimizer settings for many tasks;
- GA does not depend on the problem being investigated, since it works with encoded parameters;
- GA is successfully used for a wide range of tasks;

- GA is applicable for non-formalizable tasks when the objective function is not clearly formulated or even absent.

Among the disadvantages, the following:

- due to the heuristic nature of genetic algorithms, there is no guarantee that there are optimality criteria for the resulting solution;
- relatively high computational complexity. It is determined by the assessment of both prospects and retrospectives for the process of modeling evolution;
- low efficiency of genetic algorithms in the final phases of the search, because there are no mechanisms aimed at quickly reaching the local optimum.
- using GA, it is difficult to find the exact global optimum;
- GA is not easy to configure to find all solutions to the problem;
- not for all tasks it is possible to find the optimal coding of parameters;
- GA is hard to apply to isolated functions. Isolation ("finding a needle in a haystack") is a problem for any optimization method, since the function does not provide any information that tells you in which area to look for the maximum. Only a random hit of an individual in the global extremum can solve the problem;
- GA has difficulty optimizing highly noisy functions. The additional noise greatly affects the convergence of many evolutionary methods, and therefore often slows down the search for GA solutions.

The main difficulty in solving engineering optimization problems with a large number of local optima is the preliminary convergence of the algorithms. It consists in the fact that for a multimodal function, the solution may be a local optimum, which delivers a nonoptimal value. At the level of the genetic algorithm, the convergence problem is partially solved by using various options for parameters and their modifications.

The general trend in the study of these algorithms is increasingly the use of combined parameter methods. In this case, it is important to combine a preliminary knowledge of the tasks to be solved and preliminary results, as with the reverse descent algorithm.

The main advantage of genetic algorithms lies precisely in the fact that they can significantly overcome these difficulties.

Population-based multi-agent algorithms work with a set of potential solutions. Each decision is gradually improved and evaluated, thus, each potential decision influences how other solutions will be improved. Most population methods borrowed this concept from biology: the process of finding the best solution "copies" a certain natural process or the behavior of certain animal species, and their species characteristics are taken into account.

The class of complex systems, referred to as flock type algorithms, is also often used the term "bionic algorithms", a rich source of non-standard numerical methods with which you can solve complex problems when there is not enough information about the optimized function.

The list of common swarm algorithms

- Ant colony algorithm;
- Particle Swarm Algorithm;
- Stochastic Diffusion Search;
- Bacteria swarm algorithm;
- Electromagnetic search;
- Bee Algorithm;
- Algorithm for the dynamics of river formation;
- Search for schools of fish;
- Monkey algorithm;
- Intelligent Water Drop Algorithm;
- Bat algorithm;
- Gravity Search Algorithm;
- Firefly Algorithm;
- Wolf Pack Algorithm.

It should also be noted that most flock type algorithms originally designed to solve unconditional optimization problems with real variables, and they all have some parameters that it is necessary to select when solving a particular problem (the most significant "parameter" in this sense is the size of the population of potential solutions). However, studies have shown that it is impossible to determine in advance which it is from these algorithms that should be applied to solve one or another optimization task, in addition, as already mentioned, for each the algorithm you need to choose in advance the size of the population of potential solutions, which in turn is also a challenge.

Population search engine optimization algorithms in comparison with classical algorithms have undeniable advantages, especially when solving high-dimensional problems, multimodal and poorly formalized problems. Under these conditions, population algorithms can provide a high probability of localization of the global extremum of the optimized function.

It is also important that population-based algorithms make it possible to more efficiently find classical sub-optimal (close to optimal) solutions than classical algorithms. Such a solution is often sufficient.

Population algorithms will almost certainly lose to the exhaustive search algorithm when solving the optimization problem of low dimension. In the case of a smooth and unimodal optimizable function, population algorithms are usually less efficient than any classical gradient method. The disadvantages of population algorithms include the strong dependence of their efficiency on the values of free parameters, the number of which in most algorithms is quite large.

Since population algorithms are stochastic, their effectiveness, as a rule, varies widely depending on the success of the initial approximation obtained at the initialization stage of the population. In this regard, to evaluate the effectiveness of these algorithms, multiple runs of the algorithm are used, based on different initial approximations (multistart method). The main criteria for the effectiveness of population-based algorithms are the reliability of the algorithm—an estimate of the probability of localization of a global extremum, as well as its convergence rate—an estimate of the mathematical expectation of the required number of tests (calculations of the value of the optimized function).

The advantages of the gradient descent method are, first of all, the ease of implementation, as well as the fact that the method is guaranteed to converge to a global or local minimum for convex and non-convex functions, respectively. However, there are many disadvantages of this method, due to which this method is rarely used in real practice.

1. Gradient descent can be very slow on large data sets, because at each iteration, the gradient for all vectors of the training set is calculated.
2. It does not allow updating the model "on the fly" and adding new examples of the training sample in the learning process also because the weights of the objective function are updated immediately for the entire initial data set.
3. For non-convex functions, there is a problem of falling into local minima, because the method guarantees an exact solution only for convex error objective functions.
4. Choosing the best learning speed can be a difficult problem. Too low a learning speed can lead to very slow convergence, on the contrary, a high learning speed can prevent convergence, and as a result, the error function will fluctuate around a minimum and will not reach it.
5. Uniform updating of all parameters with the same learning speed leads to a deterioration in the quality of training if the initial data set is not balanced, that is, the sample contains classes represented by fewer objects.

Due to the constancy of the learning speed parameter λ during the model training process, it is not possible to get the process out of the local minimum if λ small enough, and vice versa, it is impossible to stop at an acceptable optimum if λ big enough.

The problem is also the fact that in a complex model, different parameters may vary in different ways: some require a smaller learning step, while others, on the contrary, more. This problem is especially relevant when inputs are not normalized and scaled.

These defects make it impossible to apply the pure gradient method to deep neural network models.

In this work, evolutionary (genetic) and swarm algorithms are used to solve the problems of structural-parametric synthesis of hybrid neural networks.

2.3 Problem Statement of Multicriteria Optimization

In general, a multicriteria optimization problem involves a set of N parameters (variables), a set of objective functions K and a set of restrictions M [6–8].

Thus, under solving a multicriteria problem, it is necessary to find the optimum for K criteria, and the task itself is formally written as follows:

$$y = f(x) = (f1(x), f2(x), \ldots, fk(x)) \rightarrow opt,$$

where $x = (x1, x2, \ldots xN)^T X$ is the vector of solutions that satisfies m restrictions; $g(x) = (g1(x), g2(x), \ldots, gm(x)) \geq 0$; $y = (y1, y2, \ldots, yk)$ is the vector of objective functions.

In this case, X denotes the space of solutions, and Y—criteria space.

Restrictions $S = \{(\vec{x}_i, y_i) : i = 1 \ldots n\}$ determine the set of admissible problem solutions.

The admissible set D is defined as the set of vectors-solutions x which satisfying the restrictions of $g(x)$

$$D = \{x \ X | g(x) \geq 0\}$$

Then the admissible area in adjectives space is the image D, is denoted by

$$Y_f = f(D) = \bigcup_{x \epsilon D} \{f(x)\}$$

The multicriteria problems turn out to be a special class of problems where the usual heuristics often lead to contradictions, since they have not any universal meaning of "optimum" as in single-criteria optimization problems, which makes it difficult to compare one method of multicriteria optimization with others. And all because the solution of this kind of problem is not the only optimal solution, but a set of compromise solutions, more commonly known as *Pareto-optimal* (*effective*) solutions.

The presented definition of an effective solution can be specified by introducing the concept of Pareto domination, which is the basis in the theory of multicriteria choice and more accurately reveals the essence of the formulation "one solution is better than another".

The concept of Pareto domination

We assume that the minimization problem for N criteria is solved. Then the concept of Pareto domination for any two vectors is defined by three possible variants:

1. Solution a dominates the solution b : $a < b$, *if* $f_i(a) < f_i(b)\ \forall\ i = \overline{1, N}$.
2. Solution a weakly dominates solution b : $a \geq b$, *if* $f_i(a) \leq f_i(b)\ \forall\ i = \overline{1, N}$.
3. Solution a and b are incomparable: $a \approx b$, if $f_i(\mathrm{a}) \nless f_i(\mathrm{b}) \wedge f_i(\mathrm{a}) \ngtr f_i(\mathrm{b})$ $\forall\ i = \overline{1, N}$.

Thus, if the vector x non-dominant relatively D, then it is called Pareto-optimal, i.e. if it does not exist $y \in D : y > x$, that $\mu^*_{uv}(y_u)$ is Pareto optimal.

Pareto set and front

The set of all effective points is called ***the Pareto set*** in the space of variables (alternatives), and their image in the criterion space—***Pareto front***.

From all of the above, we can draw the following conclusion. For any admissible point that is outside the Pareto set, it will be found the point the Pareto set which gives for all objective functions the value not worse than at this point and at least by one objective function—strictly better. It follows that the solution of a multicriteria optimization problem is advisable to choose from Pareto set, because any other one can obviously be improved some Pareto point by at one criterion without degrading the other criteria.

From the point of view of mathematics, some solution from the Pareto set cannot be recognized as the best for another solution from the Pareto set, so after the formation of the Pareto set, the problem of multicriteria optimization can be considered mathematically solved.

One of the most promising methods for solving the problem of multicriteria optimization is genetic algorithms.

2.4 Multicriteria Genetic Algorithms

2.4.1 The General Design of a Genetic Algorithm

Genetic algorithms belong to the class of evolutionary algorithms and have a number of characteristics that make them better than classical optimization methods [9–11]:

- the search of effective solutions with help of genetic algorithms does not need the specific knowledge about the same task of parameters that go into it;
- in genetic algorithms, instead of deterministic ones, it is used the stochastic operators, which have shown themselves to be quite stable in the noisy environment;
- the inherent parallelism of genetic algorithms—simultaneous consideration of a large number of population individuals—makes them less sensitive to local Optima and the influence of noises.

The use of genetic algorithms for solving problems of multicriteria optimization allows to lose of the main disadvantages of classical methods, since genetic algorithms are suitable for problems of large dimension and are able to capture Pareto-optimal points even with a single start of the algorithm. By supporting population solutions and applying the concept of Pareto optimality, genetic algorithms can find different Pareto optimal solutions in parallel.

Thus, unlike most classical approaches to solving problems of multicriteria optimization, when to obtain each individual point it is necessary to carry out a separate start of the Pareto-optimal solutions search algorithm, applying an evolutionary approach to vector optimization, due to the inherent parallelism in genetic algorithms, it is possible to obtain different points of the Pareto set in one run of the algorithm. This circumstance is an obvious advantage of the evolutionary approach to solving problems of multicriteria optimization over traditional methods of their solution.

By analogy with nature evolution, in genetic algorithms, candidate solutions are called *individuals*, a set of candidates-solutions-*population*. Each individual defines a possible solution of the problem, while, however, it is not itself a solution vector, but rather encodes it based on the corresponding solution encoding structure. In genetic algorithms, this structure is determined by a vector-a vector of bits or a vector of real numbers-a set of genes that form *chromosomes*. The set of all possible vectors forms a space of individuals or a population.

The quality of the individual under primisation solution is determined by a scalar value, the so-called *fitness function*.

In the **selection** process, which can be both stochastic and deterministic, the worst solutions—the unfitted individuals are eliminated from the population, while the individuals with greater fitness—the fittest-undergo reproduction. The goal is to intensify the search in certain areas of the search space and increase the average "quality" within the population.

Crossover and mutation aim to create new solutions by modifying existing solutions within the search space (Figs. 2.2 and 2.3). The recombination operator generates a certain number of subsidiaries by crossing a certain number of parents. To simulate the stochastic nature of evolution, the recombination operator is associated with the probability of crossing. In contrast, the mutation operator modifies individuals by changing minor parts of according to them vectors to a given degree of mutation. Both of these operators (mutation and crossover) work directly with individuals (their genotypes), that is, they act in the space of individuals, and not with decoded solutions vectors (phenotypes of individuals).

Based on the above concepts, natural evolution is modeled by an iterative computational process. First, randomly (according to a given scheme) it is created the initial population, which is the starting point of the evolutionary process. Next, a certain number of times the cycle is repeated, that is consisted of the steps of evaluation

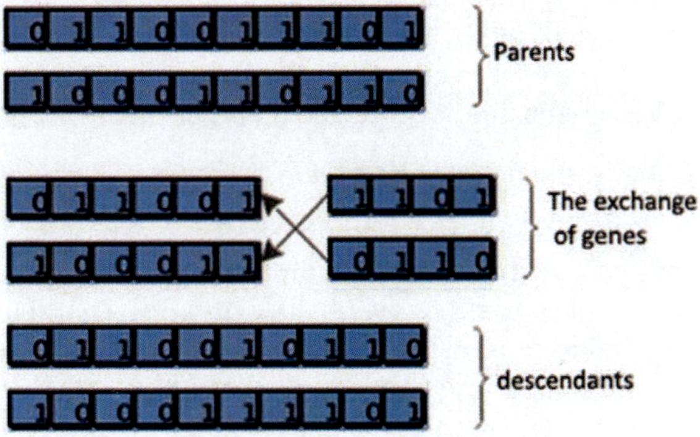

Fig. 2.2 Scheme of crossover

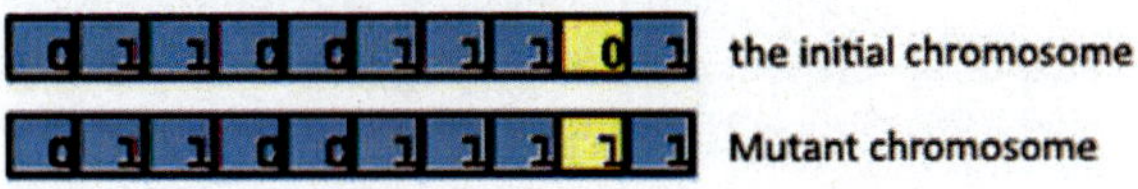

Fig. 2.3 Scheme of mutation

(fitness assignment), selection, crossing and/or mutation. Each such cyclic iteration is called a generation and, often, a predetermined maximum number of generations serves as a criterion for stopping the entire cycle. In addition, other conditions, that the stop criterion can be other conditions such as stagnation or the existence of an individual with satisfactory fitness, can serve as a criterion for stopping. Finally, the best individual(s) in the final population or found during the hole evolutionary process is the result of an evolutionary algorithms.

2.4.2 The Choice of Adaptive Probabilities of Crossover and Mutation

Practical experiments show that the crossing and mutation have a critical meaning with help of genetic algorithms [2, 12, 13]. Determination of probabilities of crossover and mutation should be used, as a rule, is executed with help of using the method of trials and errors. Optimal values of these probabilities vary between different tasks and even at different stages of genetic search.

In this book it is used an adaptive approach to determine the probability of crossing and mutation. These probabilities are adapted according to the evaluation results of the respective descendants in each generation. The experimental results show that the proposed scheme significantly improves the performance of genetic algorithms and exceeds the previous proposed methods.

As stated above, the operation of crossing happened only with probability p_c. When chromosomes can't be cross, they remain unchanged. Similarly, the mutation operator is used to change certain some elements in some individuals with probability p_m, resulting in additional genetic diversity.

The papers [13, 14] provides recommendations for setting probability values p_c and p_m. Typical value p_c are in the range of 0.5–1.0, while typical values p_m are in the range of 0.001 to ~ 0.05. These General principles were taken from empirical researches on a fixed set of test problems and are insufficient because the optimal value of probabilities is p_c and p_m is specific for each task.

Scheme of adaptive selection of crossing and mutation probabilities

In a classical genetic algorithm, genetic operators such as crosses and mutations are performed with constant probability. Different values of crossing and mutation probabilities can, however, better or worse contribute to the research of different search directions in the state space, thereby affecting the performance of the applied genetic algorithms.

In fact, the overall productivity of a genetic algorithms depends with support of an acceptable level of productivity during the evolution process. Thus, it is optimal to use a genetic algorithm that adapts to the search productivity at each iteration and "adapts" its crossing and mutation probabilities.

The essence of the proposed approach is as follows: dynamically adjust the parameters of the genetic algorithm (the probability of crossing and mutation) according to the degree of efficiency of each operator at this stage of the search. In order to evaluate the effectiveness of the genetic algorithms, its analyses ability to produce descendants with better fitness.

We consider the descendants obtained from two parents after crossing, then the scheme of adaptive selection of crossing and mutation probabilities can be represented as follows.

Step 1. As initial values of p_c and p_m we will choose the following typical recommended values [15]:

- probability of crossing $p_c = 0.9$;
- probability of mutation p_m=0.05.

Step 2. Let's define the crossover progress value CP as

$$CP = fsum_{child} - fsum_{parent},$$

where $fsum_{child}$ are values of fitness functions of two descendants sum; $fsum_{parent}$ are values sum of parent individual's fitness functions.

Step 3. Determine the average value of the crossover progress $\overline{CP}$ (measures the overall productivity of crossover operator on a given search iteration) for a generation that has experience n_c crossover operations

$$\overline{\tilde{N}D} = \frac{1}{n_{\tilde{n}}} \sum_{i=1}^{n_{\tilde{n}}} \tilde{N}P$$

Step 4. Let's determine the value of mutation progress MP

$$MP = f_{new} - f_{old}$$

where f_{new} is the value of the fitness function of the new individual (after mutation); f_{old} is the fitness of the original individual.

Step 5. Let's determine the average value of mutation progress $\overline{MP}$ for a generation that is experiencing n_m operations of mutation, the average progress value of the mutation

$$\overline{MD} = \frac{1}{n_m} \sum_{i=1}^{n_m} MP$$

Step 6. We correct the crossing and mutation probabilities, respectively, to the average values of their progress. The operator that performed better on a given iteration (with a higher average of progress) should participate more often (increase probability) on the next iteration, and vice versa.

The adjustment is performed as shown below:

$$p_c = p_c + \alpha \cdot p_{ci} p_m = p_m - \alpha \cdot p_m, \ if \ > \ W = \{\omega_1, \omega_2, \ldots, \omega_s\},$$
$$p_c = p_c - \alpha \cdot p_c \ and \ p_m = p_m - \alpha \cdot p_m, \ if \ < \ W = \{\omega_1, \omega_2, \ldots, \omega_s\}, \mu_{ij} = (x_i)$$

where α is the adaptation rate factor, α belongs [0.10; 0.15].

Note that after each adjustment, we need to make sure that the crossing and mutation operators are able to work continuously.

For this reason, we set the minimum allowable values for the crossing and mutation probabilities to 0.25 and 0.01 respectively.

The specified sequence of steps is repeated on each iteration.

2.4.3 *Selection of Population Size and Number of Iterations*

Some problems have very large solution spaces (i.e. set of variables, with a large range of admissible values for these variables). In such cases, a population of 100 chromosomes is probably small, since such a number of individuals simply will not be able to plausibly represent a sufficiently large sample of the solutions space. The following imperical rules of thumb will help to determine how large a population is needed to solve the given problem.

To do this we introduce the concept of scheme and order [15].

Under the scheme we will understand a subset of the space of genotypes G. If the elements of G are binary strings x, then, allowing some components of the string to take arbitrary values, and the rest only 0 or 1, we get a scheme or pattern. The concept scheme was introduced to define a set of chromosomes that have some common properties, that is similar to each other. For example: 1**0. The elements of the subset that this pattern represents will then be 1000, 1010, 1100, and 1110. Geometrically the scheme is a hyperplane in the solution search space.

When considering schemes, it is convenient to use the extended binary alphabet, in which, in addition to 0 and 1, an additional character is entered *, denoting any valid value, that is 0 or 1; the * symbol in a particular position means "don't care". *Allele*—this is the value that a gene on a particular chromosome takes. O*rder*—this is the number of specified fixed elements in a gene. That is, the number of alleles that take the value 0 or 1 in a scheme, then this scheme is a set of chromosomes containing zeros and ones at some predetermined positions.

Example:

Scheme	Orders
***	0
101	3
*11	2
1**	1

The first rule of population size selection is based on [16]

$$population\ size = order\left[\frac{length \cdot 2^{chromosize}}{chromosize}\right]$$

where *length* is the number of binary bits that each individual contains; *chromosize* is the average number of bits per chromosome (that is, *length* is divided by the number of parameters that each individual encodes), rounded to the nearest integer. If we take 120 chromosomes', length 6 bits each, and substitute in the formula, we will get: $[(120 \cdot 2^6)/6] = 1280$. Then we translate from the decimal system 1280 to binary **101**00000000, and in the end, we get the order 3. Thus, the main idea is to create a population of individuals, each of which is represented as a chromosome.

The second of the rules is based on statistical calculations and provides another possible approach for calculating the population size.

In the theory of cluster analysis, there is a relationship between the quality of evaluation of cluster statistics, the number of samples in the cluster and the number of parameters that define the cluster element. If we consider each chromosome as a vector, then to prevent the bias from accumulating the errors of solutions, the population (number of samples) must increase with increasing vector size. This principle is described by the following relation [9]:

$$population\ size = order\left[\left(1 + B^{-1}\right) \cdot (length + 2)\right]$$

where *length* is the number of binary bits that each individual contains; *B* is the value of the confidence interval statistical evaluation (as usual 0.05 or 0.1). For example, consider a chromosome of length 6 bits each, *B* exactly 0.1 and the order is 3 and substitute in the formula, we get: $population\ size = 3\left[\left(1 + 0, 1^{-1}\right) \cdot (6 + 2)\right]$ then in the end we get the population size equal to 264.

Usually, a "simple" genetic algorithm with selection, crossing and low probability of mutation of the population converges for "several" generations (for example, 30–50 iterations for a population of size 1000). In paper [16] it is shown that the average number of iterations required for search convergence reaches $O(\log N)$ generations', where *N* is population size. Thus, the search of a solution using a genetic algorithm help is finished quite quickly.

There are a number of population convergence estimates. The most popular among them is "bit-wise average convergence measure", what was proposed in [14].

This approach consist of the estimation of population convergence based on the values of the parameters that are presented in the individuals: when the number of individuals having the same value of a certain of the encoded parameters exceeds a certain specified limit (for example, 90% of the individuals of the population), it can be approved that the search process converged and it can be completed.

2.4.4 Analysis of Existing Multicriteria Optimization Algorithms

As the basis of the multicriteria problems solution algorithms in the evolutionary algorithm's theory, it is used the procedure of the general evolutionary algorithm. At the same time, the difference of all methods of multicriteria optimization by genetic algorithms consist of the modification of the stages of fitness assignment (stage 2) and selection (stage 3). These steps are implemented by different methods, but in such a way as to direct the search to the Pareto-optimal set and solve the problem of diversity support in population, to prevent premature convergence and achieve a well-distributed (representative) set of non-dominant solutions.

The most common modifications of genetic algorithms that implement different schemes of fitness assignment and selection include the following:

1. VEGA—Vector Evaluated Genetic Algorithm [17];
2. FFGA—Fonseca and Fleming's Multiobjective Genetic Algorithm [18];
3. NPGA—Niched Pareto Genetic Algorithm [19];
4. SPEA—Strength Pareto Evolutionary Algorithm [20];
5. NCGA—Neighborhood Cultivation Genetic Algorithm [21];
6. SPEA2—an improved version of the algorithm SPEA [22].

It is existed the next multicriteria optimization evolutionary algorithms:

1. adjective functions are considered separately;
2. a generalized criterion is constructed;
3. the concept of Pareto domination is used.

There are the following approaches to the appointment of fitness and selection.

The switching of the objective functions (NCGA)

In evolutionary algorithms of this class, instead of combining the adjective functions to obtain of scalar fitness value, the phase of selection it is executed the switch in the objective functions. Each time an individual is selected for reproduction, the decision to admit him to the intermediate population is made in relation to different adjective functions. For example, you can fill an intermediate population with equal portions relative to different adjective functions. Change the order or randomly select the order of revision of the adjective functions.

Aggregation (convolution) of adjective functions

In this case the adjective functions, as classical approaches are convolved in one parametric adjective function but this function parameters aren't changed after different optimizes, but varied under the moment during one stat.

Some approaches use a weighted method where each individual is evaluated using a specific weight combination. Thus, all members of the population are evaluated by different adjective functions, and optimization is carried out simultaneously in many directions.

Using the concept of Pareto domination (SPEA, SPEA2)

All non-dominating individuals are assigned the rank 1 and they are eliminated temporarily removed from the population. Then the next non-dominant individuals are assigned rank 2 and they also eliminated from consideration. This process continues until all individuals are assigned the certain rank relative to other members of the population. As a result, the rank of an individual determines its values of fitness.

Thus, fitness is connected with the hole population, while in other approaches, the value of an individual's fitness is calculated independently of other individuals.

To well approximate a Pareto-optimal set in one run, it is necessary to perform a polymodal search to find a representative set of solutions. Therefore, implementation the diversity of the population is one of the most important aspects of multicriteria optimization by genetic algorithms. Unfortunately, a simple genetic algorithm leads to a single solution, that is, it does not provide such a possibility, so it was designed at it has been designing the approaches to increase the scatter of points in the search space (population diversity).

The most common approaches are as follows:

The distribution of the total

Subpopulations, so-called **niches**, are formed and maintained, in which individuals must share resources. The more individuals are in the immediate neighborhood of a particular individual, the more significantly the value of his fitness will decrease.

A **niche**—is a group of individuals which have the same fitness. In the case of multicriteria optimization for fitness decreasing of individuals, which got into the same niche, it is usually the approach which called commonly General fitness division. This is done for order that to prevent the population convergence to one solution. Thus, it can be created stable subpopulations, each of which corresponds to different optimums of the function.

A limited crossing

Two individuals are allowed to cross only if they are at some distance from each other. As under fitness distribution, the distance between two individuals can be determined with different metrics use. It should be noted that this approach has not become widespread in practice.

Isolation by distance

In this mechanism, each individual is assigned a specific position in the population. In general, it is differed two approaches: either it is determined spatial structure of one population is determined in such a way that spatial niches are formed; or the population is divided into separate subpopulations which are randomly exchanged by individuals, with so-called migrations of random individuals.

Redefinition

According to this method, individuals are composed from active and nonactive parts: the first one determines the encoded vector of solutions and the inactive part is not used. Thus, the information can be "hidden" in the individual, as under since during the evolution the active and nonactive parts can be changed places.

Restart

Another approach which is used to prevent premature convergence consist in "restart" of the whole or the part of the population after a certain period of time, or in case of the search get into stagnation (no improvement for several generations).

In this case, under "restart" means the replacement of all or part of population individuals on a randomly generated one, i.e., the repetition of population initial initialization or its part.

Consolidation (Clustering)

Here, the new individuals (descendent) are replace similar individuals in the population. In this case, unlike the general evolutionary algorithm, selection, recombination, and mutation not all of population participates but only a small part of individuals.

Along with support, the important part places the concept of elitism, the basic idea of which is always to include the best individuals in the next population, that don't lose the good features due to action of genetic operators.

Use of elitism to support population diversity

The use of elitism guarantees that the maximum population fitness will not decrease from generation to generation, but can only increase. In contrast to the methods discussed earlier, the use of elitism concept in genetic algorithm always contributes to a faster convergence of the population. Therefore, this approach is good under unimodal functions optimization, and under multicriteria optimization problems solving can cause premature convergence.

Accordingly, unlike single-criterion optimization, the use of elitism concept under multi-criteria optimization is a more complicated procedure associated with the concept of dominance. Instead of one best individual, there is created the elitist set whose size can be compared with sizes of the population itself. This raises the following questions:

In general, there are two major elitist approaches:

1. Directly copy non-dominant individuals to the next population. Sometimes, it is used the variants with restrictions when only N individuals optimal for a particular function are passed into the next population regardless of values by other criteria.
2. It is often the concept of external individuals set (**archieve**) is used. Thus, in each generation, a certain percentage of the population is filled or replaced by randomly selected archives, or according to some criterion, for example, after the time elapsed for the individual to remain in the archive.

In multicriteria genetic algorithms, for the background it is taken the general evolutionary algorithm that consist of presented previously basic components. However, under specific multicriteria problems solutions methods design the basic attention is devoted to the modification of stages of fitness and selection assignment with population diversity support.

Based on a comparative analysis of the accuracy of these algorithms [21], the best results are demonstrated by the NCGA and SPEA2 algorithms.

2.4.5 The Transition from Conditional to Unconditional Multicriteria Task

The above evolutionary algorithms do well with the unconditional problems of multicriteria optimization, but under solving problems with restrictions, the obtained solutions are not always satisfactory [9, 23, 24]:

- they may not contain the conditional maximum point;
- the resulting points can be scattered in the search space;
- part from found solutions may locate outside the permitted area.

In this regard, we conclude that evolutionary algorithms are not well adapted to solve constrained problems and require some modification to take into account the specifics of conditional problem of optimization.

Turning to the problem of conditional optimization, we can distinguish the following approaches to its solution [25]:

The first approach is based on reducing the problem of conditional optimization to the problem of unconditional optimization:

- penalty functions algorithm. For GA, special algorithms for this approach include the "deadly" penalty method, the static penalty method, and the dynamic penalty method;
- rolling tolerance algorithm.

The second approach includes algorithms that do not use the reduction of conditional optimization problem to the unconditional optimization problem:

- each restriction is converted to a separate objective function;
- "treatment" procedure by a local search;
- Orvosh-Davis reduction method;
- modified algorithm of complexes.

Hybrid methods also exist. For example, the method of behavioral memory, "treatment" + deadly penalties.

The quality of the obtained solution largely depends on the choice of the step of implementing the multi-criteria optimization algorithm, on which the "treatment" procedure is applied.

For the solution of conditional MCO given problem, it is developed and researched the next algorithms:

1. *The removal of* unsatisfying to restrictions solutions after the algorithm learning finishing.
2. *The removal of* unsatisfying to restrictions solutions after every population generation.
3. *Taking into account* the degree of individuals restrictions violation of under Pareto dominance.
4. *There restrictions transformation into* additional criteria and multicriteria problems with additional criteria solution.
5. *Converting the restrictions* into criteria, multicriteria problems with additional criteria solution, and "treatment" solutions that do not satisfy the restrictions after the algorithm is finished.
6. *Multicriteria problem* optimization solution without taking into account the restrictions and the "treatment" of obtained solutions which do not satisfy the restrictions after the algorithm is finished.
7. *Converting the restrictions* into criteria, multicriteria problems with additional criteria solution, and "treatment" solutions that do not satisfy the restrictions after each iteration.
8. "*Treatment*" *of unsatisfied* to solutions restrictions after each iteration.

For more detailed research and comparison of the proposed algorithms, a detailed description of each algorithm and its pseudo-code are given below.

Consider in detail each algorithm.

1. *The removal of un satisfying to restrictions solutions after the algorithm learning finishing*

This method is the simplest. After all generations of populations, it is assumed the deletion of all final individuals which don't satisfied to the restrictions. As a result, it can be received quite a few solutions satisfy to the restrictions. The algorithm pseudocode (added operations distinguish by italics):

```
Procedure SPEA2 with removal at the end
Input: population size, max generations, f_k(X) criteria, g_k(X) constraints
begin
  Population = Initialize Population;
  Compute fitness for each individual in Population;
  for i = 1 to max generations do
     Parents = Binary tournament selection on Population;
     Offspring = Apply crossover and mutation to Parents;
     Copy Population to Offspring;
     Clear Population;
     Compute fitness for each individual in Offspring;
     Copy all non-dominated solutions from Offspring into Population;
     If Population size > population size then Truncate Population;
     If Population size < population size then
       Add to the Population the best dominated solutions;
   end
   Remove solutions from the Population that violate constraints;
end
```

2. *The removal of unsatisfying to restrictions solutions after every population generation*

This method is a modification of the previous method.

It is intended the delation of individuals which do not satisfy the restrictions after each generation. But, using the selection operator, every time, still going on the generation of new population of the original size. Thus, a new population is already generated only from feasible solutions. Under large numbers of iterations, the number of final solutions will be closed to the original population size.

Algorithm pseudocode (added operations distinguish by italics):

```
Procedure SPEA2 with removal after each iteration
Input: population size, max generations, f_k(X) criteria, g_k(X) constraints
begin
  Population = Initialize Population;
  Compute fitness for each individual in Population;
  for i = 1 to max generations do
    Parents = Binary tournament selection on Population;
    Offspring = Apply crossover and mutation to Parents;
    Copy Population to Offspring;
    Clear Population;
    Compute fitness for each individual in Offspring;
    Copy all non-dominated solutions from Offspring into Population;
    If Population size > population size then Truncate Population;
    If Population size < population size then
      Add to the Population the best dominated solutions;
    Remove solutions from the Population that violate constraints;
  end
end
```

3. *Taking into account the degree of individuals restrictions violation of under Pareto dominance*

After each iteration, before fitness function calculation, each individual of population is examined their satisfaction restrictions. If an individual does not satisfy the any of restrictions, then it is marked as not feasible and it is fixed the degree of its unsatisfactoriness (the number of restrictions which it does not satisfy).

It is performed the modification of individuals comparison by Pareto under fitness function formation. The change is that initially under two individual's comparison, it is examined their satisfaction to the restrictions. If one of the individuals does not satisfy any of them, then in this comparison it is considered dominant regardless of the values of its criteria vector. If both do not satisfy, then the dominant one will be is chosen the individual which does not satisfy a greater number of restrictions. If both are feasible, then the usual Pareto-comparison is carried out according to the values of the criteria vector.

In the process of population generation, a number of individuals which do not satisfy to the restrictions is decreases and, with large numbers of generations, finally tends to 0. However, after training, it can be stayed the unsatisfied individuals which are marked as not feasible and can be easily deleted from the final obtained result.

Algorithm pseudocode (added operations distinguish by italics):

```
Procedure SPEA2 with dominated not feasible solutions
Input: population size, max generations, f_k(X) criteria, g_k(X) constraints
begin
  Population = Initialize Population;
  Compute fitness for each individual in Population;
  for i = 1 to max generations do
    Parents = Binary tournament selection on Population;
    Offspring = Apply crossover and mutation to Parents;
    Assign a constraint violation for each individual in the Offspring;
    Copy Population to Offspring;
    Clear Population;
     Compute fitness for each individual in Offspring;
    Copy all non-dominated solutions from Offspring into Population;
    If Population size > population size then Truncate Population;
    If Population size < population size then
      Add to the Population the best dominated solutions;
  end
end
```

```
Function Pareto dominance with dominated not feasible solutions
Input: sol1, sol2
begin
  if sol1 constraint violation < sol2 constrain violation do
    return -1
  else if sol1 constraint violation > sol2 constrain violation do
    return 1
  end
  return Simple Pareto dominance sol1 sol2
end
```

4. *The restrictions transformation into additional criteria and multicriteria problems with additional criteria solution*

In this method it is used restrictions transform into additional criteria, i.e. the initial task is represented as the set of criteria: given objective functions plus additional criteria which correspond to existed restrictions. Thus, the problem of conditional multicriteria optimization takes the following:

- initial task: objective functions—$F(X) \rightarrow$ opt, restrictions—$G(X) < B$;
- transformed task: objective functions: $F(X) \rightarrow$ opt, $|G(X) - B| \rightarrow$ min.

The usual algorithm SPEA2 is used next.

Algorithm pseudocode (added operations distinguish by italics):

```
Procedure SPEA2 with transformation of constraints into criteria
Input: population size, max generations, f_k(X) criteria, g_k(X) constraints
begin
  Transform g_k(X) constraints into criteria format;
  Add transformed constraints to the criteria;
  # the criteria look like [ f_1(X), f_2(X), ..., f_|F|(X), g*_1(X), g*_2(X), ..., g*_n(X) ]
  # where g*_k(X) is transformed constraint g_k(X)
  # n is number of constraints
  Population = Initialize Population;
  Compute fitness for each individual in Population;
  for i = 1 to max generations do
    Parents = Binary tournament selection on Population;
    Offspring = Apply crossover and mutation to Parents;
    Copy Population to Offspring;
     Clear Population;
     Compute fitness for each individual in Offspring;
    Copy all non-dominated solutions from Offspring into Population;
    If Population size > population size then Truncate Population;
    If Population size < population size then
      Add to the Population the best dominated solutions;
  end
end
```

5. *Converting the restrictions into criteria, multicriteria problems with additional criteria solution, and "treatment" solutions that do not satisfy the restrictions after the algorithm is finished*

The main difference between the obtained solution of unconditional multicriteria task from the task with restrictions—is necessary that final solutions not only belong to the Pareto set but wherein are located in admissible area. Therefore, it is additionally used the procedure which permits pull solutions to an acceptable area.

This method also uses procedure solutions "treatment" at the end of the computational procedure, but the population learning is carried out not only by initial criteria, but by additional criteria corresponding to restrictions, as described in method 4.

The "Treatment" occurs by the Pareto local search method [26].

The Pareto local search is an original method of local descent in binary variables space with the transition by the first improvement. Its hallmark is method of best solution determination—the comparison is neighboring points. In this case, the decision either the point obtained in this stage is better than previous one received before it is based on Pareto principle, i.e., the transition neighboring point is performed if it dominant the current one by the criteria totality.

The Pareto local search algorithm has the following form.

The main idea of local search is the existence of surroundings that represents some set of solutions close to the current solution.

The iterative procedure of local search (in the case of a minimization task) can be described as follows.

1. Select some initial admissible point X.
2. Select any point Y from surroundings of the point X, such that $F(Y) < F(X)$.
3. If this point Y doesn't exist, then stop. Point X—minimum point.
4. Otherwise, put $X = Y$. Go to step 2.

This algorithm starts from an arbitrary initial admissible point and scans all points around point X for search the best solution around point X. If succeeded find better solution, it is remembered and the search is continued by surroundings of found best solution. If the best solution isn't found then the current point is the local minimum.

Thus, the search of optimal solution in local search algorithm is realized with help of enumeration of surroundings points, variety components of solutions boolean vector (solution vector mutation). The decision about search transference into new point is made after a numeration the closest nearest surroundings points. There are 2 ways of viewing the surroundings:

1. the fastest descent when viewing the entire surroundings and it is realized the transition to the point with the best value of the function;
2. the transition after the first improvement is an obvious alternative of the fastest descent, which should be used in circumstances when it is absent the specific information which gives reason to hope that the "fastest" descent method will be indeed the fastest.

In this method of conditional multicriteria problem solution, it is used the learning of all populations without restrictions, and after learning, it is realized "treatment" of all points which aren't satisfied to the restrictions.

Algorithm pseudocode (added operations distinguish by italics):

```
Procedure SPEA2 with transformation of constraints into criteria and treatment
Input: population size, max generations, f_k(X) criteria, g_k(X) constraints
begin
  Transform g_k(X) constraints into criteria format;
  Add transformed constraints to the criteria;
  # the criteria look like [ f_1(X), f_2(X), ..., f_|F|(X), g*_1(X), g*_2(X), ..., g*_n(X) ]
  # where g*_k(X) is transformed constraint g_k(X)
  # n is number of constraints
  Population = Initialize Population;
  Compute fitness for each individual in Population;
  for i = 1 to max generations do
    Parents = Binary tournament selection on Population;
    Offspring = Apply crossover and mutation to Parents;
    Copy Population to Offspring;
     Clear Population;
     Compute fitness for each individual in Offspring;
    Copy all non-dominated solutions from Offspring into Population;
    If Population size > population size then Truncate Population;
    If Population size < population size then
      Add to the Population the best dominated solutions;
  end
  Treatment of each solution from the Population that violates constraints;
end
```

6. *Multicriteria problem optimization solution without taking into account the restrictions and the "treatment" of obtained solutions which do not satisfy the restrictions after the algorithm is finished*

This method uses the algorithm of "treatment" solutions which do not satisfy the restrictions after the algorithm is finished.

Algorithm pseudocode (added operations distinguish by italics):

```
Procedure SPEA2 with treatment of not feasible solutions at the end
Input: population size, max generations, f_k(X) criteria, g_k(X) constraints
begin
  Population = Initialize Population;
  Compute fitness for each individual in Population;
  for i = 1 to max generations do
    Parents = Binary tournament selection on Population;
    Offspring = Apply crossover and mutation to Parents;
    Copy Population to Offspring;
    Clear Population;
    Compute fitness for each individual in Offspring;
    Copy all non-dominated solutions from Offspring into Population;
    If Population size > population size then Truncate Population;
    If Population size < population size then
      Add to the Population the best dominated solutions;
  end
  Treatment of each solution from the Population that violates constraints;
end
```

```
Function Treatment
Input: solution, k, visited = []
begin
  for all neighbor in N(solution,k) do
    if solution ⊁ neighbor and neighbor passes constrains
          and neighbor not in visited do
      Add neighbor to visited;
      return Treatment(neighbor, k, visited);
  return solution;
end
```

where $\mathcal{N}(\mathbf{solution}, \mathbf{k})$ are all 1-, 2-, …, k-neighbors of solution

7. *Converting the restrictions into criteria, multicriteria problems with additional criteria solution, and "treatment" solutions that do not satisfy the restrictions after each iteration*

This method it is assumed that the population learning only on adjectives functions but treatment procedure is executed after each iteration.

Algorithm pseudocode (added operations distinguish by italics):

```
Procedure SPEA2 with treatment after each iteration
Input: population size, max generations, f_k(X) criteria, g_k(X) constraints
begin
  Population = Initialize Population;
  Compute fitness for each individual in Population;
  for i = 1 to max generations do
    Parents = Binary tournament selection on Population;
    Offspring = Apply crossover and mutation to Parents;
    Copy Population to Offspring;
    Clear Population;
    Compute fitness for each individual in Offspring;
    Copy all non-dominated solutions from Offspring into Population;
    If Population size > population size then Truncate Population;
    If Population size < population size then
      Add to the Population the best dominated solutions;
    Treatment of each solution from the Population that violates constraints;
  end
end
```

8. *"Treatment" of unsatisfied to solutions restrictions after each iteration*

The last method uses the "treatment" procedure after each iteration, and the restrictions are converted into additional criteria.

Algorithm pseudocode (added operations distinguish by italics):

```
Procedure SPEA2 with constraints transformation and treatment after each iteration
Input: population size, max generations, f_k(X) criteria, g_k(X) constraints
begin
  Transform g_k(X) constraints into criteria format;
  Add transformed constraints to the criteria;
  # the criteria look like [ f_1(X),f_2(X),...,f_|F|(X),g*_1(X),g*_2(X),...,g*_n(X) ]
  # where g*_k(X) is transformed constraint g_k(X)
  # n is number of constraints
  Population = Initialize Population;
  Compute fitness for each individual in Population;
  for i = 1 to max generations do
    Parents = Binary tournament selection on Population;
    Offspring = Apply crossover and mutation to Parents;
    Copy Population to Offspring;
     Clear Population;
     Compute fitness for each individual in Offspring;
    Copy all non-dominated solutions from Offspring into Population;
    If Population size > population size then Truncate Population;
    If Population size < population size then
      Add to the Population the best dominated solutions;
    Treatment of each solution from the Population that violates constraints;
  end
end
```

For results analyzes and the comparison describe above methods it is solved the follow conditional multi-criteria optimization problem:

$$Minimize = \begin{cases} f_1(x, y) = 2 + (x - 2)^2 + (y - 1)^2 \\ f_2(x, y) = 9x - (y - 1)^2 \end{cases}$$

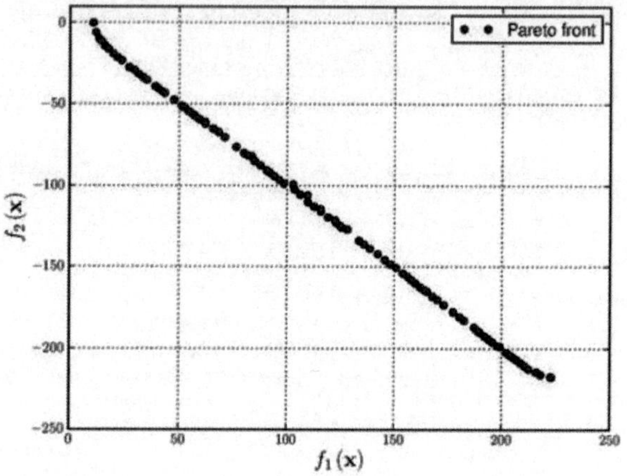

Fig. 2.4 Real Pareto front of solved problem

$$Restrictions = \begin{cases} g_1(x, y) = x^2 + y^2 \le 225 \\ g_2(x, y) = x - 3y + 10 \le 0 \end{cases}$$

$$Variables\ constraints = \begin{cases} -20 \le x \\ y \le 20 \end{cases}$$

Real Pareto front has the form shown in Fig. 2.4.

First, we will provide detailed results and examples of how the SPEA2 algorithm works with the 3rd method to solve the proposed problem. The number of individuals in the population is 5, the number of generations is 20. For each variable, 5 bits are allocated. The integer values of the variables are encoded first in bit form, and then by Gray code.

Let us present an example of one SPEA2 algorithm iteration functioning. For some further generations, a table of separate individuals values will be presented.

Initially, a random population was generated that looks like:

10100 11111 01110 11001
11101 00001 00010 0110011111 00101

At the beginning of each iteration, with the help of a selection operator, which is represented in SPEA2 as binarytournament, parents are selected for a new population.

Binarytournament #1

Two individuals were randomly selected: 11111 00101 and 11111 00101. In this case, the same individuals were randomly selected, so there is no better among them. This individual will be the first parent.

Binarytournament #2

The individuals were selected: 11111 00101 и 00010 01100. The second individual dominates the first, so it is chosen as the second parent.

Using the crossover operator, we get two off springs: 01111 01100 и 10110 00101.

Thus, repeating several times the above algorithm, we generate 6 offsprings and mutate them with some probability. We calculate the values of their objective functions and restrictions. The intermediate results are shown in Fig. 2.5 The first two values are the values of the variables, the second are the values of the objective functions, the third are the values of the restrictions.

Add to the current population, we get a total of 11 individuals. We calculate the values of fitness functions and select a new population of 5 individuals. First, non-dominated individuals are added, that is, the value of the fitness function for which is less than 1. Next, if the population size is less than the specified one, then we add the dominated elements with the best value of the fitness function, otherwise we will drop the worst individuals from the new population by the value of the fitness function. A new population is presented in Fig. 2.6.

Carrying out operations and generating new generations using the SPEA2 algorithm, in the end we get only non-dominated individuals. Examples of populations at 5, 10, 15 and 20 (last) generations are presented in Figs. 2.7, 2.8, 2.9 and 2.10 respectively.

```
Offspring
 ['10001', '00111'], [102, 54], [125, 35]
 ['10100', '11101'], [127, -85], [160, -22]
 ['01000', '01110'], [51, -45], [26, 2]
 ['10101', '11101'], [132, -76], [169, -21]
 ['11011', '00101'], [43, -43], [20, 20]
 ['00100', '00101'], [252, -142], [185, 9]
```

Fig. 2.5 Descendant of the 1st generation

	Gray binary x	x	Gray binary y	y	f1(x, y)	f2(x, y)	R(i)	D(i)	F(i)	g1(x, y)	g2(x, y)	Passes
1	10100	4	11111	11	106	-64	0	0.012994891130348065	0.012994891130348065	137	-19	True
2	01110	-9	11001	7	159	-117	0	0.012994891130348065	0.012994891130348065	130	-20	True
3	11101	2	00001	-9	102	-82	8	0.014516323497263173	8.014516323497263	85	39	False
4	00010	-17	01100	-2	372	-162	9	0.00352606129951738	9.003526061299517	293	-1	False
5	11111	1	00101	-4	28	-16	6	0.009885677864257758	6.009885677864258	17	23	False

Fig. 2.6 New 2nd population

	Gray binary x	x	Gray binary y	y	f1(x, y)	f2(x, y)	R(i)	D(i)	F(i)	g1(x, y)	g2(x, y)	Passes
1	11110	0	11100	13	150	-144	0	0.02631578947368421	0.02631578947368421	169	-29	True
2	10100	4	11111	11	106	-64	0	0.5	0.5	137	-19	True
3	10100	4	11110	10	87	-45	0	0.034637963349705855	0.034637963349705855	116	-16	True
4	10100	4	11101	12	127	-85	0	0.03154724921393508	0.03154724921393508	160	-22	True
5	10100	4	11111	11	106	-64	0	0.5	0.5	137	-19	True

Fig. 2.7 5th population

	Gray binary x	x	Gray binary y	y	f1(x, y)	f2(x, y)	R(i)	D(i)	F(i)	g1(x, y)	g2(x, y)	Passes
1	11100	3	11110	10	84	-54	0	0.08705619744773416	0.08705619744773416	109	-17	True
2	10110	7	11000	6	52	38	0	0.010860387694094743	0.010860387694094743	85	-1	True
3	11110	0	11100	13	150	-144	0	0.015308177932089431	0.015308177932089431	169	-29	True
4	10100	4	11101	12	127	-85	0	0.03154724921393508	0.03154724921393508	160	-22	True
5	10100	4	11111	11	106	-64	0	0.034637963349705855	0.034637963349705855	137	-19	True

Fig. 2.8 10th population

	Gray binary x	x	Gray binary y	y	f1(x, y)	f2(x, y)	R(i)	D(i)	F(i)	g1(x, y)	g2(x, y)	Passes
1	11101	2	11101	12	123	-103	0	0.09045365561714636	0.09045365561714636	148	-24	True
2	11100	3	11011	8	52	-22	0	0.024321896743916053	0.024321896743916053	73	-11	True
3	11000	-4	11101	12	159	-157	0	0.013500849901138536	0.013500849901138536	160	-30	True
4	10100	4	11000	6	31	11	0	0.5	0.5	52	-4	True
5	11100	3	11110	10	84	-54	0	0.01707401204471959	0.01707401204471959	109	-17	True

Fig. 2.9 15th population

	Gray binary x	x	Gray binary y	y	f1(x, y)	f2(x, y)	R(i)	D(i)	F(i)	g1(x, y)	g2(x, y)	Passes
1	11010	-1	01011	3	15	-13	0	0.08132970726457843	0.08132970726457843	10	0	True
2	11000	-4	11001	7	74	-72	0	0.04587368524026632	0.04587368524026632	65	-15	True
3	11010	-1	11001	7	47	-45	0	0.038338870065251024	0.038338870065251024	50	-12	True
4	11101	2	11101	12	123	-103	0	0.014947700322802236	0.014947700322802236	148	-24	True
5	11000	-4	11101	12	159	-157	0	0.008182760127489483	0.008182760127489483	160	-30	True

Fig. 2.10 20th population

You can also notice that, using this method, in the end, all individuals began to satisfy the restrictions. In Fig. 2.11 is a graphic representation of the real Pareto front and on it 5 points—our trained individuals.

As you can see, even a small number of iterations allows to obtain excellent solutions of multicriteria conditional optimization using a small modification of the algorithm SPEA2.

Next, consider the example of the work and the results of each method described in part 4.

1. After training a population of 100 individuals at the end, only 30 met the constraints and 70 were discarded. In Fig. 2.12 shows all individuals, in Fig. 2.13 only the remaining 30.
2. Removal of individuals who do not satisfy the constraints after each iteration. At the end of the training, they received 41 correct decisions. Preview on Fig. 2.14.
3. This method allows you to obtain a population whose points meet the constraints, but only 42 are unique. Data presented on Fig. 2.15.
4. Simple conversion of constraints into criteria gives very poor results. 12 non-dominated points according to 4 criteria (2 real + 2 restrictions) that pass the restrictions. Of these, only 7 are non-dominated by two target criteria. Thus 93

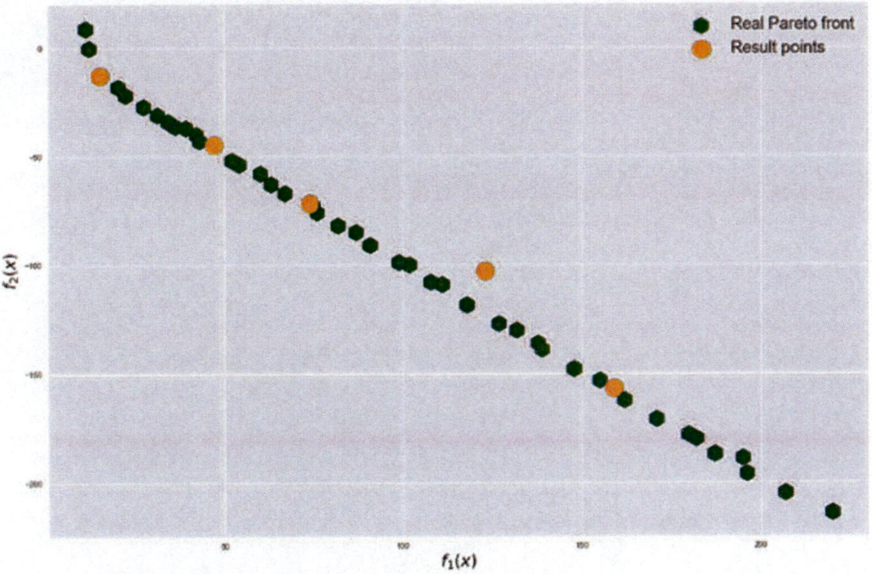

Fig. 2.11 Real Pareto front and five result individuals

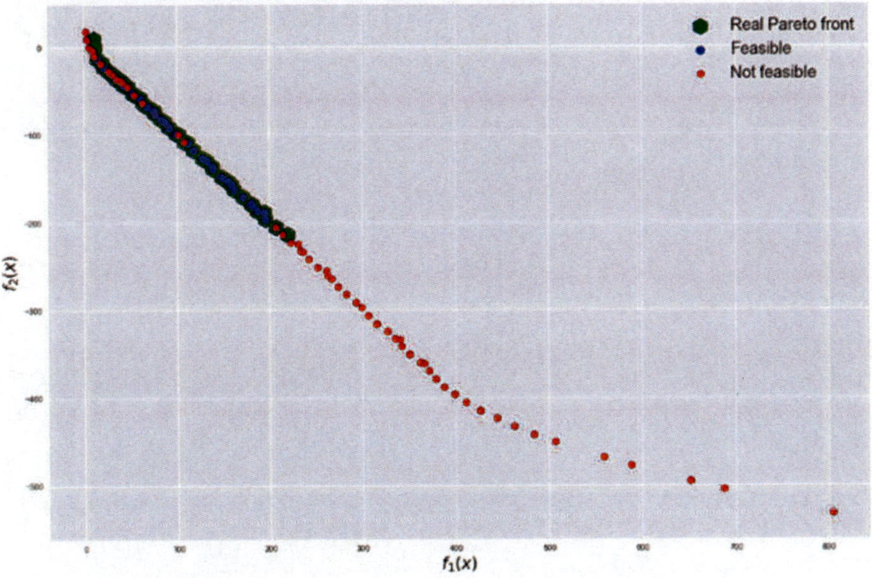

Fig. 2.12 The whole population

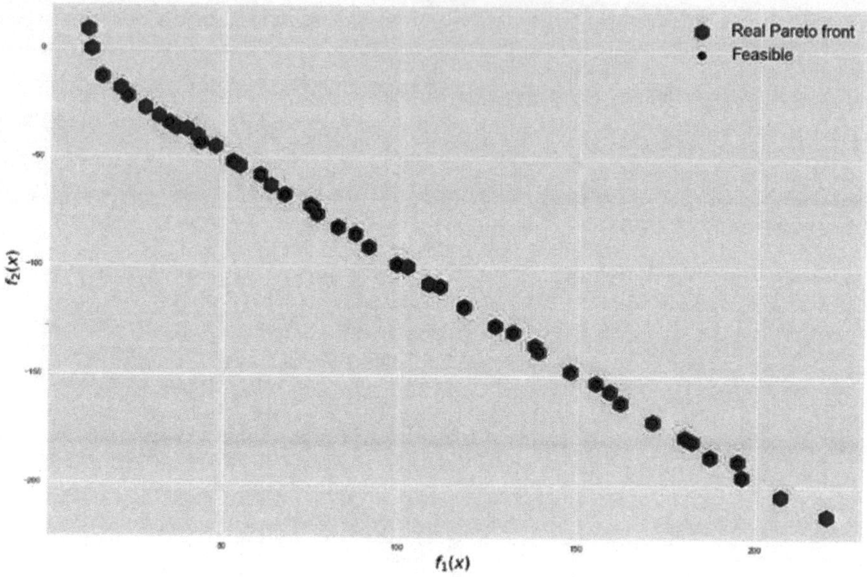

Fig. 2.13 Resulting individuals satisfying constraints

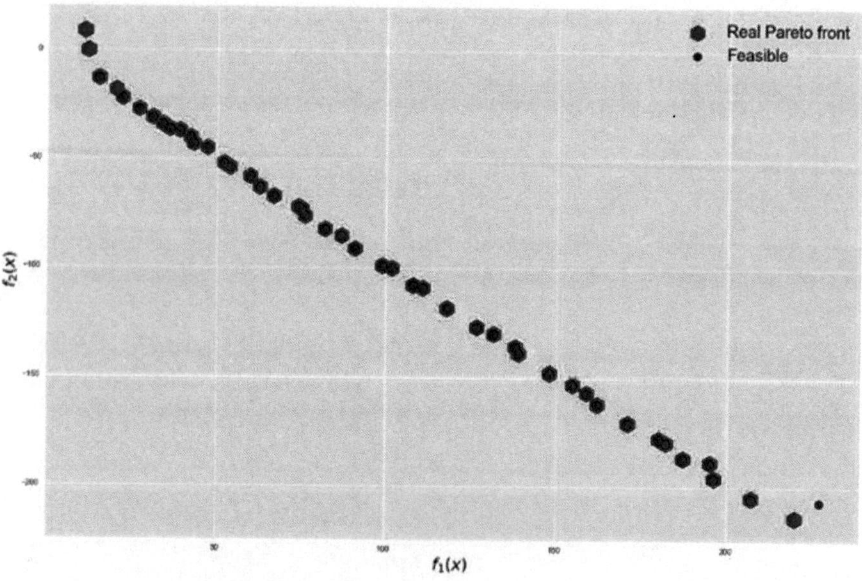

Fig. 2.14 Resulting individuals

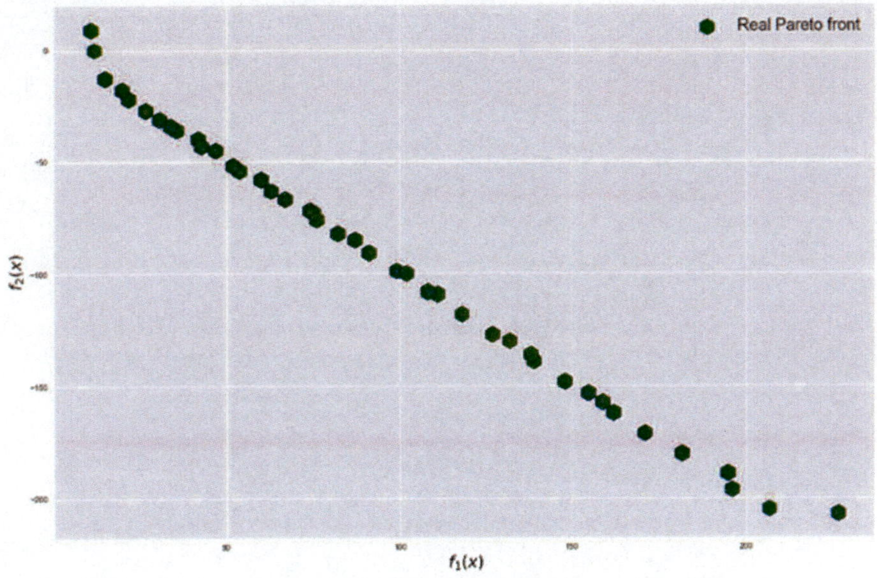

Fig. 2.15 Resulting population

results were thrown back. Visualization of the complete population in Fig. 2.16, the final solutions in Fig. 2.17.

5. In this case, the additional transfer of restrictions to the criteria, the treatment procedure allows you to get slightly better results than the 4: 17 method, 83 are discarded. In Fig. 2.18 represented individuals who do not meet the restrictions and cured individuals. In Fig. 2.19 final decisions. Using the treatment procedure, all not feasible individuals were "cured". But some of them converged at the same points, so there are fewer unique cured points than the original not feasible points.

In Fig. 2.20 is a visual representation of the transition of points that are outside the bounds of the constraints into an admissible field.

6. In this case, the treatment procedure allows you to get points which satisfy the restrictions, but for the most part they will not be better than the set of points that satisfy the restrictions initially. We got 24 results. Details in Figs. 2.21 and 2.22.
7. Learning without regard to treatment restrictions after each iteration allows you to get pretty good results: 41 final solutions. Visualization in Fig. 2.23.
8. Training with the transformation of restrictions into the criteria of "cure" After each iteration, it allows to obtain the entire population that is not dominated by four criteria, but only two individuals will be targeted by two criteria (Fig. 2.24).

Based on the results obtained, it can be seen that of the presented methods, one of the simplest methods with the removal of individuals that do not pass the restrictions after each iteration is the best. This method is quite simple and fast, does not require strong modification SPEA2.

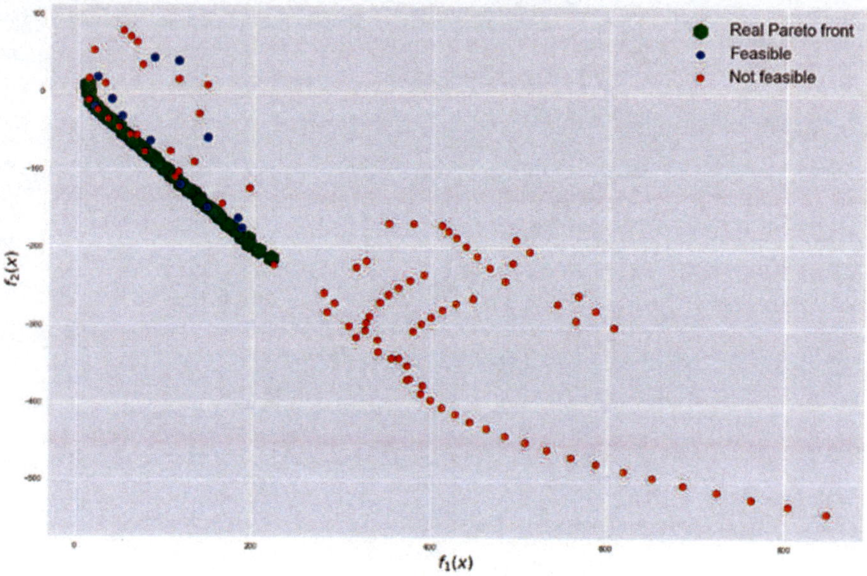

Fig. 2.16 Resulting population

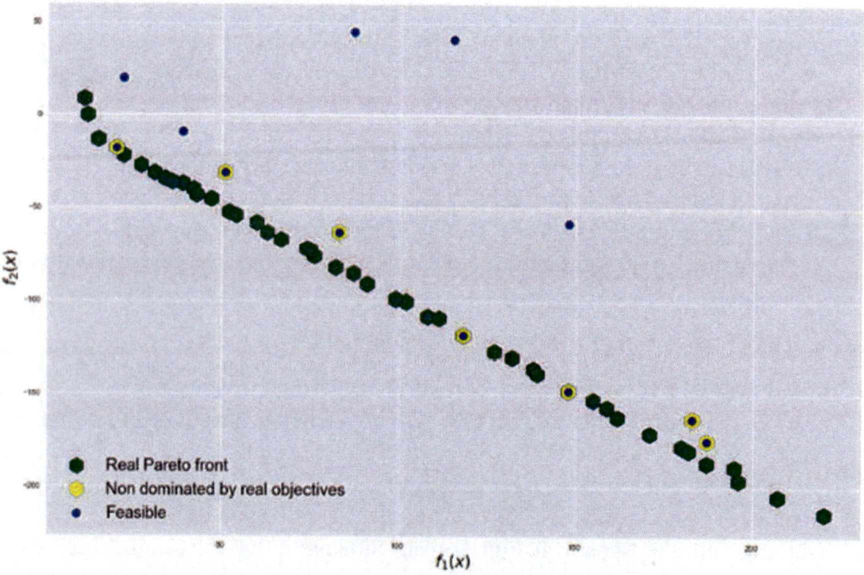

Fig. 2.17 Resulting nondominated individuals

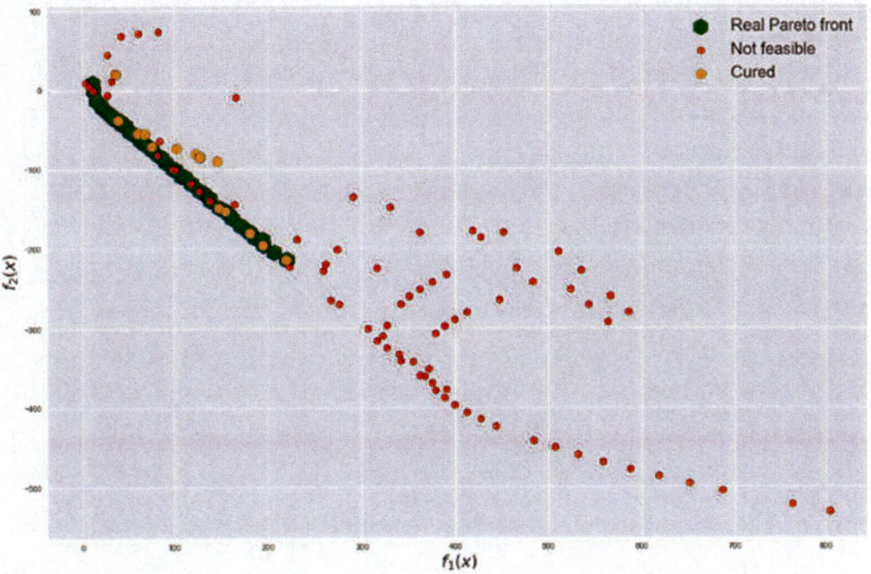

Fig. 2.18 Not feasible and cured individuals

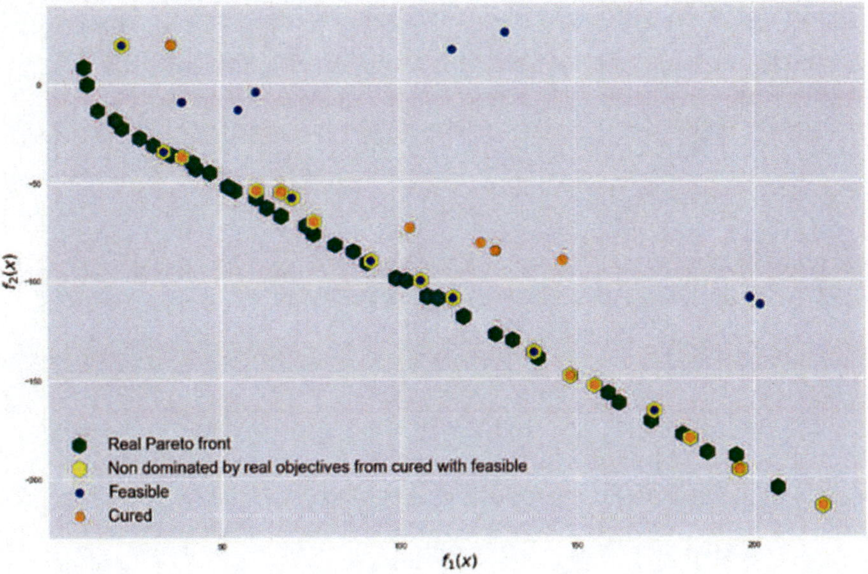

Fig. 2.19 Nondominated resulting individuals

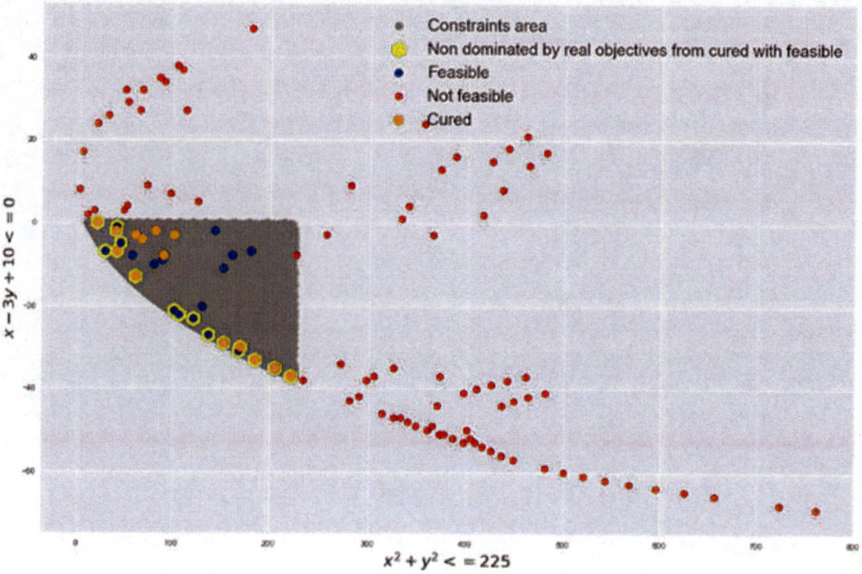

Fig. 2.20 Example of curing

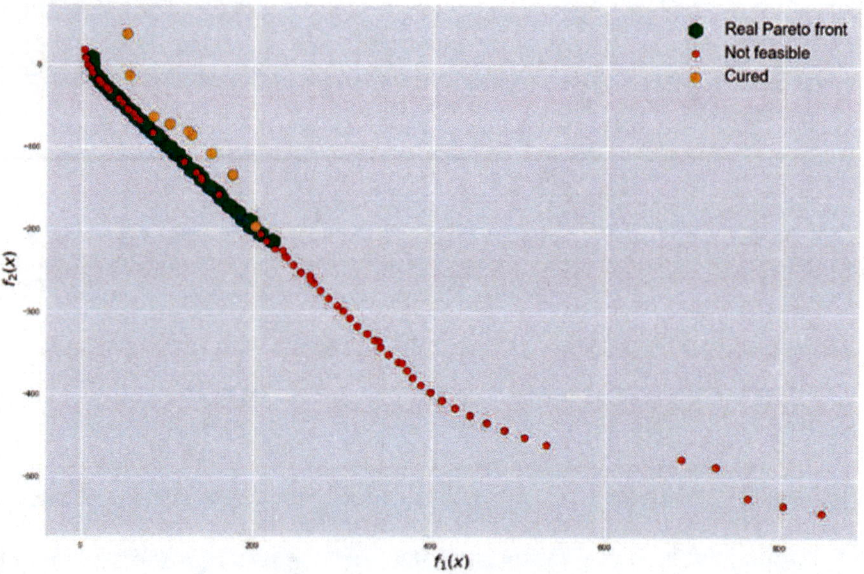

Fig. 2.21 Not feasible and cured individuals

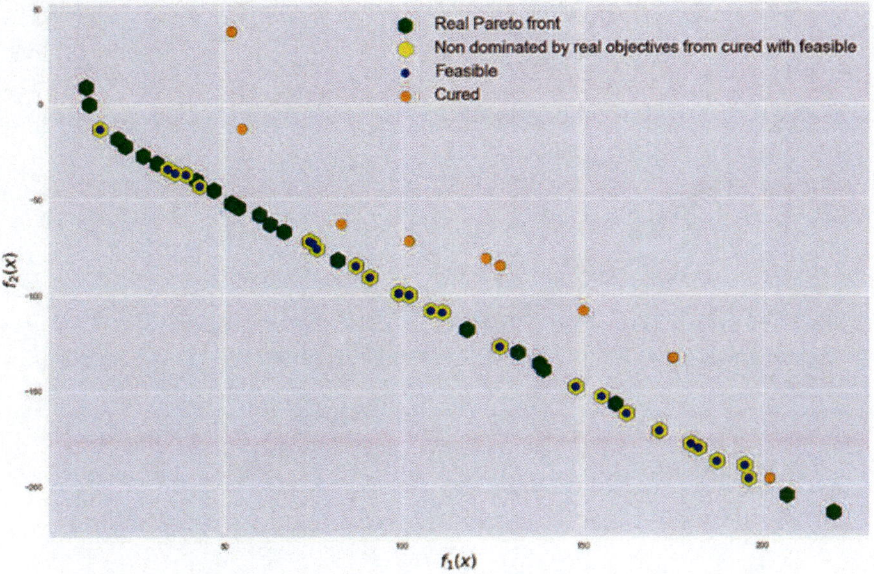

Fig. 2.22 Resulting individuals

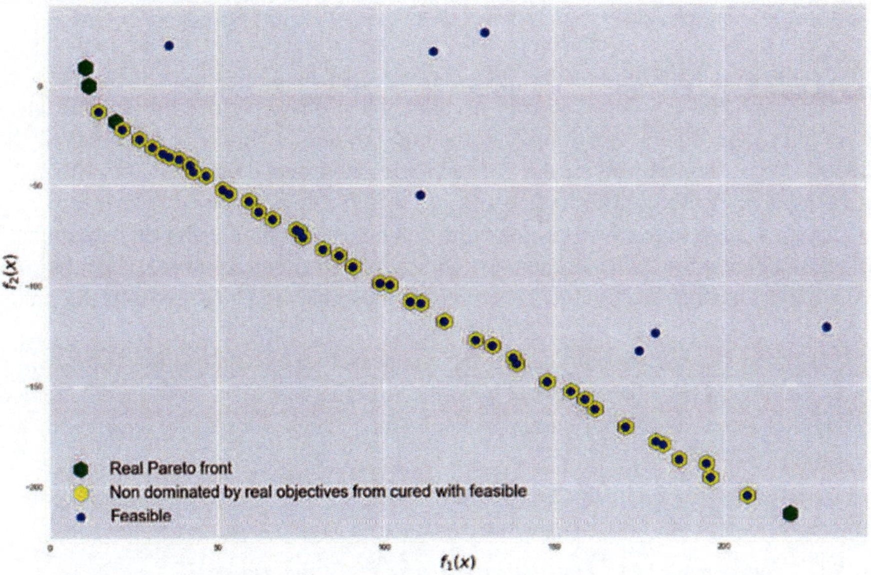

Fig. 2.23 Resulting nondominated individuals

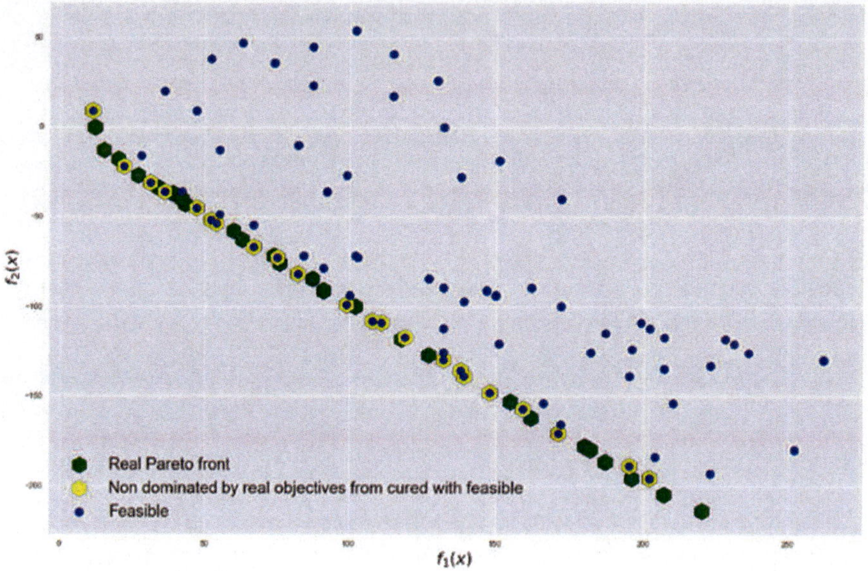

Fig. 2.24 Resulting individuals

The method obtained the same results by marking individuals who do not meet the restrictions with the feasible flag and preserve their degree of unsatisfactory, so that later this can be taken into account when checking the dominance when calculating the fitness function. This method requires a small modification of the SPEA2 algorithm. However, the changes are small and do not greatly affect the running time of the algorithm.

Another method received similar results. This is a method using the "treatment" of individuals after each iteration. It requires much more additional calculations, which greatly affect the time it takes to compute solutions. But the results are pretty good.

Based on the results obtained, it can be seen that translating restrictions into criteria does not provide additional benefits. And the described "treatment" method requires a lot of additional computational time, which does not pay off with the resulting solutions.

2.4.6 Refinement ("Treatment") of Decision Points

The use of treatment algorithms after completion of the search algorithm for solutions to the unconditional optimization problem is proposed. One of these algorithms are genetic algorithms, for example, SPEA2.

To solve this problem, we propose the use of two Pareto local searches.

Let's consider each algorithm in more detail.

Algorithm for changing the components of a Boolean solution vector.

The main idea of a local search is the existence of a neighborhood, which is a set of solutions that are close to the current solution in a sense [3, 25, 27].

Solutions to optimization problems must belong to an admissible region, that is, X ∈ D. If the objective functions F(X) are given on the set D, then we have the problem (D, F(X)). In this case, the system of neighborhoods of N is the mapping of the set D into the set 2n (the set of all subsets of the set D) defined for each problem. Moreover, two points belonging to an admissible region will be called k-neighboring if they differ from each other by the values of their k coordinates.

The iterative procedure for local search (in the case of the minimization problem) can be described as follows [25].

(1) Choose some starting point X.
(2) Perform the procedure IMPROVEMENT (X):

$$\text{IMPROVEMENT (X)} = \begin{cases} any\ Y \in N(X),\ where\ F(Y) < F(X),\ if\ Y\ exist \\ \text{NO},\ \ if\ Y\ doesn\text{'}t\ exist \end{cases}$$

(3) IF IMPROVEMENT (X) = "NO", so end. Point X—point of minimum.
(4) contionue ***X*** = IMPROVEMENT (X). Go to step 2.

This algorithm starts from an arbitrary initial admissible point and uses the procedure IMPROVEMENT (X) to search for the best solution in the neighborhood of point X. If you can find the best solution, then it is remembered and the search continues in the vicinity of this best solution. If the best solution could not be found, then the current point is the local minimum.

Thus, the search for the optimal solution in the local search algorithm is carried out by looking at the points of the neighborhood, changing the components of the Boolean solution vector. And the decision to transfer the search to a new point is made after viewing the nearest points in the neighborhood. There are 2 ways to view the neighborhood: (1) the steepest descent, when the entire neighborhood is viewed and a transition to the point with the best value of the function is performed, and (2) the first improvement transition is a clear alternative to the steepest descent, which should be used in conditions where there is no specific information that gives reason to hope that the "fastest" descent will be really fastest [27].

In the case of multicriteria problems, we are dealing with the so-called Pareto local search (PLS). Pareto local search is an original method of local descent in the space of binary variables with the transition to the first improvement. Its distinguishing feature is the way to determine the best solution—comparing the neighboring points. In this case, the determination of whether the point obtained at this stage is better than the previous one obtained before is based on the Pareto principle, i.e. the transition to a neighboring point is carried out if it dominates the current one in terms of a set of criteria.

1. Algorithm with a change in an integer vector of values

This algorithm is similar to the previous one, however, the neighborhood points are chosen not by changing the Boolean solution vector, but by moving to a neighboring point in the space of integers. All other concepts of a k-neighbor, options for viewing neighborhoods and determining the best solution using the Pareto principle remain unchanged. The advantage of this simpler algorithm is that in integer form we have fewer values in the neighborhood, therefore, the algorithm will work faster.

Results

To analyze the results and to compare the aforementioned methods, the following conditional multicriteria optimization problem was solved:

$$Minimize = \begin{cases} f_1(x, y) = 2 + (x - 2)^2 + (y - 1)^2 \\ f_2(x, y) = 9x - (y - 1)^2 \end{cases}$$

$$Constraints = \begin{cases} g_1(x, y) = x^2 + y^2 \leq 225 \\ g_2(x, y) = x - 3y + 10 \leq 0 \end{cases}$$

$$Variables\ constraints = \begin{cases} -20 \leq x \\ y \leq 20 \end{cases}$$

Real Pareto front looks as shown on Fig. 2.25.

Initially, the SPEA2 algorithm solved the problem of unconditional optimization. The initial population size is 100.

After training without taking into account restrictions, 34 individuals who satisfy the restrictions received, 30 of which are unique, and 66 who do not satisfy and are subject to "treatment" Fig. 2.26.

Two described treatment algorithms were used: Pareto Boolean and integer local search (Fig. 2.27).

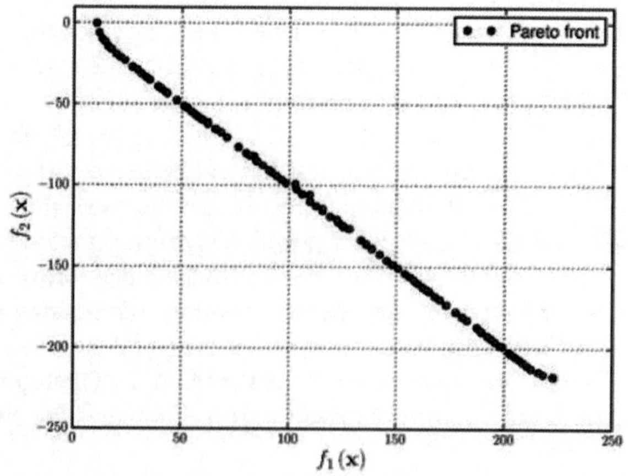

Fig. 2.25 Real Pareto front of the solved problem

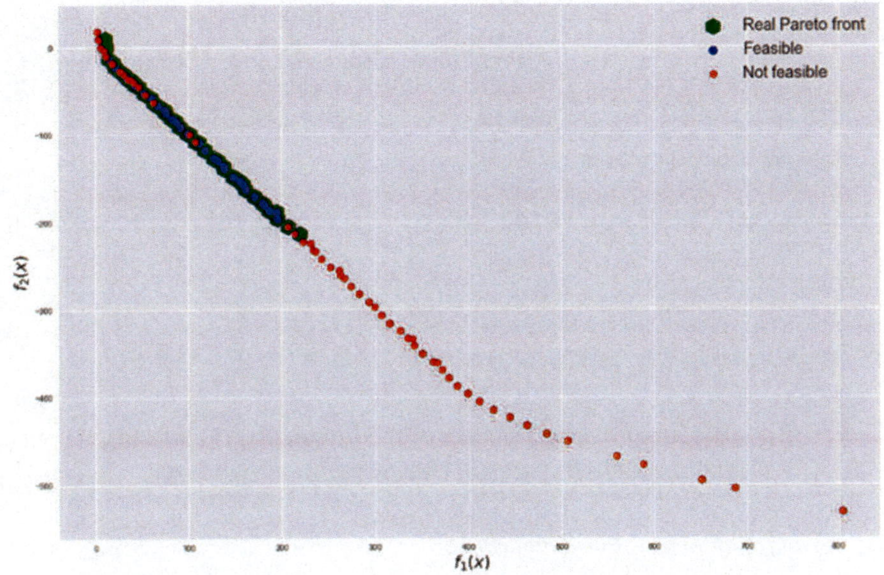

Fig. 2.26 The whole population

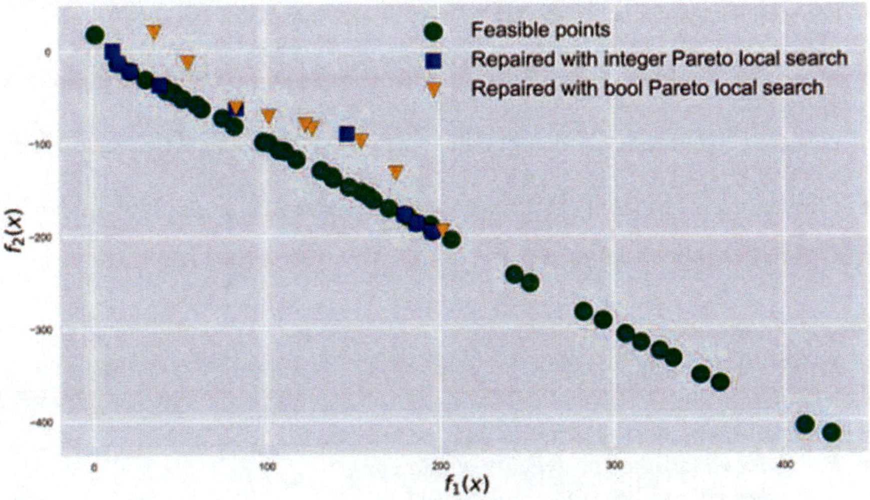

Fig. 2.27 Repaired points

	Pareto Boolean local search	Pareto integer local search
The number of cured points	66 (all)	34

(continued)

(continued)

	Pareto Boolean local search	Pareto integer local search
The number of unique points from the cured	24	10
The number of undominated unique points from the cured	9	9
The number of final non-dominated unique solutions	30	32
Did they add anything new	No	Yes

2.4.7 *Schematic of a Hybrid Genetic Algorithm for Solving Conditional Multicriteria Optimization Problems*

(1) *Features of the developed algorithm:*

- transition from restrictions to additional criteria;
- use the concept of Pareto-optimality, dominance and density of solutions to calculate the adaptability of the solution;
- adaptive determination of the probability of crossing and mutation;
- use of a special type of crossover—neighborhood crossover;
- support of an external multitude of individuals (archive) for the realization of elitism;
- clustering to eliminate clumps of points and increase the representativeness of the solution;
- "chromosome treatment" in the area of initial constraints to clarify solutions.

(2) *Description of the algorithm*

Inputs:

- *N (population size),*
- *N_A (archive size),*
- *T (maximum number of generations),*
- *P_c (initial value of the probability of crossing),*
- *P_m (initial value of mutation probability).*

Output: A (a plurality of non-dominant individuals).

Step 0. Initialization: Initial population is generated P_0, for this:

(1) Randomly (according to the uniform distribution law), N of the possible solutions of the problem in the area of feasible solutions is chosen. If the solution is a certain vector $X(x_1 \dots x_k)$, then each component of this vector is randomly selected.

(2) The resulting N random resolutions are encoded into a binary code. Depending on the constraints on the minimum and maximum values of the junction (or individual components of a vector), the number of bits required to encode that value (or a given component of a vector) is determined. For example, using 10 bits, each of which corresponds to one bit, you can encode numbers from 0 to $2^{10} = 1024$. The encoded binary representation (the vector of zeros and ones) formed by the encoding is called a chromosome.

We also create an empty archive $A_0 = \emptyset$ and set the iteration number $t = 0$.

Step 1. Determination of fitness: performed based on the use of the concept of Pareto domination [18], algorithm for calculating which (fitness function F) for each of the individuals in P_t and A_t has the following form.

(1) Suppose we have a plurality of individuals that make up a population P_t and archive A_t, where each individual i is assigned a value $S(i) \in [0,1)$, called "power" (which shows how many decisions it dominates), which is proportional to the number of members of the population $j \in P_t$, for which $f(i) \geq f(j)$ in the case of multi-criteria optimization. The proportion is as follows:

$$S(i) = \frac{n}{N+1},$$

where N is the population size; n is the number of individuals who dominate the condition $f(i) \geq f(j)$.
The "power" of each individual i in many individuals of the archive A_t and populations P_t will be defined as the sum of the "force" per individual and "force" given the dominance coded by I criterion over criterion encoded by j:

$$S(i) = |\{j|j \epsilon P_t + A_t \wedge i > j\}|,$$

where $x_{\kappa+1} \ldots x_1$ is the dominance coded by i criterion over criterion encoded by j.

(2) Based on value $S(i)$ is calculated raw the value of the fitness function $R(i)$ traits i, which is calculated by summing the "forces" of all individuals j, which dominate or weakly dominate the individual and by all criteria

$$x_{l-1} \ldots x_n \; R(i) = \sum j \in P_t + A_t, j > iS(j).$$

where P_t a plurality of individuals of the population; A_t is the multiple individuals archive.

(3) The density estimation method is an adaptation of the k-th neighbor method, where density at any point is a (decreasing) function of the distance to the k-th nearest data point. A function is called a drop-down at some interval if for any argument value from that interval, a larger argument value corresponds to a smaller function value. The density inversion is taken to invert the distance to

the k-th nearest neighbor. For each individual and, distances (in the criterion space) to the individuals j in the archive and in the general sample are counted and stored in the list.

To calculate the value of the fitness function, the value of the density of the individuals is used: for each individual, the Cartesian distance from it to the other individuals is calculated j in the archive and population.

After sorting the list in ascending order kth the element gives the desired distance to the individual i, which is indicated σ_i^k. That is σ_i^k denotes the distance from the individual i to the nearest k-th the neighbor. We use k, equal to the square root of the sample size.

We calculate the density value $D(i)$ for the individual i:

$$D(i) = \frac{1}{\sigma_i^k + 2} \quad k = \sqrt{(N + N_A}, $$

where N is the size of the population; N_A is the size of the archive; k is rounded to the nearest integer.

A denominator is added to the denominator to ensure that its value is greater than zero and that $D(i) < 1$.

(4) Finally, adding $D(i)$ to the original fitness value $R(i)$ individual i gives its fitness $F(i)$. Therefore, the final value of the fitness function $F(i)$ for the individual i is defined as $F(i) = R(i) + D(i)$.

Step 2. Archive upgrade. Create an interim archive x_i.

(a) copy individuals whose decision vectors are invariant with respect to P_t in A_t*;

(b) remove those individuals from A_t*, whose corresponding decision vectors are weakly dominated by A_t*;

(c) reduce the number of individuals stored in the archive and place the resulting reduced number of individuals in the A_{t+1}. The following procedure is used to reduce the number of individuals in the archive:

(1) All nondominant individuals (values of fitness functions less than 1) from the archive and populations are copied to the archive of the next iteration:

$$A_{t+1} = \{i | i \in P_t + A_t \wedge F(i) < 1\}, \; i = \overline{1, N}.$$

where P_t is the intermediate population; A_t is the intermediate archive; A_{t+1} is the reduced archive; $F(i)$ are fitness features for the individual i; N is the population size.

(2) If the size of the created archive corresponds to the desired one ($|A_{t+1}| = N_A$), then the step is completed, otherwise two situations are possible.

(3) The size of the created archive:

(a) less than desired ($|A_{t+1}| < N_A$). In this case, the best $N_A - |A_{t+1}|$ individuals that dominate the individuals in the previous archive and the population are copied to the new archive. This can be done by sorting the combined set $P_t + A_t$ according to the values of their fitness functions and the first $N_A - |A_{t+1}|$ individuals with $F(i) \geq 1$ from the sorted set are copied to the archive $At + 1$;

(b) more than desired ($|A_{t+1}| = N_A$). In this case, the archive size reduction process begins, which iteratively removes the individuals from A_t + 1, until ($|A_{t+1}| = N_A$).

- create a blank list D;
- for each individual i its Cartesian distance is calculated $d(i, j)$ to all other individuals j. The smallest of the distances obtained d_i=min $d(i, j)$ is added to the list $D = D \in i$, where $i = \overline{1, N}$.

If there are several individuals with a minimum distance, then the problem is solved, taking into account the second shortest distance, etc.

- the values of the received list D are sorted in ascending order. The person corresponding to the lowest value in the list D is removed from the archive;
- the process will be repeated until the archive reaches the desired size ($|A_{t+1}| = N_A$).

Step 3. Ranking. Individuals in the population are sorted according to descending fitness values (from best individual to worst).

Step 4. Grouping. The individuals are divided into groups, each consisting of two individuals. These two individuals are selected from the beginning of the list of sorted individuals.

Step 5. Crossing and mutation. In each of the formed groups there is a crossover and a mutation.

Crossing For any two individuals, the need to perform the crossing operation is determined randomly:

(1) A number is randomly generated p_c* from the interval [0,1]. The obtained number is compared with the given probability of crossing. If $p_c* > p_c(t)$, where $p_c(t)$—probability of crossing on t iterations, the crossing does not occur and the parent individuals remain unchanged. Otherwise, the process of crossing occurs (see paragraph 2, step 5).

(2) This algorithm uses the single-point crossover variant—both selected "parents" are cut at a randomly selected point, after which their chromosomes exchange their fragments (see Fig. 2.2).

Mutation

The need to perform a mutation is determined similarly to a similar operation for crossing:

(3) A number is randomly generated p_m* from the interval [0, 1]. The obtained number is compared with the given mutation probability. If $p_m* > p_m(t)$, where $p_m(t)$—the probability of a mutation at t iteration, then the mutation does not occur and the parent individuals remain unchanged. Otherwise, a mutation process takes place (see paragraph 4, step 5).
(4) A mutation is associated with a random change of one or more genes in the chromosome, as shown in Fig. 2.3.

Two parent individuals are formed from two parent individuals. Parents are removed from the group.

Adaptation of probabilities of crossing over mutations

(5) Crossing progress values are calculated B_{IJ} and the progress of the mutation $X = \left(\sum_{i=1}^{n} x_i\right)/n$, (see Sect. 2.4.2). New values of crossing and mutation probabilities are calculated based on the calculated values of the progress:

$$pc(t+1)=pc(t)+\alpha{\cdot}pc(t) \text{ and } p_M(t+1)=p_M(t)-\alpha{\cdot}p_M(t), \text{ if } \overline{CP}>\overline{MP}$$

$$pc(t+1)=pc(t)-\alpha{\cdot}pc(t) \text{ and } p_M(t+1)=p_M(t)+\alpha{\cdot}\,p_M(t), \text{ if } \overline{CP}>\overline{MP}$$

where α is the adaptation rate; α belongs [0.10; 0.15].

Step 6. All the children are grouped together into a new population P_t.

Step 7. End. Put $Pt + 1 = At$ and $t = t + 1$. If $t \geq T$ or some other stop criterion is then fulfilled A_t—there is a multitude of resolutions you want, otherwise go to step 1.

Step 8. "Treatment of points".

To eliminate the identified shortcomings of genetic algorithms, it is proposed to carry out "treatment" (refinement) of non-dominant points obtained after stopping the genetic algorithm.

(1) Select some initial valid solution (chromosome) *X*.
(2) Form a plurality of chromosomes $Y = \{y_1, \ldots., y_k\}$, where chromosome yi is obtained by mutation of the *i*th bit of the original chromosome X, $i = \overline{1, k}$, k is the number of bits in the chromosome.
(3) We calculate the fitness function values for all chromosomes from the set *Y*, using the procedure described in step 1 and selecting among them the chromosome *Y** with the best fit value.
(4) If $F(Y^*) < F(X)$ (for the maximization problem) then the solution *X* is the best in this neighborhood. We move on to step 8, step 1, and select the next point to improve until all the solutions are cured.
(5) Otherwise, we accept $X = Y^*$. Proceed to step 2 step 8.

Step 9. Eliminate clots of dots

(1) Initialize multiple clusters C. Each individual $i \in A$ forms a separate cluster C_i.
(2) If $|C| \leq N_A$, go to step 5 of step 9, otherwise go to step 3 of step 9.
(3) Calculate the distance between all possible pairs of clusters. Remoteness d_c two clusters C_i and $C_j \in C$ is defined as the average distance between pairs of individuals belonging to these clusters:

$$d_c = \frac{1}{|C_i| \cdot |C_j|} \sum_{c_i \in C_i, c_j \in C_j} ||c_i - c_j||,$$

de operator $|| \cdot ||$ reflects the Euclidean distance (in the space of goals) between two individuals c_i and c_j, operator $|| \cdot ||$ determines the number of elements in the set.
Let $x = (x_1, x_2, \ldots, x_k)$ and $y = (y_1, y_2, \ldots, y_k)$—decisions that are relevant to the individual X and Y, then the Euclidean distance between these individuals can be defined as $||X - Y|| = ||x - y|| = \sqrt{\sum_{i=1}^{k} (x_i - y_i)^2}$.
(4) Identify two clusters C_i and $C_j \in C$ with a minimum distance d_c. These clusters merge into one larger cluster $C_{ij} = C_i \bigcup C_j$.
Go to **Step 2.**
(5) Select a representative individual for each cluster, and remove all other individuals from the cluster. (Thus, a representative individual is a centroid, a point with a minimum average distance to all other points of the cluster.) Define a reduced archive by combining representative individuals of all clusters.

The result of the proposed algorithm is the approximation of the Pareto set is representative—the points are evenly distributed, there is no condensation, there are no dominant points.

Examples of the use of the hybrid genetic algorithm of multicriteria optimization developed in the work for solving the problem of structural-parametric synthesis of the neural network of base NN modification, pretrained deep neural network with DBN, the convolutional CNN are given in the Sects. 3.6.4, 4.11.3 and 4.12.3.

2.5 Introduction Swarm Intelligence Algorithms

Swarm intelligence algorithms are a class of nature-inspired metaheuristics, which are defined as the collective intelligence behavior of self-organized and decentralized systems (artificial groups of simple agents).

Two fundamental concepts that are considered as necessary properties of SI are self-organization and division of labor. Self-organization is defined as the capability of a system to evolve its agents or components into a suitable form without any external help. Self-organization relies on four fundamental properties of positive feedback, negative feedback, fluctuations and multiple interactions. Positive and negative feedbacks are useful for amplification and stabilization respectively. Fluctuations meanwhile are useful for randomness. Multiple interactions occur when the swarms share information among themselves within their searching area. The second property of SI is division of labor, which is defined as the simultaneous execution of various simple and feasible tasks by individuals. This division allows the swarm to tackle complex problems that require individuals to work together [28].

2.5.1 The General Idea Behind Swarm Intelligence Algorithms

The main idea for the swarm intelligence algorithms is to generate an initial population of agents (possible solutions) and using algorithm specific mutation and migration operators modify the agents positions in the search space and other agents' properties (e.g. velocity, individual best score) iteration by iteration. The iteration of such algorithms is also called a generation [29]. During this iterative process, the communication between the swarm individuals may take place, which allows the agents to know the current swarm best solution, and such communication affect the whole swarm behavior. A swarm algorithm performs an optimization process until a certain termination criterion is satisfied. The termination criterion can be defined as reaching the maximum number of iterations or the optimization process stagnation. The optimization process stagnation means that the swarm's best solution doesn't change more that the specified tolerance in a specified number of iterations.

There are mainly two types of a swarm behavior. These are exploration and search intensification. The first one allows exploring the search space as much as possible, so that there is more chance to find a better solution. The exploration is achieved by using stochastic components in the algorithm. The second type of behavior is used to speed up the convergence of the algorithm. The intensification is done by all of the swarm agents gathering towards the current best global solution. Both of these behaviors are controlled by the algorithm specific set of parameters. The proper tuning of these parameters allows to find a satisfactory solution in a relatively small amount of iterations (generations).

The main idea of intelligent swarm algorithms is to generate an initial population of agents (possible solutions) and, using algorithm-specific mutation and migration operators, change the position of agents in the search space and iterate other properties of agents (for example, speed, individual best result) from iteration to iteration. An iteration of such algorithms is also called generation. During this iterative process, data exchange between swarm groups can occur, which allows agents to know the current best swarm solution, and such communication affects the behavior of the entire swarm.

The swarm algorithm performs the optimization process until a certain termination criterion is satisfied. The termination criterion can be defined as achieving the maximum number of iterations or as stagnation of the optimization process. Stagnation in the optimization process means that the best swarm solution does not change more than the specified tolerance for the specified number of iterations.

2.5.2 General Swarm Intelligence Algorithm

There are two main types of swarm behavior. This is research and search intensification. The first allows you to explore the search space as much as possible so that you have a better chance of finding the best solution. The study is achieved by using stochastic components in the algorithm. The second type of behavior is used to accelerate the convergence of the algorithm. Intensification is carried out by all swarm agents gathering in the direction of the best global solution to date. Both of these behaviors are controlled by an algorithm-specific set of parameters. Proper adjustment of these parameters allows you to find a satisfactory solution for a relatively small number of iterations (generations) [30].

The work of the SI algorithm includes the following steps.

1. Generation of the initial set of particles. Somehow, as a rule, using randomization, particles are distributed in the search space. Iteration number $j = 1$.
2. Calculate the criterion for each particle as a function of the particle position $f(X_{ij})$. For individual algorithms, such as ant colony algorithms, the criterion calculation is performed after moving the particles, so for them the second step in the first iteration is skipped.
3. Particle displacement (migration). In this step, the main feature of swarm intelligence algorithms is implemented—performing each part of its actions on the basis of:
   - individual rules and experience;
   - indirect exchange of information with other swarm particles;
   - stochastic properties.
4. Checking the completion condition. If the condition is met, the process completes, the value $X_{final^{best}}$ will be the end result, otherwise there is a transition to step 2 (with an increase in the iteration number j per one).

The termination condition may be related to the execution of a predetermined number of iterations, or the detection of a solution not worse than specified as satisfactory, or stagnation of search (X_j^{best}) does not change over a certain number of iterations of the algorithm).

The analysis of the third step of the algorithm allows to determine whether it applies to the algorithms of swarm intelligence. In swarm algorithms, in formulas that determine the behavior of particles, there is either an element associated with one or more of the best positions found in all swarms, or the weighted average position of swarm particles (center of gravity), where the weights are proportional to the criterion corresponding to the position of the particle. Yes, the particle swarm algorithm uses the best position among all particles and all iterations performed to organize the exchange of information between particles. X_j^{best}, in the bee swarm algorithm there are some of the best particle positions, in monkey search the center of gravity. In the ant colony algorithm, a graph is used for interaction, the weights of the arcs of which vary depending on what value of the criterion was obtained during the movement of particles on it.

Unlike the basic principle of information exchange, evolutionary algorithms do not have a mediated interaction, instead it uses the process of selecting the decision with the best value of the criterion.

2.6 Analysis of Swarm Intelligence Algorithms

2.6.1 *Particle Swarm Optimization*

2.6.1.1 Intuitive Interpretation

Particle Swarm Optimization algorithm (*PSO*) is based on a social and psychological behavior model of some groups of organisms (swarm). The algorithm research was also inspired by such problems, as simulating the birds flock or fishes school movement [31]. The goal of the research was to figure out the principles, which help birds in the flock to coordinate their movement direction [32–37].

So the basic idea is formulated as follows. The algorithm at each iteration has a fixed-sized population (swarm) of candidate solutions. These solutions (particles) of the swarm are moved around in the search space with some velocities. This movement is iterative and determined by three components: inertial, social and cognitive components. The inertial component is responsible for changing the velocity, which particle on the previous iteration (deceleration, acceleration or leaving without change), the other two components determine the inclination to move either to the individual's best place it has ever found or to the global (population) best one. The movement has stochastic characteristics as well. After performing such movements, if improved positions are being discovered, than these will then come to guide the swarm. The process is repeated until, hopefully, a satisfactory solution is discovered or the termination condition is satisfied [38].

2.6.1.2 Algorithm Description

Let's describe how the algorithm is applied to the problem of unconstrained continuous global optimization in a D-dimensional search space. The first step of the algorithm is the generation of initial population of N particles. Each particle is a potential solution to the problem. The coordinates of the particle i at the iteration t

are defined as a real-number D-dimensional vector $X_i^t = \left(x_{i,1}^t, x_{i,2}^t, \ldots, x_{i,D}^t\right)$. Each particle i moves in a search space with a velocity $V_i^t = \left(v_{i,1}^t, v_{i,2}^t, \ldots, v_{i,D}^t\right)$.

The update rule for each particle's coordinates at the iteration $(t+1)$:

$$V_i^{t+1} = \omega V_i^t + c_1 U_D(0;1) \otimes \left(X_i^{pbest} - X_i^t\right) + c_2 U_D(0;1) \otimes \left(X^{gbest} - X_i^t\right),$$
$$X_i^{t+1} = X_i^t + V_i^{t+1},$$

here $\otimes$ is the element-wise vector product operator; ω, c_1, c_2—algorithm parameters; $U_D(0;1)$—D-dimensional vector of real-number values uniformly distributed at the interval [0; 1]; X_i^{pbest}—coordinates vector which corresponds to the best solution found by particle i for all of the iterations; X^{gbest}—coordinates vector which corresponds to the best solution found globally by all particles for all of the iterations. The pseudo code for the Particle swarm optimization algorithm:

begin
    Objective function f(X),
    Define the algorithm's parameters
    for *each particle i = 1,...,N* **do**
        Initialize the particle's position X_i
        Initialize the particle's best known position to its initial position
        Initialize the particle's speed to zero
    end
    Get initial best particle
    while *a termination criterion is not met* **do**
        for *each particle i = 1,...,N* **do**
            Update the particle's velocity and position by the update rule
            if $f(X_i) < f(X_i^{pbest})$ **then**
                Update the particle's best known position
                if $f(X_i) < f(X^{gbest})$ **then**
                    Update the swarm's best known position
                end if
            end if
        end for
    end while
    Results post-processing
end

2.6.1.3 Algorithm Parameters

Besides the population size and optimal number of iterations, the algorithm has 3 parameters: ω (inertial parameter), c_1, c_2—acceleration parameters, which correspond to cognitive and social components of each particle. Let's study them in great detail.

Inertial parameter ω. The parameter defines the weight of particles' inertia (if $\omega < 1$, the particle's movement slows down). Bigger values of this parameter corresponds to more extensive search and smaller values make the search more local and intensified. It's also known as a good practice to gradually decrease the value of ω along with the optimization process.

Cognitive parameter c_1 *and social parameter* c_2. The parameters are responsible for the weights of particles' cognitive and social components. If $c_1 = c_2 = 0$, the particles' trajectories become straight lines going to infinity. If $c_1 > 0$, $c_2 = 0$, the particles try to reach the minimum of the fitness function independently of one another. The success of the solution is determined by the value of social parameter, because it coordinates the movement of the particles in a swarm. If both parameters have small values, the particles trajectories are smooth. The bigger the values, the more stochastic the movement of particles becomes.

2.6.2 *Firefly Algorithm*

2.6.2.1 Intuitive Interpretation

The Firefly Algorithm is inspired by the way the fireflies communicate with each other. Fireflies produce short and rhythmic flashes. Different firefly species have unique flashing patterns. The fundamental functions of such flashes are to attract mating partners, and to attract potential prey. In addition, flashing may also serve as a protective warning mechanism [39].

The main principle is similar to the one in PSO: we have a population of solutions which are moved around the search space based on the algorithm's rules until a solution is found (or the termination criterion is met).

The algorithm associates the flashing light with the objective function to be optimized and is based on idealized model of fireflies' behavior. This model has the following rules:

- all fireflies are unisex so that one firefly is attracted to other fireflies regardless of their sex;
- attractiveness is proportional to their brightness, thus for any two flashing fireflies, the less brighter one will move towards the brighter one. The attractiveness is proportional to the brightness and they both decrease as their distance increases. If no one is brighter than a particular firefly, it moves randomly;

- the brightness or light intensity of a firefly is affected or determined by the landscape of the objective function to be optimized.

2.6.2.2 Algorithm Description

For minimization problem we define the brightness to be proportional to the objective function.

Similarly to the PSO algorithm, each firefly i position at the iteration t is defined a real-number D-dimensional vector $X_i^t = \left(x_{i,1}^t, x_{i,2}^t, \ldots, x_{i,D}^t\right)$, where D is the search space dimension.

The attractiveness of the firefly j to firefly i is equal to:

$$\beta_{i,j} = \beta_0 \exp\left(-\gamma r_{i,j}^2\right), i, j \in [1 : N], i \neq j,$$

where $r_{i,j}$ is the distance between firefly i and firefly j (in case of this research, the Cartesian distance); β_0 is the attractiveness at zero distance ($r = 0$); γ is the light absorption coefficient.

Actually, instead of using $\exp\left(-\gamma r_{i,j}^2\right)$ for decreasing the attractiveness as the distance between fireflies increases, it is possible to use other monotonically decreasing functions, for instance:

$$\beta_{i,j} = \frac{\beta_0}{\left(1 + \gamma r_{i,j}^2\right)}, i, j \in [1 : N], i \neq j.$$

Using this definition of attractiveness may simplify and speed-up the calculations, because we no longer need to compute the exponential function.

The other alternative for attractiveness function:

$$\beta_{i,j} = \beta_0 \exp\left(-\gamma r_{i,j}^n\right), i, j \in [1 : N], i \neq j, n \geq 1.$$

The update rule for each firefly's coordinates at the iteration (t +1):

$$X_i^{t+1} = X_i^t + \beta_{i,j}\left(X_j^t - X_i^t\right) + \alpha \epsilon_i,$$

where the third term randomizes the search using D-dimensional vector ϵ_i of normally or uniformly distributed values at the interval $[-1; 1]$ and α as a randomization coefficient. To maintain a balance between diversification and intensification of the search it is recommended to decrease the randomization coefficient with a growth of the iteration number. This process can be performed via the formula:

$$\alpha = \alpha_\infty + (\alpha_0 - \alpha_\infty)e^{-t},$$

where α_0 is the initial value of α; α_∞ is the final value of α.

The update rule can be modified to improve the convergence by adding a component, which takes in consideration global best position:

$$\lambda\sigma_i \otimes \left(X_i^t - X^{gbest}\right),$$

where λ is the parameter similar to α; $\boldsymbol{\sigma}_i$ is the similar to the randomization vector used in the initial update rule variant.

So the final version of the update rule:

$$X_i^{t+1} = X_i^t + \beta_{i,j}\left(X_j^t - X_i^t\right) + \alpha\epsilon_i + \lambda\sigma_i \otimes \left(X_i^t - X^{gbest}\right).$$

The pseudo code for the Firefly Search algorithm:

begin
  Objective function f(X),
  Generate initial population of fireflies
  Light intensity I_i at x_i is determined by $f(X_i)$
    and in case of minimization the less the value of $f(X_i)$,
    the more intensive the light I_i is
  Define light absorption coefficient γ
  while *a termination criterion is not met* **do**
    for $i = 1 : N$ *(all N fireflies)* **do**
      for $j = 1 : N$ *(all N fireflies, inner loop)* **do**
        if $I_i < I_j$ **then**
          Move firefly i towards j
        end if
        Evaluate new solutions and update light intensity
      end for
    end for
    Rank the fireflies and find the current global best X^{gbest}
  end while
  Results post-processing
end

2.6.2.3 Algorithm Parameters

Besides the population size and optimal number of iterations, the algorithm has 5 parameters: zero distance attractiveness β_0, light absorption coefficient γ, initial value of randomization coefficient α_0, final value of randomization coefficient α_∞, randomization coefficient λ.

Zero distance attractiveness β_0. The parameter is used to calculate attractiveness between two fireflies at different distances. Bigger values of this parameter correspond to bigger steps of fireflies towards the more attractive ones.

Light absorption coefficient γ. The parameter γ characterizes the variation of the attractiveness, and its value is very important in determining the speed of the convergence and how the algorithm behaves.

Initial and final value of randomization coefficient α_0 *and* α_∞. The parameters are used to compute and gradually decrease the size of the α step factor. The values are drawn from the interval [0; 1].

Randomization coefficient λ. Determines the weight of the third component of the update rule, which coordinates the movement towards the global best solution.

2.6.3 Cuckoo Search Algorithm

2.6.3.1 Intuitive Interpretation

Cuckoos have a remarkable reproduction strategy. Lots of species engage the obligate brood parasitism [40] by laying their eggs in the nests of other host birds (often other species). Some host birds can engage direct conflict with the intruding cuckoos. If a host bird discovers the eggs are not their owns, they will either throw these alien eggs away or simply abandon its nest and build a new nest elsewhere. Some cuckoo species have evolved in such a way that female parasitic cuckoos are often very specialized in the mimicry in color and pattern of the eggs of a few chosen host species [41]. This reduces the probability of their eggs being abandoned and thus increases their reproduction [42, 43].

In addition, the timing of egg-laying of some species is also amazing. Parasitic cuckoos often choose a nest where the host bird just laid its own eggs. In general, the cuckoo eggs hatch slightly earlier than their host eggs. Once the first cuckoo chick is hatched, the first instinct action it will take is to evict the host eggs by blindly propelling the eggs out of the nest, which increases the cuckoo chick's share of food provided by its host bird. Studies also show that a cuckoo chick can also mimic the call of host chicks to gain access to more feeding opportunity.

The Cuckoo Search algorithm associates the quality of eggs and nests' position with the objective function. The idea is to have a population of solutions (nests with eggs) and to have a cuckoo, which tries to find new better nests (solutions) in the search space and dump its egg in there by following specific rules described below.

2.6.3.2 Algorithm Description

For minimization problem we define the fitness to be proportional to the objective function.

The algorithms is based on idealized model of cuckoos' behavior. This model has the following rules:

- each cuckoo lays one egg at a time in a randomly chosen nest;
- the best nests with high-quality eggs (with the best value of the objective function) will carry over to the next generation;
- the number of nests is fixed and each egg laid by a cuckoo can be discovered by the host bird with probability $p_a \in [0; 1]$. In this case, the discovered nest is replaced by the new one (with new random solution).

Each nest *i* position at the iteration *t* is defined a real-number *D*-dimensional vector $X_i^t = (x_{(i,1)}^t, x_{(i,2)}^t, \ldots, x_{(i,D)}^t)$, where *D* is the search space dimension.

The update rule for all of the solutions:

$$X_i^{(t+1)} = X_i^t + \alpha \otimes Levy_D(\lambda),$$

where $\alpha > 0$ is the step size; $\otimes$ is the element-wise vector product operator; $Levy_D(\lambda)$ is the D-dimensional vector of real-number values drawn from a Levy distribution.

The pseudo code for the Cuckoo Search algorithm:

begin
    Objective function f(X)
    Generate initial population of N host nests
    while*a termination criterion is not met* **do**
        Get a cuckoo randomly by Levy flights
        Evaluate its quality/fitness F_i
        Choose a nest among N (say, j) randomly
        if $F_i < F_j$ **then**
            replace j by the new solution
        end if
        A fraction (p_a) *of worse nests*
            are abandoned and new ones are built
        Keep the best solutions (or nests with quality solutions)
        Rank the solutions and find the current best
    end while
    Results post-processing
end

2.6.3.3 Algorithm Parameters

Besides the population size and optimal number of iterations, the algorithm has 2 parameters: α is the step size for Levy flights, p_a is the probability of nest discovery and deletion.

Step size α *for Levy flights*. The purpose of the parameter is obvious. The value of the step size should be related to the scales of the concrete problem. In *the MCS algorithm* (*Modified Cuckoo Search*) α is a D-dimensional vector and it's elements are calculated taking in consideration current best nest:

$$\alpha = U_1(0; 1)(X^g best - X_i).$$

2.6.4 Bat Search Algorithm

2.6.4.1 Intuitive Interpretation

The idea of *the Bat Search algorithm* is drawn from the bats advanced capability of echolocation [3]. Bats use echolocation to detect prey, avoid obstacles, and locate their roosting crevices in the dark. These bats emit a very loud sound pulse and listen for the echo that bounces back from the surrounding objects. Their pulses vary in properties and can be correlated with their hunting strategies, depending on the species [44].

The algorithm flow can be summarized as follows: each bat, which represents a solution, flies randomly in the search space with some velocity with a varying frequency or wavelength and loudness. As it searches for its prey, it changes frequency, loudness and pulse emission rate to control the way it moves. Search is intensified by a local random walk around the current global best solution. Selection of the best continues until certain stop criteria are met.

2.6.4.2 Algorithm Description

The algorithms is based on idealized model of bats' behavior. This model has the following rules:

- bats can distinguish prey from obstacles and sense the distance to them by the means of echolocation;
- bats fly randomly with velocity V_i at position X_i with a fixed frequency $f_{\min}$, varying wavelength λ and loudness A_0 to search for prey. They can automatically adjust the wavelength (or frequency) of their emitted pulses and adjust the rate of pulse emission $r \in [0; 1]$, depending on the proximity of their target;
- although the loudness can vary in many ways, we assume that the loudness varies from a large (positive) A_0 to a minimum constant value $A_{\min}$.

Each bat is a potential solution to the problem. The coordinates of the bat i at iteration t are defined as a real-number D-dimensional vector $X_i^t = \left(x_{i,1}^t, x_{i,2}^t, \ldots, x_{i,D}^t\right)$. Each bat i moves in a search space with a velocity $V_i^t = (v_{(i,1)}^t, v_{(i,2)}^t, \ldots, v_{(i,D)}^t)$.

The pseudo code for the Bat Search algorithm:

begin
    Objective function f(X)
    Generate initial population of N bats
    while *a termination criterion is not met* **do**
        Generate new solutions by adjusting frequency
        Update velocities and locations/solutions
        if *(rand >r_i)* **then**
            Select a solution among the best solutions
            Generate a local solution around the selected best solution
        end if
        Generate a new solution by flying randomly
        if *(rand < A_i& $f(X_i) < f(X^{gbest})$)* **then**
            Accept the new solutions
            Increase r_i and reduce A_i
        end if
        Rank the bats and find the current best
    end while
    Results post-processing
end

The update rule for the solutions and velocities

$$
\begin{aligned}
f_i &= f_{\min} + (f_{\max} - f_{\min})\beta, \\
V_i^{(t+1)} &= V_i^t + (X_i^t - X^{gbest})f_i, \\
X_i^{(t+1)} &= X_i^t + V_i^{(t+1)},
\end{aligned}
$$

where β is a random vector drawn from a uniform distribution at the interval [0; 1]; X^{gbest}—coordinates' vector which corresponds to the best solution found globally by all bats for all of the iterations.

For the local search part, we generate a new solution using random walk:

$$
\begin{aligned}
X_{new} &= X_{old} + \epsilon \bar{A}^t, \\
\bar{A}^t &= 1/N \sum_{(i=1)}^{N} A_i^t,
\end{aligned}
$$

where ϵ—D-dimensional vector of uniformly distributed values at the interval [−1; 1]; $\overline{A_i^t}$—average loudness of all the bats at the iteration t as it may be assumed from its definition.

Furthermore, the loudness and the rate of pulse emission for each bat have to be updated accordingly as the iterations proceed. New values are given by

$$
A_i^{t+1} = \alpha A_i^t,\ r_i^{t+1} = r_i^0[1 - \exp(-\gamma t)],
$$

where α and γ are constants, so that for any $0 < \alpha < 1$ and $\gamma > 0$, we have

$$A_i^t \to 0, r_i^t \to r_i^0, ast \to \infty.$$

2.6.4.3 Algorithm Parameters

Besides the population size and optimal number of iterations, the algorithm has 6 parameters: minimal frequency $f_{\min}$, maximum frequency $f_{\max}$, initial loudness A_0, initial rate of pulse emission r_0, coefficients α and γ.

Minimal and maximum frequencies' values $f_{\min}$ and $f_{\max}$. These parameters are used to compute the frequencies for each solutions' update. The frequency f_i essentially controls the range of the ith bat's movement. So the values of $f_{\min}$ and $f_{\max}$ should be chosen according to the domain size of the problem.

Initial loudness A_0. Used to calculate the loudness for each bat on each iteration.

Initial rate of pulse emission r_0. Used to calculate the rate of pulse emissions for each bat on each iteration. Both loudness and rate of pulse emission basically control the intensity of the local search.

Coefficients α *and* γ. The coefficients control how do the loudness and the rate of pulse emission decrease as the iterations proceed.

2.6.5 *Wolf Pack Search Algorithm*

2.6.5.1 Intuitive Interpretation

The Wolf Pack Search algorithm is inspired by the strategies used by wolves and social work division in a pack [45]. There is a lead wolf; some elite wolves act as scouts and some ferocious wolves in a wolf pack. They cooperate well with each other and take their respective responsibility for the survival and thriving of wolf pack. The algorithm abstracts three intelligent behaviors of the wolves, which are scouting, calling and besieging, and two rules, which are winner-take-all generation rule of lead wolf and stronger-survive renewing rule of wolf pack. All of the abstractions are described below [46, 47].

2.6.5.2 Algorithm Description

If the predatory space of the artificial wolves is a $N \times D$ Euclidean space, N is the number of wolves, D is the number of variables. The position of one wolf i at iteration t is a vector $X_i^t = \left(x_{i,1}^t, x_{i,2}^t, \ldots, x_{i,D}^t\right)$. $Y = f(X)$ represents the concentration of prey's smell perceived by artificial wolves, which is also the objective function value.

The whole behavior of wolf pack (optimization process) is abstracted three intelligent behaviors, scouting, calling, and besieging behavior, and two intelligent rules, winner-take-all generating rule for the lead wolf and the stronger-survive renewing rule for the wolf pack.

The winner-take-all generating rule for the lead wolf basically implies that during each iteration the wolf with the best objective function value becomes a lead wolf.

Scouting behavior. All wolves except the lead are considered to be the scout wolves and they search the solution in predatory space. Y_i is the concentration of prey smell perceived by the scout wolf *i* and Y_{lead} is the concentration of prey smell perceived by the lead wolf. If $Y_i < Y_{lead}$, the scout wolf I becomes lead wolf and $Y_{lead} = Y_i$. Else if $Y_i > Y_{lead}$, the scout wolf *i* takes a step towards h different directions; the step length is $step_a$. The update rule for the coordinates is defined below:

$$x_{i,d}^p = x_{i,d} + \sin\left(2\pi \times \frac{p}{h}\right) \times step_a^d, p = \{1, 2, \ldots, h\}.$$

In addition, h is different for each wolf because of their different seeking ways. So h is randomly selected in $[h_{\min}; h_{\max}]$ and it must be an integer. The above process is repeated until $Y_i < Y_{lead}$ or the maximum number of repetitions for scouting behavior $T_{\max}$ is reached.

Calling behavior. The lead wolf will howl and summon other wolves to gather around the prey. Here, the position of the lead wolf is considered as the one of the prey so that the other wolves aggregate towards the position of lead wolf. The position of the wolf *i* at the iteration *t* is updated according to the following equation:

$$X_i^{t+1} = X_i^t + step_b \frac{\left(X^{lead} - X_i^t\right)}{\left|X^{lead} - X_i^t\right|},$$

where $step_b$ is the step length; X^{lead} is the position of the lead wolf at the iteration *t*; $\left|X^{lead} - X_i^t\right|$—the Euclidean distance between the lead wolf and the *i*-th wolf.

If $Y_i < Y_{lead}$, the *i*th wolf becomes the lead wolf and takes the calling behavior. If $Y_i > Y_{lead}$, the *i*-th wolf keeps on aggregating towards the lead wolf with a fast speed until $\left|X^{lead} - X_i^t\right| < L_{near}$ after that the wolf takes besieging behavior. L_{near} is the distance determinant coefficient as a judging condition, which determine whether wolf i changes state from aggregating towards the lead wolf to besieging behavior.

Besieging behavior. After large-steps running towards the lead wolf, the wolves are close to the prey, then all wolves except the lead wolf will take besieging behavior for capturing prey. Now, the position of lead wolf is considered as the position of prey. In particular, X^{lead} represents the position of prey at the *t*th iteration. The position of wolf *i* is updated according to the following equation:

$$X_i^{t+1} = X_i^t + \lambda step_c \left|X^{lead} - X_i^t\right|,$$

where λ—a random *D*-dimensional vector of uniformly distributed values at the interval [−1; 1]; $step_c$—the step length of wolf *i* when it takes besieging behavior.

Y_{i0} is the concentration of prey smell perceived by the wolf I and Y_{it} represents the one after it took this behavior. If $Y_{i0} < Y_{it}$, the position Xi is updated; otherwise it remains unchanged.

The stronger-survive renewing rule for the wolf pack. The prey is distributed from the strong to the weak, which will result in some weak wolves dead. The algorithm will generate *R* wolves while deleting *R* wolves with bad objective function values. Specifically, with the help of the lead wolf's hunting experience, in the *d*-th variable space, position of the *i*-th one of *R* wolves is defined as follows:

$$X_i = X^{lead} U_D(0;\ 1), i = \{1, 2, \ldots, R\},$$

where X^{lead} is the position of artificial lead wolf; $U_D(0;\ 1)$ is a random number uniformly distributed at the interval $[-0.1; 0.1]$.

The pseudo code for the Wolf Pack Search algorithm:

```
begin
   Objective function f(X)
   Generate initial population of N wolves
   whilea termination criterion is not met do
      do
         Perform scouting behavior
      whileY_i<Y^lead or T >T_max
      calling_behavior_flag:
      do
         Perform calling behavior
      whileY_i<Y^lead
      ifL(i, lead) >L_near then
         gotocalling_behavior_flag
      endif
      Perform besieging behavior
      Renew the position of the lead
      Renew wolf pack
   end while
   Results post-processing
end
```

2.6.5.3 Algorithm Parameters

Besides the population size and optimal number of iterations, the algorithm has 6 parameters: *step,* the distance determinant coefficient L_{near}, the maximum number of repetitions for scouting behavior $T_{\max}$, the population renewing proportional coefficient β, the minimum and the maximum values for number of direction during scouting behavior $h_{\min}$, $h_{\max}$.

Step. There are 3 step coefficients for each intelligent behavior $step_a$, $step_b$, $step_c$, which determine the movement pace of each wolf during these behaviors. All of these step coefficients should have the following relationship:

$$step_a = \frac{step_b}{2} = 2 \cdot step_c = step.$$

The distance determinant coefficient L_{near}. This parameter serves as a judging condition, which determine whether the *i*-th wolf changes state from aggregating towards the lead wolf to besieging behavior. The different value of L_{near} will affect algorithmic convergence rate.

The maximum number of repetitions for scouting behavior $T_{\max}$. The bigger the value, the more explorative the search is, but the convergence speeds suffers.

The population renewing proportional coefficient β. It is used to calculate the fraction of wolves with bad objective functions that will be replaced by the newly generated ones. The number R of wolves to replace is an integer and randomly selected at the interval $[N/(2\beta); N/\beta]$.

The minimum and the maximum values for number of direction during scouting behavior $h_{\min}$, $h_{\max}$. The parameters are used to get the number of directions in which the *i*th wolf does steps during the scouting behavior.

2.6.6 Ants Algorithm

2.6.6.1 Intuitive Interpretation

The most studied and commonly used feature of ants' behavior is their ability to find the shortest path to a food source. Biologists have found that ant trails, which can be found near any anthill, owe their appearance to specific odorous substances—pheromones, which are released by each ant. The more ants have passed this way, the stronger the smell left behind by them and the more other ants it attracts. To amplify this effect, scientists have introduced a non-existent condition: pheromones evaporate as quickly as the longer path traveled by an ant. The result of this method is to find the shortest routes to all food sources, as well as the dynamic redistribution and optimization of these routes.

2.6.6.2 Algorithm Description

Let us formalize the use of this algorithm to solve the optimization problem—to provide

$\min_{x \in U} J(x)$, where

$J(x) : R^D \to R$—objective function (possibly discrete), which allows to characterize the quality of the agent's position on the basis of D parameters (in this case—ants).

$x \in R^D$—parameter vector, which we will consider as an agent.

The components of this vector specify the position of the agent in the search space U, and functional J on the basis of this provision characterizes the state of the agent. Parameters have simple limitations $x \in U \subseteq R^D$, setting the maximum and minimum values for each, $b_{low} \leq x \leq b_{up}$

The colony consists of N ants. The area of definition of each variable is divided into equal parts-sections—each of which has its own pheromone value. The number of sections is specified by a parameter d_{prec}. To find the best position of the colony, a certain segment of discrete time is divided into To find the best position of the colony, a certain segment of discrete time is divided into N_{iter} moments (also, the limit can be set to the maximum time to find a solution T_{max}).

The algorithm is as follows:

(1) Divide each variable into sections—equal intervals of one variable:

$$b_{ilow} \leq x_{i1} < x_{i2} < \cdots < x_{id_{prec}-1} < x_{id_{prec}} \leq b_{iup};$$

(2) Initialize the initial pheromone value for each section according to the parameter φ_0;

(3) For each section, calculate the probability of ants getting into it:

$$p_i^d(t) = \frac{\varphi_i^\alpha(t)\mu_i^\beta(t)}{\sum_{j=0}^{d_{prec}} \varphi_{jd}^\alpha(t)\mu_{jd}^\beta(t)}, \tag{2.1}$$

$$\sum_{i=0}^{d_{prec}} p_i^d(t) = 1,$$

where φ is the value of the pheromone section, μ—some heuristic parameter (varies depending on the specific task) α, β is the relevant coefficients that govern the significance of the pheromone and the heuristic parameter when selecting a section.

(4) For each ant to each variable, select the appropriate section, based on the probability obtained in step (4), and randomly select the coordinates of the ants within the selected sections:

$$X_k = (x_1, x_2, \ldots x_i)$$

$$x_i(t+1) = \text{rand}\left(b_i^{r(t)_i}, b_i^{r(t)_i+1}\right), i \in D, \tag{2.2}$$

where $r(t)_i$—the number of the selected section for the i-th measurement at time t.

(5) We get the value of fitness function for each ant;
(6) We update the best solution if an ant whose fitness value was greater than the fitness of the previous best ant was found among the generation;
(7) We update the pheromone values for each section:

$$\varphi_{ij}(t+1) = w * \varphi_{ij}(t) + \sum_{k=1}^{N} \begin{cases} fit(J(X_k(t))), if x_i \in X_k(t), x_i \in \left[b_i^j, b_i^{j+1}\right] \\ 0 \end{cases} \tag{2.3}$$

where w—Pheromone evaporation coefficient, k—number of ants, $fit()$—a fitness function that returns value [0, 1], $\left[b_i^j, b_i^{j+1}\right]$—the boundaries of the jth section;
(8) Return to step (3) if the stop criterion is not met (the required accuracy is reached or the iteration limit is reached).

```
while not stop_criterion:
 colony.init()
 for ant in ants:
  for dimension in dimensions:
   for di in dimension:
    // calculate probability of choosing this interval (1)
    ant.probs[dimension][di] = prob(dimension, di);
   // choose interval and value from it for dimension value (2)
   ant.solution[dimension] = random(pick(ant.probs));
  // calculate fitness of the solution
  ant.fit = fitness(J(ant.solution));
 for dimension in dimensions:
  for di in dimension:
   for ant in ants:
    // recalculate pheromone value for each interval (3)
    pheromone[dimension][di] = recalc_pheromone(ant);
```

2.6.6.3 Algorithm Parameters

Efficiency AA depends on a number of control parameters, which include:

$\boldsymbol{N}$—number of ants; $\boldsymbol{N_{iter}}, \boldsymbol{T_{max}}$—the maximum number of iterations or the maximum time allowed, $\boldsymbol{\varphi_0}$—initial pheromone concentration, $\boldsymbol{w}$—the rate of evaporation of the pheromone, $\boldsymbol{\alpha}$—intensification of pheromone, $\boldsymbol{\beta}$—intensification of heuristics, $\boldsymbol{d_{prec}}$—the number of sections of each dimension. In the following, we look at these options. The number of ants significantly affects the characteristics AA—obviously, a large number leads to more computational complexity. The more ants are used, the more pathways are built and more pheromone is deposited.

The less ants are used, the weaker the algorithm's ability to research and therefore less information about the search space. A small number of ants can cause premature convergence or find suboptimal solutions. Also, it should be noted that the optimal number of ants depends on the number of sections and should be responsible for achieving optimal study of the whole area (one way of comparison can be considered $N = d_{prec}^D$, where D—number of variables).

Maximum number of iterations

N_{iter} Maximum number also plays an important role in finding optimal solutions. With a small N_{iter} ants may not have enough time to build the best path. On the other hand, if N_{iter} too large, extra calculations will be performed. Initial concentration values φ_0 also affect the characteristics AA. Initial arc initialization is usually assigned a small or positive value φ_0, or a small random value of their range $[0, \varphi_0]$. Great value φ_0 choosing at random can lead to an initial choice of an unpromising solution. Number of sections d_{prec} is also an important parameter—too large sections will prevent convergence, too small sections will lead to premature convergence 2.

2.6.7 *Stochastic Diffusion Search*

2.6.7.1 Intuitive Interpretation

The idea of the algorithm is inspired by a famous restaurant game formulated as follows. A group of delegates attends a long conference in an unfamiliar town. Every night each delegate must find somewhere to dine. There is a large choice of restaurants, each of which offers a large variety of meals. The problem the group faces is to find the best restaurant, that is the restaurant where the maximum number of delegates would enjoy dining. Even a parallel exhaustive search through the restaurant and meal combinations would take too long to accomplish. To solve the problem delegates decide to employ a stochastic diffusion search.

Each delegate acts as an agent maintaining a hypothesis identifying the best restaurant in town. Each night each delegate tests his hypothesis by dining there and randomly selecting one of the meals on offer. The next morning at breakfast every delegate who did not enjoy his meal the previous night, asks one randomly selected colleague to share his dinner impressions. If the experience was good, he also adopts this restaurant as his choice. Otherwise he simply selects another restaurant at random from those listed in 'Yellow Pages'. Using this strategy it is found that very rapidly significant number of delegates congregate around the 'best' restaurant in town.

2.6.7.2 Algorithm Description

The algorithm is based on 4 strategies:

1. At first each agent generates a hypothesis about the best solution (initialization).
2. Each agent tests it's hypothesis and proves or rejects his previous guess.
3. Agents exchange their hypotheses and their test results (a diffusion takes place).
4. Based on the above behaviors, agents form new hypotheses, test them and communicate with each other. The process continues until the convergence condition is satisfied.

Let's describe these strategies application to solving the unconstrained continuous global minimization problem.

Initialization. A population of random solutions

X_i, $i \in [1; |S|]$ is generated and their fitness $f(X_i)$ is calculated. The recommended population size |*S*| is relatively large—it ranges from 100 to 1000 agents.

Test. The fitness of the *i*th agent is compared to the fitness of a random agent *j*, $i, j \in [1 : |S|], j \neq i$. If $f(X_i) \leq f(X_j)$, then the *i*th agent becomes active, and inactive otherwise.

Diffusion. For each of the agents we perform the following operations:

(1) If *i*th agent is inactive, then we choose a random agent j. If the *j*th agent is active, then the *i*th agent assigns a new hypothesis, which is drawn from some neighborhood $d(X_j)$ of the *j*th agent:

$$S = \frac{1}{2}\left|\sum_{i=0}^{n-1}(x_i(y_{i-1} - y_{i+1}))\right|, \ (x_0, y_o) = (x_n, x_n).$$

$$l_{o,e} : o \in N$$

$$A = \{a_i\}, \ i = \overline{1, n}$$

where *b*—a parameter in range (0; 1) and usually is equal to 0,1.

(2) Else if the *i*th agent is active, then we randomly or systematically choose a new hypothesis for the agent. It is recommended to choose the hypothesis in the center of the current unexplored areas of the search space according to the scheme:

- sort the current coordinates of all the agents by each dimension;
- find the largest interval $\{x_i, y_i\}$, $i = \overline{1, n}$ between 2 neighboring solutions for each dimension;
- Change the *i*th agent's coordinates as follows:

$$B = \{b_i\}, \ i = \overline{1, n}$$

Convergence. The algorithm may use different termination criteria. Besides the common ones used for all of the swarm intelligence algorithms, there are so called strong and weak halting criteria.

The strong halting criterion takes in consideration the fact that a hypotheses clustering takes place during the optimization process. It means that agents form sets, which use same hypotheses. According to the criterion, the iterations stop when the size of the biggest cluster exceeds some threshold or stays the same for the defined number of iterations (cluster stagnation).

The weak halting criterion. The optimization process stops when the number of active agents is bigger than some predefined value or this number doesn't change for a specified number of iterations (stagnation of the number of active agents).

The pseudo code for the Stochastic diffusion search:

begin
  Objective function f(X)
  Initializepopulation
  while *a termination criterion is not met* **do**
    Generating and testing hypotheses
      Determining agents ' activities (active / inactive)
    Diffusing hypotheses
  end while
  Find the current best solution
  Results post-processing
end

2.6.7.3 Algorithm Parameters

Besides the population size and optimal number of iterations, the algorithm has 2 parameters: **a** parameter **b** used in a diffusion and a number of iterations to satisfy the weak halting criterion.

2.6.8 Harmony Search

2.6.8.1 Intuitive Interpretation

Metaphorically, each candidate solution is a harmony vector produced by a single musician, and the population itself describes the harmony memory. The measure of a sound harmony is defined by the fitness-function.

The whole optimization process is based on an idealized improvisation process, performed by a musician. There are 3 ways to improvise:

- to play any famous piece of music (a series of pitches in harmony) exactly from musician's memory;
- to play something similar to a known piece (thus adjusting the pitch slightly);
- to compose new or random notes.

The above components are called usage of harmony memory, pitch adjusting and randomization. So the goal is to find the best sound harmony (or, at least, a good enough).

2.6.8.2 Algorithm Description

Let's describe the algorithm's principle for continuous minimization problem in a D-dimensional search space. Similarly to other population algorithm, the initial population of solutions is generated.

The usage of harmony memory ensures that the best harmonies will be carried over to the next generation harmony memory. In order to regulate the acceptance of harmonies to the new memory, a parameter r_{accept} called harmony memory accepting or considering rate is used.

The pitch adjusting is responsible for generating slightly different solutions. It is determined by a pitch bandwidth b_{range} and a pitch adjusting rate r_{pa}. The pitch is adjusted linearly as follows:

$$C = \{c_k\},\ k = \overline{1, m}$$

where X^{old} *is* the existing solution from the harmony memory; X^{new} is the new pitch (solution) after the adjustment; ε—a random number generator in the range of [−1; 1].

The randomization increases the search diversification. Although adjusting pitch has a similar role, but it is limited to certain local pitch adjustment and thus corresponds to a local search. The use of randomization can drive the system further to explore various diverse solutions so as to find the global optimum.

The pseudo code for the Harmony Search algorithm:

```
begin
        Objective function f(X)
        Generate initial harmonics (real number arrays)
        Define pitch adjusting rate (r_pa), pitch limits and bandwidth
        Define harmony memory accepting rate (r_accept)
        while a termination criterion is not met do
                Generate new harmonics by accepting best harmonics
                Adjust pitch to get new harmonics (solutions)
                if (rand >r_accept) than
                        choose an existing harmonic randomly
                else if(rand >r_pa) than
                        adjust the pitch randomly within limits
                else
                        generate new harmonics via randomization
                end if
                Accept the new harmonics (solutions) if better
        end while
        Find the current best solutions
        Results post-processing
end
```

2.6.8.3 Algorithm Parameters

Besides the population size and optimal number of iterations, the algorithm has 3 parameters: memory accepting rate r_{accept}, pitch bandwidth b_{range} and pitch adjusting rate r_{pa}.

Memory accepting rate r_{accept} is used to determine whether to accept the harmony to the new harmony memory or not. If this rate is too low, only few best harmonies are selected and it may converge too slowly. If this rate is extremely high (near 1), almost all the harmonies are used in the harmony memory, then other harmonies are not explored well, leading to potentially wrong solutions. Therefore, typically, we use $r_{accept} = 0.7$–0.95.

Pitch bandwidth b_{range} is a step of pitch adjustment.

Pitch adjusting rate r_{pa} allows us to control the degree of adjustment. A low pitch adjusting rate with a narrow bandwidth can slow down the convergence of the search because of the limitation in the exploration of only a small subspace of the whole search space. However, a very high pitch-adjusting rate with a wide bandwidth may cause the solution to scatter around some potential optima as in a random search. Thus, we usually use $r_{pa} = 0.1$–0.5 in most applications.

2.6.9 Gravitational Search

2.6.9.1 Intuitive Interpretation

The algorithm defines it's candidate solutions as objects and their fitness is measured by their masses. All of the objects in a population attract each other by the gravity force, thus causing a global movement towards the heavier objects (with better fitness score). Hence, masses cooperate using a direct form of communication, through gravitational force. The heavy masses move more slowly than lighter ones, which guarantees the exploitation step of the algorithm.

In GSA, each object has four specifications: position, inertial mass, active gravitational mass, and passive gravitational mass. The position of the mass corresponds to a solution of the problem, and its gravitational and inertial masses are determined using a fitness function.

The GSA could be considered as an isolated system of masses. The masses obey the following laws:

- *Law of gravity:* each particle attracts every other particle and the gravitational force between two particles is directly proportional to the product of their masses and inversely proportional to the distance between them, R (here R is used instead of R^2, because according to the experiment results, obtained by the authors, R provides better results than R^2 in all experimental cases).
- *Law of motion:* the current velocity of any mass is equal to the sum of the fraction of its previous velocity and the variation in the velocity. Variation in the velocity or acceleration of any mass is equal to the force acted on the system divided by mass of inertia.

2.6.9.2 Algorithm Description

At the tth iteration the force acting on the object i from object j is defined by the following formula:

$$F_{ij}(t) = G(t)\frac{M_i^{passive}(t) \times M_j^{active}(t)}{R_{ij}(t) + \varepsilon}\left(X_j^t - X_i^t\right),$$

where M_j^{active} is the active gravitational mass related to particle j;

$M_i^{passive}$ is the passive gravitational mass related to agent i; $G(t)$ is the gravitational constant at the iteration t; ε is a small constant; and $R_{ij}(t)$ is the Euclidean distance between two particles i and j.

It is supposed that the total force that acts on object i in a dimension d be a randomly weighted sum of dth components of the forces exerted from other objects:

$$F_i^d(t) = \sum_{j=1, j \neq i}^{N} rand_j F_{ij}^d(t).$$

Random weights gives the algorithm a stochastic characteristic, thus making the search more exploratory.

Hence, by the law of motion, the acceleration of the particle i at the iteration t, and in the d-th direction is given according to the formula:

$$a_i^d(t) = \frac{F_i^d(t)}{M_i^{inertial}(t)},$$

where

$M_i^{inertial}$ is the inertial mass of the ith particle.

Furthermore, the next velocity of an agent is considered as a fraction of its current velocity added to its acceleration. Therefore, *its position and its velocity could be calculated as follows*:

$$V_i^{t+1} = rand_i \times V_i^t + a_i^t,$$
$$X_i^{t+1} = X_i^t + V_i^{t+1},$$

where $rand_i$—a uniform random variable in the interval [0; 1].

The gravitational constant, G, is initialized at the beginning and will be reduced with time to control the search accuracy:

$$\frac{dm}{dt} = \sum_i m_i$$

For the implementation, the following formula was used:

$$c_p m \frac{dT}{dt} - A_d Z \frac{dP}{dt} = Q + \sum_i h_i m_i$$

where α is a parameter, which is used to regulate the speed of reduction; t is the current number of iterations passed; T is the total number of iterations.

Gravitational and inertia masses are simply calculated by the fitness evaluation. A heavier mass means a more efficient agent. This means that better agents have higher attractions and walk more slowly. Assuming the equality of the gravitational and inertia mass, the values of masses are calculated using the map of fitness. The gravitational and inertial masses are updated by the following equations:

$$p = \rho RT,$$

The pseudo code for the Gravitational Search algorithm:

begin
Objective function f(X)
Generate initial population
while *a termination criterion is not met* **do**
Evaluate the fitness for each object
Update the G, best and worst of the population
Calculate M and a for each object
Update velocities and positions
end while
Find the current best solution
Results post-processing
end

2.6.9.3 Algorithm Parameters

Besides the population size and optimal number of iterations, the algorithm has 2 parameters: the initial value of gravitational constant G_0, reduction regulation parameter α. Both regulate the step of the algorithm. Bigger values for G0 and smaller values for α correspond to bigger steps (thus, increase the exploration) and vice versa.

2.6.10 Benchmarks

The most commonly used functions to test the genetic algorithms were selected to test the algorithm, namely: *Sphere, Rastrigin, Easom, Lévi function N.13* etc. Their graph and options are listed below [48]:

Sphere function

$f(x) = \sum_{i=1}^{n}(x_i^2)$;
$\alpha = 1$,
$\varphi_0 = 0.1, w = 0.75$
$x_i \in [-10, 10]$ (Fig. 2.28);

Rastrigin function

$f(x) = 10n + \sum_{i=1}^{n}(x_i^2 - 10A * \cos(2\pi x_i))$;
$\alpha = 1, \varphi_0 = 0.1, w = 0.75$
$x_i \in [-5.12, 5.12]$ (Fig. 2.29);

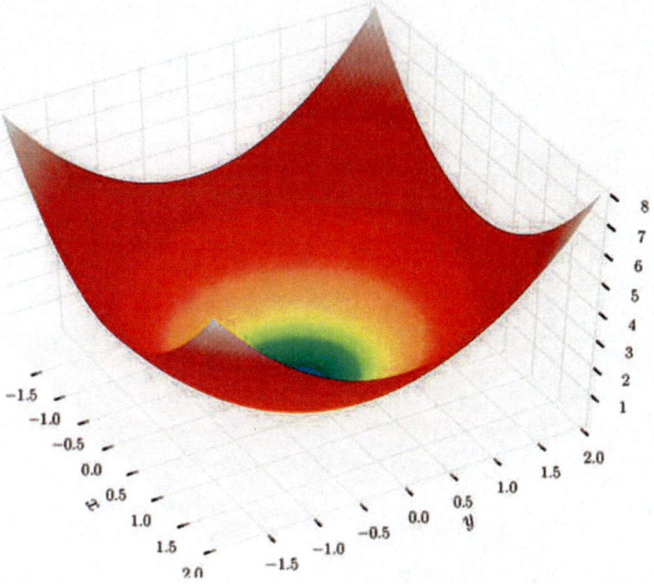

Fig. 2.28 Sphere function graph

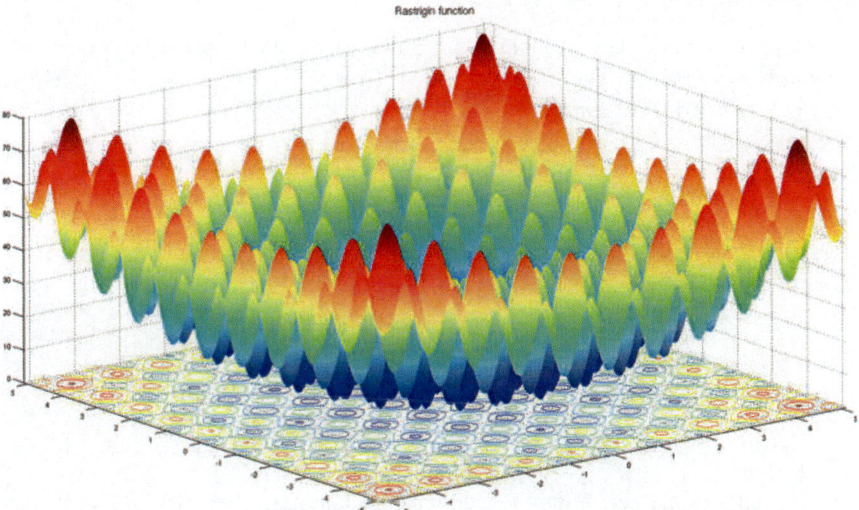

Fig. 2.29 The Rastrigin function graph

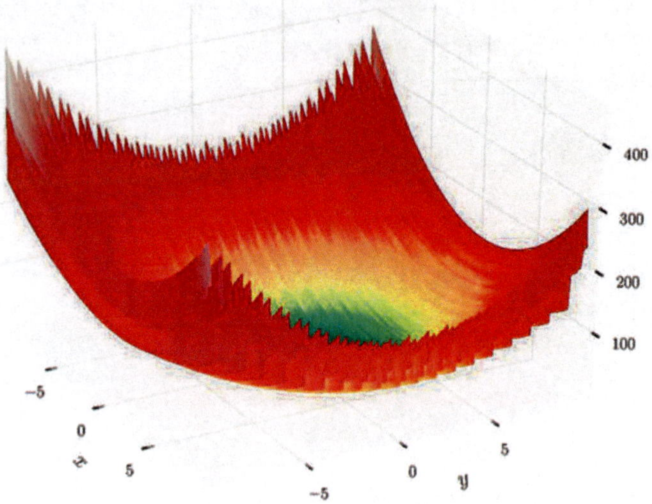

Fig. 2.30 The Lévi function N.13 graph

Lévi function N.13

$f(x, y) = sin^2 3\pi x + (x-1)^2\left(1 + sin^2 3\pi y\right) + (y-1)^2\left(1 + sin^2 2\pi y\right);$
$\alpha = 1, \varphi_0 = 0.1, w = 0.75$
$x_i \in [-10, 10]$ (Fig. 2.30);

Easom function

$f(x, y) = -\cos(x)\cos(y)\exp\left(-\left((x-\pi)^2 + (y-\pi)^2\right)\right);$
$\alpha = 0.1,$
$\varphi_0 = 0.2, w = 0.25$
$x_i \in [-100, 100]$ (Fig. 2.31);

$$f(x) = \sum\nolimits_{i=1}^{n-1}\left[100\left(x_{i+1} - x_i^2\right)^2 + (x_i - 1)^2\right]$$

Here, n represents the number of dimensions and x_i in [−5, 10] for i = 1, …, n (Figs. 2.32, 2.33 and 2.34).

$$f(x) = 418.9829d - \sum_{i=1}^{d} x_i \sin\left(\sqrt{|x_i|}\right)$$

$$f(x, y) = -20\exp\left[-0.2\sqrt{0.5\left(x^2 + y^2\right)}\right] - \exp\left[0.5(\cos 2\pi y)\right] + e + 20$$

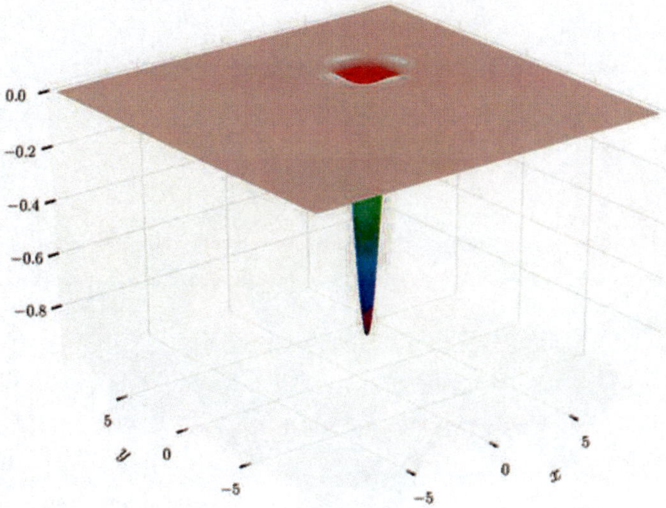

Fig. 2.31 The Easom function graph

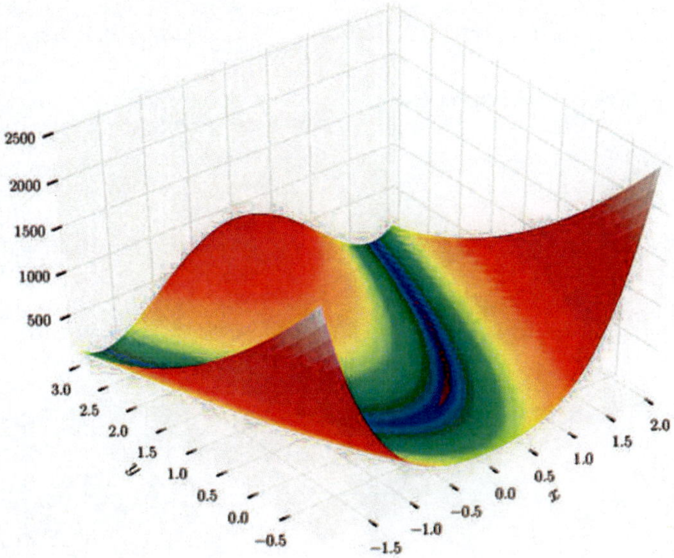

Fig. 2.32 The Rosenbrock function graph

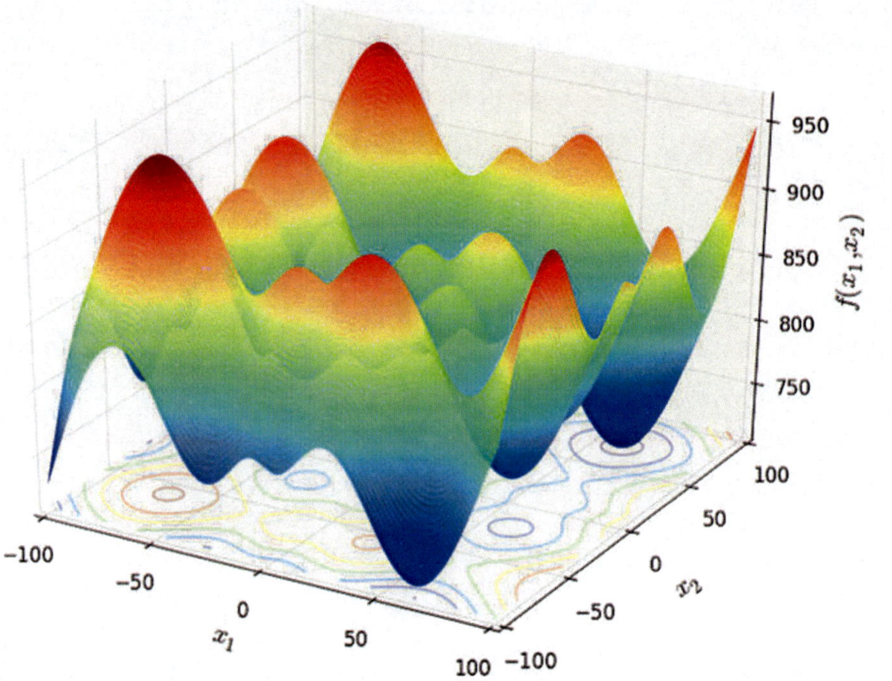

Fig. 2.33 The Schwefel function graph

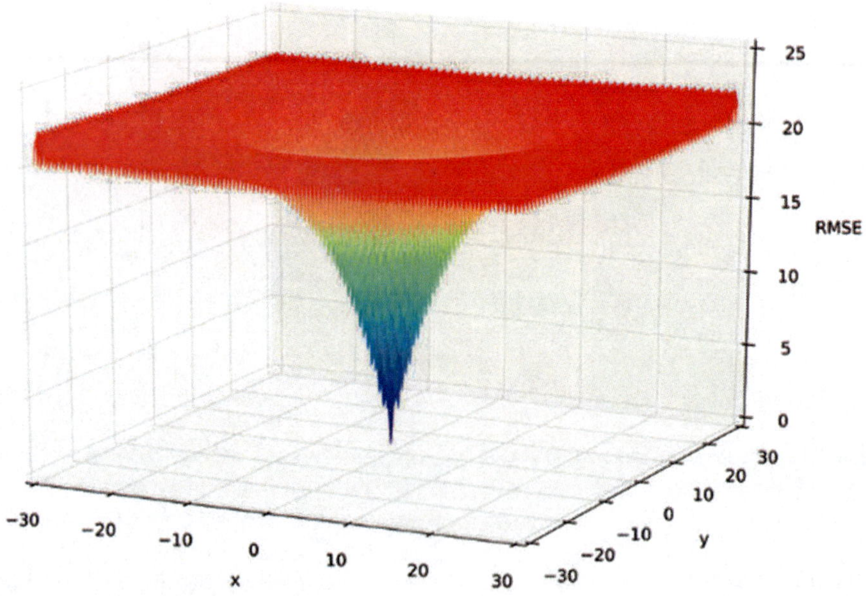

Fig. 2.34 The Ackley function graph

Table 2.1 Particle swarm optimization algorithm benchmarks ($D = 2$)

Functions	Best score	Average score	Scores' std	Success rate (%)
Sphere	1.03271296e−70	1.22448926762e−13	1.218272e−12	100.0
Rosenbrock	5.24879696e−15	7.6237213735e−07	4.071194e−06	100.0
Schwefel	2.54551324e−05	18.9648683582194	43.41460173	77.0
Ackley	4.4408921e−16	0.02579938388778	0.25669954	99.0
Griewank	0.0	0.01028899766	0.015153223	28.0

Benchmark results with stagnation criteria (δ_t=100, tolerance = 10^{-4}) for: Particle swarm, Firefly, Cuckoo search, Bat Search, Wolf Pack algorithms (Tables 2.1, 2.2, 2.3, 2.4 and 2.5).

Benchmark results for ant algorithm

Because the accuracy of the ant colony algorithm depends directly on the parameter d_{prec}, then its work will be analyzed for some range of values of this parameter.

Table 2.2 Firefly algorithm benchmarks ($D = 2$)

Functions	Best score	Average score	Scores' std	Success rate (%)
Sphere	2.1051625e−16	4.612786e−14	4.5004979e−14	100.0
Rosenbrock	5.9709494e−13	4.4055256e−11	4.55276813e−11	100.0
Schwefel	2.5458132e−05	2.6289436e−05	9.5750548e−07	100.0
Ackley	5.6310876e−07	8.7877072e−06	4.9782492e−06	100.0
Griewank	2.2209367e−08	0.0029744485	0.00361346663	57

Table 2.3 Cuckoo search algorithm benchmarks ($D = 2$)

Functions	Best score	Average score	Scores' std	Success rate
Sphere	7.8686077e−23	4.60546119e−18	1.1815461e−17	100.0
Rosenbrock	1.70531275e−11	3.1626659e−08	8.9698615e−08	100.0
Schwefel	2.5455133e−05	7.9037787e−05	0.00036309616	97.0
Ackley	1.0812373e−10	5.8392246e−08	9.6923468e−08	100.0
Griewank	9.984764e−09	0.00486636915	0.00473441488	22.0

Table 2.4 Bat search algorithm benchmarks ($D = 2$)

Functions	Average score	Best score	Scores' std	Success rate
Sphere	0.0	0.0	0.0	100.0
Rosenbrock	4.5215364e−07	0.00779121477	0.01663758005	40.0
Schwefel	2.5579336e−05	194.809620718	137.984441023	19.0
Ackley	1.0309895e−05	0.0309207168	0.30695096001	76.0
Griewank	0.0	0.0	0.0	100.0

Table 2.5 Wolf Pack algorithm benchmarks ($D = 2$)

Functions	Average score	Best score	Scores' std	Success rate
Sphere	0.0	1.668157e−301	0.0	100.0
Rosenbrock	6.9663433e−07	0.00368076855	0.02425397674	57.9999999999
Schwefel				
Ackley	4.4408921e−16	4.4408921e−16	0.0	100.0
Griewank	0.0	0.0	0.0	100.0

Also, along with the parameter d_{prec}, as already mentioned when considering parameters, it makes sense to change the parameter N—number of ants. Destination rule *N choose*

$N = d_{prec}^D$, where is the number of variables D, in our case, equal to 2 (Tables 2.6, 2.7, 2.8 and 2.9; Figs. 2.35, 2.36, 2.37, 2.38, 2.39, 2.40, 2.41 and 2.42).

2.7 Hybrid Swarm Optimization Algorithms

To increase the accuracy of swarm algorithms, hybrid algorithms are used, which are built as a result of using the work of individual swarm algorithms. There are two approaches to their construction, namely:

1. Collective, when individual algorithms included in the hybrid work simultaneously. After completing each step, the algorithm that has achieved the best result is determined. The values of the variables of the best algorithm serve as the initial conditions for all algorithms in the next step.
2. Built-in algorithms, consisting of two separate swarm algorithms, one of which is the basic and the second meta-algorithm.

2.7.1 Cooperative Algorithm

The main idea is to generate three populations with each of the algorithms used in this case PSO, Firefly Algorithm, Cuckoo Search algorithm, (one for each metaheuristics), which would collectively solve the optimization problem based on competition and cooperation.

The main parameter that requires tuning for all algorithms is the size of the population or the number of individuals/particles. The task of choosing the number of individuals is rather complicated in itself, since it is necessary to determine their number so that for a certain number of calculations of the objective function an optimal solution is achieved, on the one hand, and so that these calculations are as few as possible, on the other. In addition, the number of iterations/generations required by the algorithm to achieve the optimal solution with a given accuracy depends

Table 2.6 Sphere function result

d_{prec} and N	Accuracy							
	10e−2	10e−3	10e−4	10e−5	10e−6	10e−7	10e−8	10e−9
1, 1	1432 1432	–	–	–	–	–	–	–
2, 4	189 1512	1762 14,096	–	–	–	–	–	–
3, 9	28 756	69 1863	460 12,420	–	–	–	–	–
5, 25	8 1000	24 3000	98 12,250	293 36,625	–	–	–	–
10, 100	3 3000	6 6000	98 98,000	136 136,000	–	–	–	–
20, 400	1 8000	3 24,000	3 24,000	20 160,000	176 1,408,000	–	–	–
30, 900	1 27,000	2 54,000	5 135,000	19 513,000	19 513,000	514 13,878,000	556 15,012,000	–
50, 2500	1 125,000	1 125,000	1 125,000	6 750,000	22 2,750,000	178 22,250,000	252 31,500,000	–
100, 10,000	1 1,000,000	1 1,000,000	3 3,000,000	4 4,000,000	6 6,000,000	24 24,000,000	177 177,000,000	177 177,000,000
200, 40,000	1 8,000,000	1 8,000,000	2 16,000,000	3 24,000,000	6 48,000,000	15 120,000,000	221 1,768,000,000	–
300, 90,000	1 27,000,000	1 27,000,000	1 27,000,000	2 54,000,000	5 135,000,000	8 216,000,000	53 1,431,000,000	96 2,592,000,000

Table 2.7 Rastrigina function result

d_{prec} and N	Accuracy							
	10e−2	10e−3	10e−4	10e−5	10e−6	10e−7	10e−8	10e−9
1, 1	–	–	–	–	–	–	–	–
3, 9	–	–	–	–	–	–	–	–
5, 25	686 85,750	874 109,250	–	–	–	–	–	–
11, 121	104 138,424	254 338,074	948 1,261,788	–	–	–	–	–
21, 441	9 83,349	84 777,924	97 898,317	181 1,676,241	–	–	–	–
31, 961	10 297,910	26 774,566	26 774,566	83 2,472,653	–	–	–	–
51, 2601	5 663,255	13 1,724,463	34 4,510,134	134 17,775,234	203 26,928,153	–	–	–
101, 10,201	3 3,090,903	6 6,181,806	8 8,242,408	29 29,878,729	31 31,939,331	–	–	–
201, 40,401	3 24,361,803	5 40,603,005	5 40,603,005	6 48,723,606	10 81,206,010	35 284,221,035	282 2,290,009,482	–
301, 90,601	2 54,541,802	4 109,083,604	4 109,083,604	7 190,896,307	9 245,438,109	12 327,250,812	65 1,772,608,565	306 8,344,895,706

Table 2.8 Levi function N13 result

d_{prec} N	Accuracy							
	10e−2	10e−3	10e−4	10e−5	10e−6	10e−7	10e−8	10e−9
1, 1	–	–	–	–	–	–	–	–
2, 4	2069 16,552	–	–	–	–	–	–	–
3, 9	83 2241	83 2241	83 2241	–	–	–	–	–
5, 25	43 5375	211 10,375	1049 131,125	2068 258,500	–	–	–	–
10, 100	5 5000	5 5000	403 403,000	–	–	–	–	–
20, 400	4 32,000	9 72,000	19 152,000	521 4,168,000	–	–	–	–
30, 900	2 54,000	5 135,000	5 135,000	46 1,242,000	297 8,019,000	–	–	–
50, 2500	2 250,000	3 375,000	8 1,000,000	15 1,875,000	272 34,000,000	519 64,875,000	–	–
100, 10,000	2 2,000,000	2 2,000,000	2 2,000,000	9 9,000,000	31 31,000,000	236 236,000,000	–	–
200, 40,000	1 8,000,000	1 8,000,000	3 24,000,000	3 24,000,000	8 64,000,000	8 64,000,000	171 1,368,000,000	–
300, 90,000	1 27,000,000	2 54,000,000	2 54,000,000	5 135,000,000	6 162,000,000	15 405,000,000	28 756,000,000	203 5,481,000,000

Table 2.9 Easome function result

d_{prec} N	Accuracy							
	10e−2	10e−3	10e−4	10e−5	10e−6	10e−7	10e−8	10e−9
1, 1	–	–	–	–	–	–	–	–
2, 4	–	–	–	–	–	–	–	–
3, 9	–	–	–	–	–	–	–	–
5, 25	1952 244,000	–	–	–	–	–	–	–
10, 100	765 765,000	–	–	–	–	–	–	–
20, 400	31 248,000	–	–	–	–	–	–	–
30, 900	92 2,484,000	–	–	–	–	–	–	–
50, 2500	6 750,000	–	–	–	–	–	–	–
100, 10,000	103 103,000,000	232 232,000,000	–	–	–	–	–	–
200, 40,000	2 16,000,000	154 1,232,000,000	154 1,232,000,000	–	–	–	–	–
300, 90,000	2 54,000,000	29 783,000,000	105 2,835,000,000	157 4,239,000,000	157 4,239,000,000	–	–	–

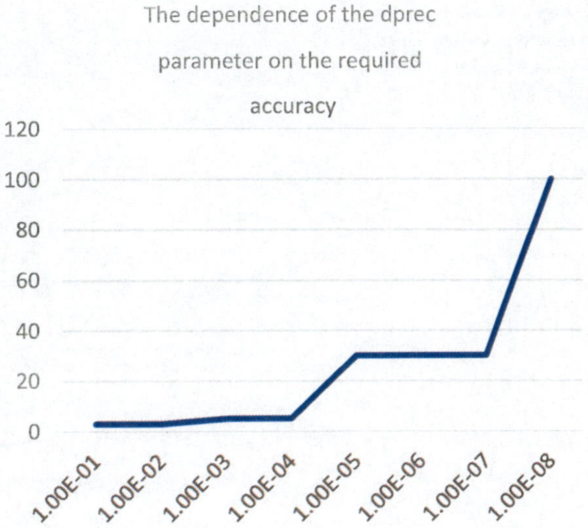

Fig. 2.35 The graphs of dependencies of the d_{prec} parameter

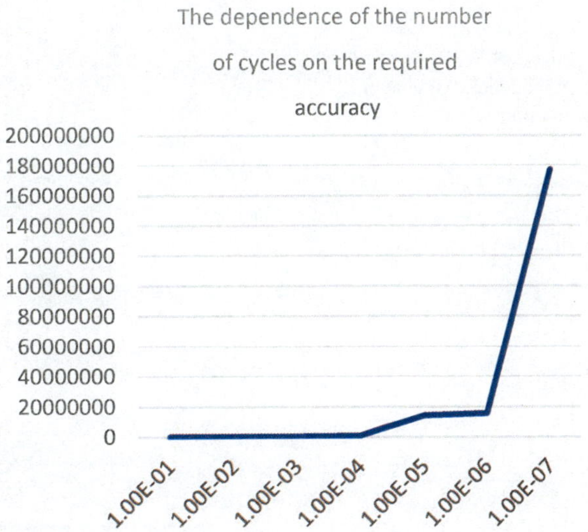

Fig. 2.36 The graphs of dependencies of the cost of accuracy (sphere)

on the size of the population. Together, these parameters determine the number of calculations of the objective function during the operation of the algorithm, on which the execution time of one program start (run) depends.

Therefore, it was decided to automate the process of setting the size of the population so that the number of individuals for each population is determined during the operation of the algorithm. Namely, the size of each population can both increase and decrease depending on how the value of the fitness function changes. In other words, if at the *t*-th iteration, the average fitness of individuals of the *k*th population

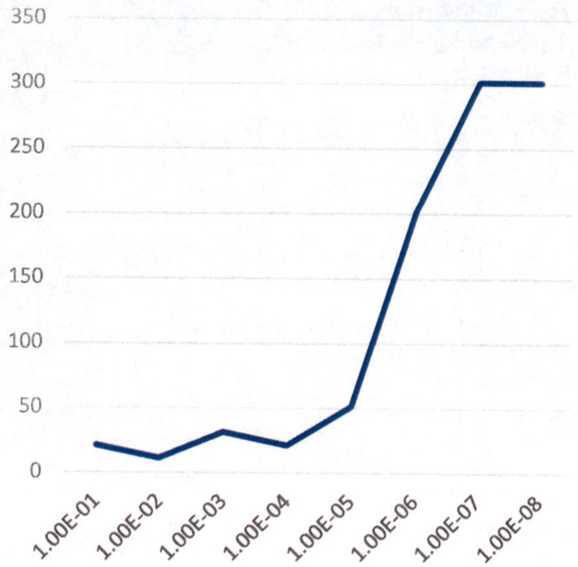

Fig. 2.37 The graphs of dependencies of the d_{prec} parameter

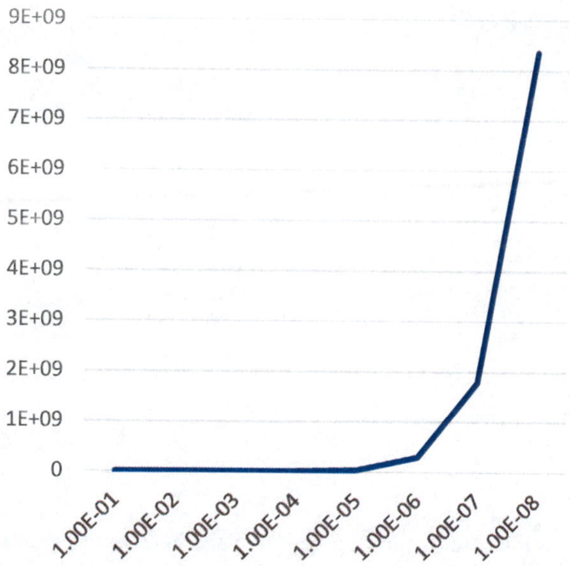

Fig. 2.38 The graphs of dependencies of cost of accuracy

m_i is better than the average fitness of other populations, then the kth population is considered the "winner", and all the rest are "losers". Individuals are removed from the "losing" populations, the rest are added to the "winning" population. Thus, the best algorithm for the task at each iteration is determined.

On the other hand, the total number of all individuals can also either increase or decrease. If over a given number of generations the value of the fitness function

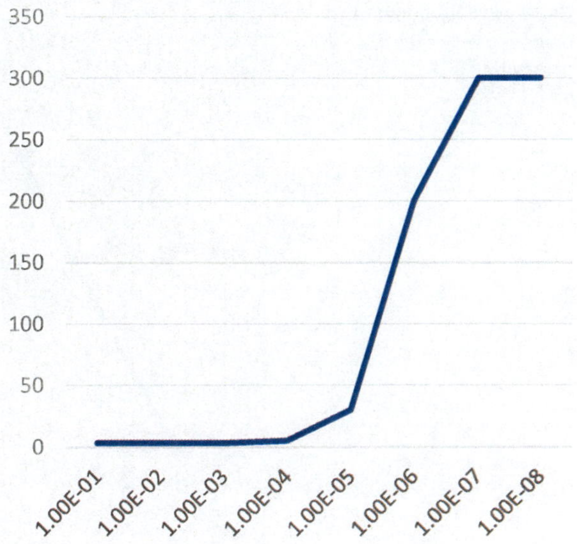

Fig. 2.39 The graphs of dependencies of the d_{prec} parameter

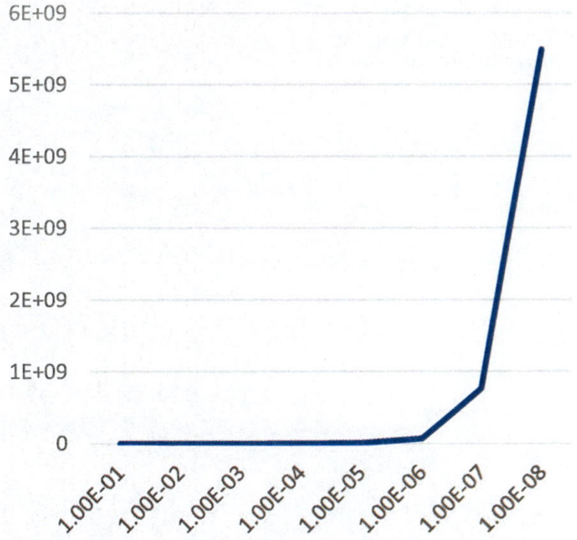

Fig. 2.40 The graphs of dependencies of cost of accuracy (Levi function N13)

does not improve, then the size of all populations increases. And vice versa, if over a certain given number of generations the value of the fitness function only improves, then the size of all populations decreases.

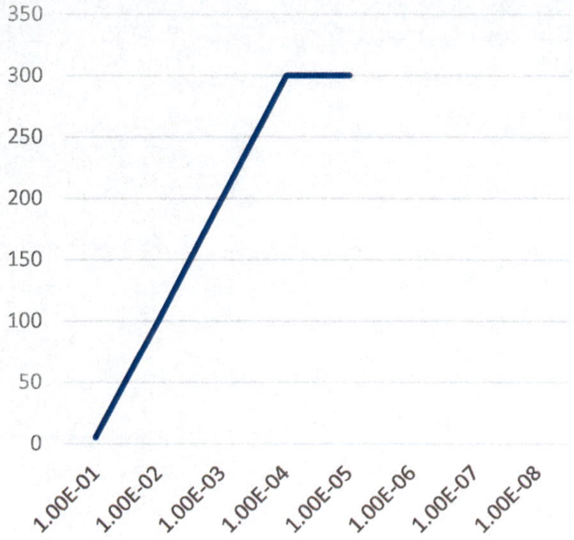

Fig. 2.41 The graphs of dependencies of the d_{prec} parameter

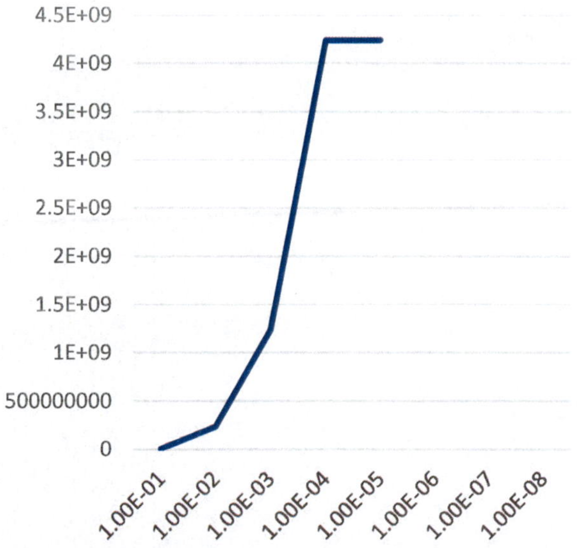

Fig. 2.42 The graphs of dependencies of the cost of accuracy (Easoma)

In addition, all populations cooperate with each other. They exchange individuals: the worst individuals of one population are replaced by the best individuals of other populations, thereby transmitting information about the best solutions received by the whole team of algorithms in general.

Cooperation of Biology Related Algorithms

Set the minimum size of future populations
Set the maximum size of future populations
initialization 3 populations **P,F, C** minimum size randomly
initialization the corresponding parameters for each algorithm
GBest definition
until *the stop criterion is satisfied*
*execute PSO algorithm for population **P***
*execute firefly algorithm for population **F***
*execute Cuckoo search algorithm for population **C***
calculation of the average suitability of a population: the population with maximum suitability is the "winner", other populations are the "losers"
*for **each** "loser" population*
it is necessary to reduce the size of the "losing" population by 10% (but so that its size is not less than the minimum)
end of cycle
until *the size of the winning population and the entire population as a whole exceeds the maximum allowable value*
increase the size of the winning population by the number of individuals removed from the losing populations
end of cycle
if *absolutely Q calculations of the objective functionmigrate the best individuals*
if *changed generations*
if GBest*did not improve*
until *the size of all populations as a whole exceeds the maximum increase*
*the size of each population by **K** individuals*
end of cycle
if GBest*was improving all the time*
while *the size of all populations is greater than the minimum*
reduce the size of each population by K individuals
end of cycle
*update value**GBest***
end of cycle

The number of calculations of the objective function for the migration between populations, as well as the number of iterations to increase or decrease the size of all populations were empirically selected.

The results are given for cooperative algorithm, which consists of Cuckoo Search, Firefly Algorithm and Particle Swarm Optimization, and for these algorithms separately. In all cases, the population size is equal to 20 particles, each algorithm's termination criterion was reaching the maximum number of iterations, which is 100, each optimization was run for 100 times for statistically significant results. Success rate is the percent of optimization results, which score is less or equal 0.0001.

Object functions and parameters used for testing:

1. Sphere (bounds = [−5.12; 5.12]):
   (1) Firefly algorithm parameters: (2, 1, 0.1, 0.5, 0.8);
   (2) Cuckoo search parameters: (0.1, 2);
   (3) PSO parameters: (0, 1.7, 1.7).
2. Rosenbrock (bounds = [−2.048; 2.048]):
   (1) Firefly algorithm parameters: (1, 0.3, 0.7, 0.5, 0.5);
   (2) Cuckoo search parameters: (0.1, 1);
   (3) PSO parameters: (0.5, 0.8, 2).

Table 2.10 Each algorithm contribution in cooperative algorithm ($D = 2$)

Functions	Firefly algorithm contribution (%)	Cuckoo search contribution (%)	PSO contribution (%)
Sphere	15.2	18.2	66.6
Rosenbrock	7.28	33.18	59.54
Schwefel	83.24	6.07	10.69
Ackley	63.14	15.69	21.17
Griewank	71.52	8.08	20.4

Table 2.11 Cooperative algorithm benchmarks ($D = 2$)

Functions	Best score	Avg. score	Scores' std	Success rate (%)	Avg. time (s)
Sphere	**2.456447e−70**	**8.555204e−59**	4.83125e−58	100.0	1.154136
Rosenbrock	6.280511e−09	0.00664737	0.018780827	34.0	1.140975
Schwefel	2.5455178e−05	9.535774707	46.715439442	96.0	1.207859
Ackley	**4.4408921e−16**	2.8893053e−05	0.000174777	97.0	0.806806
Griewank	**0.0**	0.00483512878	0.0040016077	37.0	1.404220

3. Schwefel (bounds = [−500; 500]):

   (1) Firefly algorithm parameters: (5, 0.3, 1, 5, 1.8);
   (2) Cuckoo search parameters: (0.1, 1);
   (3) PSO parameters: (0.3, 3, 0.8).

4. Ackley (bounds = [−32.768; 32.768]):

   (1) Firefly algorithm parameters: (1, 0.3, 1, 1, 1.2);
   (2) Cuckoo search parameters: (0.25, 1);
   (3) PSO parameters: (0, 1.50315925, 1.74936669).

5. Griewank (bounds = [−600; 600]):

   (1) Firefly algorithm parameters: (5, 0.3, 1, 3, 1.1);
   (2) Cuckoo search parameters: (0.05, 1);
   (3) PSO parameters: (0.3, 2, 1.1) (Tables 2.10, 2.11, 2.12, 2.13 and 2.14; Figs. 2.43, 2.44, 2.45, 2.46 and 2.47).

2.7.2 *Static Parametric Meta-optimization for Swarm Intelligence Algorithms*

Let's call the vector of behavioral parameters of the swarm algorithm *a strategy of the algorithm*. The strategy determines the effectiveness of the algorithm. There are

Table 2.12 Firefly algorithm benchmarks ($D = 2$)

Functions	Best score	Avg. score	Scores' std	Success rate (%)	Avg. time (s)
Sphere	1.9166451e−15	1.3999496e−12	1.95501e−12	100.0	0.786515
Rosenbrock	**3.15129e−13**	**1.436699e−10**	2.04545e−10	100.0	0.762923
Schwefel	2.54558e−05	4.767902043	33.375121736	98.0	0.867649
Ackley	1.10167724e−06	**9.1209655e−06**	4.68323e−06	100.0	0.889470
Griewank	9.7697073e−09	**0.0043019781**	0.003668169	41.0	0.882320

Table 2.13 Cuckoo algorithm benchmarks ($D = 2$)

Functions	Best score	Avg. score	Scores' std	Success rate (%)	Avg. time (s)
Sphere	7.0085866e−22	4.1172033e−18	1.277876e−17	100.0	0.161709
Rosenbrock	5.8655307e−13	9.90969471e−08	3.534285e−07	100.0	0.218852
Schwefel	2.5455133e−05	**1.18441752149**	11.784464625	96.0	0.169399
Ackley	1.0988512e−09	4.6477204e−08	7.504856e−08	100.0	0.257644
Griewank	3.6686684e−09	0.0058587958	0.005169419	20.0	0.221452

Table 2.14 PSO algorithm benchmarks ($D = 2$)

Functions	Best score	Avg. score	Scores' std	Success rate (%)	Avg. time (s)
Sphere	1.0918948e−42	9.7713431e−09	9.7181174e−08	100.0	0.028481
Rosenbrock	1.0326349e−11	0.00185575366	0.0184448903	99.0	0.050412
Schwefel	**2.5455132e−05**	17.875215852	42.559664035	82.0	0.037682
Ackley	**4.4408921e−16**	0.07739818617	0.4401026094	97.0	0.065973
Griewank	**0.0**	0.01217487735	0.014619663	18.0	0.052823

several ways one can obtain the rational set of parameters for a specific algorithm to solve a specific task. One way is to determine strategy by hand-tuning, but as the number of parameters increases, the number of possible combinations of parameters' values grows exponentially, and evaluating all of such combinations becomes a very hard computational task. That's way we need another approach for searching for the best strategy. The one which is covered in the article is called meta-optimization and it is implemented with additional layer of optimization. Basically, one algorithm's parameters are tuned by another algorithm. One type of such optimization problem is called parameter adaptation problem.

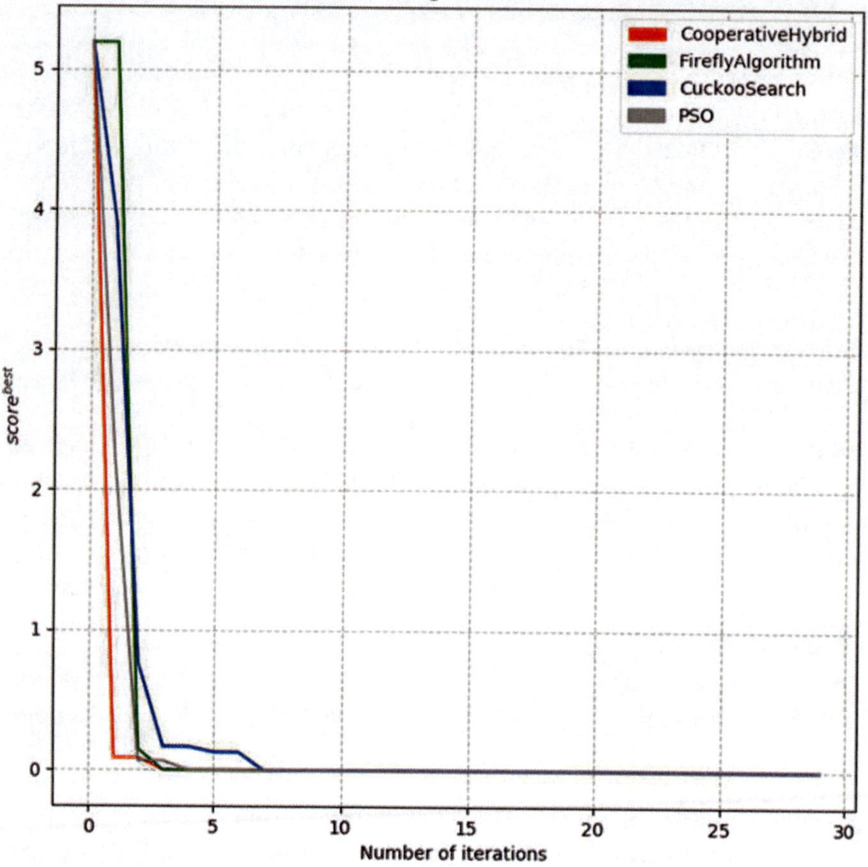

Fig. 2.43 Score changes with iterations for sphere

2.7.2.1 Parameter Adaptation Problem Statement

The first important step to solve the meta-optimization problem is to formulate it using mathematical notation.

The parameter adaptation problem implies that the optimal strategy for the tuned algorithm is static and is searched for a specific optimization problem.

Let function $f(X)$ be an optimization problem, which maps vector of parameters X to a fitness measure:

$$f : \mathbb{R}^n \to \mathbb{R},$$

where n is the dimension of vector X.

The optimization problem itself is formulated as finding the optimal solution $X^{\min}$, which satisfies the equation:

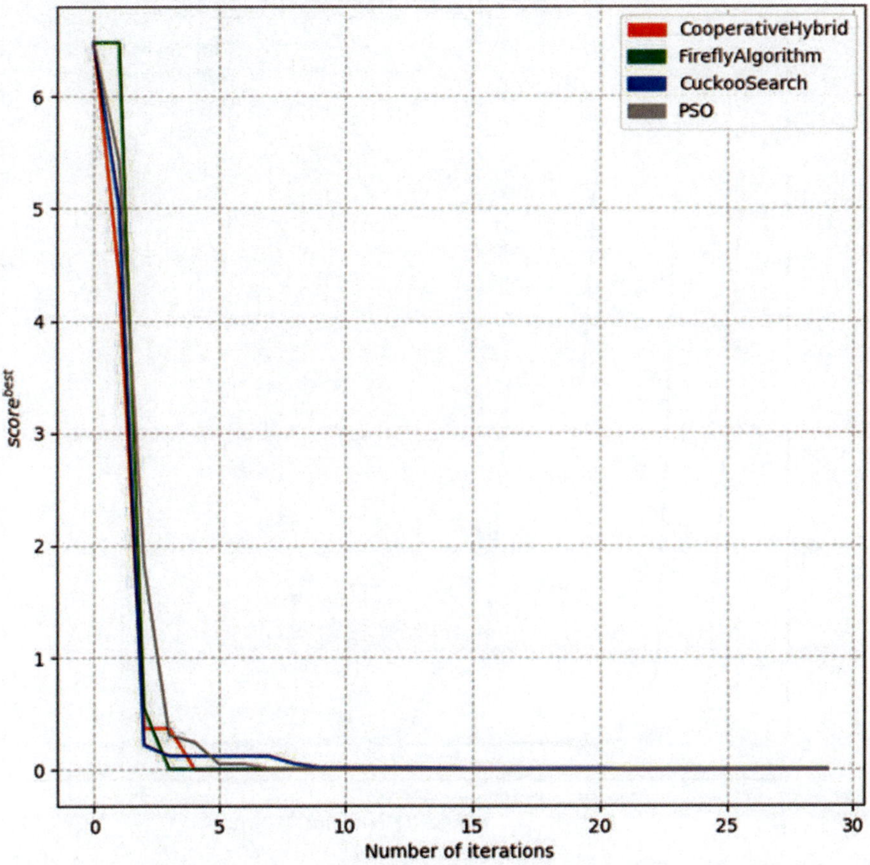

Fig. 2.44 Score changes with iterations for Rosenbrock

$$\forall Y \in \mathbb{R}^n : f\left(X^{\min}\right) \le f(Y).$$

Let $a(B)$ be a base algorithm used to solve the aforementioned optimization problem, where $B = \left(b_1, b_2, \ldots, b_{|B|}\right)$—strategy of the base algorithm. Let's also define the upper and lower bounds b_i^{lower}, b_i^{upper} for each element of the B. The bounds are specific to the optimization problem. Than D_B is a set of possible strategies of the algorithm $a(B)$:

$$D_B = \{b_i | b_i^{lower} \le b_i \le b_i^{upper}, i \in [1 : |B|]\}.$$

The meta-optimization (parameter adaptation) problem is formulated as finding such strategy $B^{best} \in D_B$ of the algorithm $a(B)$ for a problem $f(X)$, which minimizes the object meta-function $\mu(f, B)$:

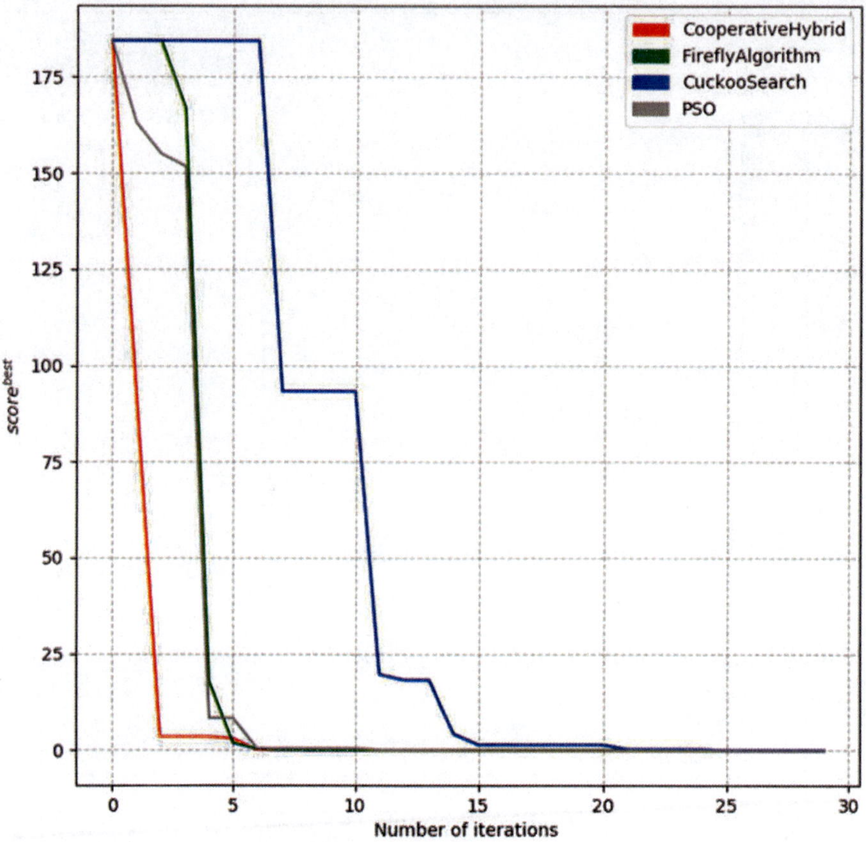

Fig. 2.45 Score changes with iterations for Schwefel

$$\min_{B \in D_B} \mu(f, B) = \mu\left(f, B^{best}\right)$$

2.7.2.2 Meta-function Variants for Parameter Adaptation Problem

In this paper, it is proposed to use the following performance and efficiency meta-functions for the meta-optimization problem. The first suggested meta-function:

$$\mu_1(f, B) = score(B),$$

where *score*(*B*) is the value of the objective function at the solution provided by *a*(*B*).

The second suggested meta-function:

$$\mu_2(f, B) = evals(B),$$

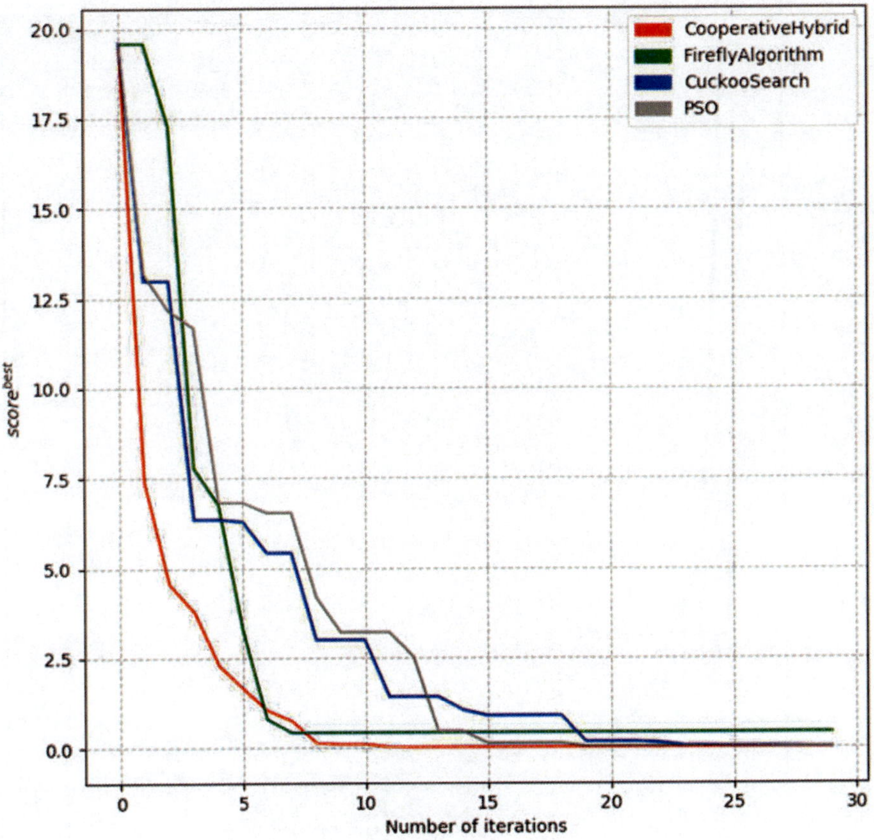

Fig. 2.46 Score changes with iterations for Ackley

where *evals*(*B*) is the number of objective function evaluations for the whole optimization process with algorithm a using strategy B. Note that this meta-function should be used when the termination criterion is not just surpassing some maximum number of iterations, but when we also specify some tolerance to be reached. One way to check if the satisfactory tolerance is achieved is to use a stagnation criterion.

The meta-function, which will be used in meta-optimization process, is represented by the additive convolution of the above ones:

$$\mu(f, B) = \lambda_1 \mu_1(f, B) + \lambda_2 \mu_2(f, B),$$

where $\lambda_{1,2}$ *are* weight coefficients for each estimation.

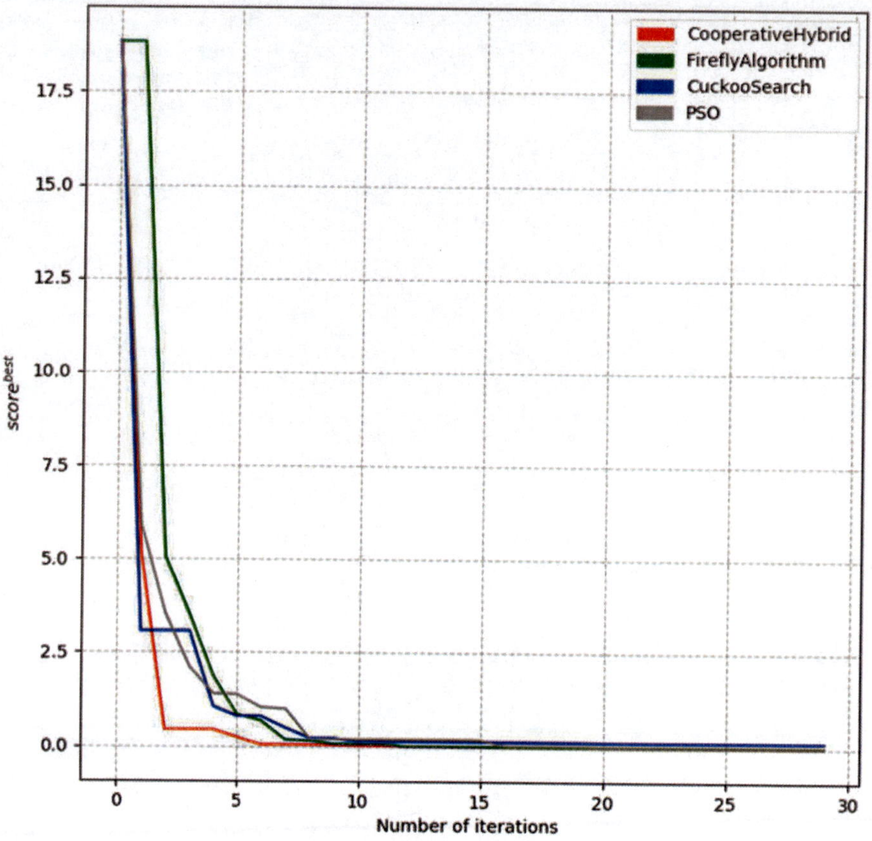

Fig. 2.47 Score changes with iterations for Griewank

2.7.2.3 Description of the Swarm Intelligence Algorithms with Meta-optimization

In this paper, 3 algorithms with parameters adaptation problem are described. Those are PSO meta-optimized with PSO; PSO meta-optimized with Cuckoo Search; PSO meta-optimized with Bat Search [49].

The basic idea for the algorithms is that they consist of the base algorithm, which is used to solve the main optimization problem and the meta-algorithm used to find an optimal strategy for the base one. The object function for the base algorithm is an object function of the general optimization problem. The meta-algorithm evaluates the meta-function defined in the 2-nd chapter at each generation and tries to minimize it by searching for better strategies for the base algorithm. It may be a good practice to set some parameters of the base algorithm before the meta-optimization process, so that the search space for the meta-algorithm is smaller, thus the computational time is shorter. The strategy for the meta-algorithm is set manually.

Now let's describe the implementation of the aforementioned algorithms one by one in details.

1. *PSO meta-optimized with PSO.*

   Base algorithm: PSO.
   Meta-algorithm: PSO.

The algorithm moves a set of potential solutions around the search space till a termination condition is satisfied. This movement is iterative and determined by three components: inertial, social and cognitive components. The inertial component is responsible for changing the velocity, which particle on the previous iteration (deceleration, acceleration or leaving without change), the other two components determine the inclination to move either to the individual's best place it has ever found or to the global (population) best one. The set of candidate solutions is called a population. For the base algorithm, the search space is the space of the potential solutions to the main optimization problem and for the meta-algorithm, the search space is the space of possible strategies. In general, there is a D-dimensional search space and the first step of the algorithm is the generation of initial population of N particles. Each particle is a potential solution to the problem. The coordinates of the particle i at the iteration t are defined as a real-number D-dimensional vector $X_i^t = \left(x_{i,1}^t, x_{i,2}^t, \ldots, x_{i,D}^t\right)$. Each particle i moves in a search space with a velocity $V_i^t = \left(v_{i,1}^t, v_{i,2}^t, \ldots, v_{i,D}^t\right)$.

The update rule for each particle's coordinates at the iteration $(t+1)$:

$$V_i^{t+1} = \omega V_i^t + c_1 U_D(0;1) \otimes \left(X_i^{pbest} - X_i^t\right) + c_2 U_D(0;1) \otimes \left(X^{gbest} - X_i^t\right),$$
$$X_i^{t+1} = X_i^t + V_i^{t+1},$$

where $\otimes$ is the element-wise vector product operator; ω, c_1, c_2 are algorithm parameters; $U_D(0;1)$ is the D-dimensional vector of real-number values uniformly distributed at the interval [0; 1]; X_i^{pbest} are coordinates vector which corresponds to the best solution found by particle i for all of the iterations; X^{gbest} are coordinates vector which corresponds to the best solution found globally by all particles for all of the iterations.

Base algorithm parameters (strategy): ω (inertial parameter), c_1, c_2 are acceleration parameters, which correspond to cognitive and social components of each particle.

Meta-algorithm parameters: the same as for the base one.

The pseudo code for the PSO parameters adaptation with PSO:

```
begin
  Meta-function μ(B)
  Define the meta-algorithm's parameters
  N – meta-algorithm population size
  for each particle of meta-algorithm swarm i = 1,...,N do
    Initialize the meta-particle's position B_i
    Initialize the meta-particle's best-known position to its initial position
    Initialize the meta-particle's speed to zero
  end
  Get initial best meta-particle (initial best strategy B^gbest)
  while a termination criterion is not met do
    for each particle i = 1,...,N do
      Update the meta-particle's velocity and position by the update rule
      if μ(B_i) < μ(B_i^pbest) then
        Update the meta-particle's best-known position
        if μ(B_i) < μ(B^gbest) then
          Update the meta-swarm's best-known position
        end if
      end if
    end for
  end while
  Results post-processing
End
```

2. *PSO meta-optimized with Cuckoo Search.*

   Base algorithm: PSO.
   Meta-algorithm: Cuckoo Search.

The PSO algorithm was already described, so let's describe the Cuckoo Search principles. The Cuckoo Search algorithm associates the quality of eggs and nests' position with the objective function. The idea is to have a population of solutions (nests with eggs) and to have a cuckoo, which tries to find new better nests (solutions) in the search space and dump its egg in there by following specific rules described below.

Similarly to the PSO (and every other swarm algorithm), the Cuckoo Search changes the positions of the candidate solutions by following a special behavior:

- each cuckoo lays one egg at a time in a randomly chosen nest;
- the best nests with high-quality eggs (with the best value of the objective function) will carry over to the next generation;
- the number of nests is fixed and each egg laid by a cuckoo can be discovered by the host bird with probability $p_a \in [0; 1]$. In this case, the discovered nest is replaced by the new one (with new random solution).

Each nest i position at the iteration t is defined a real-number D-dimensional vector $X_i^t = \left(x_{i,1}^t, x_{i,2}^t, \ldots, x_{i,D}^t\right)$, where D is the search space dimension.

The update rule for all of the solutions:

$$X_i^{t+1} = X_i^t + \alpha \otimes Levy_D(\lambda),$$

where $\alpha > 0$ is the step size; $\otimes$ is the element-wise vector product operator; $Levy_D(\lambda)$ is the D-dimensional vector of real-number values drawn from a Levy distribution.

Base algorithm parameters (strategy): ω (inertial parameter), c_1, c_2 are acceleration parameters, which correspond to cognitive and social components of each particle.

Meta-algorithm parameters: α is the step size for Levy flights, p_a is the probability of nest discovery and deletion.

The pseudo code for the PSO parameters adaptation with Cuckoo Search:

begin
  Meta-function μ(B), define the meta-algorithm's parameters
  Generate initial population of N host nests for the meta-algorithm
  while*a termination criterion is not met* **do**
    Get a cuckoo randomly by Levy flights
    Evaluate its quality/fitness μ(B$_i$)
    Choose a nest among N (say, j) randomly
    if *μ(B$_i$) <μ(B$_j$)***then**
      replace j by the new solution
    end if
    A fraction (p$_a$) of worse nests
      are abandoned and new ones are built
    Keep the best solutions (or nests with quality solutions)
    Rank the solutions and find the current best
  end while
  Results post-processing
end

3. *PSO meta-optimized with Bat Search.*

   Base algorithm: PSO.
   Meta-algorithm: Bat Search.

The PSO algorithm was already described, so let's describe the Bat Search principles. The algorithm flow can be summarized as follows: each bat, which represents a solution, flies randomly in the search space with some velocity with a varying frequency or wavelength and loudness. As it searches for its prey, it changes frequency, loudness and pulse emission rate to control the way it moves. Search is intensified by a local random walk around the current global best solution. Selection of the best continues until certain stop criteria are met.

The algorithms is based on idealized model of bats' behavior. This model has the following rules:

- bats can distinguish prey from obstacles and sense the distance to them by the means of echolocation;
- bats fly randomly with velocity V_i at position X_i with a fixed frequency $f_{\min}$, varying wavelength λ and loudness A_0 to search for prey. They can automatically adjust the wavelength (or frequency) of their emitted pulses and adjust the rate of pulse emission $r \in [0; 1]$, depending on the proximity of their target;
- although the loudness can vary in many ways, we assume that the loudness varies from a large (positive) A_0 to a minimum constant value $A_{\min}$.

Each bat is a potential solution to the problem. The coordinates of the bat i at iteration t are defined as a real-number D-dimensional vector $X_i^t = \left(x_{i,1}^t, x_{i,2}^t, \ldots, x_{i,D}^t\right)$. Each bat i moves in a search space with a velocity $V_i^t = \left(v_{i,1}^t, v_{i,2}^t, \ldots, v_{i,D}^t\right)$.

The update rule for the solutions and velocities:

$$\begin{aligned} f_i &= f_{min} + (f_{max} - f_{min})\beta, \\ V_i^{t+1} &= V_i^t + \left(X_i^t - X^{gbest}\right)f_i, \\ X_i^{t+1} &= X_i^t + V_i^{t+1}, \end{aligned}$$

where β is a random vector drawn from a uniform distribution at the interval [0; 1]; X^{gbest} are coordinates' vector which corresponds to the best solution found globally by all bats for all of the iterations.

For the local search part, we generate a new solution using random walk:

$$\begin{aligned} X_{new} &= X_{old} + \varepsilon \overline{A^t}, \\ \overline{A^t} &= \frac{1}{N}\sum_{i=1}^{N} A_i^t, \end{aligned}$$

where ϵ is the D-dimensional vector of uniformly distributed values at the interval $[-1; 1]$; $\overline{A_i^t}$ is the average loudness of all the bats at the iteration t as it may be assumed from its definition.

The loudness and the rate of pulse emission for each bat have to be updated accordingly as the iterations proceed. New values are given by

$$A_i^{t+1} = \alpha A_i^t,\ r_i^{t+1} = r_i^0\left[1 - \exp(-\gamma t)\right].$$

where α and γ are constants, so that for any $0 < \alpha < 1$ and $\gamma > 0$.

Base algorithm parameters (strategy): ω (inertial parameter), c_1, c_2 are acceleration parameters, which correspond to cognitive and social components of each particle.

Meta-algorithm parameters: minimal frequency $f_{\min}$, maximum frequency $f_{\max}$, initial loudness A_0, initial rate of pulse emission r_0, coefficients α and γ.

The pseudo code for the PSO parameters adaptation with Bat Search:

begin
Meta-function $\mu(B)$
Define the meta-algorithm's parameters
Generate initial population of N bats for the meta-algorithm
while*a termination criterion is not met* **do**
Generate new solutions by adjusting frequency
Update velocities and locations/solutions
if *(rand $>r_i$)* **then**
Select a solution among the best solutions
Generate a local solution around the selected best solution
end if
Generate a new solution by flying randomly
if *(rand $<A_i \& \mu(B_i) < \mu(B^{gbest})$)* **then**
Accept the new solutions
Increase r_i and reduce A_i
end if
Rank the bats and find the current best
end while
Results post-processing
end

4. *ACO meta-optimized with PSO.*

Base algorithm: ACO.
Meta-algorithm: PSO.

The parameters of the particle swarm will be the values of the AA (particle) of the particle swarm, and the accuracy of the AA and the execution time will be used as the evaluation criteria. Since each parameter is a separate variable, with the addition of each of them, the tuning time is significantly increased (O (nD…), where D is the number of measurements). Therefore, only the 3 most significant MA parameters were selected for tuning:

- α—Pheromone Significance Factor;
- d_{prec}—number of measurement sections;
- N—number of ants.

Other options can be customized based on the overall evaluation of the optimized feature:

- φ_0—initial pheromone concentration;
- w—the rate of evaporation of the pheromone.

Given this information, we specify the meta-algorithm. Its optimization problem takes the following form:

Find one $X^*\left(\alpha^*, d^*_{prec}, N^*\right)$, that $ACO(X^*) \rightarrow min$, where ACO—basic algorithm (Ant Colony Optimization). In fact, the parameters should also be

φ_0 and w, but as noted above, we take the values of these parameters constant, so we ignore them when writing.

The formulas for obtaining new parameter values at each iteration of the meta-algorithm take the form:

$\alpha(t+1) = \alpha(t) + c_{in}v_\alpha(t) + c_{cog}(r_\alpha(t) - \alpha(t)) + c_{soc}(p_\alpha(t) - \alpha(t))$;

$d_{prec}(t+1) = d_{prec}(t) + c_{in}v_{d_{prec}}(t) + c_{cog}\left(r_{d_{prec}}(t) - d_{prec}(t)\right) + c_{soc}\left(p_{d_{prec}}(t) - d_{prec}(t)\right)$;

$N(t+1) = N(t) + c_{in}v_N(t) + c_{cog}(r_N(t) - N(t)) + c_{soc}(p_N(t) - N(t))$;

Pseudocode for *ACO meta-optimized with PSO*

The hybrid algorithm pseudocode is, for the most part, a pseudocode of an algorithm that acts as a meta-algorithm, except for one feature: to obtain a function value based on a solution (the result of the basic algorithm), to avoid randomly biased solutions, and bringing the result to an acceptable appearance, the basic algorithm is executed several times, and the obtained results are averaged (in our case the number of iterations was equal to 10).

```
particles.init();

while notstop_criterion:
forparticle in particles:
// move each particle (1)
 particle.move(particle.vector);
 // calculate average results
 foriin iterations:
 particle.result += ACO(particle.vector);
 particle.result /= iterations;
 forparticle in particles:
 // update global and locals best solutions
 particle.update_bests();
 forparticle in particles:
 // update vectors (3)
 particle.update_vector();
```

2.7.2.4 Meta-optimization and Base Algorithm Choice

One of the main problems of parameter adaptation problem is to choose the appropriate meta and base algorithms. The optimization for the above meta-function requires a lot of calculations, as we basically do a full optimization process at each generation of the meta-algorithm. So, it is desired to achieve the satisfactory result in the least possible amount of time by using proper meta and base algorithms. The way we make a choice among other available algorithms is based on several criteria. The suggested criteria for the base algorithm choice:

1. Has the best average score among all of the other available algorithms at some specified amount of iterations $T_{\max}$ on a set of test functions for optimization.
2. Has the shortest evaluation time for the $T_{\max}$ iterations on a number of test functions for optimization.
3. Has a small number of parameters to optimize over. The less dimensions the search space has, the faster is the optimization process.

The suggested criteria for the meta-algorithm are formulated as follows:

1. Obviously, the meta-algorithm has to improve the best average score of the base algorithm at some specified amount of iterations $T_{\max}$ on a set of test functions for optimization.
2. Has the shortest evaluation time for the $T_{\max}$ iterations on a number of test functions for optimization.

2.7.2.5 Meta-algorithms Parameters' Initialization

Another problem of meta-optimization is to properly set the meta-algorithm's population size and parameters (strategy). So, here are some recommendations how to do it.

Increasing the population size may improve the meta-optimization solutions, but it also makes the computations more complex. The population is chosen based on the dimension of the base algorithm strategy search space. For instance, for the PSO algorithm as a meta-algorithm and search space dimension $D = 2$ the population size may vary from 10 to 20; for $D = 3$ we may want to increase the number of particles, so that the size is about 20–30.

As for the strategies, the choice is based on the bounds of the base algorithm strategy search space. All of the swarm intelligence algorithms have parameters, which control the particles step and exploration/exploitation rate. So, for search spaces with small ranges of parameters, we may want to set the step size to be relatively small.

2.7.2.6 Benchmarks

See Tables 2.15, 2.16, 2.17, 2.18, 2.19 and 2.20.

ACO meta-optimized with PSO algorithm benchmarks

The tests were performed for different functions. The research results are as follows:

Sphere function

Setting result $\boldsymbol{\alpha}$: 1.69254 (versus 1) (Table 2.21; Fig. 2.48)

The average cost improvement is **34%**.

Rastrigina function

Setting result $\boldsymbol{\alpha}$: 0.979081 (versus 1) (Table 2.22; Fig. 2.49)

The average cost improvement is **7%**.

Table 2.15 PSO algorithm meta-optimized with PSO benchmarks ($D = 2$, base $T_{max} = 50$, meta $T_{max} = 50$, $\lambda_1 = 1$, $\lambda_2 = 0$, runs = 50, $e = 0.0001$)

Functions	Best score	Average score	Scores' std	Success rate (%)
Sphere	4.579894e−36	2.7442816e−15	1.9209971e−14	100.0
Rosenbrock	2.975853e−11	0.0207741020318	0.0568023382377	40.0
Schwefel	2.5455132e−05	47.282485557384	65.74944906774	62.0
Ackley	4.4408921e−16	0.1038777139018	0.505441686471	90.0
Griewank	2.746102e−09	0.012221675206	0.0091683181564	4.0

Table 2.16 PSO benchmarks without Meta-optimization with same seeds as in Table 2.15 ($D = 2$, $T_{max} = 50$, runs = 50, $e = 0.0001$)

Functions	Best score	Average score	Scores' std	Success rate (%)
Sphere	2.4283547e−35	1.942099114e−15	1.35946938e−14	100.0
Rosenbrock	5.8145632e−10	0.0102757514902	0.037354238614	62.0
Schwefel	0.0232000338	39.843647023503	63.391744250035	0.0
Ackley	0.1753981864	1.9807641209579	1.0071862930836	0.0
Griewank	0.0216966632	0.239982543552	0.1480157446776	0.0

Table 2.17 PSO algorithm meta-optimized with Bat search benchmarks ($D = 2$, base $T_{max} = 50$, meta $T_{max} = 50$, $\lambda_1 = 1$, $\lambda_2 = 0$, runs = 50, $e = 0.0001$)

Functions	Best score	Average score	Scores' std	Success rate (%)
Sphere	8.2332659e−37	1.11337232e−05	7.64479426e−05	98.0
Rosenbrock	9.1337927e−12	0.0229007174695	0.0432846635779	24.0
Schwefel	2.5455132e−05	62.461548636026	91.421709972633	56.0
Ackley	4.4408921e−16	0.3937889491266	1.1506795904082	68.0
Griewank	4.4408921e−16	0.0196794031218	0.0161688499722	12.0

Table 2.18 PSO benchmarks without meta-optimization with same seeds as in Table 2.15 ($D = 2$, $T_{max} = 50$, runs = 50, $e = 0.0001$)

Functions	Best score	Average score	Scores' std	Success rate (%)
Sphere	2.952518e−37	1.43445097e−25	9.60690187e−25	100.0
Rosenbrock	1.010342e−10	0.00534722521118	0.0266727070985	72.0
Schwefel	0.0075808915	42.179153709893	60.720004286041	0.0
Ackley	0.1596013097	1.7731788320129	0.9614073035587	0.0
Griewank	0.0124720353	0.2443062602677	0.1536610721152	0.0

Table 2.19 PSO algorithm meta-optimized with Cuckoo search benchmarks ($D = 2$, base $T_{\max} = 50$, meta $T_{\max} = 50$, $\lambda_1 = 1$, $\lambda_2 = 0$, runs = 50, $e = 0.0001$)

Functions	Best score	Average score	Scores' std	Success rate (%)
Sphere	2.8678282e−36	0.0001281595957	0.0008892359418	98.0
Rosenbrock	2.0477848e−12	0.0666932042432	0.3185117804953	46.0
Schwefel	2.5455132e−05	75.803006475174	85.16940000324	44.0
Ackley	4.4408921e−16	0.2588922741186	0.773690395221	82.0
Griewank	2.553513e−15	0.0193188257938	0.0199278674243	14.0

Table 2.20 PSO benchmarks without meta-optimization with same seeds as in Table 2.15 ($D = 2$, $T_{\max} = 50$, runs = 50, $e = 0.0001$)

Functions	Best score	Average score	Scores' std	Success rate (%)
Sphere	1.403173e−36	4.86048752e−08	3.400996523e−07	100.0
Rosenbrock	6.5509675e−11	0.010576433067	0.039559987715	62.0
Schwefel	0.0098750526	35.185656742725	57.689149184387	0.0
Ackley	0.01657186	1.7674575921894	1.0762303972009	0.0
Griewank	0.0140190785	0.2348410206461	0.1486528359596	0.0

Function Levi N 13
Setting result $\boldsymbol{\alpha}$: 1.261801 (versus 1) (Table 2.23; Fig. 2.50)
The average cost improvement is **10%**.
Easom function
Setting result $\boldsymbol{\alpha}$: 0.254911 (versus 0.1) (Table 2.24; Fig. 2.51)
The average cost improvement is **70%**.

2.8 Modern Gradient Algorithms

Due to the continuous improvement of existing neural network algorithms differing in their properties and implementation features often the problem arises of determining a more effective method, minimizing the error function, guaranteeing better results when solving a task problem.

It is known that the most used method for training neural networks is the error back propagation algorithm, in which the objective function is minimized by the gradient descent method [50], that is, at each iteration, the weights change occurs according to the formula

$$w_{N+1} = w_N - \alpha \nabla_w E(w),$$

where E is the objective error function, depending on the parameters: w are the weights of the neural network, α is the learning rate. Thus, in this method, the

Table 2.21 Working result algorithm by function Sphere

d_{prec} and N	Accuracy							
	10e−2	10e−3	10e−4	10e−5	10e−6	10e−7	10e−8	10e−9
1, 1	1069 1069	–	–	–	–	–	–	–
2, 4	70 560	1536 12,288	–	–	–	–	–	–
3, 9	21 567	54 1458	346 9342	–	–	–	–	–
5, 25	6 750	8 1000	64 8000	216 27,000	–	–	–	–
10, 100	1 1000	3 3000	71 71,000	108 108,000	–	–	–	–
20, 400	1 8000	1 8000	3 24,000	13 104,000	50 400,000	–	–	–
30, 900	1 27,000	2 54,000	3 81,000	14 378,000	13 351,000	355 9,585,000	–	–
50, 2500	1 125,000	1 125,000	1 125,000	4 500,000	15 1,875,000	178 3,125,000	252 31,500,000	1000 125,000,000
100, 10,000	1 1,000,000	1 1,000,000	1 1,000,000	3 3,000,000	4 4,000,000	18 18,000,000	177 177,000,000	395 395,000,000
200, 40,000	1 8,000,000	1 8,000,000	2 16,000,000	2 16,000,000	3 24,000,000	9 72,000,000	221 1,768,000,000	251 2,008,000,000
300, 90,000	1 27,000,000	1 27,000,000	1 27,000,000	2 54,000,000	4 108,000,000	7 189,000,000	53 1,431,000,000	96 2,592,000,000

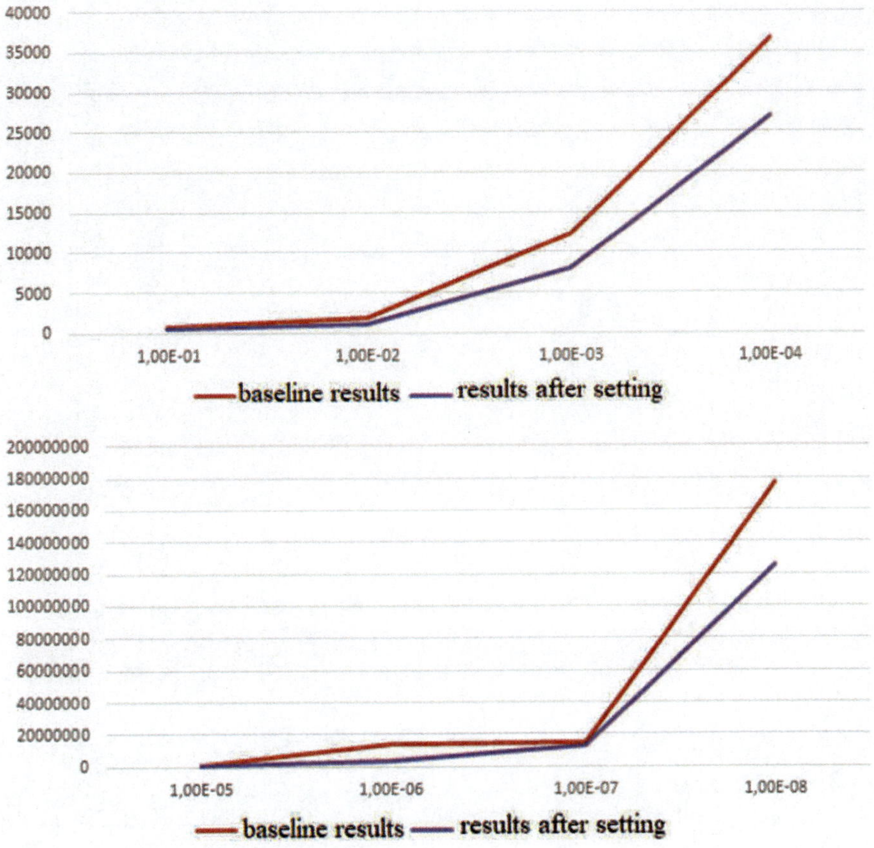

Fig. 2.48 The graphs of dependence of algorithm parameter on accuracy

weights of the neural network are updated in the opposite direction to the gradient of the objective function with a step determined by the learning speed.

Due to the constancy of the parameter of the learning speed
α during the model training process, it is impossible to make the processes go from the local minimum if α it is small enough, and on the contrary, it is impossible to stop at an acceptable optimum if it is large enough.

Another problem is the fact that in a complex model, different parameters may vary in different ways: some require a smaller step of learning, while others, on the contrary, more. This problem is especially relevant when the input is unnormalized and unscaled.

Modern machine learning packages use different variations of the classic gradient descent method, which have higher performance and accuracy in real practical tasks due to the solutions implemented in them problems (see Sect. 2.2). However, often the user of already implemented methods of learning neural networks uses built-in

Table 2.22 Working result algorithm by function **Rastrigina**

d_{prec} and N	Accuracy							
	10e−2	10e−3	10e−4	10e−5	10e−6	10e−7	10e−8	10e−9
1, 1	–	–	–	–	–	–	–	–
3, 9	–	–	–	–	–	–	–	–
5, 25	686 85,750	874 109,250	–	–	–	–	–	–
11, 121	104 138,424	254 338,074	948 1,261,788	–	–	–	–	–
21, 441	9 83,349	84 777,924	97 898,317	181 1,676,241	–	–	–	–
31, 961	10 297,910	26 774,566	26 774,566	83 2,472,653	819 24,398,829	–	–	–
51, 2601	4 530,604	10 1,326,510	30 3,979,530	125 16,581,375	199 26,928,153	–	–	–
101, 10,201	3 3,090,903	5 5,151,505	7 7,212,107	24 24,727,224	27 31,939,331	–	–	–
201, 40,401	2 16,241,202	5 40,603,005	5 40,603,005	6 48,723,606	9 81,206,010	31 265,448,645	279 2,265,647,679	–
301, 90,601	2 54,541,802	4 109,083,604	4 109,083,604	7 190,896,307	8 245,438,109	12 327,250,812	58 1,579,835,665	287 7,815,596,786

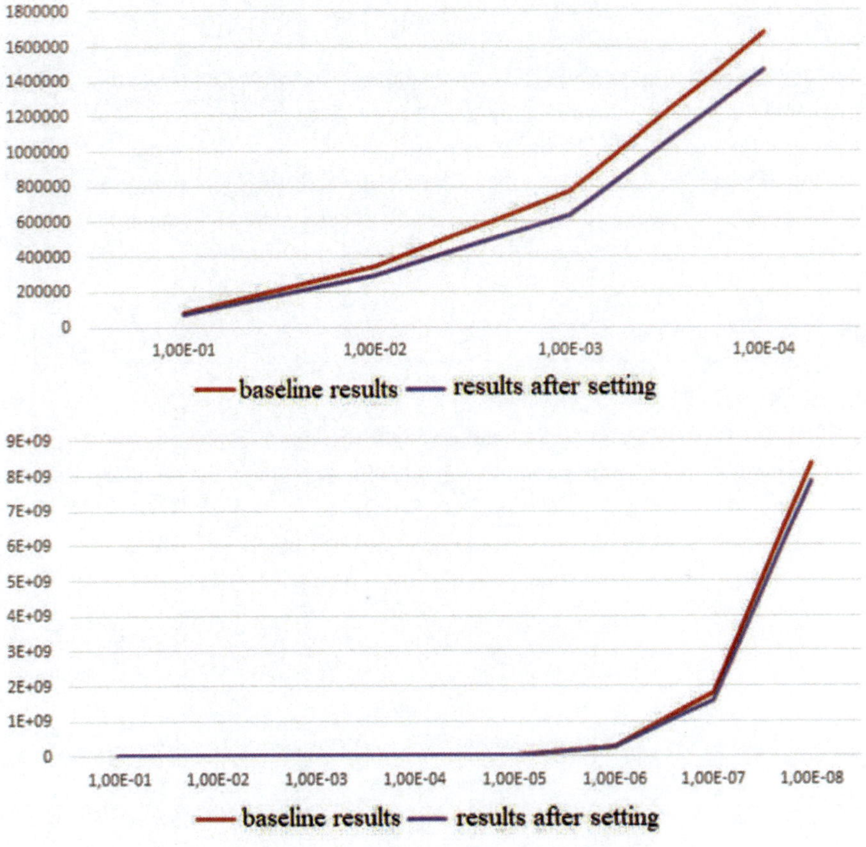

Fig. 2.49 The graphs of dependence of algorithm parameter on accuracy

optimization algorithms in "black box" mode, i.e. does not have enough information about the behaviors available for use methods on the data set in question.

To avoid problems and improve the results of the gradient descent method, a number of powerful upgraded optimizer methods are used, including Stochastic Gradient Momentum Descent (SGDM), Nester Accelerated Gradient (NAG), Adaptive Gradient Algorithm (AdaGrad, Medium, RMSProp), Adaptive Moment Evaluation (AdaM), Nesting Adaptive Moment (NAdaM)

Stochastic Gradient Descent with Moment (SGDM)

This method is the closest development of the gradient method, but it shows much better optimization results due to the introduction of two principle heuristics:

(1) splitting training data into equal subsets of data (minibatch) and randomly mixing them at each training epoch;
(2) the use of Momentum during determination of parameter update values based on their previous values.

Table 2.23 The results of the algorithm on the function **Levi N 13**

d_{prec} N	Accuracy							
	10e−2	10e−3	10e−4	10e−5	10e−6	10e−7	10e−8	10e−9
1, 1	–	–	–	–	–	–	–	–
2, 4	2069 16,552	–	–	–	–	–	–	–
3, 9	83 2241	83 2241	83 2241	–	–	–	–	–
5, 25	43 5375	211 10,375	884 110,500	–	–	–	–	–
10, 100	5 5000	5 5000	401 401,000	–	–	–	–	–
20, 400	4 32,000	9 72,000	19 152,000	478 3,824,000	–	–	–	–
30, 900	2 54,000	5 135,000	5 135,000	46 1,242,000	297 8,019,000	902 24,354,000	–	–
50, 2500	2 250,000	3 375,000	8 1,000,000	15 1,875,000	272 34,000,000	519 64,875,000	–	–
100, 10,000	2 2,000,000	2 2,000,000	2 2,000,000	9 9,000,000	31 31,000,000	236 236,000,000	–	–
200, 40,000	1 8,000,000	1 8,000,000	3 24,000,000	3 24,000,000	8 64,000,000	8 64,000,000	156 1,248,000,000	172 1,376,000,000
300, 90,000	1 27,000,000	2 54,000,000	2 54,000,000	5 135,000,000	6 162,000,000	15 405,000,000	26 702,000,000	155 4,185,000,000

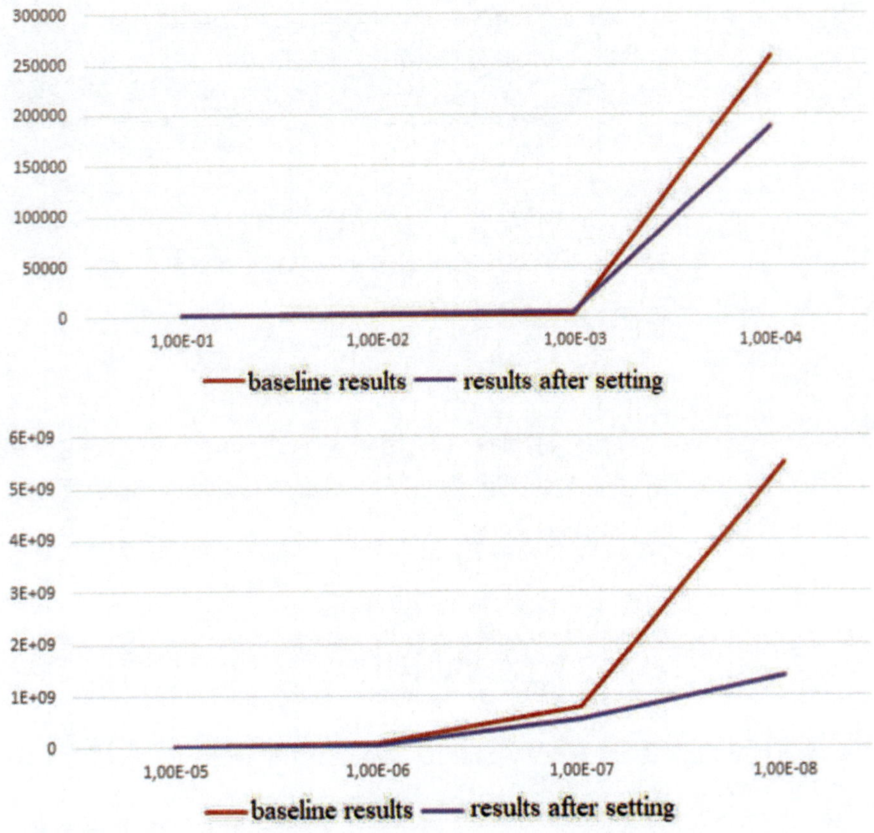

Fig. 2.50 The graphs of dependence of algorithm parameter on accuracy

The idea of stochasticity of this method consists in updating of parameters of a network after passing of a signal of each data sample taken at random from the general training dataset [16, 51]. Such training is called on-line training, but it has the significant disadvantage of considerable time cost in the case of large training datasets for passing the signal in both directions and gradients calculation and performing parameter updating operations for each sample separately.

$$\theta^t = \theta^{t-1} - \lambda \nabla E_k(\theta), \tag{2.4}$$

where, k is the separate pattern of data, ∇E is the gradient of loss function calculated by expression.

To avoid these disadvantages without losing the useful properties of the algorithm, use the partitioning of the datasets into groups of samples (minibatches), which were previously randomly mixed [17].

Table 2.24 Results of the algorithm on the Easom function

d_{prec} N	Accuracy							
	10e−2	10e−3	10e−4	10e−5	10e−6	10e−7	10e−8	10e−9
1, 1	–	–	–	–	–	–	–	–
2, 4	–	–	–	–	–	–	–	–
3, 9	4689 126,603	–	–	–	–	–	–	–
5, 25	1952 244,000	–	–	–	–	–	–	–
10, 100	765 765,000	–	–	–	–	–	–	–
20, 400	31 248,000	–	–	–	–	–	–	–
30, 900	92 2,484,000	–	–	–	–	–	–	–
50, 2500	6 750,000	–	–	–	–	–	–	–
100, 10,000	103 103,000,000	232 232,000,000	–	–	–	–	–	–
200, 40,000	2 16,000,000	37 296,000,000	50 400,000,000	366 2,928,000,000	485 3,880,000,000	–	–	–
300, 90,000	2 54,000,000	29 783,000,000	50 1,345,000,000	50 1,345,000,000	74 1,982,000,000	–	–	–

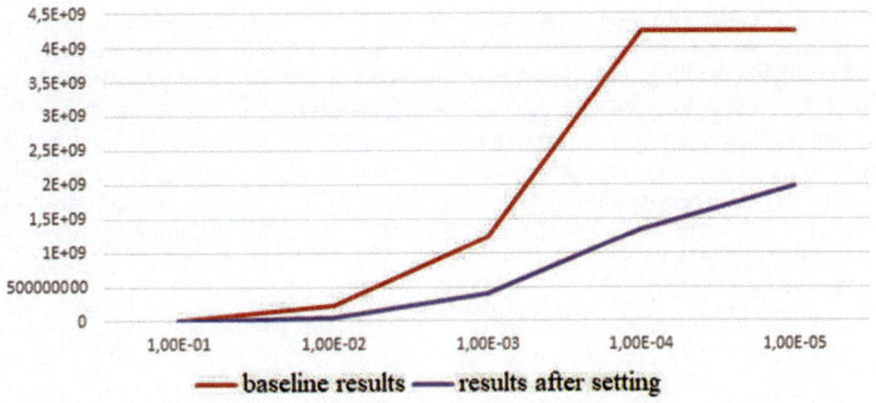

Fig. 2.51 The graphs of dependence of algorithm parameter on accuracy

In this case ∇E_k is the k-th minibatch by size $1 \ll m \ll M$ (M—total volume of the dataset) which is chosen relatively large and is usually controlled by the size of available memory of the computing device.

This approach is well-paralyzed since each minibatch can be processed individually and use distributed computing systems to train.

The second component of SGDM-Moment is an additional parameter of calculating functions of network weights updated values that takes into account the previous weights change values [52].

$$\Delta\theta^t = v^t = \mu\Delta\theta^{t-1} - \lambda\nabla E(\theta) \quad (2.5)$$

$$\theta^t = \theta^{t-1} + v^t \quad (2.6)$$

or

$$\theta^t = \theta^{t-1} - \lambda\nabla E(\theta) + \alpha\Delta\theta^{t-1} \quad (2.7)$$

where μ is the moment coefficient, $\Delta\theta^{t-1}$ is the previous value of updating the network settings, v^t is the can be interpreted as the velocity of the parameter point on the surface of the objective function.

The name "moment" is taken as an analogy of "moment of inertia" in mechanics: the parameter tensor w wanders the landscape of the objective function, receiving the acceleration from the gradient of this function ("force"). Unlike a simple gradient descent, this allows to continue moving with some inertia in the same direction, avoiding oscillations.

Nesterov Accelerated Gradient (NAG)

Methods the development of the previous method. Invented in 1983 by Nesterov [37], it is nowadays quite popular in machine learning. SGDM first calculates the gradient at the current point and then realizes jump in the direction of the updated accumulated gradient. Nesterov accelerated gradient algorithm first "jumps" in the direction of the previous jump at the rate stored in the previous step, and only then calculates new gradients. In other words, NAG implements the strategy "Jump → correction" (Fig. 2.52).

It looks like this:

$$v^t = \mu v^{t-1} + \lambda \nabla E\left(\theta - \mu v^{t-1}\right), \tag{2.8}$$

$$\theta^t = \theta^{t-1} - v^t. \tag{2.9}$$

This method of moment calculating avoids the undesirable high rate of parameters change, that can lead to explosive gradients. It seems to predict the parameters change, and if the move was too large and leads to a worse state of the objective function, it calculates the gradient by the parameter with taking into account made move. Also, NAG is even better at combating unnecessary oscillations around the optimum point [21], it proves better (compared to SGDM) works on a large number of tasks, especially with recurrent neural networks. Both methods work satisfactorily at relatively large values of μ and λ, often within limits $\mu \in [0.9, 1)$ and $\lambda \in [0.01, 0.001]$ [10].

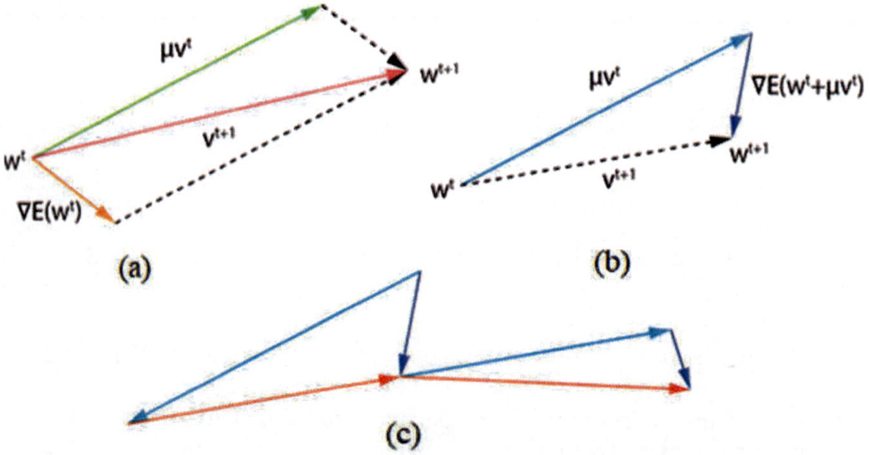

Fig. 2.52 Comparison of ways to calculate and update network weights for SGDM (**a**) and NAG (**b**) methods, as well as two steps of NAG (**c**) where the red line shows the accumulated gradient

Adaptive Gradient Algorithm

Otherwise the algorithm of adaptive gradients (AdaGrad, AdaDelta) [50, 53]—development of NAG algorithm, which aims to control such optimization parameter as learning rate λ. In all previous methods, it is unchanged and is applicable for all network parameters at all training epoches. This becomes a problem for deep neuron networks and wide-layer networks when the number of parameters reaches millions. The first reason is that parameters are more or less sensitive, that is, they change less or more, more often or less frequently. The second reason of the problem is the depth of the network and gradients attenuation in the layers distant from its output, so that such gradients need to be amplified enough for a useful update of the deep layers parameters.

AdaGrad adaptively changes λ for each parameter separately, reducing it for parameters that changed very often and increasing for those that are changed less frequently. That is, it tries to solve two problems of gradient methods at once: attenuation and explosion of gradients, wherein individually for each parameter.

$$\theta^{t+1} = \theta^t - \frac{\lambda}{\sqrt{\varepsilon + G^t}} \cdot g^t \tag{2.10}$$

where λ is the initial learning rate, ε is the very small value that prevents division by zero at $G^t = 0$, usually $\varepsilon = 1.0e - 10$,

$$g^t = \nabla E(\theta^t) \tag{2.11}$$

$$G^t = \sum_{\tau=1}^{t} g^\tau g^{\tau T} \tag{2.12}$$

where g^τ is the objective function gradient by parameter, G^t is the cumulative gradient by this parameter for all periods of learning, which is a diagonal matrix of squares g^τ.

Thus, AdaGrad accumulates the information about the change of gradients by parameter and adapts λ according to this information.

This method performs much better results on very deep models with hundreds millions of parameters than the previous ones, and, importantly, even for the non-convex functions which often encountered when using neural network models.

Root Mean Square Propagation (RMSProp)

This optimization method was invented by Hinton to solve the problems of the previously listed methods, including AdaGrad. Like AdaGrad, its work is to adaptively change the learning rate 00 λ, but these algorithms do it differently [54, 55].

This method is based on another called RProp, which is to "stimulate" gradients that change in one direction several epochs in a row, and to "suppress" those that change the sign often. Because different gradients vary differently, it is appropriate to use the learning step individually as well. RProp only works with a full dataset, without the minibatch division, and uses only a gradient sign, so all parameters are trained with the same gradient amplitude, that permit to avoid the plateau for very small gradient values. If, for example, in the last two learning steps, the gradient sign by the parameter w_i did not change, then for it the learning step λ is increased with the coefficient $k = 1.2$, otherwise it decreases with $k = 0.5$. However, since the sign change estimation by parameter is executed by averaged over the entire dataset, it makes no sense to use RProp with minibatchas, since the sign of the parameter gradient can vary from minibatch to minibatch.

To apply this problem, it was developed RMSProp, in which the gradients remain the same as in the classical gradient method, and the rate or step of training is controlled by the accumulated rate of gradient change in the previous steps.

Let

$$d\theta = \nabla E(\theta) \tag{2.13}$$

Then

$$v_{d\theta} = \beta v_{d\theta} + (1 - \beta) d\theta^2 \tag{2.14}$$

$$\theta = \theta - d\theta \frac{\lambda}{\sqrt{v_{d\theta} + \varepsilon}} \tag{2.15}$$

where ε is the null division insurance at $v_{d\theta} = 0$.

Therefore, RMSProp helps to avoid gradient oscillations by the parameter and speed up for the optimum search, given that it allows to use sufficiently large values of the learning step $\leq \delta t$.

Adaptive Moment Assessment (AdaM)

The optimization method that combines the approaches of AdaGrad and RMSProp methods. It also changes the learning steps individually for the parameters, taking into account the "history" of gradients changes on them [56].

Let

$$v = \beta_1 v + (1 - \beta_1) d\theta \tag{2.16}$$

$$s = \beta_2 s + (1 - \beta_2) d\theta^2 \tag{2.17}$$

$$\widehat{v_t} = \frac{v_t}{1 - \beta_1^t} \tag{2.18}$$

$$\widehat{s_t} = \frac{s_t}{1 - \beta_2^t} \tag{2.19}$$

where v and s called the first and second moment estimates, $\widehat{v_t}$ and $\widehat{s_t}$ are adjusted for parameters displacement of the moment estimate.

Then the parameters change as follows:

$$\theta = \theta - \lambda \frac{\widehat{v_t}}{\sqrt{\widehat{s_t} + \varepsilon}} \tag{2.20}$$

The use examples of described above algorithms are represented in Chaps. 3 and 4.

For many particularly deep neural networks, this algorithm performs better than other optimizers, although for shallow networks it does not give much advantage over AdaGradch NAG.

Its advantages:

- The actual learning step in each era is closely related to a given step;
- The upgrade step does not directly depend on the current gradient value, which avoids plateaus that are a problem for SGD, which is quickly stuck;
- AdaM was designed to combine the benefits of AdaGrad, which works well with sparse gradients, and RMSProp, which works well on small minibatch sizes. This method is also easily combined with NAG (NAdaM) and SGDM.

However, it has drawbacks. Despite its very rapid convergence [54], it has been experimentally found that AdaM sometimes converges to poorer optima than SGDM or NAG. In particular, the CIFAR family newsletters Studies have been conducted that have shown the possibility of replacing AdaM after a certain amount of training epochs with SGDM, which has reached a better optimum [57].

References

1. Kingma, D.P., Ba, J.L.: Adam: a method for stochastic optimization. In: International Conference on Learning Representations. Published as a conference paper at ICLR 2015, 30 Jan 2017, pp. 1–15 (2017)
2. Akhmedova, Sh.A.: Study of the effectiveness of the "flock" algorithm for multicriteria optimization problems. In: Youth and Science: A Collection of Materials of the VIII All-Russian Scientific and Technical Conference of Students, Graduate Students and Young Scientists Dedicated to the 155th Anniversary of the Birth of K.E. Tsiolkovsky (2012)
3. Arai, M.: Bounds on the number of hidden units in binary-valued three-layer neural networks. Neural Netw. **6**(6), 855–860 (1993)
4. Mantegna, R.N.: Fast, accurate algorithm for numerical simulation of Lévy stable stochastic processes. Phys. Rev. E **49**, 4677–4683 (1994)
5. Eiben, A.E., Smith, J.E.: Introduction to Evolutionary Computation. Berlin (2003)
6. Gaft, M.G.: Decision Making Under Many Criteria (1979)
7. Reznikov, B.: Methods and optimization algorithms on discrete models of complex systems, p. 250. VIKI Named After Mozhaisky (1983)
8. Semenkina, H.: Optimization of management of complex systems by the method of generalized local search. In: Semenkina, H., Zhidkov, V. (eds.) p. 215. MAKS Press (2002)

9. Coello Coello Carlos, A.: An empirical study of evolutionary techniques for multiobjective optimization in engineering design. Ph.D. thesis, Department of computer science, Tulane University, New Orleans, LA (1996)
10. Duchi, J., Hazan, E., Singer, Y.: Adaptive subgradient methods for online learning and stochastic optimization. J. Mach. Learn. Res. **12**, 2121–2159 (2011)
11. Goldberg, D.E.: A Comparative Analysis of Selection Schemes Used in Genetic Algorithms. In: Goldberg, D.E., Deb, K. (eds.) pp. 36–57 (1991)
12. Goldberg, D.E.: Critical Deme Size for Serial and Parallel Genetic Algorithms. In: Goldberg, D.E., Kargupta, H., Horn, J., Cantu-Paz, E. (eds.) pp. 365–452. IlliGAL Report №95002, Illinois Genetic Algorithms Laboratory, University of Illinois, Urbana, Illinois (1995)
13. Schaffer, J.D.: Multiple objective optimization with vector evaluated genetic algorithms. In: Grefenstette, J.J. (ed.) Proceedings of an International Conference of Genetic Algorithms and Their Applications, Pittsburgh, PA, pp. 93–100 (1985)
14. Akhmedova, Sh.A.: Formation of an optimal investment portfolio of an enterprise with a binary "flock" algorithm. Information Technology, System Analysis and Management. In: Proceedings of the IX All-Russian Scientific Conference of Young Scientists, Graduate Students and Students, vol. 2, pp. 218–220. Publishing House of TTI SFU, Taganrog (2011)
15. Rastrigin, L.: Random search. Knowledge (45) (1979)
16. Bottou, L., Bousquet, O.: The tradeoffs of large scale learning. Adv. Neural Inf. Process. Syst. **20**, 161–168 (2008)
17. Ning, Q.: On the momentum term in gradient descent learning algorithms. Neural Netw. Off. J. Int. Neural Netw. Soc. **12**(1), 145–151 (1999)
18. Zitzler, E.: SPEA2: Improving the performance of the strength Pareto evolutionary algorithm. In: Zitzler, E., Laumanns, M., Thiele, L. (eds.) pp. 236–296. Technical Report 103, Computer Engineering and Communication Networks Lab (TIK), Swiss Federal Institute of Technology ETH) Zurich (2001)
19. Horn, J.: A niched Pareto genetic algorithm for multiobjective optimization. In: Horn, J., Nafpliotis, N., Goldberg, D.E. (eds.) Proceedings of the First IEEE Conference on Evolutionary Computation, vol. 1, pp. 82–87, Piscataway (1994)
20. Bengio, Y., Boulanger-Lewandowski, N., Pascanu, R.: Advances in optimizing recurrent networks. In: 1988 International Conference on Acoustics, Speech, and Signal Processing, ICASSP-88 (1988). https://doi.org/10.1109/icassp.2013.6639349
21. Sutskever, I., Martens, J., Dahl, G., Hinton, G.: On the importance of initialization and momentum in deep learning. In: 30th International Conference on Machine Learning, ICML 2013, pp. 1139–1147 (2013)
22. Sutton, R.S.: Training with Reinforcement, p. 399. BINOM, Knowledge Laboratory (2012)
23. Vorozheikin, Yu.A., Gonchar, T.N., Panfilov, I.A., Sopov, E.A., Sopov, S.A.: About one modification of the probabilistic genetic algorithm for solving complex problems of conditional optimization. Bulletin of the Siberian State Aerospace University Named After Academician M.F. Reshetneva (4), 79–84 (2009)
24. Zvonkov, V.: A Comparative Study of Genetic Algorithms and a Flock Optimization Algorithm, pp. 786–797. Siberian State Autonomous University Krasnoyarsk (2012). [Electronic resource] access mode: http://www.ssc.smr.ru/media/ipuss_conf/14/7_14.pdf
25. Lin, C.-J.: Design of neuro-fuzzy systems using a hybrid evolutionary learning algorithm. In: Lin, C.-J., Xu, Y.-J. (eds.). J. Inf. Sci. Eng. (23), 463–477 (2007)
26. Akhmedova, Sh.A.: Study of the effectiveness of the "flock" algorithm for problems of conditional optimization. In: Youth and Science: A Collection of Materials of the VII All-Russian Scientific and Technical Conference of Students, Graduate Students and Young Scientists Dedicated to the 50th Anniversary of the First Manned Space Flight (2011)
27. Hinton, G.E.: Reducing the dimensionality of data with neural networks. In: Hinton, G.E., Salakhutdinov, R. (eds). Science **313**(5786), 504–507 (2006)
28. Akhmedova, Sh.A., Semenkin, E.S.: Development and Research of the Efficiency of the Flock Optimization Algorithm, p. 67. LAP LAMBERT Academic Publishing, Germany. ISBN 978-3-659-37617-7

29. Eberhart, R.C., Hu, X.: Human tremor analysis using particle swarm optimization. In: Proceedings of the IEEE Congress on Evolutionary Computation (CEC 1999), pp. 1927–1930 (1999)
30. Krishnanand, K.N., Ghose, D.: Glowworm swarm optimization for simultaneous capture of multiple local optima of multimodal functions. Swarm Intell. **3**(2), pp. 87–124 (2009)
31. Kennedy, J., Eberhart, R.: Particle swarm optimization. In: Proceedings of IEEE International Conference on Neural Networks, vol. IV, pp. 1942–1948 (1995)
32. Fukuyama, Y., Takayama, S., Nakanishi, Y., Yoshida, H.: A particle swarm optimization for reactive power and voltage control in electric power systems. In: Proceedings of the Genetic and Evolutionary Computation Conference 1999 (GECCO 1999), pp. 1523–1528 (1999)
33. Kennedy, J.: Small worlds and mega-minds: effects of neighborhood topology on particle swarm performance. In: Proceedings of IEEE Congress on Evolutionary Computation (CEC 1999), pp. 1931–1938 (1999)
34. Moore, J., Chapman, R.: Application of Particle Swarm to Multiobjective Optimization. Department of Computer Science and Software Engineering, Auburn University (1999)
35. Shi, Y., Eberhart, R.C.: Parameter selection in particle swarm optimization. In: Proceedings of Evolutionary Programming VII (EP98), pp. 591–600 (1998)
36. Yang, S., Wang, M., Jiao, L.: A quantum particle swarm optimization. In: Proceedings of the Congress on Evolutionary Computing, vol. 1, pp. 320–324 (2004)
37. Akhmedova, Sh.A., Semenkin, E.S.: Particle swarm optimization algorithms development and effectiveness investigation. Bulletin of the Siberian State Aerospace University Named After Academician M.F. Reshetneva **4**(44), 78–79 (2012)
38. Kennedy, J., Eberhart, R.C.: A discrete binary version of the particle swarm algorithm. In: Proceedings of the International Conference on Computational Cybernetics and Simulation, pp. 4104–4108 (1997)
39. Yang, X.S.: Firefly algorithms for multimodal optimization. In: Proceedings of th 5th Symposium on Stochastic Algorithms, Foundations and Applications, pp. 169–178 (2009)
40. Yang, X.S., Deb, S.: Cuckoo search via Levy flights. In: Proceedings of World Congress on Nature & Biologically Inspired Computing, pp. 210–214 (2009)
41. Payne, R.B., Sorenson, M.D., Klitz, K.: The Cuckoos. Oxford University Press (2005)
42. Tuba, M., Subotic, M., Stanarevic, N.: Modified Cuckoo search algorithm for unconstrained optimization problems. In: Proceedings of the 5th European Computing Conference (ECC'2011), pp. 263–268 (2011)
43. Valian, E., Mohanna, S., Tavakoli, S.: Improved Cuckoo search algorithm for global optimization. Int. J. Commun. Inf. Technol. IJCIT **1**(1), 31–44 (2011)
44. Yang, X.S.: A new metaheuristic bat-inspired algorithm. In: Nature Inspired Cooperative Strategies for Optimization, SCI 284, pp. 65–74 (2010)
45. Yang, C., Tu, X., Chen, J.: Algorithm of marriage in honey bees optimization based on the Wolf Pack search. In: Proceedings of the International Conference on Intelligent Pervasive Computing, pp. 462–467 (2007)
46. Shoghian, Sh., Kouzehgar, M.: A comparison among Wolf Pack Search and four other optimization algorithms. World Acad. Sci. Eng. Technol. **6**(72), 418–423 (2012)
47. Zahadat, P., Schmickl, T.: Wolf Pack-inspired evolutionary algorithm and a reaction-diffusion-based controller are used for pattern formation. In: Proceedings of the 2014 Conference on Genetic and Evolutionary Computation (GECCO'2014), pp. 241–248 (2014)
48. Molga, M., Smutnicki, Cz.: Test Functions for Optimization Need, 5 April 2005
49. Wei, J., Wang, Y., Wang, H.: A hybrid particle swarm evolutionary algorithm for constrained multi-objective optimization. Comput. Inform. **29**, 701–718 (2010)
50. Hinton, G., Tieleman, T.: Lecture 6.5—RMSProp: divide the gradient by a running average of its recent magnitude. COURSERA: Neural Netw. Mach. Learn. **4**(2), 26–31 (2012)
51. Duchi, J., Hazan, E., Singer, Y.: Adaptive subgradient methods for online leaning and stochastic optimization. In: COLT (2010)
52. Nesterov, Y.Y.: Method for Minimizing Convex Functions with Convergence Rate O $(1/k^2)$, pp. 543–547. Report, AS USSR, T. 269, rel. 3 (1983)

53. Zeiler, M.D.: Adadelta: An Adaptive Learning Rate Method, 22 Dec 2012, pp. 1–6. [Електронний ресурс] Режим доступу: https://arxiv.org/pdf/1212.5701v1
54. Kingma, D., Ba, J.: Adam: a method for stochastic optimization. In: International Conference on Learning Representations (2015). [Electronic resource] access mode: https://arxiv.org/pdf/1412.6980
55. Ruder, S.: An Overview of Gradient Descent Optimization Algorithms (2016). [Електронний ресурс] Режим доступу: https://ruder.io/optimizing-gradient-descent/
56. Reddi, S., Kale, S., Kumar, S.: On the Convergence of Adam and Beyond (2018). [Електонний ресурс] Режим доступу: https://arxiv.org/pdf/1904.09237
57. Xavier, G., Bengio, Y.: Understanding the difficulty of training deep feedforward neural networks. J. Mach. Learn. Res. Proc. Track. **9**, 249–256 (2010)

Chapter 3
Formation of Hybrid Artificial Neural Networks Topologies

3.1 Problem Statement of Optimal Artificial Neural Network Synthesis

Let us set the problem of neural network synthesis as follows.

A finite set $J = \{(R_j, Y_j)\},\ j = 1, \ldots, P$ pairs the "attribute value" type, where $R_j Y_j$ input and output vectors of NN, respectively.

It is necessary to synthesize such optimum NN based on training sample J, which would provided effective solution of the applied problem (classification, approximation, forecasting). Vector optimality criterion is defined as

$$I = \{I_1(x), I_2(x)\} \to \text{opt},$$

where $I_1(x) = E_{gen}(x)$ is the error of the generalization defining the solution error value of given task on the test sample; $I_2(x) = S(x)$ is the complexity of neural network (number of inter neuronal connections); $X = (x_{11}, x_{12}, \ldots, x_{1*2}, x_2, x_3, x_4)^T$ is the vector that defines the topology, the structure and parameters of the network, where x_{1i} is the number of neurons in the ith hidden layer; $i = \overline{1, x_2}$, x_2 is the number of hidden layers; x_3 is the number of inter neuronal connections; x_4 is the set of w_{ij} weight coefficients values; i is the hidden layer number; j is the neuron number.

M. Zgurovsky et al., *Artificial Intelligence Systems Based on Hybrid Neural Networks*, Studies in Computational Intelligence 904,
https://doi.org/10.1007/978-3-030-48453-8_3

3.2 Synthesis Methodology of Hybrid Neural Networks Based on One Topology Neural Networks Using Neurons of Different Types

Synthesis of hybrid neural networks, which are used for the solution of a specific application problem on the basis of a given training sample and a neural network of single topology with neurons of different types is a complex problem which consists in the preliminary definition of NN initial topology (further this network will be considered as base) according to the classification presented in Fig. 1.4 for given class of applied tasks (for example, classification, approximation, decision making, forecasting by the training samples. Presented on test sites) (see Sect. 3.3.1), in optimal choice of layers number, types of neurons, number of neurons in layers, order of layers alternation with different neurons, determination of weight coefficients values, parameters of activation functions (neighbourhood functions).

Based on the above the synthesis methodology of hybrid neural networks can be represented as such a sequence of actions.

1. Optimal choice of topology of the base neural network.
2. Structural-parametric synthesis of HNN based on the optimal choice of the base NN with the determination of the optimal number of layers, types of neurons, number of neurons in the layers, the order of alternation of layers with different neurons, the values of the weighting coefficients, the parameters of the activation functions (neighbourhood functions).

To solve the problem of structural-parametric synthesis of HNN based on the optimal choice of base NN, it is advisable to create a two-stage optimization algorithm: the genetic algorithm will be the most effective procedure in the initial stage of a solution search, which determines the area of global extremum existence; the second stage of the search will be related to the refinement of the minimum based on the local optimization algorithm (see Sect. 3.6.4). The task is divided into two sub-tasks: search of the optimal structure (number of hidden layers, types of neurons, number of neurons in layers, order of alternation of layers with different neurons) and adjustment of weight coefficients and parameters of activation (neighbourhood) functions (sub-task of parametric optimization). To solve both sub-problems, we use a two-stage optimization algorithm, the first stage of which uses a hybrid multicriteria evolution algorithm to locate the area of search for the optimal structure and weight coefficients of the NN. In the second stage, the determination of optimal values of the NN weight coefficients is performed on the basis of the error backpropagation and gradient methods (steepest descent, of stochastic gradient descent, stochastic gradient descent with moment (SGDM), the Nesterov accelerated gradient (NAG), Adaptive Gradient Algorithm (AdaGrad, AdaDelta), root mean square propagation (RMSProp), adaptive moment estimation (AdaM), Nesterov adaptive moment (NAdaM)) (see Sect. 2.8).

The results of HNN structural-parametric synthesis problem solution are considered in the examples (see Sect. 1.11).

3.3 Synthesis Methodology for Hybrid Neural Networks Based on Neural Networks of Different Topologies Using Single-Type Neurons

Synthesis of hybrid neural networks that are used to solve a specific application task based on a given training sample is a complex problem, which is to pre-determine the initial NN topology (hereafter referred to as the base network) according to the classification shown in Fig. 1.4 for this class of application tasks (e.g., classification, approximation, decision making, forecasting by training samples presented on test sites) (see Sect. 3.3.1), further modification of the base NN by the training sample corresponding to this application task (see Sect. 3.6)) and the formation of a hybrid structure by creating a module that incorporates a base NN (see Sect. 3.7) or a module ensemble (see Sect. 3.8), which provides new opportunities to increase the efficiency of the task solution [1–3]. In Sect. 3.6, the problem of modification the base NN is reduced to solving the conditional problem of multicriteria optimization based on the developed hybrid genetic algorithm. In Sect. 3.7 it is substantiated the feasibility of creating hybrid NN of module structure. In Sect. 3.8 it is proposed an approach to further complicate the hybrid NN topology by creating a module ensemble.

Based on the above the synthesis methodology of hybrid neural networks can be represented as such a sequence of actions.

1. Optimal choice of topology of the base neural network.
2. Modification based on the solution of multicriteria optimization problem by changing the number of hidden layers, neurons in them, cross-connections, determining the weights values.
3. Structural—parametric synthesis of the neural networks module.
4. Structural—parametric synthesis of an ensemble consisting of NN modules

Each of the points in the proposed methodology is a separate task, so let's consider them separately. New methods and algorithms have been developed to solve each of the above tasks, so the proposed approach can be considered as methodology.

3.3.1 Optimal Selection of Base Artificial Neural Network Topology

3.3.1.1 Problem Statement of Base Neural Network Topology Optimal Selection

The set of neural networks $\{NN\}_i$, $i = \overline{1, n}$ is given and the applied task (approximation, classification, forecasting, decision making, control, etc.) is determined.

It is necessary from the set $\{NN\}_i$, $i = \bar{1}, \bar{n}$ for the given problem according to the learning sample of the same type of tasks, presented on the site [4], to choose a topology of base NN according to the criterion of E_{gen}.

3.3.1.2 Algorithm of Base Neural Network Topology Optimal Choice

The problem of optimal selection of base NN topology is solved by using the method of brute force. The algorithm is as follows.

(1) By the task class, to which apply the given task, it is defined the training samples that are on the site of learning tasks.
(2) The selected learning sample is divided in accordance: 80%—training sample, 20%—test one.
(3) In accordance to the classification of NN (see Fig. 1.4), it is performed learning sequentially for each NN according to the algorithm specified for it.
(4) Using test sample for each NN the E_{gen} is determined.
(5) As the base NN is selected that one, which has the smallest E_{gen}.

Consider examples of base NN topology optimal choice for tasks, respectively: approximation, classification and forecasting.

3.3.2 *Examples of Optimal Selection of Base Neural Network Topology*

As a result of the brute force method use based on numerical experiments and testing various NN topologies (see Fig. 1.6) for solution of approximation problem according to the criterion of the E_{gen} selected ANFIS and NEFPROX.topologies.

In solving the task of approximation, the ANFIS and NEFPROX topologies showed the best results according to this criterion.

Learning sample format is as follow:

- 10 clear numerical parameters;
- learning sampling size of 200 values;
- test sampling size of 50 values.

The results of experiments are shown in Table 3.1.

The results of the ANFIS network with the test sample are shown in Table 3.2.

The results of NEFPROX Network training are shown in Table 3.3.

The results of the NEFPROX network with the test sample are shown in Table 3.4.

Table 3.1 ANFIS network learning results

Training time (s)	Learning error	Number of rules	Number of training cycles
6	0.001	24	200

Table 3.2 Results of the ANFIS network functioning on the test sample

Work time (s)	Error during operation
9	0.04

Table 3.3 NEFPROX learning outcomes

Number of membership functions	Training time (s)	Learning error	Number of rules	Number of training cycles
10	4	0.001	17	200
20	9	0.001	24	200

Table 3.4 The results of the NEFPROX network with the test sample

Number of belonging functions	Work time (s)	Error during operation
10	2	0.18
20	6	0.06

As base topologies for solving the problem of classification, the best ones were the topologies of fuzzy neural networks, which are presented in Table 3.5.

A set of "Wine" [4, 5] data will be used to work with the data of different types, in which there are three types of data.

1. **Clear variables**. The following characteristics are given: persistent acidity, citric acid, residual sugar, chloride, all sulfur dioxide, sulfate.
2. **Fuzzy variables**. Given: volatile acidity, the content of free sulfur dioxide, pH.
3. **Linguistic variables**. Each term of linguistic variables corresponds to a fuzzy subset of that uses a triangular function as a function of belonging [6].

- Density $T = \{\text{high, average, low}\}$.
- The alcohol content of $T = \{\text{high, medium, low}\}$.

According to the setting of the task (Sect. 3.1), using the package Matlab Neural Network Toolbox and the educational sample of Wine, will solve the problem of choosing the optimal topology of NN. The results of the problem solving are shown in the Table 3.6.

Based on the results obtained, we can conclude that the best topology is ANFIS and NEFCLASS.

As a result of numerical experiments and testing of various neural network topologies, ANFIS, NEFPROX and TSK topologies are selected for prediction task solution. To compare the efficiency of different topologies during the solution of the prediction task the following task was chosen: determination of the age of shellfish by its physical attributes [4]. Usually, to solve a similar problem, the shellfish is simply cut, and then it is counted the number of rings on its body under a microscope. However, there is another, easier and faster method based on the analysis of the physical parameters of the shellfish.

The description of sampling parameters is given in Table 3.7. As input, the settings are 1–7. The result of prediction is option 8. To accelerate learning, all the parameters of the numerical type were numbered by division 200.

Table 3.5 Comparison of topologies

Topology	Number of layers	Output rules	Learning algorithm	Scope of						
				Classification	Clustering	Approximation	Prognosis	Optimization	Memory	Control
ANFIS	5	Takagi-Sugeno, Tsukamoto	Gradient descent			+	+			
TSK	5	Takagi-Sugeno	Hybrid (2 stages)			+	+			
Wang-Mendel	4		Hybrid (2 stages)			+	+			
NEFCON			Fuzzy sets method							+
NEFPROX	3		Fuzzy sets method			+				
NEFCLASS	3		The method of fuzzy sets, genetic algorithms	+						

Table 3.6 Results of the problem solution by different topologies

Topology	Training time (s)	Error on training sample	Work time (s)	Error on test sample
ANFIS	6	0.001	9	0.04
TSK	6	0.002	11	0.12
Wang-Mendel	7	0.002	12	0.14
NEFCON	5	0.003	10	0.15
NEFPROX	4	0.002	9	0.11
NEFCLASS	9	0.001	6	0.0598

Table 3.7 Data set description

#	Name	Data type	Units of measurement	Description
1	Gender	Binary	–	0—Female, 1—Male
2	Length	Numerical	mm	Largest linear size
3	Diameter	Numerical	mm	Perpendicular length size
4	Height	Numerical	mm	Height
5	Full weight	Numerical	g	Weight of whole shellfish
6	Body weight	Numerical	g	Weight without shells
7	Dried weight	Numerical	g	Weight of dried shellfish
8	Number of rings	Entire	–	The number of rings indicates the shellfish's age

The training sample contains 350 records. The test sample that contains 50 records is used to test the accuracy of the work. The learning sample fragment is presented in Table 3.8.

The results of training are given in Table 3.9.

The results of the ANFIS network on the test sample are shown in Table 3.10.

The results of NEFPROX network training are shown in Table 3.11.

Table 3.8 Fragment of the testing sample

Gender	Length	Diameter	Height	Full weight	Body weight	Dried weight	
1	0.455	0.365	0.095	0.514	0.2245	0.15	15
1	0.35	0.265	0.09	0.2255	0.0995	0.07	7
0	0.53	0.42	0.135	0.677	0.2565	0.21	9
1	0.44	0.365	0.125	0.516	0.2155	0.155	10
0	0.53	0.415	0.15	0.7775	0.237	0.33	20
0	0.545	0.425	0.125	0.768	0.294	0.26	16
1	0.475	0.37	0.125	0.5095	0.2165	0.165	9
…	…	…	…	…	…	…	…

Table 3.9 ANFIS network learning results

Network type	Training time (s)	Learning error	Number of rules	Number of training cycles
Standard topology	6	0.001	24	200
Modified topology	6.2	0.005	24	200

Table 3.10 The results of the ANFIS network with the test sample

Network type	Duration of work (s)	The error on the test sample when working
Standard topology	9	0.04
Modified topology	9.5	0.1

Table 3.11 NEFPROX learning results

Network type	Number of membership functions	Training time (s)	Learning error	Number of rules	Training cycles
Standard topology	10	4	0.001	17	200
Modified topology	10	4.1	0.007	17	200
Standard topology	20	9	0.001	24	200
Modified topology	20	9.3	0.008	24	200

The results of the NEFPROX network on the test sample are shown in Table 3.12.
The results of TSK Network training are shown in Table 3.13.
The results of the TSK network with the test sample are shown in Table 3.14.
Based on the experiments conducted, you can make the following conclusions:

Table 3.12 Results of the NEFPROX network with the test sample

Network type	Membership functions	Duration of work (s)	Error during operation
Standard topology	10	2	0.18
Modified topology	10	2.05	0.14
Standard topology	20	6	0.06
Modified topology	20	6.1	0.12

Table 3.13 TSK network learning results

Network type	Training time (s)	Learning error	Number of rules	Number of training cycles
Standard topology	8	0.001	24	270
Modified topology	8.5	0.003	24	270

Table 3.14 Results of the TSK network on the test sample

Network type	Duration of work (s)	Error during operation
Standard topology	10	0.03
Modified topology	11	0.1

1. The NEFPROX network operates fastest of all considered topologies, but has the smallest precision.
2. The TSK network has the greatest accuracy, but it is more complex to learn and needs more time for learning and work.
3. The ANFIS network is optimal in the ratio of accuracy/rate. In addition, it is quite simple in structure and quickly trained.

It is necessary to learn these topologies to work with data of different types. Since default the given HNN working with clear numerical inputs, it is necessary to bring all inputs to a uniform clear format. For this it will be use the converted block at the input.

Networks modification for possibility to work with different types of input data leads to an accuracy decrease up to 5–7%, compared to standard networks, regardless of topologies. In this regard, there is a need to synthesize new, more complicated integrated topologies.

Consider a case of the selected optimal technology NN modification. To do this, analyze the connections of various types that can be set between the neurons of different layers in a multi-layer neural network (MLNN).

3.4 Structure and Training of Networks with Neurons of Type Sigm_Piecewise

Considering the problems of multilevel architectures of neural networks and the relative complexity of the model of the neuron itself sigm_piecewise, it is advisable to use the architecture of Perceptron with one hidden layer (Fig. 3.1).

Since the neuron function with the model sigm_piecewiseis is an approximation of a piecewise linear function piecewise_linear, then the sum of these neurons is of the form in turn will be an approximation of the corresponding sum of the function piecewise_linear.

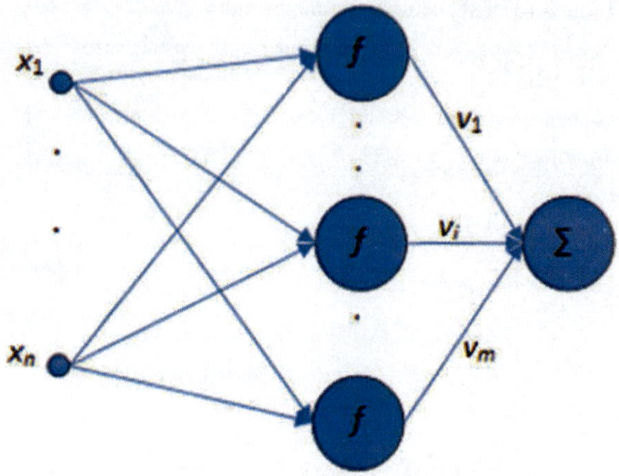

Fig. 3.1 The perceptron architecture with one hidden layer

$$\sum_{j=1}^{n} v_j \text{ sigm\_}piecewise\left(\vec{x};\ \vec{w}_{+}^{(j)};\ \vec{w}_{-}^{(j)};\ \vec{h}^{(j)}\right), \tag{3.2}$$

Since the function piecewise_*linear* $\left(\vec{x};\ \vec{w}_{+}^{(j)};\ \vec{w}_{-}^{(j)};\ \vec{h}^{(j)}\right)$ splits the space into two subspaces by a hyperplane $\vec{h}$, and gives in each subspace its own linear function, then the sum of such functions can already divide the space into a large number of subspaces, depending on the number of functions in the sum and in each subspace can set a certain linear function, that is, with a sufficient number of functions in the sum, you can set a piecewise linear function of any complexity. Hence, even this simple architecture with one hidden layer should be enough for approximation of any continuous functions.

It is known that when training the multilayer neural networks with a large number of neurons in layers and/or a large number of layers, it is very difficult to find a parameters vector at which the global minimum of the error function is achieved. This is because the error function, that needs to be minimized, in the case of complex neural networks is also a complex nonlinear function of the parameters vector that are optimized and the minimization of such functions in general is an unsolved problem. Usually, under minimization of such functions, gradient methods are used, which are very sensitive to local minima and "plateau"—regions in the space of parameters vectors, where the gradient vector of the function is close to zero, under falling into such regions the gradient methods tend to "stuck". Because of this, more and more attention is paid to so-called pre-learning methods that attempt to initialize the network parameters vector in such a way that:

- not get into the so-called "problem" regions of the parameter space;
- to begin training already with some "satisfactory" value of an error.

The idea of the proposed method is to gradually increase the number of neurons in the hidden layer, with the addition of a new neuron, only its parameters and the parameter responsible for the weight of this neuron in the neuron-adder of the output layer are learnt—all other parameters are fixed.

To use this method, the condition must be satisfied that for any training sample $\langle X^{(T)}, y^{(T)} : X^{(T)} \to \Re \rangle$ and for any given error value $\varepsilon > 0$ to be achieved, there are such a number of neurons in the hidden layer m, and such vectors of parameters $\vec{w}, \ldots, \vec{w}_m; \vec{v}; \vec{v} \in R^m$, that the network error of the described type with such parameters will be less than or equal to ε:

$$\forall \langle X^{(T)},\ y^{(T)} : X^{(T)} \to R \rangle,\ \ \varepsilon > 0 : \exists m \in N,\ \vec{w}_1,\ \ldots,\ \vec{w}_m, \tag{3.3}$$

$$\vec{v} : E(X^{(T)}, y^{(T)}, \mathrm{Net}(\vec{x}) = \sum_{i=1}^{m} v_i f(\vec{x}; \vec{w}_i) \le \varepsilon \tag{3.4}$$

If this condition is not met, it is possible that adding and training any number of new neurons will not reduce the network error on the training sample. For example, this condition does not hold for a network with linear neurons in a hidden layer, that is, when $f(\vec{x}; \vec{w}) = \vec{w}^T \vec{x}$ since the linear combination of any number of such neurons will always remain a linear function of $\vec{x}$ so if the examples in the training sample are not perfectly described by some linear function, then there is some error limit $\varepsilon_{threshold}$, less than which can't be obtained with any network configuration with linear neurons in a hidden layer.

Having a training and validation samples of the form and $\langle X^{(V)}, y^{(V)} : X^{(V)} \to \Re \rangle$ accordingly, the pre-training of a single-layer network of the described type according to the proposed method consists of the following steps:

1. At the first iteration, a network of the described type with one neuron in the hidden layer is trained.
2. After i iterations we have a network with i neurons in the hidden layer.
3. At the iteration $i + 1$, another neuron is added to the hidden network layer, after which the network training is repeated, but all network parameters except the parameters of this neuron and its weight in the output layer adder neuron are fixed.
4. At each iteration i are calculated:
   - the current value of the network error on the validation sample-denote it as $E^V(i)$;
   - the minimum value among all errors in the previous iterations-denote it as $E^V_{\min}(i) = \min\limits_{j=1,\ldots,i=1} E^V(j)$.
5. The addition of new neurons is performed as long as the condition is satisfied that during the I last iterations the minimum value of the network error on the validation sample has been sufficiently improved, that is:

$$E^V_{\min}(i) \le a \cdot E^V_{\min}(i - I),\ a \in (0, 1], \quad I \in N,\ \forall i > I, \tag{3.5}$$

where a is a constant that specifies the desired level of improvement of the minimum error value for I iterations—usually it is chosen from a set of numbers $\{0.9,\ 0.99,\ 0.999\ \ldots\}$ that is when $a = 0.9$ for I iterations the minimum error of the network on validating sample should decrease by at least 10% at $a = 0.9$—not less than 1% and so on.

6. After stopping, the network returns to the iteration where the minimum error on the validation sample was reached.

The use of this method for pre-learning networks described type is appropriate, when for the reaching satisfactory value of network error on the training sample it is t enough the network with a relatively small number of neurons in the hidden layer from 10 to 10000 (depending on the processing power of a computer that is used for training, and time allocated for training), since at each iteration it is learnt the parameters of only one neuron. This situation usually occurs either when the unknown approximated function is fairly simple, or when the model of neurons used in the hidden layer is a fairly complex function.

3.5 Analysis of Different Types Connections

Different types of connections can be established between neurons of different layers in MLNN (Fig. 3.2).

In general, the output signal of the j-layer neuron is the input signal of the layer numbered $j + S$. The following types of connections between neurons are possible: sequential ($S = 1$); cross ($S > 1$); lateral ($S = 0$); inverse ($S < 1$).

In the case of NN with feed-forward serial connection, the output signal of neurons of some layer is the input signal of neurons of the layer following it. Lateral connections are established between neurons of the same layer, cross (or connections through the layer) are established between neurons of some layer and subsequent layers (but not the next layer).

Feedback loops are connections between neurons of some layer and previous layers. Examples of single-layer and multi-layer neural networks with serial connections (in the case of multi-layer also with cross-connections) are single-layer and multi-layer perceptrons, with lateral—Kohonen networks, with inverse-Hopfield, Hemming networks, Boltzmann machine.

In the presence in a multilayer neural network of serial, reverse, cross and lateral connections, the concept of a layer neuron degenerates, because it is difficult and even impossible to determine the number of neurons that receive a signal at the same time.

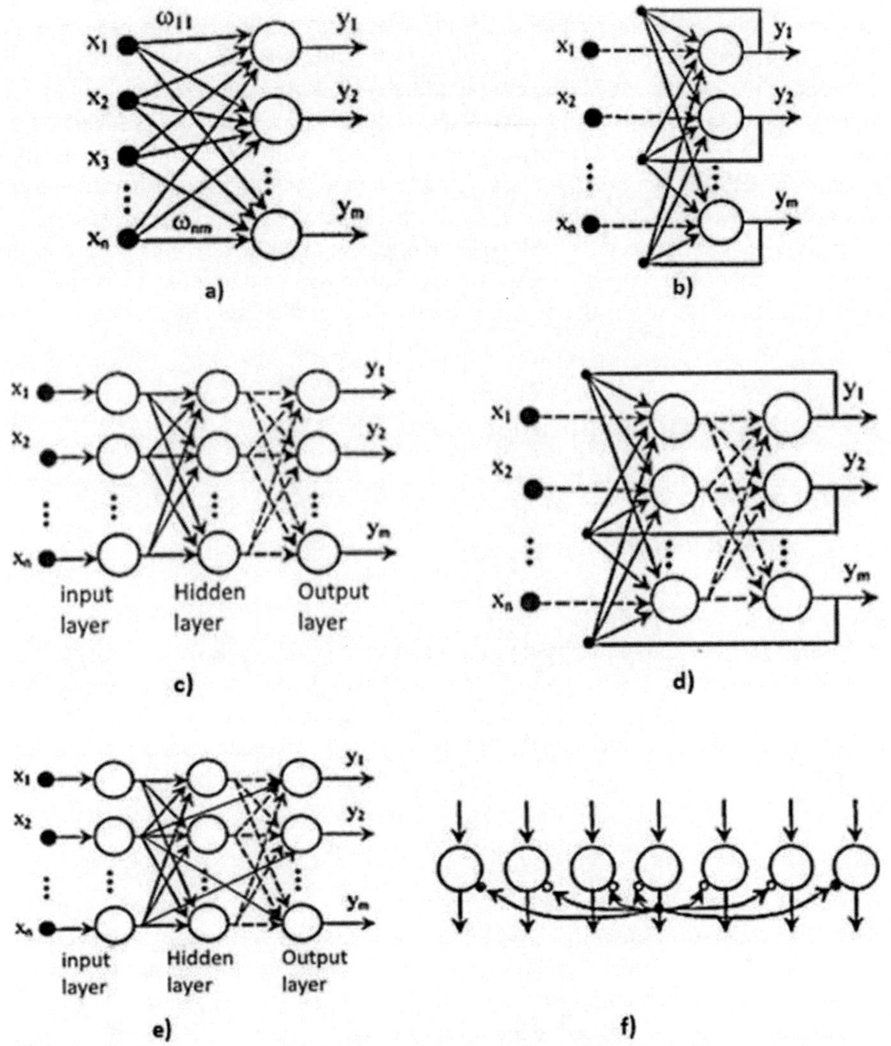

Fig. 3.2 The main types of connections between neurons of different layers in the model NN connectionist type: **a** is the single-layer network with direct communication; **b** is the multilayer network with direct connection; **c** is the multilayer network with cross-connection; **d** is the single-layer recurrent network (feedback); **e** is the multilayer recurrent network; **f** is the network with lateral connections

In addition to the considered main types of connections, in practice, pyramidal and layered NN are also used, in which different types of neurons and activation functions can be present.

The complexity of NN in the work is determined by the total number of connections between all neurons NN and, in turn, sets the computational complexity of the network as the number of computer operations, required to convert the input signal to the output. On the base of the given classification of different types of NN and the problems solved by them in this work, in accordance with Sect. 3.1, the following NN are chosen: NN without feedbacks, in which there are only feed-forward (serial and possibly cross) or lateral connections. This type of networks occupies a fairly wide class of NN, makes it possible to build effective combined NN, as well as integrate with other technologies. According to the proposed methodology of hybrid NN (Sect. 3.2), the next task is the optimal modification of the base NN.

3.6 Suboptimal Modification of the Base Neural Network

3.6.1 Problem Statement of Base Neural Network Modification

Given a finite set of attribute-value $J = \{(R_j, Y_j)\}$, $j = 1, \ldots, P$ pairs, where $R_j Y_j$ are the input and output vectors, respectively.

It is necessary to optimally modify the base NN on the training sample of this problem (structure and parameters) of the selected optimal topology on the reference sample of this type of problems. The vector criterion is taken as the optimality criterion

$$I = \{I_1(x), I_2(x)\} \to \text{opt}, \tag{3.6}$$

where $I_1(x) = E_{gen}(x)$ is the generalization error, determines the magnitude of the error of solving the problem; $I_2(x) = S(x)$ is the neural network complexity (number of interneuron connections); $x = (s, p, q, w)^T$ s is the number of hidden layers; $q = \{q_{ir,if}\}$ is the cross connection; $i \neq j, i, j$ are layer numbers, $j, i = \overline{1, s}$; r, f is the numbers of neurons in the hidden layers i and j, respectively, $r = \overline{1, p_i}$, $\overline{1, p_j}$:

$$q_{tr,if} = \begin{cases} 1, & \text{cross connection is present,} \\ 0, & \text{cross connection is absent,} \end{cases}$$

$w = \{w_{ij,k}\}$, $i = \overline{1, s}$, $j = \overline{1, p_j}$, $k = \overline{1, g_{ij}}$ is the values of weight coefficients; g_{ij} is the number of inputs of the ith neuron of the jth layer.

This problem belongs to the class of multicriteria optimization problems.

3.6.2 *Topology Variants Formation of Modified Base Artificial Neural Network*

The main problem in implementing a modified hybrid algorithm (Sect. 2.4.7) is encoding potential solutions. One way involves encoding a pre-selected sequence of network transformation rules in a chromosome: adding a layer, adding connections of different types, adding neurons (Fig. 3.3) with the subsequent calculation of the vector of weight coefficients by means of the chosen algorithm of training and check of an error of generality.

The proposed method of encoding is to specify in the chromosome the structure of the neural matrix of connections between neurons, that is, encoding all possible paths in the neural network in the chromosome: each gene of the chromosome represents some path in the acyclic graph NN.

The procedure for gradually increasing the network topology by adding the appropriate components (layers, neurons, individual connections: interlayer, cross) is shown in Fig. 3.3.

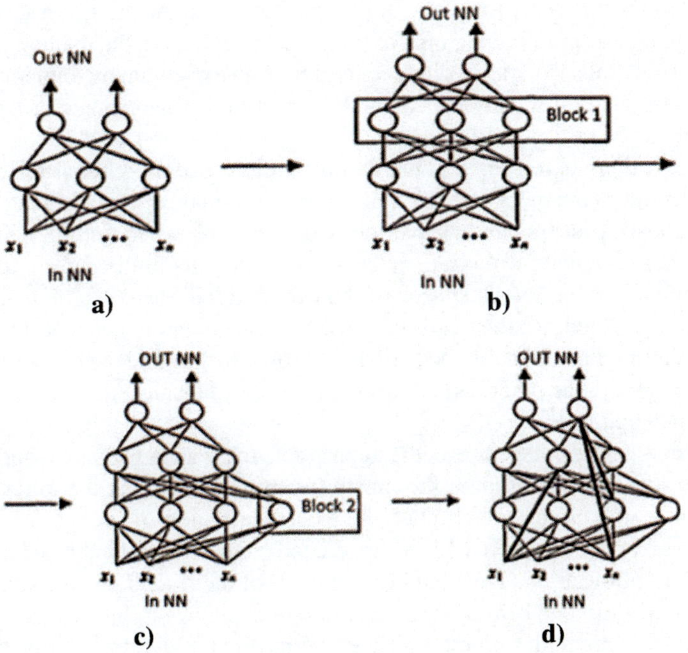

Fig. 3.3 Variants of topologies that are formed in the NN: **a** is the original topology; **b** is the adding a hidden layer; **c** is the adding elements of the first hidden layer; **d** is the adding cross-links

3.6.3 *The Method of Optimal Base Neural Network Modification Based of Hybrid Multicriteria Evolutionary Algorithm and Adaptive Merging and Growing Algorithm*

3.6.3.1 Analysis of the Evolutionary Algorithm Efficiency

The results obtained by the author indicate a low percentage of experiments in which the extremum value was obtained with a given accuracy for the hybrid genetic algorithm (HGA), methods of random, gradient search for optimization of multiextremal functions. However, when optimizing unimodal functions, a good approximation of the found extremums to the true extremum is observed for all methods, and when optimizing multiextremes ones—for the methods of HGA and random search [7, 8].

The reason for this is the property of the HGA and the random search method, which is associated with the rapid localization of the zone of existence of the extremum.

The gradient algorithm is characterized by a sequential study of the search zone, which allows in most experiments to find a local extremum with a given accuracy, but it is not suitable for finding a global extremum. In contrast to the method of random search in HGA mechanisms of directed motion to an extremum by implementing the algorithm of "natural selection", so HGA gives a higher percentage of localization of a global optimum.

So HGA, there is, on the one hand, quite time consuming, requires the user to specify certain parameters, determining a set of optimal control parameters so that an evolutionary process can balance the search and use when finding good quality solutions (for example, if the rate of crossover and mutation is chosen too high, a significant part of the search space will be explored, but there is a high probability of losing good solutions, the inability to use existing solutions), detects the inability during training of neural networks with high probability to find the exact value of the extremum, and on the other hand, makes it possible to localize the area of existence of the global extremum.

According to the conclusions, it is appropriate to create a two-stage optimization algorithm: genetic algorithm is the most effective procedure at the initial stage of a solution search, which determines the existence region of global optimum; the second phase of the search will be associated with the refinement of minimum, based local optimization algorithm. The analysis of the results shows that the use of a local optimization algorithm in the last iterations allows to increase the accuracy of finding the extremum for both single-extremal and multi-extremal functions. The development and implementation of this algorithm is considered below. The use of a two-stage optimization algorithm in the NN training procedure will simultaneously solve two problems: to increase the rate of convergence of the algorithm due to the properties of the genetic algorithm, to explore the entire search space as a whole and to increase the accuracy of finding the extremum by using effective methods of local optimization.

As mentioned above, the problem of optimal modification of the base neural network falls into two subtasks: the search for the optimal structure (the number of hidden layers, neurons in them and cross-connections between neurons) and the adjustment of weight coefficients (the subtask of parametric optimization). For the solution of both subproblems use a two-step optimization algorithm in the first stage, of which it is applied a hybrid multiobjective evolutionary algorithm, which localizes the search area of the optimal structure and weight coefficients of NN [9, 10]. In the second stage the determination of the optimal values for weight coefficients NN is performed on the base of backpropagation and steepest descent methods use (stochastic gradient descent), and the optimal values determination of the number of hidden layers and neurons in them will be carried out using the adaptive merging and growing algorithm, when for each neuron of the hidden layer is calculated significance, and in the case where its value is less than the threshold, label these neurons with subsequent calculation the correlation coefficients of these neurons with unmarked neurons of this layer. Each labeled hidden neuron will be merged with its most correlated from this layer unmarked analog. In the case that the error of learning in the process of merging operations execution stops to decrease, the operation of growing is performed, that is, adding one neuron to this hidden layer, etc.

3.6.3.2 Adaptive Merging and Growing Algorithm

The adaptive merging and growing algorithm looks like this.

(1) Create an initial population of base ANN of size M, which is pre-calculated. The number of hidden layers, neurons of each hidden layer, the values of the weight coefficients are uniformly randomly generated in the middle of certain ranges, the size of which is determined by the given restrictions.

(2) Initialize counters of the epoch $\mu_{ij} = 0$, (i is the number of the hidden layer; j is the number of the neuron of the hidden layer; $i = \overline{1, n_1}$, $j = \overline{1, n_{2i}}$; n_1 is the number of hidden layers; n_{2i} is the number of neurons in the ith hidden layer) for each hidden neuron h_{ij}. This counter is used to count the number of epochs during which the hidden layer neurons learn until everything is well. The number of epochs is a parameter that is pre-calculated.

(3) Learn each ANN using a two-step optimization algorithm during training for a certain number of training epochs.

(4) Increase the value of era counters for $i = \overline{1, n_1}$; $j = \overline{1, n_{2i}}$;

$$\mu_{ij_k} = \mu_{ij_k} + \tau, \tag{3.7}$$

where k is the epoch number $k = \overline{1, N}$; τ is the epoch number.

(5) Evaluate each ANN according to a predefined fitness function on the test sample, if the stopping criterion is fulfilled, then the evolutionary process is

stopped and proceed to step 6. Otherwise, the evolutionary process is continued and proceed to step 15.

(6) Calculate the error of the ANN on the test sample. If the stop criterion is met, the learning process is stopped and the current network architecture is the final ANN. Otherwise, continue.

(7) Delete the neuron label of each hidden layer, if it exists, and calculate the significance $\eta_{ij}, i = \overline{1, n_1}; j = \overline{1, n_{2i}}$ of each neuron of the hidden layers. For this purpose, for each neuron, it is determined the percentage of data for which this neuron was not activated (the output was close to 0) by the formula n_0/n, where n_0 is the number of data for which the neuron was not activated; n is the total sample volume.

(8) If the significance of one or more neurons of the hidden layers is less than the calculated threshold value η^* (defined by user), label these neurons with S and continue.

(9) Calculate the correlation between each S-labeled hidden neuron and other non-marked neurons of the same hidden layer by examples from the training sample.

(10) Merge each S-labeled hidden neuron with its most correlated non-binary analog from this layer, this new neuron does not contain any label and the algorithm initializes a new counter epoch for it with an initial zero value.

(11) Repeat the training of the modified ANN, which is obtained after merging the hidden neurons, until its previous level of error is reached. If the modified ANN is able to reach its previous error level, continue. Otherwise, restore not modified ANN and to move to p. 13.

(12) Update the era counter for each hidden neuron of the modified ANN and go to step 6. The era counter is updated as follows:

$$\mu_{ijk} = \mu_{ijk} + \tau_r, i = 1, 2, \ldots, N, \tag{3.8}$$

where τ_r is the number of epochs for which the modified ANN is re-trained after the merging operation.

(13) Test the criterion of adding a neuron that monitors progress in reducing learning error. If this criterion is satisfied, continue. Otherwise go to step 3 for further training. It is assumed that since the merging operation is considered unsuccessful (or can not be applied) and the criterion of adding a neuron is not satisfied, the performance (functioning) of the ANN can be improved only by training.

(14) Add one neuron to the current ANN architecture and proceed to step 3. Since the error of the ANN after training is not significantly reduced, and the merging operation is recognized as unsuccessful (or can not be applied), the performance (functioning) of the ANN can be improved by adding hidden neurons h_{ij}. A hidden neuron is added by splitting (splitting) an existing hidden neuron ANN. The split (split) operation produces two new hidden neurons from an existing hidden neuron. The epoch counters are initialized by dividing the μ_i

by two, where μ_i is the number of epochs for which a hidden h_{ij}, neuron is learnt still.

(15) Select ANN for evolution operations execution.

(16) Apply evolutionary operators such as crossover and/or mutation to ANN architectures and weights to produce offspring.

(17) Get a new General sample from the "parents" and "offspring" for the next generation, then go to step 3.

To test the algorithm we use Python language and libraries Keras, Tensorfow. The test task is to recognize images of numbers. The sample is taken from the Mnist database. It has already defined a training sample (60,000 images) and a test sample (10,000 images).

The initial model takes the input image-a matrix of pixels 28 × 28, which is transformed into an auxiliary layer vector with 784 numbers. The neural network is a perceptron with 3 hidden layers of 50 neurons each and an output layer with 10 outputs.

The structure of the model in the program

Layer (type)	Output shape	Param #
flatten_1 (Flatten)	(None, 784)	0
hidden_1 (Dense)	(None, 50)	39,250
hidden_2 (Dense)	(None, 50)	2550
hidden_3 (Dense)	(None, 50)	2550
output (Dense)	(None, 10)	510
Total	params:	44,860
Trainable	params:	44,860
Nontrainable	params:	0

flatten_1 is the auxiliary layer for transformation of a matrix of pixels into a vector of numbers

hidden_1/2/3 is the layer of hidden neurons 44,860

output is the output layer of the perceptron

The 2nd column shows the number of outputs of each layer. For hidden layers this number corresponds to the number of neurons in the layer.

After the first iteration of joins the model structure takes the following form:

Layer (type)	Output shape	Param #
flatten_1_input (Input Layer)	(None, 28, 28)	0
flatten_1 (Flatten)	(None, 784)	0
hidden_1 (Dense)	(None, 43)	33,755
hidden_2 (Dense)	(None, 42)	1848
hidden_3 (Dense)	(None, 39)	1677

(continued)

(continued)

Layer (type)	Output shape	Param #
output (Dense)	(None, 10)	400
Total	params:	37,680
Trainable	params:	37,680
Nontrainable	params:	0

After the first iteration, the hidden layers have 43, 42, and 39 neurons each, respectively. After that it is necessary to train the model again:

Train on 60,000 samples, validate on 10,000 samples
Epoch 1/5
−2s – loss: 0.3937 – acc: 0.9026 – val_loss: 0.2503 – val_acc: 0.9320
Epoch 2/5
−2s – loss: 0.1849 – acc: 0.9477 – val_loss: 0.1915 – val_acc: 0.9480
Epoch 3/5
−2s – loss: 0.1464 – acc: 0.9576 – val_loss: 0.1650 – val_acc: 0.9530
Epoch 4/5
−2s – loss: 0.1274 – acc: 0.9626 – val_loss: 0.1587 – val_acc: 0.9573
Epoch 5/5
−2s – loss: 0.1142 – acc: 0.9659 – val_loss: 0.1595 – val_acc: 0.9550

Model loss after retraining: 0.15952704775482415.

Let's compare the error of the obtained model with the error of the original model (Fig. 3.4).

As you can see, we have received a significant improvement. This abrupt change in error is due to the fact that Keras does not modify the topology of the model in training, so the first iteration of merging additionally rejects all artifacts due to the suboptimal initial topology.

After the fourth iteration, we got a deterioration in the error. Restore the model from the 3rd iteration and stop the program. Structure of the final model:

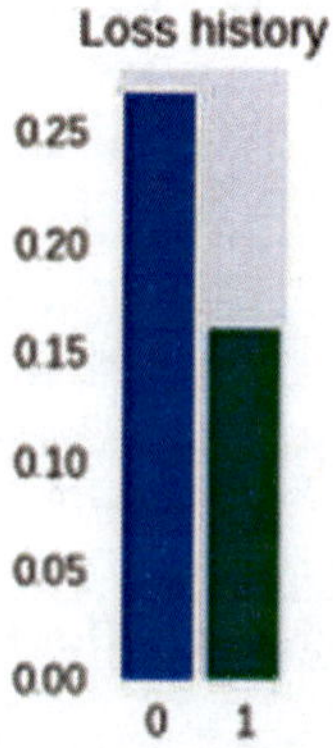

Fig. 3.4 The comparison of the obtained model error with the error of the original model; the blue bar is the error of the initial model, green bar—the error of the resulting model

Final		Model
Layer (type)	Output Shape	Param #
flatten_1_input (InputLayer)	(None, 28, 28)	0
flatten_1 (Flatten)	(None, 784)	0
hidden_1 (Dense)	(None, 26)	20,410
hidden_2 (Dense)	(None, 29)	783
hidden_3 (Dense)	(None, 26)	780
output (Dense)	(None, 10)	270
Total	params:	22,243
Trainable	params:	22,243
Non-trainable	params:	0

The final model has 3 hidden layers with 26, 29 and 26 neurons each. The number of neurons was reduced by 48%, 42% and 48% respectively relative to the original model. The total number of hidden neurons was reduced by 46%.

3.6.4 Two-Level Algorithm of Parameter Synthesis of Base Neuron Network Optimal Modification

As a more optimal algorithm for tuning neural network parameters, it is proposed to use a two-level algorithm of parametric synthesis, that supposes sequential tuning of the neural network using the genetic algorithm on the first stage and traditional neural network optimizers on the second stage [11, 12]. The purpose of such optimization algorithm use is the acceleration of learning process and increase of the neural network functioning accuracy after learning by this algorithm, compared to existed alternatives [13].

The scheme of hybrid algorithm functioning of the neural network tuning consists of the following steps.

(1) Parameters setting for neural network learning. At this stage, it is given the fixed topology of neural network, training sample, conditions of learning stop and tuned a genetic algorithm.
(2) Initial adjustment of network parameters. At this stage, by applying the hybrid genetic algorithm (see Chap. 2), It is happened the search of optimal weight coefficients and biases of neural network which give the highest accuracy and the lowest classification error. The result of performing this stage is a set of weight coefficients and biases.
(3) Further adjustment of neural network parameters. At this stage, it is formed the new neural network model with the initial coefficients obtained at the previous stage. After that, the neural network is learnt by the selected algorithm of gradient optimization up to learning stop condition execution.

The use of a genetic algorithm at the first stage of the neural network optimization will allow to expand the area of possible solutions search and to increase the approaching chances to the point of the global minimum. At the second stage the use of optimization algorithm which is based on the method of steepest descent will permit the guarantee to search the point of global minimum with higher probability [14].

To investigate the efficiency of the algorithm functioning it was carried out a series of experiments with use of different optimizers for neural network parameters tuning for two types of tasks: classification and approximation problems [15].

The first task of the neural networks used in the experiments was the task of clothing images classification. The networks were learned on Zalando's Fashion-MNIST database, which is positioned as the most sophisticated alternative of the common MNIST database for handwriting recognition [16]. This sample consists of 60,000 images for learning and 10,000 images for testing of classification effeciency. The samples consist of black and white images of clothes 28×28 pixels divided into 10 classes (T-shirts, pants, sweaters, dresses, coats, sandals, shirts, sneakers, bags and boots).

The training sample was divided into two parts—training (80%, i.e. 48,000 images) which was used by the optimizers under network parameters adjustment the, and test (20%, i.e. 12,000 images), which was used for network classification accuracy estimation during the learning. Network learning was performed up to classification accuracy achievement of 85%, so the base comparison characteristic of different optimization algorithms use is the number of learning epochs required for given accuracy achievement [17, 18].

For the experiments conducting it was used fully connected feedforward neural networks with a stable architecture consisting of: an input layer that contained 784 neurons, two hidden layers of 512 neurons each, and an output layer of 10 neurons [19, 20].

In two-level optimization algorithm it was used the following genetic algorithm settings: population size—25 individuals, archive size—25 individuals, number of iterations—10, crossing probability—80%, mutation probability—20%. The purposes of the genetic algorithm were neuron network classification error value minimization and classification accuracy value maximization [21].

During the experiments, it was used the following tuning optimizers: for the steepest descent algorithm, Adagrad, RMSProp, and Adam a training coefficient was 0.01, for an accelerated Nesterov gradient used a training coefficient of 0.01 and a impulse coefficient of 0.9, for Adadelta optimizer used a training coefficient of 1. The values of coefficients were chosen experimentally with the purpose of algorithm functioning efficiency increasment [22].

For learning results comparison execution using each optimization algorithm, the experiments were performed ten times, after which the average number of learning epochs required to achieve 85% accuracy was calculated. The comparison of optimization algorithms use with two-level algorithm is given in Table 3.15.

Table 3.15 Number of epochs to achieve 85% accuracy

Optimization algorithm	Number of epochs	
	Under tuning by only the optimizer	Under tuning by two-level algorithm
The steepest descent	3666	2876
Accelerated Nesterov gradient	1732.5	1425.5
Adagrad	1257	1037
RMSProp	309	295.2
Adadelta	203.2	183.4
Adam	52.7	48.7

Table 3.16 Comparison of two-level algorithm optimizers

Optimization algorithm	Performance increase of two-level algorithm (%)	
	Compared to the steepest descent algorithm	Compared to a one-level optimization algorithm
The steepest descent	–	27.47
Accelerated Nesterov gradient	157.17	21.54
Adagrad	253.52	21.21
RMSProp	1141.87	4.68
Adadelta	1898.91	10.79
Adam	7427.72	8.21

In the Table 3.16 it is given the percentages of efficiency increase of two-level algorithm function efficiency in comparison with optimization algorithms and with gradient descent in particular.

In Table 3.17 it is given the comparison of the application of the two-level optimization algorithm with the steepest descent algorithm at the second stage and two-level optimization algorithms with others (Accelerated Nesterov Gradient, Adagrad,

Table 3.17 Comparison of two-level algorithms with different optimizers

Optimization algorithm in the second stage	Acceleration of work in comparison with two level steepest descent algorithm	
	In the epochs	Percentage (%)
Accelerated Nesterov gradient	1450.5	101.75
Adagrad	1839	177.34
RMSProp	2580.8	874.25
Adadelta	2692.6	1468.16
Adam	2827.3	5805.54

RMSProp, Adadelta, Adam) in the second stage of neural network setup. In table it is represented the comparison of number of epochs would need to achieve 85% classification accuracy and the percentage value of algorithms acceleration.

Thus, the use of a two-level algorithm for neural network parameters optimization allowed to increase the efficiency of learning in comparison with the use of only singular optimization algorithms. Such effect of efficiency increasment is noticeable even under using a small number genetic algorithm iterations at the first stage of the two-level algorithm [23].

The second task under two-level algorithm functioning research was the task of the building cost function approximation. As a data set, it was used the information about buildings in agglomeration around Boston and the surrounding area for the 1970. The sample consists of 506 records, of which 404 (80%) were used for learning and 102 (20%) were used for test of approximation accuracy. Each record characterizes the area or surrounding of Boston and the buildings within it. The sample provides 14 parameters for each record, which includes [24]:

- crime rate in the city per capita;
- percentage of residential land allocated for plots over 25,000 square feet;
- percentage of land for business not related to retail;
- indicator of the boundary of the district or city with the Charles River;
- the concentration of nitrogen oxides in ten millionths of a fraction;
- average number of rooms in buildings;
- percentage of buildings built before 1940;
- weighted average distance to five job centers in Boston;
- highway accessibility index;
- real estate tax at the cost of the building at $10,000;
- the ratio of the number of students to the number of teachers;
- percentage of African-American population in the city or area;
- percentage of the population belonging to the working class;
- the average cost of a residential building, measured in thousands of dollars.

The task of neural networks was the approximation of the building cost (the last parameter in the sample) based on the remaining thirteen parameters. For neural networks functioning efficiency estimation was used the value of mean square error and mean absolute error. For prediction realization a set of 12 parameters is applied to the input of a neural network and one output value is expected that corresponds to the predicted cost of the building.

The learning of artificial neural networks was carried out before reaching the value of the mean absolute error $ME = 3$, which corresponds to the deviation of the predicted value of the building cost from the actual $3000. To compare the efficiency of the optimizers functioning, it is used the value of epochs number required for given error achievement error.

For experiments conduction it is used a fully connected feedforward network with a stable architecture consisting of: an input layer containing 13 neurons, one hidden layer with 16 neurons, and an output layer with 1 neuron.

In two-level optimization algorithm it was used the following genetic algorithm settings: population size—25 individuals, archive size—25 individuals, number of iterations—10, crossing probability—80%, mutation probability—20%. The purpose of the genetic algorithm was the value minimization of the mean square error and the mean absolute error.

For comparison realization the learning with each algorithm use was realized 10 times, after that it was calculated the average number of learning epochs required for desired accuracy achievement. The comparison of the optimization algorithms use with the two-level algorithm is given in Table 3.18.

During the experiments, the following optimizer settings were used: for the steepest descent algorithm it was used a training coefficient of 0.01, the accelerated Nesterov Gradient used a training coefficient of 0.01 and a impulse coefficient of 0.5, for Adagrad, RMSProp, and Adam it is used the coefficients of 0.1, for Adadelta was used value 1. The values of the coefficients were selected experimentally in order to improve the efficiency of the algorithms.

In the Table 3.19 it is shown the percentages of function efficiency increasment of two-level algorithm in comparison with optimization algorithm and with steepest descent method. The negative values mean a decrease in the speed of training.

Table 3.18 The number of epochs required to reach the error of $3000

Optimization algorithm	Number of epochs to reach the specified error	
	Under tuning by only optimizer	Under tuning by two-level algorithm
The steepest descent	59.5	42.8
Accelerated Nesterov gradient	33.1	23.4
Adagrad	149.3	141.9
RMSProp	110.2	93.8
Adadelta	663.8	615
Adam	53.2	46.6

Table 3.19 Comparison of two-level algorithm optimizers

Optimization algorithm applied in the second stage	Efficiency increase of two-level algorithm (%)	
	Compared to the steepest descent algorithm	Compared to a one-level optimization algorithm
The steepest descent	–	39.01
Accelerated Nesterov gradient	154.27	41.45
Adagrad	−58.07	5.21
RMSProp	−36.57	17.48
Adadelta	−90.33	7.93
Adam	27.68	14.16

The best result from all the optimizers was Nesterov's Accelerated Gradient, and the use of Adagrad, RMSProp and Adadelta optimization algorithms reduced the learning speed in comparison with the steepest descent algorithm use. However, two-level algorithm use for neural network parameters optimization was permitted to increase the efficiency of training in comparison with only optimization algorithms use, including the use of Nesterov Accelerated Gradient.

Thus, the efficiency of using a two-level it was demonstrated the efficiency of neural network adjustment under classification and approximation tasks solution. The use of a two-level algorithm in conjunction with any of the researched optimizers allowed to increase the learning of the neural network and to achieve faster the desired prediction accuracy.

In case if the modification of ANN topology which is chosen for given task solution does not give good results, it is necessary asynthesis of new topology. In this work for new topology synthesis task solution it is proposed a module principle of hybrid neural networks organization.

If the modification of the ANN topology, which is chosen to solve the problem does not give tangible results, the synthesis of a new topology is necessary. In this paper, a modular principle of hybrid neural networks (HNN) organization is proposed to solve the problem of synthesis of a new topology.

3.7 Results of Neural Network Learning Using SWARM

3.7.1 Intelligence Algorithms

Section 2.4 deals with the classification, construction and analysis of genetic algorithms, and Sect. 3.6.3 addresses the adaptation and use of genetic algorithms for training neural networks.

Sections 2.5, 2.6, 2.7 and 2.8 deal with the classification, construction, and analysis of swarm algorithms.

Section 2.9 discusses the classification, construction, and analysis of modern gradient methods, which, as part of hybrid algorithms, together with genetic algorithms, are an effective tool for training neural networks (see Sect. 3.6.4).

In recent years, a large number of works [25] have been devoted to the use of swarm algorithms for training NN, however, it should be noted that these are mainly articles devoted to the particle swarm algorithm [26, 27].

Compared to the particle swarm method (PSM), genetic algorithms, for example, require a large number of calculations for low-efficient solutions, as a result of which the solution search process is slow [28], and in some cases they may have lower accuracy than the MPP [1]. Also, in genetic algorithms, the amount of memory required is proportional to the size of the population; therefore, in the case of ANN, in which the number of weights and threshold values is large, the population size is limited, which reduces the efficiency of the algorithm [2]. Unlike the GA, the PSO algorithm has no complicated evolutionary operators such as crossover and mutation [29–33].

In relation to neural networks, particle movement occurs in the hyperspace of network weights, each i-th particle is essentially a separate possible configuration of ANN, whose position is determined by the vector of values of weights $(w_1, w_2, \ldots, w_D)$, where D is the number of ANN weights. During training—minimizing the error function—these various configurations move in the search space, and, influencing each other, seek to find the optimal position.

The search for the optimum in parallel in several places can significantly reduce the likelihood of a jam in the local optimum. This type of behavior seems ideal when exploring large search spaces, especially with a relatively large maximum rate parameter. Some particles will explore far beyond the current minimum, while the population still remembers the global best. This seems to solve one of the problems of gradient-based algorithms.

In [34], the particle swarm method was compared in the ability to train ANN with an algorithm that is a combination of the back propagation method of the error and the genetic algorithm (GA was used to select the optimal parameters of the back propagation method of error). Comparison of the results was carried out for several classification problems, the PSM showed approximately the same accuracy as the combined method ($\pm 10\%$ in different experiments), but at the same time spent on training from 1.5 to 3 times less time.

Several methods were used in [35]: the PSM, the back-propagation error method, and the evolutionary algorithm were used to train neural networks for solving various types of classification and approximation problems. According to the results for classification problems, the particle swarm method found on average 2 times greater accuracy of finding the minimum of errors than the evolutionary algorithm, and 4 times more accuracy than the method of back propagation of errors. For approximation problems, the PSM showed the accuracy of finding the minimum on average 10% better than the evolutionary algorithm, and 16% better than the backpropagation method.

In [36], the efficiency of the PSM, the error back propagation method, and the hybrid combination of these two methods were compared: the use of the PSM for global search, with the switch to the back propagation method for local search [37], in areas where the global optimum is supposedly located. The training was carried out for a neural network with one hidden layer, and the number of hidden layer neurons varied from 4 to 16 for each of three tasks: two for classification and one for approximation.

Various modifications of the method are also possible, for example, the introduction of the inertia factor, which makes it possible to slow down the particle motion over time in order to more accurately find the optimum. It was shown in [4] that, for the particle swarm method, there is a range of values of the algorithm parameters such that its convergence is ensured, while relations are determined that allow the boundaries of this region to be calculated. It was shown in [28] that simple modifications of the canonical PSM provide convergence to a local minimum, and in [38] a modification of the PSM is presented for which convergence to a global minimum is proved.

Swarm Intelligent Algorithms have a disadvantage of easily getting into a local optimum, because they have several parameters to be adjusted [39, 40]. If these parameters are not appropriately set, the search will become very slow near the global optimum.

After suitable adjustment of the parameters for the SA, the rate of convergence can be speeded up and the ability to find the global optimum can be enhanced. It can be achieved by the construction of hybrid swarm algorithm (see Sect. 2.7).

3.7.2 Results of Intelligence Algorithms Use NN Learning

Let us consider the use of various swarm and hybrid swarm algorithms developed by the authors and presented in Sect. 3.2 for training neural networks.

Neural network architecture

Number of hidden layers—3, each having 5 neurons. The activation function for the neurons is hyperbolic tangent.

Datasets used for testing

The neural network was used for classification problems on the following datasets:

(1) *Iris dataset*

- classes: 3;
- samples per class: 50;
- samples total: 150;
- dimensionality: 4;
- features: real, positive.

(2) *Breast cancer dataset*

- classes: 2;
- samples per class: 212(M), 357(B);
- samples total: 569;
- dimensionality: 30;
- features: real, positive.

(3) *Wine dataset*

- classes: 3;
- samples per class: [41];
- samples total: 178;
- dimensionality: 13;
- features: real, positive.

Table 3.20 Iris dataset results

Algorithm	Training set score (%)	Test set score (%)	Avg. time (s)
PSO	94.025	91.033	1.64518
Cuckoo search	90.958	89.833	2.17550
Cooperative	95.192	93.033	8.54292
SGD	82.942	79.600	0.79717
LBFGS	97.233	93.533	0.22204
Adam	97.125	95.400	0.60204

Optimizers tested

The swarm optimizers used for training: Particle Swarm Optimization, Cuckoo Search, Cooperative Algorithm (PSO, Cuckoo Search, Firefly Search) [42, 43].

The other optimizers for comparing the results: Stochastic Gradient Descent, LBFGS, Adam.

Parameters for PSO: $N = 40$, [0.5, 1.5, 1.5];

Parameters for Cuckoo Search: N = 40, [0.1, 1];

Parameters for Cooperative algorithm:

(1) PSO: $N = 30$, (0.3, 3, 0.8);
(2) Cuckoo Search: $N = 30$, (0.1, 1);
(3) Firefly Algorithm: $N = 30$, (2, 0.3, 1, 3, 1.1);
(4) alpha = 0.7;
(5) steps = 2.

Parameters for the other optimizers are set to default for scikit-learn implementation:

https://scikit-learn.org/stable/modules/generated/sklearn.neural_network.MLPClassifier.html.

Termination criterion for the swarm algorithms (except for cooperative): maximum number of generations—200, solution stagnation generations—100; for cooperative: maximum number of generations—50.

Maximum number of iterations/epochs for other algorithms—2500.

The results were obtained by performing 100 different training runs (Tables 3.20, 3.21 and 3.22).

3.7.3 Meta-Algorithm for Neural Network Training Results

Meta-algorithm description

Meta-optimizer: PSO.

Base algorithm: PSO.

Table 3.21 Wine dataset results

Algorithm	Training set score (%)	Test set score (%)	Avg. time (s)
PSO	62.296	60.389	1.80075
Cuckoo search	58.986	58.222	3.03049
Cooperative	66.148	64.333	14.33499
SGD	41.894	41.056	0.08989
LBFGS	41.147	39.917	0.02272
Adam	54.768	52.194	0.22785

Table 3.22 Cancer dataset results

Algorithm	Training set score (%)	Test set score (%)	Avg. time (s)
PSO	91.503	89.631	3.01254
Cuckoo search	90.890	90.167	5.56119
Cooperative	91.662	90.105	27.14380
SGD	63.556	63.859	0.12833
LBFGS	68.255	68.167	0.10831
Adam	77.369	76.667	0.64681

Parameters used for tests

(1) *Iris dataset*:

- meta-optimizer's behavioral parameters: (0; 1.5; 1.5);
- meta-optimizer's population size: 40;
- base algorithm population size: 40;
- meta-optimizer's termination criterion: maximum number of iterations is 20;
- base algorithm termination criterion: maximum number of iterations is 50 (same for the PSO without meta-optimization).

(2) *Wine dataset*:

- meta-optimizer's behavioral parameters: (0; 1.5; 1.5);
- meta-optimizer's population size: 40;
- base algorithm population size: 40;
- meta-optimizer's termination criterion: maximum number of iterations is 20;
- base algorithm termination criterion: maximum number of iterations is 50 (same for the PSO without meta-optimization).

(3) *Cancer dataset*:

- meta-optimizer's behavioral parameters: (0.5; 1.5; 1.5);
- meta-optimizer's population size: 40;
- base algorithm population size: 40;
- meta-optimizer's termination criterion: maximum number of iterations is 20;

Table 3.23 Iris dataset results

Algorithm	Training set score (%)	Test set score (%)	Avg. time (s)
PSO	97.575	95.567	1.02969
Meta-algorithm	99.042	95.833	418.79328

Table 3.24 Wine dataset results

Algorithm	Training set score (%)	Test set score (%)	Avg. time (s)
PSO	69.894	64.306	1.30222
Meta-algorithm	81.866	78.472	552.10273

Table 3.25 Cancer dataset results

Algorithm	Training set score (%)	Test set score (%)	Avg. time (s)
PSO	90.956	88.860	0.91160
Meta-algorithm	92.18	90.570	377.80936

- base algorithm termination criterion: maximum number of iterations is 20 (same for the PSO without meta-optimization).

Results (20 optimization runs for each dataset) (Tables 3.23, 3.24 and 3.25).

3.8 Method of Structural-Parametric Synthesis of the Module of Hybrid Neural Events

3.8.1 Determination of the Module Structure of Hybrid Neural Networks

The module topology involves the sequential combination of several different neural network architectures. In general, the module operates in the same way as an individual NN. Its advantage is the combination of various data transformations, which allows to obtain more accurate results. The module topology is presented in Fig. 3.5.

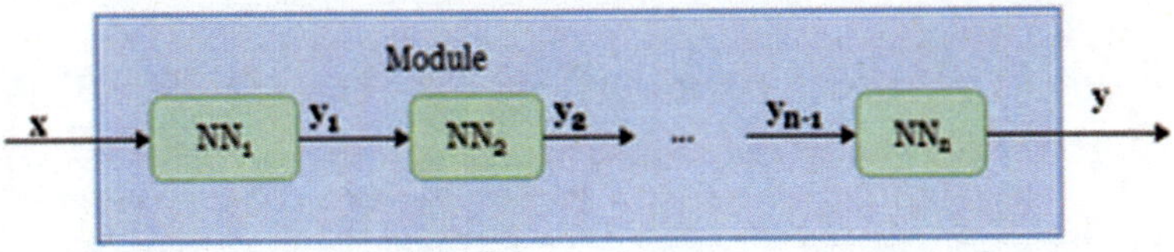

Fig. 3.5 Module topology

Modern technologies allow to operate with huge networks as with simple elements for building something larger, like LEGO bricks. The abstraction level of modern software is very high, everything is at an intuitive and understandable level, so the practical implementation of NN modules is a fairly easy task.

It is proposed to choose a modified base neural network as the main neural network that is part of the module (see Sects. 3.1–3.6). Thus, it is necessary to determine the topology, number and location of other neural networks besides base one, which are part of the module.

The main problem of modules composition is their learning algorithm. Inside the module there are several neural networks, therefore, there are different ways of module learning:

- all networks are trained together;
- some networks are trained together, some are trained separately;
- each network is trained separately on the training sets; then they are combined into a module.

Three variants of neural networks module are possible:

- neural network that is learnt without a teacher (Kohonen network)—a basic neural network—base neural network;
- base neural network—GMDH (group method of data handling)—neural network;
- neural network that is trained without a teacher (Kohonen network)—base neural network—GMDH—neural network.

3.8.2 *Module Based on Kohonen Network and Base Neural Network*

The block-diagram of such hybrid network is shown in Fig. 3.6.

Since the Kohonen neural network is trained without a teacher, the learning algorithm of such module has the following form [44]:

1. The Kohonen neural network is trained by input.
2. For the reference inputs of the trained Kohonen NN, its outputs, which are the reference inputs for the base neural network, are determined.
3. Base NN training with the teacher is performed.

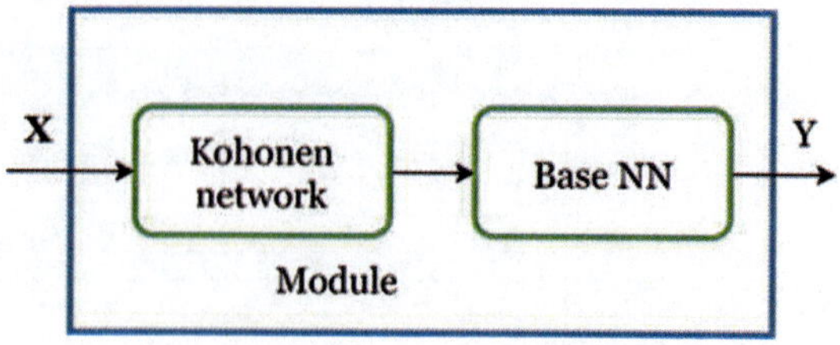

Fig. 3.6 Kohonen + base NN module topology

At the same time it is necessary to take into account the peculiarities of the Kohonen NN, which has two modes of operation: accreditation and interpolation.

There is a situation when several times the same neuron becomes the “winner”. As a result, only one neuron in the next layer belonging to another network as a part of hybrid NN can be activated. This mode of operation is called the accreditation mode. It has limited accuracy, because it is actually the result of only one neuron functioning.

The most common methods for solving this problem are [45]:

1. Feelings of justice: in order to prevent lack of training with one of the neurons, a “sense of justice” is introduced. If the neuron most often wins the “competition”, that is, receives maximum output more often than in one of the M cases, then its output value is artificially reduced to allow other neurons to win. This includes all the network neurons in the learning process.
2. Weights correction is proportional to the output: in this modification, the weights of not only the “winner” neuron are corrected, but also of all others. The normalization is performed by the maximum value of the layer output or by its average value. This method also eliminates “dead” neurons and improves weights density distribution.

This mode is called interpolation mode. The number of neurons that become “winners” neurons should be selected depending on the task, provided that the optimal value is in the range from 2 to 1/2 of the number of neurons in the Kohonen layer. This method is more accurate than the accreditation method because it is capable of establishing more sophisticated matches.

As a result, we get that choosing not one neuron-“winner”, but a group of neurons-“winners”, it is increased the accuracy of the hybrid neural network [28].

The number of neurons-“winners” depends on the task.

The algorithm of neurons-“winners” choice:

1. It is set a certain value C (count), that will be equal to the required number of neurons-“winners” for the given task.
2. We calculate the value of the output signals for the Kohonen layer neurons by the formula

$$\mathrm{OUT}_k = \sum_{n}^{n=1\ldots M} w_{nk} * x_n, \tag{3.9}$$

where OUT_k is the k-neuron output of the Kohonen layer; $x_1 \ldots x_n$ are signals of input vector $\mathbf{X}$.

Let us move on to the key feature and, at the same time, the problem of the Kohonen network—the choice probability of the same neuron as the neuron-“winner”, that will lead to the primitive disregard for other neurons learning. The method of this problem solution is the method of “justice”. In this case, it is optimal precisely because of its implementation simplicity and, as a result, homogeneous training of all neurons in the Kohonen layer. It is given by the formula

$$pass = \frac{1}{M}, \tag{3.10}$$

where M is the number of iterations, for which the neuron can only be "winner" once, *pass* is the value of change in the weight of the neural network.

The rule of this method use is: "For a neuron that has been awarded" winner "status more than once in M iterations, the weight is decreased, but the activation of the neuron is not canceled." Thus, for the neuron, that became a winner for the second time in a certain number of iterations, it is decreased the weight, thus it is reduced its "rank", that allows to train other neurons.

$$w_{k(i+1)} = w_{k(i)} - pass, \tag{3.11}$$

where $w_{k(i+1)}$ is the new weight of the neuron-"winner"; $w_{k(i)}$ is the old weight of the neuron-"winner"; i is the iteration number. With this method, other neurons on the background of such neuron-"winner" are "risen" in their "rank".

From the above we see, that it is not advisable to use a low value of M, because then the weight of the neuron will be severely reduced, which will sharply reduce its potential to become a neuron-"winner" in further network learning.

The NN module with such a topology is mainly used to solve the classification problem. The Kohonen network in this case will play the role of an initial classifier, and also will reduce the vector dimension of the input data of base network, which will refine and correctly distribute the results of the Kohonen network.

This module version can also be used to solve the approximation problem. In this case, the quality of the Kohonen network as a classifier is lost, but it will reduce the vector dimension of the input data, for example, as Principal Component Analysis (PCA). Reducing the dimension of the data will permit to simplify the architecture of the subsequent base network and, accordingly, to reduce the training time. But there is a problem of output neurons number determination in the Kohonen network. In case of the classification problem, it is equal to the number of classes; for the approximation problem, this number is determined empirically.

Let us consider various variants for the formation of such module topology.

To compare and evaluate the effectiveness of the presented modules, for example, the classification problem and the approximation problem were solved. The solution of the classification problem by this version of the module was described in [46], therefore, here we present only the obtained results.

As a dataset for approximation, the data of building prices in Boston were used. The data has 13 factors and 1 output—the price of the building is a real number, 506 copies. 25% of the dataset was used on a test set, the rest on a training set.

Perceptron

The module topology is shown in Fig. 3.7.

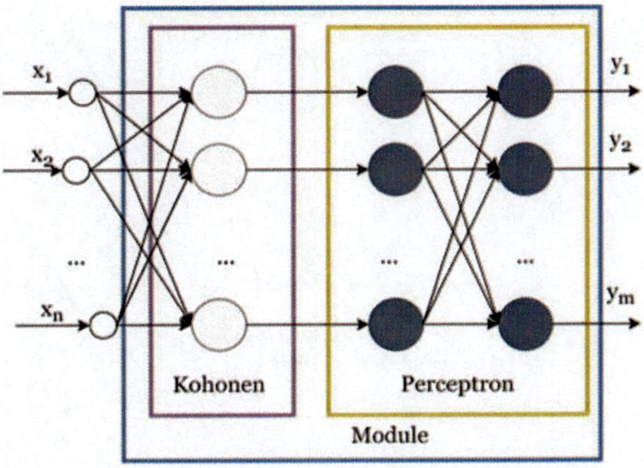

Fig. 3.7 Kohonen network + perceptron module topology

To solve the approximation problem: the number of output neurons of the Kohonen network was taken equal to 5, the number of neurons of the hidden perceptron layer was 10, the activation functions of ReLU and one output neuron.

Counter Propagation Network (CPN)

The module topology is shown in Fig. 3.8.

To solve the approximation problem: the number of output neurons in the Kohonen network was taken equal to 5, the number of internal neurons in the network of counterpropagation is 15.

Radial Base Function Network (RBFN)

The module topology is shown in Fig. 3.9.

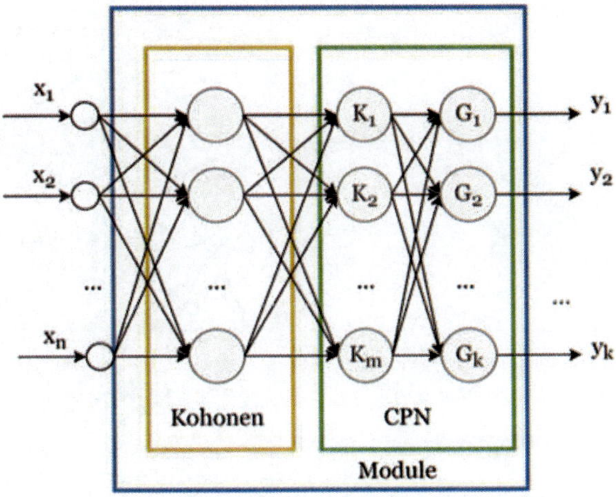

Fig. 3.8 Kohonen network + CPN module topology

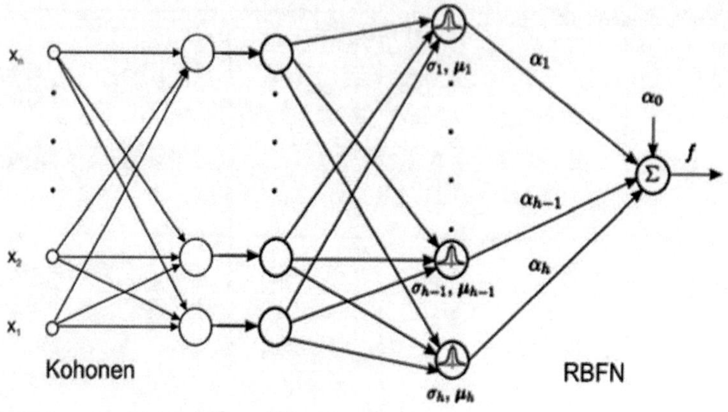

Fig. 3.9 Kohonen network + RBFN module topology

Probabilistic Neural Network (PNN)

The module topology is shown in Fig. 3.10.

NEF ClassM

The module topology is shown in Fig. 3.11.

Modified Network TSK

The module topology is shown in Fig. 3.12.

To solve the approximation problem: the number of output neurons of the Kohonen network was taken equal to 5, the number of inputs into neurons TSK—2, the number of inputs into neurons of the polynomial part—2, the degree of polynomials—2.

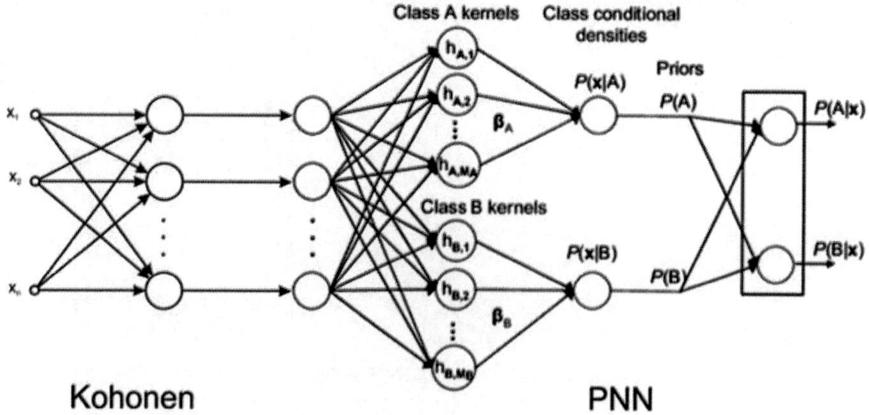

Fig. 3.10 Kohonen network + PNN module topology

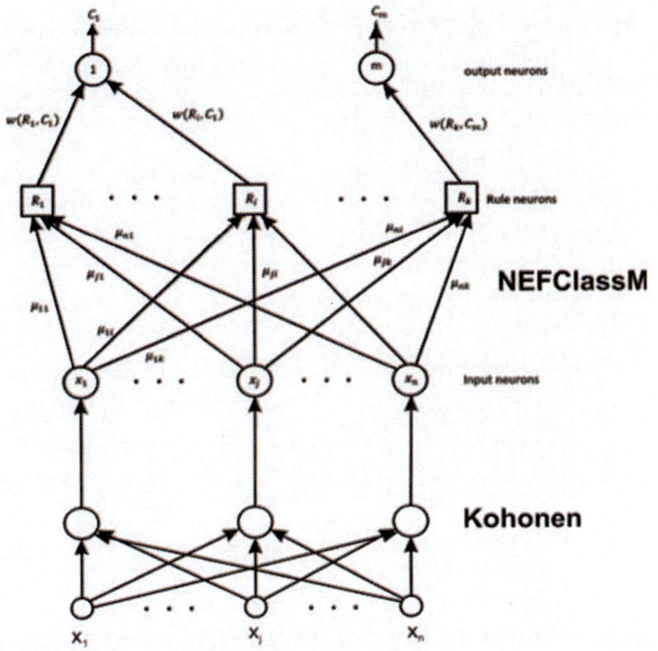

Fig. 3.11 Kohonen network + NefClassM module topology

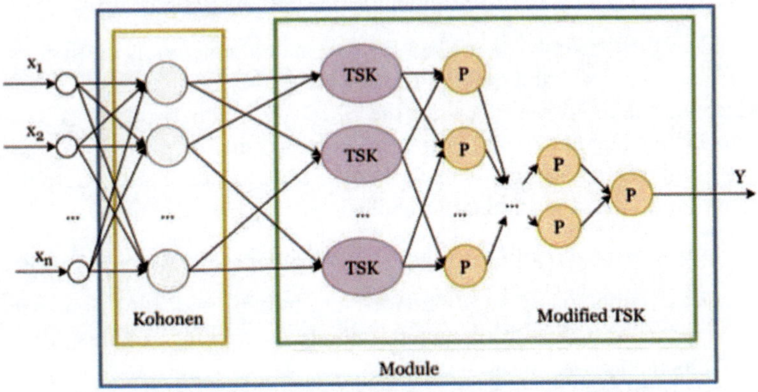

Fig. 3.12 Kohonen network + Modified TSK module topology

The results of the classification problem solution with the module topology are presented in Table 3.26. The percentage of correctly classified samples as a percentage was used as an estimate.

The obtained results of the approximation problem solution with the module topology are presented in Table 3.27. As an estimate of the approximation error, we used mean absolute percentage error (MAPE).

Table 3.26 Percentage of correctly classified samples of individual networks and module

Base NN	Without Kohonen		With Kohonen	
	Train (%)	Test (%)	Train (%)	Test (%)
Perceptron	33.8	22.22	95.77	100
RBFN	38.73	38.88	96.47	100
CPN	70.42	72.22	85.9	94.44
PNN	88.02	75	95	97.22
NEFClassM	80.28	80.55	94.36	100

Table 3.27 MAPE when using separate networks and the Kohonen network + base network module

Base NN	Without Kohonen		With Kohonen (Module)	
	Train (%)	Test (%)	Train (%)	Test (%)
Perceptron (10)	23.11	24.92	24.2	23.87
CPN	29.07	28.55	86.38	89.17
Modified TSK	15.3	19.6	17.94	20.43

3.8.3 *Module Based on Base and GMDH-Neural Networks*

The module topology of such a configuration is shown in Fig. 3.13.

The learning algorithm of such a module has the following form:

(1) Based on the existing training set, the base neural network is trained.
(2) For the reference inputs of the trained base NN, its outputs are determined, which are the reference inputs for the GMDH neural network.
(3) The GMDH neural network is trained with a teacher on the basis of the generated input training sample (step 2) and the corresponding initial output sample according to the algorithm described below.

The learning algorithm of GMDH neural network has the following form:

(1) The entire sample is randomly divided into training and validation in the ratio of 0.7–0.3 (i.e. 70% of the examples fall into the training sample, 30%—in the validation sample).
(2) It is learnt C_k^2 models of follow structure

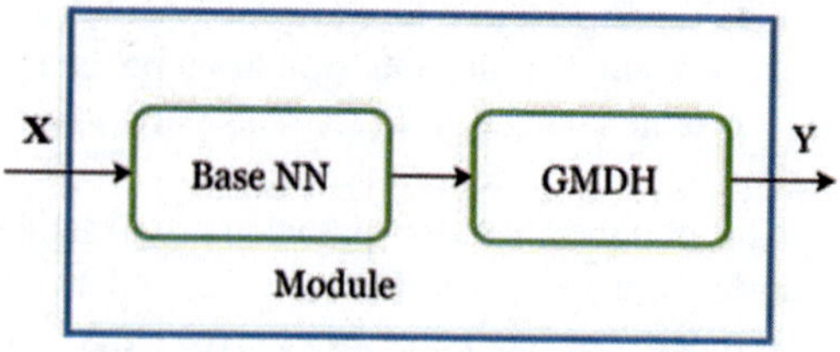

Fig. 3.13 Base network + GMDH module topology

$$f\left(x_i, x_j\right) = a_{i1}x_i + a_{j1}x_j + a_{ij}x_ix_j + a_{i2}x_i^2 + a_{j2}x_j^2, \quad (3.12)$$

using the training sample.

(3) For each model f its validation error is calculated

$$E(f) = \sum_{(\vec{x},y)\in(X_v,\vec{y}_v)} (f(\vec{x}) - y))^2, \quad (3.13)$$

(4) The models s_l with the smallest error are selected, where $(l - s)$ is the number of the series (often, for simplicity of algorithm implementation $s_l = k$ is chosen).
(5) The outputs of these models for each sample training sample form a new input matrix $X^{(1)}$ for the next row (layer) of models:

$$X^{(1)} = \begin{pmatrix} f_1(\vec{d}_1) & \dots & f_{sl}(\vec{d}_1) \\ \vdots & \ddots & \vdots \\ f_1(\vec{d}_{N-k}) & \dots & f_{sl}(\vec{d}_{N-k}) \end{pmatrix}, \quad (3.14)$$

where $\vec{d}_p = [d_p, d_{p+1}, \ldots, d_{p+k-1}]^T$, and the record $f_m\left(\vec{d}_p\right)$ means that only elements under the indexes i and j, that match the model f_m will be taken from the vector $\vec{d}_m$;

(6) Among the errors in the validation sample of all models of the current row, the minimum one is selected, and if it is less than the minimum error of the models from the previous row (or if it is the first row), it is executed the transition on the next row (beginning from step 2), otherwise—the algorithm stops, and the best model of the previous row is selected as "final"; thus, the stopping criterion of the GMDH algorithm can be written as:

$$\begin{cases} l > 1, \\ \min\limits_{f\in F_l} E(f) \geq \min\limits_{f\in F_{l-1}} E(f), \end{cases} \quad (3.15)$$

where l is the number of current row; F_l is the set of all models of current row.
(7) At the output of this stage, we obtain the GMDH neural network of the following form (Fig. 3.14).

Since GMDH is at the end, this module can only be used to solve the approximation or prediction problems, although the general architecture and principle of operation can be applied with other neural networks to solve the classification problem. In the topology of this module, the output neurons of the base network are the inputs of the GMDH network. However, a similar problem arises here, as with the Kohonen network. The number of output neurons of the core network for the approximation problem is unknown, therefore, it should be determined empirically.

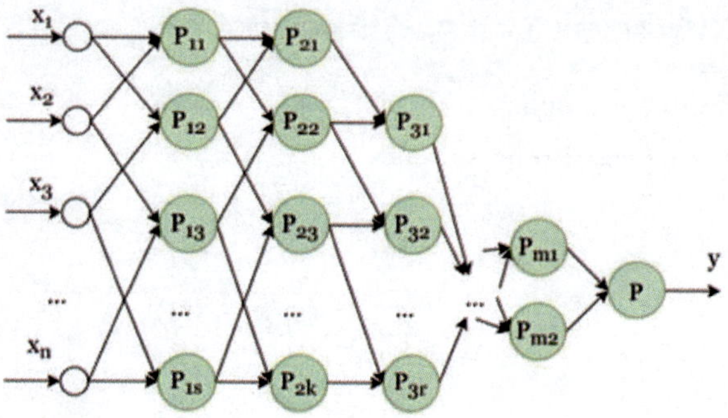

Fig. 3.14 Polynomial neural network

Let's consider different variants of a module topology construction: base neural network and GMDH network.

Perceptron

The module topology is shown in Fig. 3.15.

In this case it is considered the number of perceptron hidden layer neurons 20, ReLU activation functions and 5 output neurons. The number of inputs for GMDH neurons is 2. The degree of the polynomial is 2.

Counter Propagation Network

The module topology is shown in Fig. 3.16.

The number of internal neurons of the counter propagation network is 15, and the weekend is 5.

The number of inputs for GMDH neurons is 2, the degree of polynomial is 2.

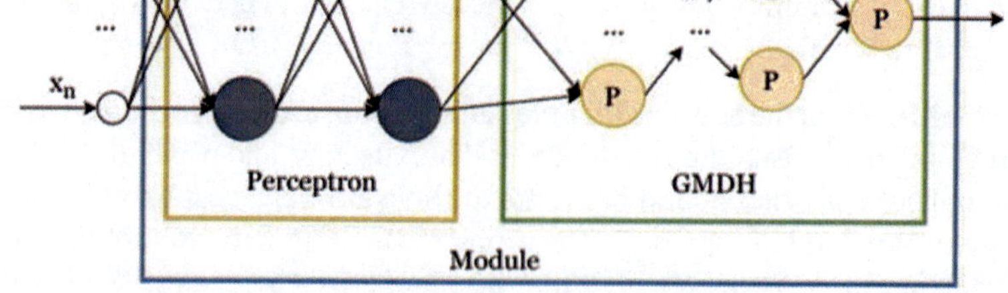

Fig. 3.15 Perceptron + GMDH module topology

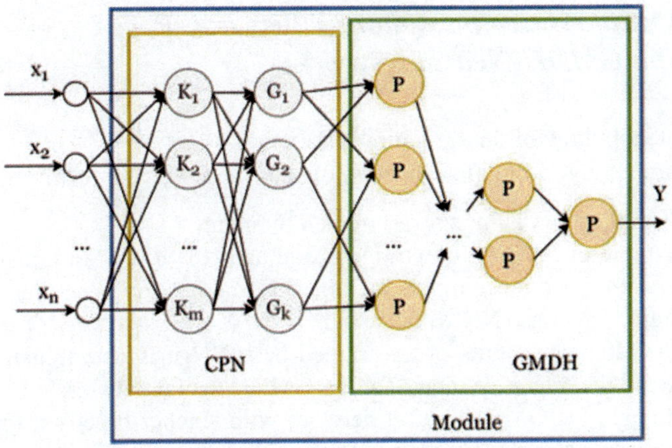

Fig. 3.16 CPN + GMDH module topology

Modified Network TSK

The module topology is shown in Fig. 3.17.

The number of inputs for the neurons of TSK is 2. The number of inputs for the neurons of the polynomial part is 2, the degree of polynomials is 2. The number of inputs for GMDH neurons is 2, the degree of polynomial is 2.

The obtained results of solving the approximation problem with the module topology are presented in Table 3.27. As an estimate of the approximation error, we used mean absolute percentage error (MAPE).

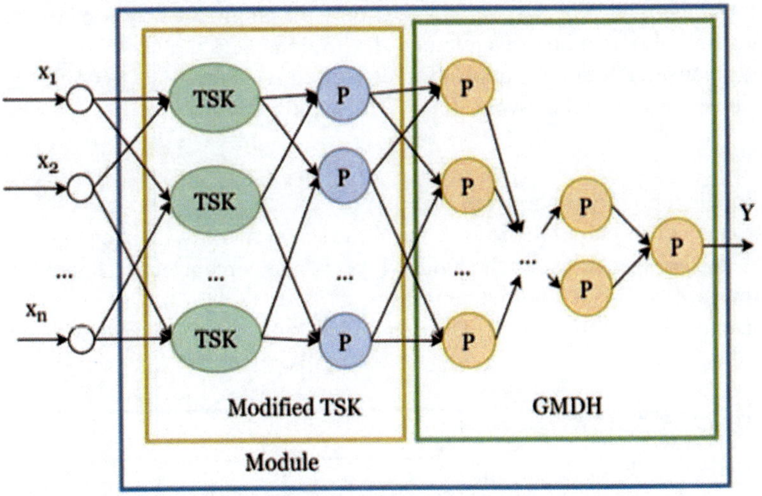

Fig. 3.17 Modified TSK + GMDH module topology

3.8.4 Module Based on Kohonen Network, Base and GMDH-Neural Networks

The module topology of this structure is shown in Fig. 3.18.

The learning algorithm of such a module has the following form:

(1) The Kohonen neural network is trained by input.
(2) For the reference inputs of a trained Kohonen NN, it is determined its outputs, which are the reference inputs for a base neural network.
(3) It is trained the base NN with teacher.
(4) For the reference inputs of the trained base NN it is determined its outputs, which are the reference inputs for the GMDH neural network.
(5) It is trained the GMDH neural network with teacher based on the generated input training sample (step 4) and the corresponding to it initial output sample.

This variant of the module can be considered as an upgraded variant of the second module type. As stated earlier, the module with the topology of the base neural network + GMDH neural network can be used to solve the approximation problem, therefore, the first NN (Kohonen) plays the role of a dimensionality reducer of the input data incoming to the base NN. This will simplify the structure of the base NN and the GMDH neural network, that will have a positive effect for the training time. But here the two problems described above are already connected: the determination of the number of output neurons of the Kohonen network and the number of output neurons of the base network. Different algorithms can be used to determine these values, for example, genetic, but this will greatly increase the complexity of the solution and the learning time. If you have some ideas of the dataset that is being worked on, it's possible to make some assumptions about the possible number of clusters in the data. The last variant is simply an empirical way to select the best combination of output neurons number of neural networks.

Let's consider various options for constructing a topology module: network Kohonen + base neural network + GMDH neural network.

Perceptron

The module topology is shown in Fig. 3.19.

The number of output neurons of the Kohonen network was taken equal to 5.

The number of neurons in the hidden layer of the perceptron is 10, the activation functions of ReLU and 5 output neurons.

The number of inputs for GMDH neurons is 2, the degree of polynomial is 2.

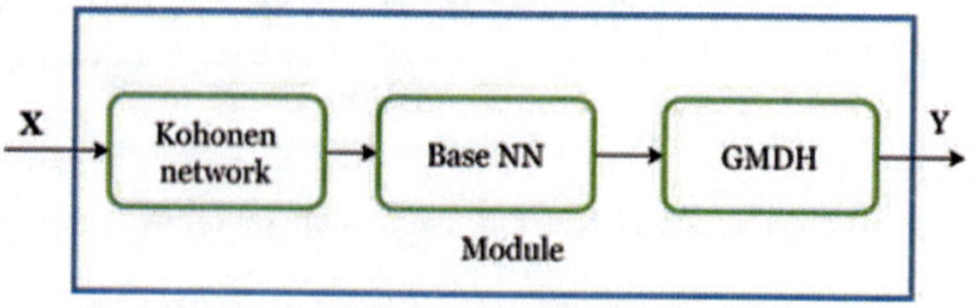

Fig. 3.18 Kohonen network + base NN + GMDH module topology

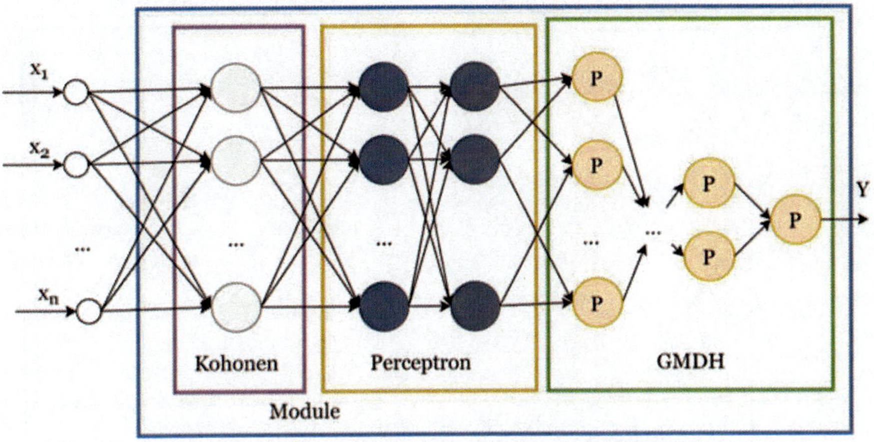

Fig. 3.19 Kohonen network + perceptron + GMDH module topology

Counter Propagation Network

The module topology is shown in Fig. 3.20.

The number of output neurons of the Kohonen network was taken equal to 5.

The number of internal neurons in the network of counter propagation is 10, and the weekend is 5.

The number of entries into GMDH neurons is 2, the degree of polynomial is 2.

Modified Network TSK

The module topology is shown in Fig. 3.21.

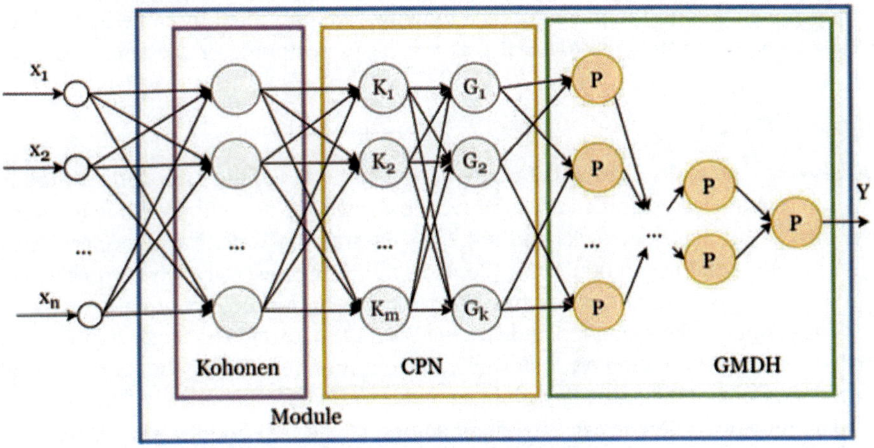

Fig. 3.20 Kohonen network + CPN + GMDH module topology

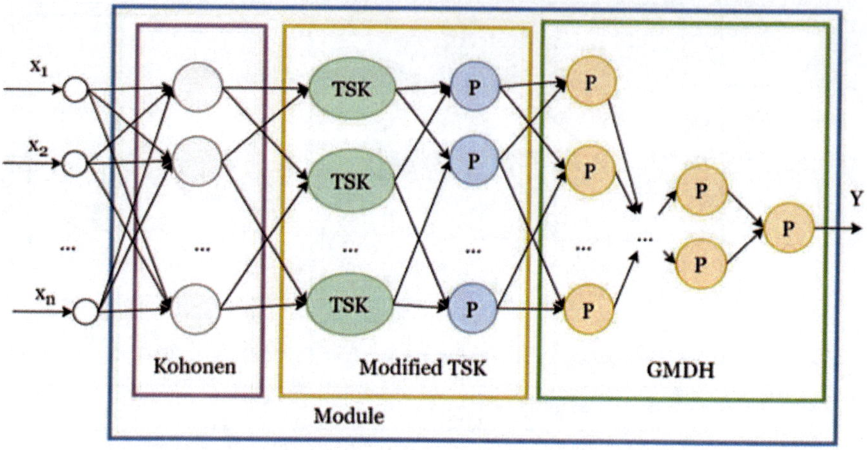

Fig. 3.21 Kohonen network + modified TSK + GMDH module topology

Table 3.28 MAPE when using the Kohonen network + base network + GMDH module

Base NN	Train (%)	Test (%)
Perceptron (10)	17.92	16.54
CPN	21.02	19.41
Modified TSK	17.25	17.73

The number of output neurons of the Kohonen network was taken equal to 5. The number of inputs for neurons TSK is 2. The number of inputs for neurons of the polynomial part is 2, the degree of polynomials is 2. The number of inputs into GMDH neurons is 2, the degree of polynomial is 2. The obtained results of the approximation problem solution with the module topology are presented in Table 3.28. Mean absolute percentage error (MAPE) was used as an estimate of the approximation error.

Results of Analysis

Analyzing the results obtained, it can be seen that the use of the Kohonen network in the first module can significantly improve the classification accuracy while reducing the size of the base networks and reducing the training time. For the approximation problem, the first module showed itself not in the best way. The error has not decreased, so the use of this combination for this task is called into question.

The results of the second module showed that the use of GMDH can significantly reduce the approximation error, however, the question remains about the number of output neurons in the core network that are fed to GMDH.

The third module consists of a combination of the first and second. As already seen, the first module is not very suitable for solving the approximation problem, and the second one works great. Thus, they partially neutralize each other, but the GMDH network still dominates. The results showed a similar trend: GMDH greatly reduced the approximation error of the first module, but it cannot be said that it is less than the error of the second module. Therefore, the use of the 3rd module with Kohonen and GMDH to solve the approximation problem is controversial.

3.9 Structural-Parametric Synthesis of an Ensemble of Modules of Hybrid Neural Networks

3.9.1 Review of Methods for Constructing Ensembles of Artificial Neural Networks

An ensemble of neural networks is a group of topologies united in a single structure, which may differ in architecture, learning algorithm, learning criteria and types of forming neurons [30, 47–49]. In another variant, the term is understood as an ensemble "combined model", the output of which is a functional combination of outputs of individual models [50].

Input data can be divided into specific groups for processing by different ANN or fed to all networks at the same time.

Forming ensembles of ANN, it is necessary to optimize two criteria—quality training of separate ANN and their optimum Association.

Known algorithms are usually divided into two classes [30]: algorithms that for new classifiers change the distribution of training examples, based on the accuracy of previous models (busting), and those in which new members of the ensemble learn independently of others (beging). The main algorithms for combining ANN into an ensemble and their disadvantages are given in Table 3.29 [50].

In contrast to [51], in this work, neural network modules are used instead of separate NN.

The need to apply the principle of modularity in hybrid NN ensemble structure is determined by the following:

- the heterogeneity of training sample data, which leads to the inability of the single-module NN to correctly approximate the required dependency;
- complexity of the algorithm of the solved problem, which requires a multimodal structure;
- differences in the characteristics of the error function on different fragments of the training sample;
- the need to accumulate knowledge of experts in the training modules of NN.

Currently, there are serial and parallel types of structures for the construction of SNN ensembles. There is currently no research on the sequential structure of module-to-ensemble connectivity. In [52], examples of sequential network building in modules are presented. An example of a parallel connection of modules is presented in [49].

Table 3.29 Characteristics of algorithms for combining expert opinions

Technology	Methodology of obtaining the result	Disadvantages
Static structure		
The averaging over ensemble	A linear combination of the output signals of ANN	The dependence of the result on the correct determination of the competence of the ANN. Increasing the complexity of the algorithm through the use of algorithms for correction of "noise emissions"
Boosting	Each new ANN is based on the results of previously constructed	The presence of a larger number of examples of learning sample. Degeneration of the ANN ensemble into a complex inefficient neural network structure, which requires a large amount of computational resources. Recent SNN learn from the "most difficult" examples
Stacking	Application of the concept learning objective	The complexity of theoretical analysis through a set of sequentially formed models. Metamodel levels may grow, which can lead to rapid depletion of computing resources
Begging	Formation of a set of ANN on the basis of a set of subsets of the training sample and the subsequent unification of the results of the ANN	Additional computational costs associated with the need to form a large number of subsets of the training sample Subsets of examples differ from each other, but are not independent, since they are all based on the same set. The algorithm requires a large amount of data to configure and train
Dynamic structure		
Mixing of ANN results	Integration of ANN knowledge through the use of a gateway network	The algorithm is demanding to computational resources when splitting the source space. It becomes possible to create a large number of areas, which will lead to excessive clustering of space and create a large group of base ANN with a complex mechanism of interaction through gateway networks. Learning and setting up a hierarchical model is a complex computational process. The stochastic gradient-based learning process is based on setting the ANN weights, first and second level gateway network, resulting in a complex algorithm of complex optimization of the entire neural network machine

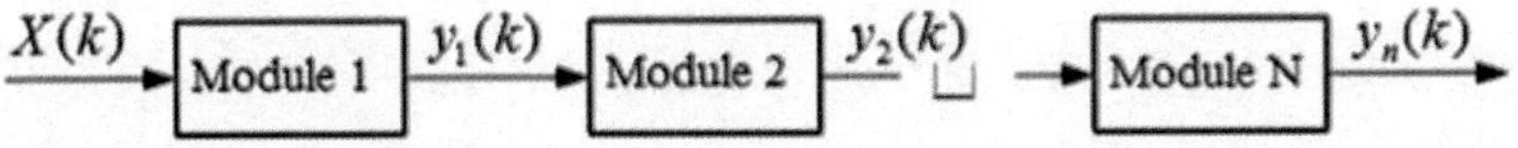

Fig. 3.22 Serial connection of modules

The main difficulty of networking in an ensemble is learning all the components to solve a problem. In order to increase the effectiveness of training, NNs are trained separately (if possible) and then combined into a single structure. However, if the tuning algorithms of the selected topologies belong to different classes of training, synchronous training of all the modules belonging to the ensemble is required, and therefore it is necessary to develop a single algorithm of tuning of all the modules of the ensemble.

Consider the principles of constructing ensembles of hybrid SNN sequential and parallel structures.

3.9.2 Serial Connection of Modules

The structure of organizing a sequential ensemble is to feed the output of one module to the inputs of another module. Such a structure is used to restore the input data or to improve their differences for the main task (approximation, classification, etc.).

The general scheme of serial connection of modules is shown in Fig. 3.22.

3.9.3 Connection of Modules in Parallel

An ensemble in which the input data is fed simultaneously to all modules that make up a hybrid neural network is called parallel. The main element in setting up such an amalgamation is the "amalgamation layer", which is responsible for the aggregation of the performance of the various components of the ensemble. The general structure of a parallel ensemble of neural network modules is shown in Fig. 3.23.

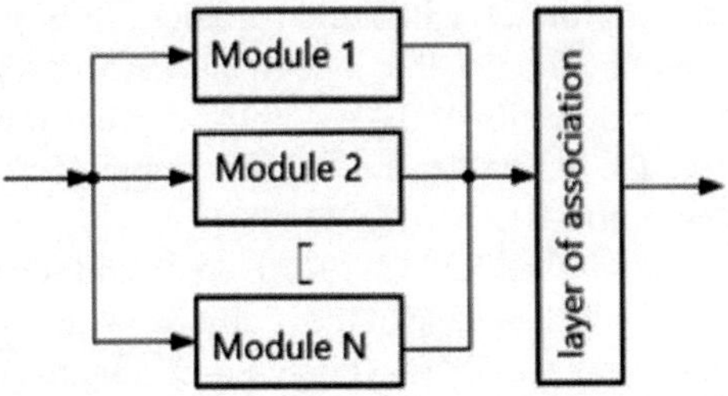

Fig. 3.23 Parallel connection of neural network modules

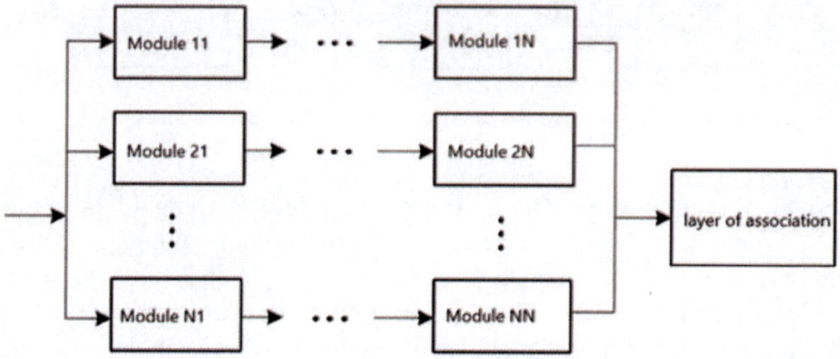

Fig. 3.24 Serial-parallel structure of the ensemble of neural network modules

The main disadvantage of using a parallel ensemble is the problem of optimum choice of topologies of NN modules that are appropriate to include in the ensemble.

3.9.4 Serial-Parallel Structure of the Ensemble of Neural Network Modules

The series-parallel ensemble of neural network modules, which is the most general structure, is shown in Fig. 3.24.

The main drawback of a sequentially parallel ensemble is the overly complex learning algorithm with probabilistic convergence.

3.9.5 Construction of Neural Network Module Ensemble Architecture

Based on the above analysis, we use a parallel ensemble structure with a layer of integration.

A necessary and sufficient prerequisite for building a module ensemble that has greater accuracy in solving the classification (prediction) problem than each individual module is to include modules that meet the criteria of accuracy and diversity in its composition. As the diversity of the ensemble decreases with increasing accuracy of the members of the ensemble, the task of creating an effective ensemble is reduced to finding a compromise.

In most ensemble methods (see Table 3.29), diversity and accuracy are achieved by manipulating training sample data. One of the problems with these approaches is that they tend to build oversized ensembles. This requires a large amount of memory to store the trained modules (classifiers). To overcome this problem, we use the Pruning procedure [53], which provides the optimum choice of a subset of individual modules (classifiers) from an already built ensemble in terms of accuracy, variety, and memory costs.

Based on the analysis of the Table 3.29, it can be shown that bagging has advantages over others:

- bagging reduces variance without increasing bias, which results in reduced error and increased stability;
- provides an increase in the accuracy of the ensemble, compared to any module (classifier) that is part of the ensemble. It may even offset the extra computational cost required to implement the bagging procedure, which involves the training of many neural networks;
- robust to noise;
- parallel training of ensemble elements.

Let $D = \{d_1, \ldots, d_N\}$ be the set of N data points where $d_i = \{(x_i, y_i) \mid i \in [1, N]\}$ is a pair of introductory features and a label representing the ith data point, $C = \{c_1, \ldots, c_M\}$ set of M-modules, where $c_i(x_j)$ gives the prediction of the ith module at the jth data point, $V = \{v^{(1)}, \ldots, v^{(N)} | v^{(i)} = [v_1^{(i)}, \ldots, v_L^{(i)}], i \in [1, N]\}$ set of vectors, where $v_j^{(i)}$ is the number of predictions for the jth label of the ith data point of the ensemble with a majority vote (majority vote), and L the number of source labels.

Required based on the accuracy and variety of classifiers $C = \{c_1, \ldots, c_M\}$ select members to form an ensemble, having a test dataset and considering that the networks are pre-trained on bootstrap samples.

According to the accuracy criterion, as shown in Sect. 3.7, the ensemble elements must be hybrid neural networks (modules). Diversity is ensured by learning the elements of the ensemble on the various datasets that can be obtained by using the bootstrap method.

The main idea behind bootstrap is to repeatedly extract repeated samples from the empirical distribution by the Monte Carlo statistical method: we take the finite set of n terms of the original sample $x_1, x_2, \ldots, x_{n-1}, x_n$, where at each step of n consecutive iterations using a random number generator evenly spaced [1, n], an “arbitrary” element is draw nx_k, which again “returns” to the original sample (i.e. can be retrieved).

Therefore, the preliminary stage in the construction of the ensemble is the creation of base classifiers, which must be independent.

These classifiers are trained on independent datasets. As a result, we have the following algorithm.

1. A set of training examples is provided $(x_1, y_1), \ldots, (x_m, y_m)$ with labels

$$y \in \{1, \ldots, k\}. \tag{3.16}$$

2. Get t bootstrap samples D_t.
3. Independently (in parallel) teach t classifiers h_t, each in their sample D_t.

Consider the concept of diversity.

Two classifiers are given c_i and c_j, where $N^{(01)}$ indicates the number of data points, incorrectly predicted c_i, but correctly predicted c_j. $N^{(10)}$ the opposite $N^{(01)}$ and indicates the number of points correctly predicted c_i, but incorrectly predicted c_j. The variety of the classifier c_i, relative to the classifier c_j, which is indicated $\mathrm{Div}_{i,j}$ is the ratio between the sum of the number of data points correctly predicted by one of the classifiers and the total amount of data and is determined by the equation

$$\mathrm{Div}_{i,j} = \frac{N^{(01)} + N^{(10)}}{N}. \tag{3.17}$$

Contribution of the diversity of the ci classifier to the designated ensemble ConDiv_i, is the sum of the differences between ci and every other classifier in the ensemble (except c_i, since according to Eq. (3.18) the difference of the classifiers is in itself equal to zero) and is determined by the equation

$$\mathrm{ConDiv}_i = \sum_{j=1}^{M} \mathrm{Div}_{i,j}. \tag{3.18}$$

For the problem of two-class classification, the contribution of the variety of the classifier c_i

$$\mathrm{ConDiv}_i = \frac{1}{N} \sum_{k=1}^{N} \left(M - v^i_{c_i(x_k)}\right), \tag{3.19}$$

where N is the number of data points; M is the total number of classifiers; $v^i_{c_i(x_k)}$ is the number of classifiers that agree with c_i the classifier in predicting the result of the classification.

Thus $M - v^i_{c_i(x_k)}$ is the number of classifiers who disagree for c_i a classifier in the prediction of the result. In the general case, the prediction of an ensemble member at one data point can be divided into four subsets in which:

1. a separate classifier predicts correctly and is in the minority group;
2. the individual classifier predicts correctly and is in the majority group;
3. the individual classifier mis-estimates and is in the minority group;
4. the individual classifier incorrectly predicts and is in the majority group.

In [48, 54], the following two rules are defined for developing a heuristic metric for estimating the individual contributions of ensemble members: (1) correct forecasts make positive contributions, wrong forecasts make negative ones; (2) correct forecasts in the minority group make more positive contributions than correct forecasts in the majority group, and incorrect forecasts in the minority group make fewer negative contributions than incorrect forecasts in the majority group.

Individual contribution of the classifier c_i is defined as follows:

$$IC_i = \sum_{j=1}^{N} IC_i^j, \tag{3.20}$$

where IC_i^j the contribution of the classifier c_i to the i-th data point d_j. IC_i^j is determined by which subset the classifier is predicting.

When $c_i(x_j)$ equal y_j, which means that c_i makes the right predictions at the point d_j, if $c_i(x_j)$ belongs to a minority group (the first subset), then IC_i^j is defined as

$$IC_i^{(j)} = 2\upsilon_{\max}^{(j)} - \upsilon_{c_i(x_j)}^{(j)}, \tag{3.21}$$

where $\upsilon_{\max}^{(j)}$ are majority of votes d_j; $\upsilon_{c_i(x_j)}^{(j)}$ is the number of predictions $c_i(x_j)$, which was previously defined.

When $c_i(x_j)$ equal y_i and $c_i(x_j)$ belongs to the majority group (in this case $\upsilon_{c_i(x_j)}^{(j)} = \upsilon_{\max}^{(j)}$) (the second subset), IC_i^j is defined as

$$IC_i^j = \upsilon_{sec}^{(j)}, \tag{3.22}$$

where $\upsilon_{sec}^{(j)}$ is the second largest number of votes in d_j tags. $\upsilon_{sec}^{(j)} = \upsilon_{\max}^{(j)}$ is an estimate of the "degree of positive contribution" in this case. Accordingly, if most of the classifiers predict correctly with the classifier on d_j, the contribution of this classifier is not very valuable, because without its prediction the ensemble would still be correct for d_j (in the absence of communication). Note that $\left(\upsilon_{sec}^{(j)} = \upsilon_{\max}^{(j)}\right)$ negative. According to the rules for the development of an individual contribution rate, all correct forecasts make a positive contribution. So, to $\left(\upsilon_{sec}^{(j)} = \upsilon_{\max}^{(j)}\right)$ the member is added $\upsilon_{\max}^{(j)}$, to normalize it so that it is always positive, which gives the Eq. (3.22). And $\upsilon_{\max}^{(j)}$ added to $\upsilon_{\max}^{(j)} - \upsilon_{c_i(x_j)}^{(j)}$ to maintain the relative order that gives the Eq. (3.21).

When $c_i(x_j)$ not equal $y_j IC_i^{(j)}$ is defined as

$$IC_i^{(j)} = \upsilon_{correct}^{(j)} - \upsilon_{c_i(x_j)}^{(j)} - \upsilon_{\max}^{(j)}, \tag{3.23}$$

where $\upsilon_{correct}^{(j)}$ is the number of votes for the correct label d_j. Like the "positive contribution", the "negative contribution" is estimated by the formula $\upsilon_{correct}^{(j)} - \upsilon_{c_i(x_j)}^{(j)}$, which is the difference between the number of votes on the correct label and the number of votes on $c_i(x_j)$. Combining the Eqs. (3.21), (3.22) and (3.23) using the Eq. (3.20), individual contribution of the classifier c_i:

$$IC_i = \sum_{j=1}^{N} \left(\alpha_{ij}\left(2\upsilon_{\max}^{(j)} - \upsilon_{c_i(x_j)}^{(j)}\right) + \beta_{ij}\upsilon_{sec}^{(j)} + \theta_{ij}\left(\upsilon_{correct}^{(j)} - \upsilon_{c_i(x_j)}^{(j)} - \upsilon_{\max}^{(j)}\right)\right), \tag{3.24}$$

where

$$\alpha_{ij} = \begin{Bmatrix} 1 \text{ if } c_i(x_j) = y_j \text{ and } c_i(x_j) \text{ is in the minority group,} \\ 0, \text{ else,} \end{Bmatrix} \quad (3.25)$$

$$\beta_{ij} = \begin{Bmatrix} 1, \text{ if } c_i(x_j) = y_j \text{ and } c_i(x_j) \text{ is in the magority group,} \\ 0, \text{ else,} \end{Bmatrix} \quad (3.26)$$

$$\theta_{ij} = \begin{Bmatrix} 1, \text{ if } c_i(x_j) \neq y_j, \\ 0, \text{ else} \end{Bmatrix} \quad (3.27)$$

According to the Eq. (3.24) formed a set of modules that are included in the ensemble. This requires the development of a module merge procedure.

We define a dynamically averaged network (DAN) via:

$$f_{AOI} = \sum\nolimits_{i=n}^{n} w_i f_i(x), \quad (3.28)$$

where $w_i \quad i = \overline{1, n}$ are determined according to the formula

$$w_i = \frac{c(f_i(x))}{\sum_{i=n}^{n} c(f_i(x))}, \quad (3.29)$$

where $c(f_i(x)) = \begin{cases} f_i(x), & \text{if } \; f_i(x) \geq 0.5, \\ 1 - f_i(x), & \text{else,} \end{cases}$ f_{DAN} is the weighted average (WA) network outputs.

The weight vector is calculated each time the original ensemble is evaluated to obtain the best solution for the particular case, rather than selecting the weights statically.

The contribution of each module to the amount is proportional to its reliability.

The general algorithm for constructing the architecture of ensembles of neural network modules is as follows:

We have a set of training examples $(x_1, y_1), \ldots, (x_m, y_m)$ with labels $y \in \{1, \ldots, k\}$.

(1) We get t bootstrap samples D_t.
(2) Independently (in parallel) teach t classifiers h_t, each in their sample D_t.
(3) Get predictions from each classifier c_i of jth data points d_j on the test sample and determine the individual contribution.

   3.1. If the prediction $c_i(x_j)$ equal y_i, that is c_i makes the right predictions in c_j
      3.1.1. If the prediction $c_i(x_j)$ belongs to a minority group, then the individual contribution is calculated by the formula (3.21).
      3.1.2. If the prediction $c_i(x_j)$ belongs to the majority group, then the individual contribution is calculated by the formula (3.22).

3.2. If the prediction $c_i(x_j)$ not equal y_i, individual contribution is calculated by the formula (3.5).

(4) Determine the individual contribution of the classifier c_i by the formula (3.24), where, depending on step 3, the appropriate ratio is set to one.
(5) Add a couple (ci, ICi) to the list OL and sort in descending order.
(6) Define the parameter p which is the desired percentage of classifiers C, which must be stored at the output of the sub-ensemble. This setting is based on existing resources such as memory, time and cost.
(7) Knowing the desired resource costs and the real costs, deduce the first p percent from the list as a shortened and ranked sub-ensemble.
(8) Determine the weighting coefficients of the modules in the ensemble by the formula (3.9).

As a result of the algorithm the ranking of selected neural networks according to their individual contribution was obtained. In practice, in ensembles a classification error usually exhibits a monotonous decrease as a function of the number of elements in the ensemble.

For large ensemble sizes, these curves reach an asymptotic constant error rate, which is usually considered the best result that can be achieved.

One of the problems with ensemble approaches is that using them leads to the creation of unreasonably large ensembles, which requires a large amount of memory to store trained modules and reduce the response time for forecasting [55]. Ensemble simplification or sample ensembles is a method that solves this problem by selecting a subset of individual modules from the prepared ensemble to form a sub-ensemble for prediction. It is necessary to carefully select modules in the sub-ensemble to ensure its minimum complexity (to reduce memory requirements and response time) and an accuracy of prediction that is the same or greater than the accuracy of the original ensemble. Thus, the next step in creating an effective ensemble is to use a simplification procedurepruning.

Usually, such methods are used to select the classifiers required Reduce-Error Pruning, Kappa Pruning, Margin Distance Minimization and Orientation Ordering. But all of these approaches are based on the selection of networks by precision or variety, which have already been considered in the individual contribution ranking. It is also known that sometimes it is not enough to take into account the variety and precision to form an effective ensemble. It is suggested to use this approach as a simplification with the help Complementary Measure, which also takes into account the interaction of the classifiers with each other.

3.9.6 Simplification Algorithm

The simplification algorithm is as follow:

(1) Get predictions from each member of the ensemble after running on the test dataset.
(2) Form a sub-ensemble S_{u-1} of outgoing members $c_i(x_j)$, which is not equal y_i (that is, they predict incorrectly).
(3) Select the qualifier with the best variety and accuracy according to the OL list obtained from the previous algorithm and add it to the sub-ensemble.
(4) Obtain the value S_u, which characterizes the impact of the best-performing classifier on the sub-ensemble, which gives incorrect predictions:

$$s_u = \arg\max_k \sum_{(x,y)\in Z_{sel}} I(y) = h_k(x), \tag{3.30}$$

$$H_{s_{u-1}}(x) \neq y, \tag{3.31}$$

where the best-performing qualifier belongs to the original ensemble $h_k(x)$ and $H_{s_{u-1}}$; $I(y)$ is the indicator function; S_u is the value by which members are selected for the reduced sub-saber.
(5) We set the threshold for the selection of the classifier. That is, if an error is made by the sub-assembly S_{u-1} more than a mistake S_u, namely, the difference of their errors exceeds a predetermined threshold value, then the classifier is considered a supplement and is selected in the reduced sub-ensemble. So,

$$S_u = \arg\max_k \sum_{(x,y)\in Z_{sel}} I\left(\left|y - H_{s_{u-1}}(x)\right| - |y - h_k(x)| > threshold\right). \tag{3.32}$$

(6) Repeat steps 3–5 with each classifier and initial sub-ensemble $S_{u-1.}$

3.10 Comparative Analysis of the Result of Solving the Classification Problem by Hybrid Neural Networks of Ensemble Topology

Here two approaches to the construction of hybrid neural networks (HNN) of ensemble topology for the classification task are considered.

In the first case, an ensemble topology based on individual neural networks is used, in particular: Perceptron, Radial-basis function network (RBFN), counterpropagation network (CPN), probability neural network (PNN), NefclassM, NaïveBayes.

In the second case, the ensemble HNN topology constructed on the basis of NN modules is considered. When solving the classification problem, NN modules include the Kohonen NN and the base NN, in which the following networks were used: Perceptron, RBFN, CPN, PNN, NefclassM, NaiveBayes. Such a module structure corresponds to the classification problem.

Table 3.30 Learning results of individual networks and modules

Base NN	Without Kohonen network		With Kohonen network (Module)	
	Train (%)	Test (%)	Train	Test (%)
Perceptron	27.46	38.89	96.48	100
RBFN	38	38.89	95.77	94.44
CPN	40.14	38.88	90.14	91.66
PNN	91.54	86.11	92.25	97.22
NEFClassM	80.28	80.55	96.47	97.22
Naïve Bayes	93.66	94.4	95.77	97.22

Table 3.31 Contribution of diversity to the ensemble of individual networks and the ensemble of modules

Base NN	Without Kohonen network	With Kohonen network (Module)
Perceptron	0.4722	0.22
RBFN	0.472	0.138
CPN	1	0.111
PNN	1.722	0.111
NEFClassM	1.55	0.16
Naïve Bayes	2	0.138

Table 3.32 Individual contribution to the ensemble of individual networks and the ensemble of modules

Base NN	Without Kohonen network	With Kohonen network (Module)
Perceptron	0	9
RBFN	0	4
CPN	38	3
PNN	51	2
NEFClassM	33	6
Naïve Bayes	69	4

Table 3.33 Accuracy of pruning ensemble of separate networks and ensemble of modules

NN	Without Kohonen		With Kohonen (Module)		Base NNs
	Train (%)	Test (%)	Train (%)	Test (%)	
PNN, CPN, Naïve Bayes, NEFClassM	88.88	95.77	98.59	100	Perceptron, RBFN, NEFClassM, Naïve Bayes

The Wine dataset was used as a training set. The number of samples in the set is 178. Each object has 13 features, represented by real numbers greater than zero, and one class label. The number of classes is 3. For the test sample, 20% of the data set was selected, respectively, 80% for training.

The Results are presented in Tables 3.30, 3.31, 3.32 and 3.33.

References

1. Fonseca, C.M., Fleming, P.J.: Multiobjective optimization and multiple constraint handling with evolutionary algorithms. Part I: a unified formulation. Technical Report 564, University of Sheffield, Sheffield, UK (1995)
2. Goldberg, D.E., Deb, K.: A comparative analysis of selection schemes used in genetic algorithms, pp. 36–57 (1991)
3. Goldberg, D.E., Kargupta, H., Horn, J., Cantu-Paz, E.: Critical deme size for serial and parallel genetic algorithms. Illigal Report № 95002, Illinois Genetic Algorithms Laboratory, University of Illinois, Urbana, Illinois, pp. 365–452 (1995)
4. Hinton, G. E.: A practical guide to training restricted Boltzmann machines. Technical Report 2010-000. Machine Learning Group, University of Toronto, Toronto, pp. 160–169 (2010)
5. Cortez, P.: Wine quality data set [Online course]. https://archive.ics.uci.edu/ml/datasets/wine+quality
6. Kruglov, V., Dli, M., Golunov, R.: Fuzzy logic and artificial neural networks. Fizmatlit, p. 221 (2001)
7. Lin, C.-J., Xu, Y.-J.: Design of neuro-fuzzy systems using a hybrid evolutionary learning algorithm. J. Inf. Sci. Eng. (23), 463–477 (2007)
8. Islam, M.M., Sattar, M.A., Amin, M.F., Yao, X., Murase, K.: A new adaptive merging and growing algorithm for designing artificial neural networks. IEEE Trans. Syst. Man Cybern. Part B Cybern. **39**(3), 705–722 (2009)
9. Coello Coello, C.A.: A comprehensive survey of evolutionary-based multiobjective optimization techniques. Laboratorio Nacional de Informatica Avanzada, Veracruz, Mexico (38) 1998
10. Coello Coello, C.A.: An empirical study of evolutionary techniques for multiobjective optimization in engineering design. Ph.D. Thesis. Department of Computer Science, Tulane University, New Orleans, LA (1996)
11. Kingma, D.P., Ba, J.L.: Adam: a method for stochastic optimization. Published as a Conference Paper at ICLR 2015 [cs.LG], pp. 1–15 (2017). arXiv:1412.6980v9
12. Duchi, J., Hazan, E., Singer, Y.: Adaptive subgradient methods for online learning and stochastic optimization. J. Mach. Learn. Res. **12**, 2121–2159 (2011)
13. Duchi, J., Hazan, E., Singer, Y.: Adaptive subgradient methods for online leaning and stochastic optimization. COLT (2010)
14. Hinton, G., Tieleman, T.: Lecture 6.5-RMSprop: divide the gradient by a running average of its recent magnitude. COURSERA Neural Netw. Mach. Learn. **4**(2), 26–31 (2012)
15. Kingma, D., Ba, J.: Adam: a method for stochastic optimization. In: International Conference on Learning Representations (2014). arXiv:1412.6980
16. Muhammad, A.I.: A review on face recognition with artificial neural network and particle swarm optimization. Int. J. Eng. Dev. Res. **7**(1), 4 (2019)
17. Zeiler, M.D.: Adadelta: an adaptive learning rate method. 1–6 [cs.LG] (2012). arXiv:1212.5701v1
18. Reznikov, B.: Methods and optimization algorithms on discrete models of complex systems. VIKI named after Mozhaisky (1983)

19. Ismail, A., Engelbrecht, A.P.: Training product units in feedforward neural networks using particle swarm optimization. In: Proceedings of the International Conference on Artificial Intelligence, Sept 1999, Durban, South Africa, vol. 40, p. 5 (1999)
20. Mirjalili, S., Hashim, S.Z.M., Sardroudi, H.M.: Training feedforward neural networks using hybrid particle swarm optimization and gravitational search algorithm. Appl. Math. Comput. (218), 11125–11137 [Electronic resource] (2012). www.elsevier.com/locate/amc
21. Nesterov, Y.Y.: Method for minimizing convex functions with convergence rate $O(1/k^2)$. Report. AS USSR. T. 269, rel. 3, pp. 543–547 (1983)
22. Reddi, S., Kale, S., Kumar, S.: On the convergence of adam and beyond (2018). arXiv:1904.09237
23. Ruder, S.: An overview of gradient descent optimization algorithms (2016). https://ruder.io/optimizing-gradient-descent/
24. Sutskever, I., Martens, J., Dahl, G., Hinton, G.: On the importance of initialization and momentum in deep learning. In: 30th International Conference on Machine Learning, ICML 2013, pp. 1139–1147 (2013)
25. Rosli, A.D., Adenan, N.S., Hashim, H., Abdullah, N.E., Sulaiman, S., Baharudin, R.: Application of particle swarm optimization algorithm for optimizing ANN model in recognizing ripeness of citrus. IOP Conf. Ser. Mater. Sci. Eng. **340**, 012015 (2018)
26. Abusnaina, A.A., Jarrar, R., Ahmad, S., Mafarja, M.: Training neural networks using salp swarm algorithm for pattern classification. In: Proceedings of the International Conference on Future Networks and Distributed Systems (ICFNDS 2018). ACM, New York, NY, USA, no. 4, p. 6 [Electronic resource]. https://core.ac.uk/download/pdf/160738309.pdf
27. Dehuria, S., Cho, S.-B.: Multi-criterion Pareto based particle swarm optimized polynomial neural network for classification: a review and state-of-the-art. Comput. Sci. Rev. (3), 19–40 (2009)
28. Chumachenko, H., Kryvenko, I.: Neural networks module learning. Electron. Control Syst. **2**(48), 76–80 (2016). NAU, Kyiv
29. Syulistyo, A.R., Purnomo, D.M.J., Rachmadi, M.F., Wibowo, A.: Particle swarm optimization (PSO) for training optimization on convolutional neural network (CNN). J. Comput. Sci. Inf. (9/1), 52–58 (2016)
30. Golovko, V.: Neural networks: training, organization and application. IPRZhR. Neurocomputers and Their Application, p. 256 (2001). Book 4
31. Vrbančič, G., Fister, Jr I., Podgorelec, V.: Swarm intelligence approaches for parameter setting of deep learning neural network: case study on phishing websites classification. In: International Conference on Web Intelligence, Mining and Semantics, June 25–27, Novi Sad, Serbia. ACM, New York, NY, USA, p. 8 [Electronic resource] (2018). https://doi.org/10.1145/3227609.3227655
32. Zhang, J.-R., Zhang, J., Lok, T.-M., Lyu, M.R.: A hybrid particle swarm optimization–back-propagation algorithm for feedforward neural network training. Appl. Math. Comput. **185**, 1026–1037 (2007)
33. Gudise, V.G., Venayagamoorthy, G.K.: Comparison of particle swarm optimization and back-propagation as training algorithms for neural networks. In: Proceedings of the 2003 IEEE Swarm Intelligence Symposium, 2003, SIS '03. Institute of Electrical and Electronics Engineers (IEEE) (2003)
34. Schaffer, J.D.: Multiple objective optimization with vector evaluated genetic algorithms. In: Grefenstette, J.J. (ed.) Proceedings of an International Conference of Genetic Algorithms and Their Applications, Pittsburgh, PA, 1985, pp. 93–100
35. Sineglazov, V., Chumachenko, E., Gorbatiuk, V.: Applying different neural network's topologies to the forecasting task. In: 4th International Conference in Inductive Modelling ICIM', pp. 217–220 (2013)
36. Tamon, C., Xiang, J.: On the boosting pruning problem. In: Proceedings of the 11th European Conference on Machine Learning, pp. 404–412 (2000)
37. Semenkina, H., Zhidkov, V. Optimization of Management of Complex Systems by the Method of Generalized Local Search. MAKS Press, p. 215 (2002)

38. Partalas, I., Tsoumakas, G., Vlahavas, I.: Focused ensemble selection: a diversity-based method for greedy ensemble selection. In: Proceeding of the 18th European Conference on Artificial Intelligence, pp. 117–121 (2008)
39. Lazzús, J.A., Salfate, I., Montecinos, S.: Hybrid neural network–particle swarm algorithm to describe chaotic time series. Neural Netw. World 601–617. (2014)
40. Hoshino, Y., Jin'no, K.: Learning algorithm with nonlinear map optimization for neural network. J. Signal Process. **22**(4), 153–156 (2018)
41. Aljarah, I., Faris, H., Mirjalili, S.: Optimizing connection weights in neural networks using the whale optimization algorithm [Electronic resource]. https://link.springer.com/article/10.1007/s00500-016-2442-1. Accessed 21 Nov 2016
42. Settles, M., Rylander, B.: Neural network learning using particle swarm optimizers. School of Engineering University of Portland (2002)
43. Yermakov, V.: Swarm particle optimization in training of artificial neural networks. Syst. Anal. 7
44. Electronic resource. http://mnemstudio.org/neural-networks-kohonen-self-organizing-maps.html
45. Sutton, R.S.: Training with reinforcement. BINOM. Knowledge Laboratory, p. 399
46. Arai, M.: Bounds on the number of hidden units in binary-valued three-layer neural networks. Neural Netw. **6**(6), 855–860 (1993)
47. Bifet, A., Holmes, G., Pfahringer, B., Kirkby, R., Gavaldá, R.: New ensemble methods for evolving data streams. In: Proceedings of the 15th ACM SIGKDD International Conference on Knowledge Discovery and Data mining, pp. 139–148 (2009)
48. Bodyanskiy, Y., Rudenko, O.: Artificial neural networks: architectures, training, applications. Teletech, Kharkov, p. 369 (2004)
49. Martınez-Muñoz, G., Hernandez-Lobato, D., Suárez, A.: An analysis of ensemble pruning techniques based on ordered aggregation. IEEE Trans. Pattern Anal. Mach. Intell. **31**(2), 245–259 (2009)
50. Martınez-Muñoz, G., Suárez, A.: Aggregation ordering in bagging. In: Proceedings of the IASTED International Conference on Artificial Intelligence and Applications. Acta Press, pp. 258–263 (2004)
51. Canuto, A.M.P., Abreu, M.C.C., de Melo Oliveira, L., Xavier, Jr., J.C., Santos, A. de M.: Investigating the influence of the choice of the ensemble members in accuracy and diversity of selection-based and fusion-based methods for ensembles. Pattern Recogn. Lett. **28**, 472–486 (2007)
52. Ma, Z., Dai, Q., Liu, N.: Several novel evaluation measures for rank-based ensemble pruning with applications to time series prediction. Expert Syst. Appl. **42**, 280–292 (2015) [Online course]
53. Martınez-Muñoz, G., Suárez, A.: Pruning in ordered bagging ensembles. In: Proceedings of the 23rd International Conference on Machine Learning, pp. 609–616 (2006)
54. Lu, Z., Wu, X., Zhu, X., Bongard, J.: Ensamble pruning via individual contribution ordering. Department of Computer Science University of Vermont, Burlington, NSW, pp. 635–745 (2007)
55. Zhang, Y., Burer, S., Street, W.N.: Ensemble pruning via semi-definite programming. J. Mach. Learn. Res. 1315–1338 (2006)

Chapter 4
Development of Hybrid Neural Networks

4.1 Deep Learning as a Mean of Improving the Efficiency of Neural Networks

The main complexity of using neural networks (NN) is the "dimension curse". In the case of increasing the size of the inputs and the number of layers, the complexity of the network and, accordingly, the learning time increases exponentially, the result is far from optimal.

Another complication of using NNs is that traditional NNs lack an interpretation of how they solve the problem. In many areas, this explanation is more important than the outcome itself (for example, medicine). The internal presentation of learning outcomes is often so complex that it cannot be analyzed, except in some of the simplest cases that are usually of no interest.

Currently, the theory and practice of machine learning are experiencing a true "deep revolution", caused by the successful application of Deep Learning methods, which are the third generation of NN. Unlike the classic (second generation) NNs of the 1980s and 1990s, new learning paradigms have eliminated a number of problems that have hampered the spread and successful use of traditional NNs.

Networks trained with deep learning algorithms not only outperformed the best alternative approaches, but also revealed some of the beginnings of understanding the meaning of the information being provided (for example, image recognition, textual information analysis, etc.).

The most successful modern industrial methods of computer vision and speech recognition are built on the use of deep networks, and giants of the IT industry such as Apple, Google, Facebook are buying up teams of researchers engaged in deep NN.

M. Zgurovsky et al., *Artificial Intelligence Systems Based on Hybrid Neural Networks*, Studies in Computational Intelligence 904,
https://doi.org/10.1007/978-3-030-48453-8_4

Deep learning theory supplements conventional machine learning technologies with special algorithms for analyzing input information at multiple levels of representation. The peculiarity of the new approach is that "deep learning" studies the subject until it finds enough informative levels of presentation to account for all factors that can influence the characteristics of the subject under study.

Thus, the NN based on this approach requires less input for training, and the trained network is able to analyze information with much higher accuracy than conventional NN [1–5].

In general, deep neural networks carry out a deep hierarchical transformation of the input space of images and represent neural networks with a large number of layers of neuronal elements [6]. There are the following deep neural networks (DNNs):

- deep belief neural networks;
- deep perceptron;
- deep convolutional neural networks: R-CNN, Fast-CNN, Faster-CNN, SSD, ResNet, etc.;
- deep recurrent neural networks;
- deep autoenconder;
- deep recurrent-convolutional neural network (deep RCNN).

Historically, the first appeared neural networks of deep trust and deep perceptron, which in the general case are multilayered perceptron with more than two hidden layers. The main difference between the deep neural network of deep perceptron is that the deep believe network in the general case is not a feed with forward propagation (feed forward neural network). Until 2006, in the scientific environment, the paradigm was a priority, that the multilayer perceptron with one, maximum of two hidden layers is more effective for the nonlinear transformation of the input space of images into output compared with the perceptron with a large number of hidden layers. It was considered that it makes no sense to use perceptron with more than two hidden layers. This paradigm was based on the theorem that a perceptron with one hidden layer is a universal approximator.

The second aspect of this problem is that all attempts to apply a back-propagation algorithm to study a perceptron with three or more hidden layers did not lead to better resolution of various tasks. This is due to the fact that the reverse error propagation algorithm is ineffective for teaching perceptrons with three or more hidden layers when using the sigmoid activation function due to the problem of vanishing gradient problem. Therefore, the use of a generalized delta rule for training perceptron with a large number of hidden layers leads to a gradient attenuation in the propagation of the signal from the last to the first layer. In 2006, J. Hinton proposed a "greedy" layer-wise algorithm, which became an effective means of learning deep neural networks. It has been shown that such deep neural network (Deep Belief Network (DBN)) has a high efficiency of nonlinear transformation and representation of data in comparison with traditional perceptron. Such a network carries a deep hierarchical transformation of the input space of images. As a result, the first hidden layer allocates a low-level space of signs of the input data, the second layer detects the space of signs of a higher level of abstraction, etc. [7].

The new learning paradigm implements the idea of learning in two stages. In the first stage, information about the internal structure of the input data is extracted from the large array of unallocated data with the help of auto-associators (by layer-by-layer training without a "teacher"). Then, using this information in a multilayered NN, it is taught with "teacher" (marked data) known methods. It is desirable to have as much unclassified data as possible. The amount of marked data can be much smaller. In our case, this is not very relevant.

4.2 Pretrained Deep Neural Networks

In general, deep learning NN (for example, Deep Belief Neural Network (DBN)), structural diagrams that include Restricted Boltzmann machines (RBM) or autoencoders are shown in Figs. 4.1 and 4.2 [1, 8–11].

This work is considered pretrained deep neural network with deep belief network.

4.3 Deep Belief Network, Structure and Usage

Deep belief network (DBN) is a deep generative neural network designed to find or reproduce the probability distribution in data, contains many layers of latent variables that are usually binary; the input layer can be either binary or continuous. It was invented by Hinton in 2006 after developing an effective algorithm for Boltzmann restricted machine learning [1, 10].

Deep belief network with n hidden layers contains n weight matrices $W^{(1)} \dots W^{(n)}$ and $b+1$ bias vectors $b^{(0)} \dots b^{(n)}$, where $b^{(0)}$ is the bias of the visible layer.

The mathematical model of DBN has the following form [6]:

$$p\left(h^{(n)}, h^{(n-1)}\right) = \exp\left(b^{(n)\mathrm{T}} h^{(n)} + b^{(n-1)\mathrm{T}} h^{(n-1)} + h^{(n-1)} W^{(n)} h^{(n)}\right), \tag{4.1}$$

$$p\left(h_i^{(k)} = 1 | h^{(k+1)}\right) = \sigma\left(b_i^{(k)} + W_i^{(k+1)\mathrm{T}} h^{(k+1)}\right) \forall i, \forall k 1, \dots, n-2, \tag{4.2}$$

in case of binary input neurons:

$$p(v_i = 1 | h^{(1)}) = \sigma\left(b_i^{(0)} W_i^{(1)\mathrm{T}} h^{(1)}\right) \forall i, \tag{4.3}$$

or, for continuous input neurons:

$$\mathrm{v} \sim N\left(v;\ b^{(0)} + W^{(1)\mathrm{T}} h^{(1)},\ \beta^{-1}\right), \tag{4.4}$$

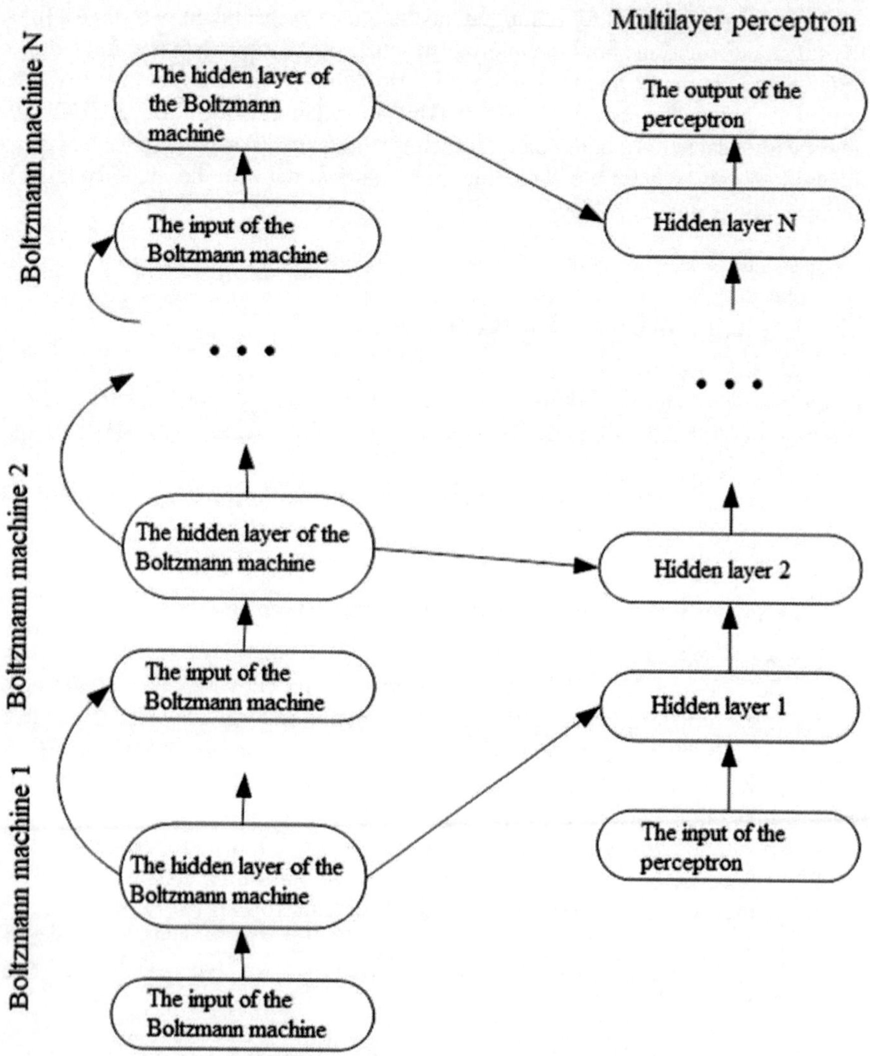

Fig. 4.1 Pretrained deep neural network with deep belief network

where $h^{(n)}$ are states of neurons of the nth hidden layer; v is the random pattern of the data; v is the generated state of visible neurons; β^{-1} is the diagonal matrix; $\sigma(\bullet)$ is the sigmoidal function.

Unlike other neural network models, the elemental unit of the DBN is not a separate layer, but a pair of neighboring layers forming a restricted Boltzmann machine (RBM), in other words, the DBN is formed by the RBM cascade. Therefore, DBN with one hidden layer degenerates into RBM.

The structure of the DBN is shown in Fig. 4.3.

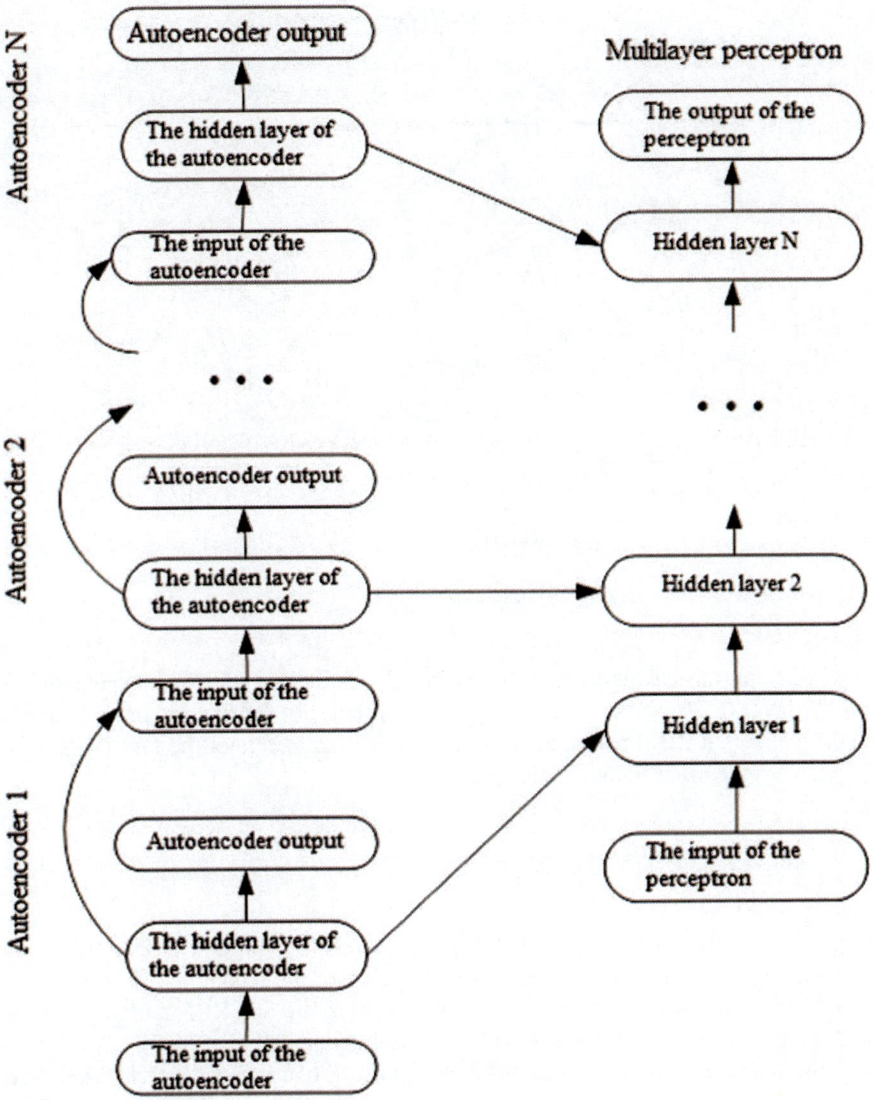

Fig. 4.2 Pretrained deep neural network with autoencoders

The work of a trained DBN consists of data generation by propagating a signal from the last hidden layer to the visible. To generate the pattern, it is necessary to create a Gibbs sample on the last pair of layers, that is, on the last RBM cascade, then the signal propagates the network to the input neurons, where the resulting vector is generated. The learning DBN is begun from the first RBM learning to maximize [6]:

$$\mathbb{E}_{v \sim p_{data}} \log p(v), \tag{4.5}$$

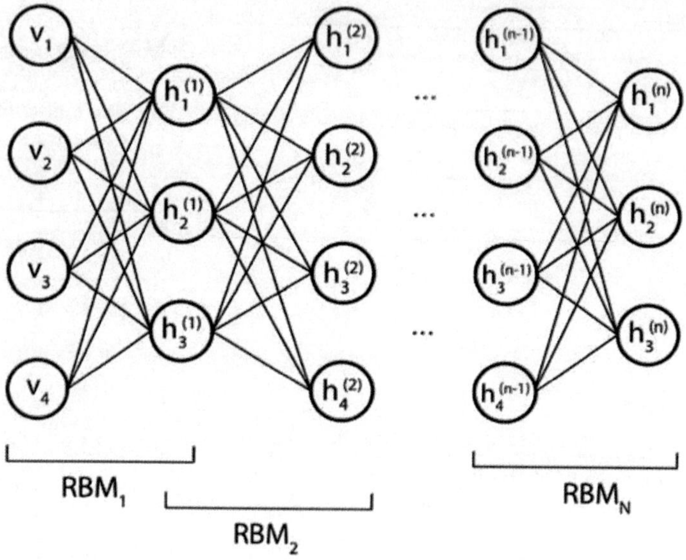

Fig. 4.3 Deep belief neural network structure

using the appropriate algorithms, where $\mathbb{E}_{v \sim p_{data}}$ is the mathematical expectation of visible layer (first), vector value; p_{data} is the probability distribution of input data v.

The obtained RBM parameters determine the parameters of the first DBN layer. Next, learn the other RBM to maximize [6]

$$P(y|x) = \frac{P(x, y)}{P(x)} = \frac{\frac{1}{Z}e^{-E(x,y)}}{\frac{1}{Z}\sum_y e^{-E(x,y)}} = \frac{e^{-E(x,y)}}{\sum_y e^{-E(x,y)}}, \tag{4.6}$$

where $p^{(1)}$ is the probability distribution, presented by the first RBM; $p^{(2)}$ is the distribution represented by the second RBM. That is, the second RBM is learned to model the distribution determined by a sample from the first RBM hidden neurons, and the first RBM works directly with the training data.

This procedure, called the greedy layer training, is continued iteratively to achieve the expected result, which satisfies the corresponding criteria of effectiveness [12].

The learned DBN can be directly used as a generative model, but it is mainly used in connection with its ability to improve the classification models. That is, it is possible to use the weights from DBN multilayer perceptron (MLP) initialization with preloading [6]:

$$h^{(1)} = \sigma\left(b^{(1)} + v^{\mathrm{T}} W^{(1)}\right), \tag{4.7}$$

$$h^{(1)} = \sigma(b^{(1)} + h^{(i-1)\mathrm{T}} W^{(i)}) \forall l \in 2, \ldots, m, \tag{4.8}$$

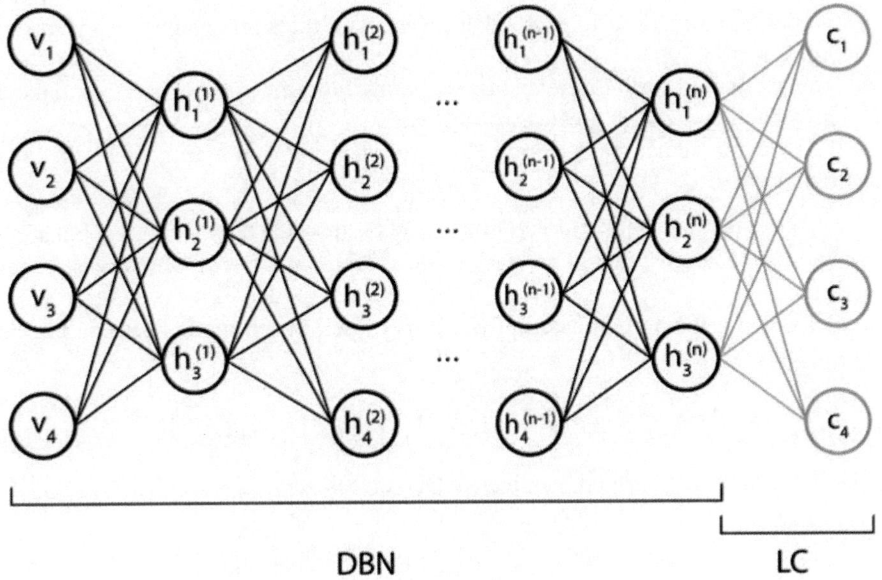

Fig. 4.4 The structure of a MLP from DBN: LC is a linear classifier

where m is the number of perceptron hidden layers.

After MLP initialization, by weights with biases, trained on the stage of primary DBN learning, the MLP are retrain to solve the classification task solution. This process is referred to as discriminant fine tunning. The structure of such hybrid network is shown in Fig. 4.4.

To test the quality of the model, it is executed the estimation of its generalized ability on the test data set.

4.4 Deep Belief Network and Semi-supervised Learning

Semi-supervised learning—is a machine learning method that is a hybrid of unsupervised learning and supervised learning methods and is used for supervised learning improvement with using unmarked training data [13]. The problem that is solved by this method is comprised in follow.

Let there be a set of data patterns $x = \{x_1, \ldots x_n | x_i \in X\}$ and a set of matching marks $x = \{y_1, \ldots y_n | y_i \in Y\}$. In addition, there is an additional set of data patterns $x = \{x_{n+1}, \ldots x_{n+m} | x_i \in Y\}$, that don't have the corresponding marks y. Using a combination of marked and unmarked data, it is necessary to achieve the best classification result compared with use of only marked data.

To apply this method, the following assumptions must be made [14]:

- the points lying close to each other are marked the same manner with greater probability;
- the data mainly form discrete clusters, points from the same cluster are marked the same manner with greater probability;
- the redundant data, that is form a diversity of much smaller dimension than the input space.

For high-volume semi-supervised learning, with great data volume application is used the generative neural network models, representative of which deep belief network.

Considering the properties of DBN, the purpose of such model learning can be expressed as follows [14]:

$$\text{argmax}_\theta\left(\log p\left(\{x_i,\ y_i\}_{i=1}^{n}|\theta\right) + \lambda \log p\left(\{x_i\}_{i+n=1}^{n+m}|\theta\right)\right), \quad (4.9)$$

where θ is the set of network parameters (W and b).

Semi-supervised learning has two variants of use [13, 14]:

1. a large amount of marked data is available for model training;
2. a large amount of unmarked data and a much smaller amount of marked data are available for model creation and training.

The semi-supervised learning procedure contains the following steps:

1. construction and training of DBN without teacher training using the full amount of training data;
2. adding a classifier layer to the DBN;
3. discriminatory retraining of the network using the marked part of the training set with model quality control.

4.4.1 Approaches to Deep Belief Networks Creating

The choice of deep learning neural networks is still an urgent problem that is being explored by scientists in machine learning area all over the world [6]. In the absence of a unified theory that would describe this problem and give a single algorithm of its solution, researchers and engineers are forced to create and use the variety heuristic techniques of deep neural network models creating, beginning from manual setting of their parameters, and up to software products creating, that execute the automized search of optimal structures.

It is practiced two approaches to DBN building:

1. a priori task of network structure;
2. optimal DBN structure search.

The first approach is applied when there is the desired structure of adjective model, such as MLP, is known, that can be caused, for example, by its application specific, limitation of computational resources, etc. Under this approach it is created RBM cascade with given parameters depth and width and layer greedy training. The purpose of model training under this approach is to achievement the highest possible quality model with given restrictions on the structure.

In the second approach, when the desired structure of the adjective model is unknown, it is executed the optimal DBN structure search by iteratively growing of RBM cascade. The purpose of training in this approach is to achieve a given threshold level of model quality.

Conducted researches [15] prove that the width of the layers can be limited by the value above:

$$w = n + 4, \tag{4.10}$$

where n is the size of the input layer of the network, or in other words, the data pattern vector size from the training set [12].

After reaching the depth of the network, which gives sufficient quality to its generalization, it is sometimes facilitated by the removal of excess connections or neurons [16].

The DBN structural synthesis task is reduced to iterative construct of restricted Boltzmann machines cascade, with step-by-step learning of them and model validation at the end of each iteration [6].

The quality criterion under DBN learning is the generalization error on the test data set. For DBN generative quality estimation is used the criterion of absolute rejection of reproduced and given data patterns:

$$e = \frac{1}{N}\sum_{i=1}^{N} |v_i - \hat{v}_i|, \tag{4.11}$$

where v is the test data pattern; $\hat{v}$ is the generative pattern; e is the playback error; N is the size of test data set.

If Deep Belief Network is used for MLP pretraining it is used the error and accuracy estimates of classification on the test data set

$$e = \frac{1}{N}\sum_{i=1}^{N} |c_i - \hat{c}_i|, \tag{4.12}$$

$$a = \frac{g}{N}. \tag{4.13}$$

4.4.2 *Problems of Deep Belief Networks Creation*

In case when the DBN structure was not set a priori, it is existing the problem of optimal structure search and adjustment of optimal parameters for such structure. Since neural networks do not yet have a universal analytical description in terms of optimal structure for different tasks, it is used empirically conditioned heuristics. Although there are many such techniques, their choice and application is more often dependent from the user, the expert. However, there is an obvious need to automate the search of DBN optimal structure, since its manual selection and learning settings are very costly and often routine [6].

It is necessary to create a software module that performs the structural-parametric synthesis of a DBN, that is, implements the following algorithms:

- DBN optimal structure searching;
- DBN optimal learning hyperparameters search and training performance;
- DBN ultimate discriminatory retraining;
- DBN transformation into objective model.

The following steps must be followed to achieve the goal:

- finding and analysis of available and effective tools for creating and training DBN and RBM;
- software development for implementation of processes neural network models creating and tuning; means of process monitoring during module operation;
- implementation for receiving, preparing and data sets storing steps, and module functioning results saving in the form of software model suitable for objective use in the production environment.

The product created must meet the following requirements:

- implement steps and algorithms that realize the purpose of this work;
- have a user-friendly and documented user interface;
- enable the user to get current results of the module functioning;
- realize the possibility of full stop of the module functioning, and work resumption from the stop point with saving the previous results, or complete restart of the process with deleting of the previous results;
- it must be enough simple to set up and deploy in a production environment.

4.5 Overview of Existing Solutions

The problem of building and learning automation of deep learning neuron network optimal model exists for a long time, so there is a lot of researches in search of its solution [6].

For today, there are at least two approaches that are directed to facilitate the construction and training of effective models by the expert.

1. automation of individual steps of model construction and learning;
2. reusing existing adjustable models of general-use with further adjustment under specific task. This approach is called "transfer learning", has its specific features, advantages and disadvantages and is described in detail in the next section.

In the direction of automatization of creation and learning neural networks it is existed certain scientific and engineering results, which were realized in the form of specialized software libraries and frameworks.

4.5.1 Google Cloud AutoML

This toolkit for building a wide variety of models for a whole host of tasks of machine learning tasks was developed by Google's Cloud AI department and implemented in the form of a cloud-based online service that allows automatically build neural network models, train them and continue to be used as part of other software systems built in the cloud ecosystem services of this company.

One of the advantages of this tool is:

- performs hypotheses creation based on input data without user intervention;
- does not require user qualification in machine learning, and has simple use;
- the result can easily be embedded into other software system that has been deployed in the same Google cloud services ecosystem.

Disadvantages:

- "Black Box"—there is no control over the process of model construction and training; it is not possible to estimate the intermediate results of functioning, etc.;
- the entire amount of input data must be marked.

The inconvenience of using the results obtained outside of Google's cloud-based online services.

4.5.2 H2O.ai

The framework designed to automate operations of the construction and training of various machine learning models, including deep neural networks. It is designed for experienced, skilled specialists, researchers of machine-learning, allows to manage a process, capable of models creating suitable for deployment in a production environment of various types.

Advantages:

- wide range of models;
- advanced configuration system;
- available control over the model of building and training process; enables the user to evaluate intermediate performance and intervene for reconfiguration, etc.;
- ability to interpret the obtained results;
- it is relatively easy to deploy in a variety of levels and types of environment (from local workstation, local server, computing cluster and to cloud services) due to an advanced set of interfaces and the use of containerization and orchestration technologies;
- fit for work with frameworks of big data processing, distributed computing;
- friendly to cloud computing services of Amazon, Google, Microsoft.

Disadvantages:

- high "entry threshold", too sophisticated tool for relatively small and simple tasks;
- it requires refinement to create, train and transform DBN;
- for work with a hybrid (unmarked + marked parts) data set requires complex configuration.

4.5.3 TPOT

The framework that positions by authors as an assistant for data researchers. Also designed to perform the full cycle of machine learning model creation: from pre-processing data, through the model search, until training of the chosen variant. It is based on genetic algorithms for optimal architectures model and their training parameters search.

Advantages:

- researcher-friendly, flexible in configuration and usage;
- transparency of monitoring processes.

Disadvantages:

- the complexity of distributed calculations uses without additional programming;
- low usability with large volumes of data, more suitable for the exploration of individual models on small data sets;
- difficulty in using with GPUs, lack of settings, need to create additional interfaces.

4.5.4 Scikit-Learn

The popular classic machine learning framework, that has a vast array of data processing tools of data processing machine learning models, rich documentation and a wealth of fans. It is very common among statisticians, researchers and students. It also contains tools for automatically finding the structure of machine learning models and hyperparameters of their learning.

Advantages:

- large toolbox, rich documentation;
- ability to automatical search of models' structures and training parameters.

Disadvantages:

- lack of convenient tools for building and deep neural networks learning;
- necessity of work results post-processing for application in real production;
- lack of ability uses calculations on graphical process without additional software.

4.5.5 *Strategy of DBN Structural-Parametric Synthesis*

Deep belief network structurally is a cascade of RBM. So, the problem of DBN creation consist of consistently addition RBM, starting from the first one that accepts input data.

The deep belief network synthesis strategy has the following steps:

1. initialization of DBN;
2. adding RBM to the existing structure. If this is the first unit of the cascade, then the input data is formed directly from the training data set, otherwise, the input data is formed by passing data of the training data set through the previous RBM in the cascade;
3. training of the current RBM;
4. adding a classifier layer and discriminating DBN training;
5. testing of all DBN for quality by the criterion of minimization of generalization error (classification accuracy) on the test data set;
6. if the generalization error exceeds the threshold, the classifier layer is deleted, the existing network parameters are fixed and steps 2–6 are repeated.

After achievement of necessary DBN quality, the process of structural-parametric synthesis is considered complete, and the model is built and trained. After validation RBM learning results, it is used generative (by error value of test pattern backing up) and classification (by the classification accuracy on the test set of marked data) model is used. When DBN is used for training on an unmarked data set, it is executed only validation for the network generative capability. The DBN parameters thus layer by layer repeat the corresponding RBM parameters, and thus the main task of DBN parameters tuning is transformed into step by step learning of each RBM in cascade.

4.6 Restricted Boltzmann Machine, Its Structure and Definitions

Restricted Boltzmann machine is a generative stochastic energy artificial neural network, that capable of learning the probability distribution in a set of input data. It is a development of the Universal Boltzmann Machine (UBM) and was created to avoid the practical restrictions of the UBM, such as the practical inefficiency of sampling from distribution, the calculation of energy values, and the like. To address these issues, structural restrictions have been imposed on UBM, which has greatly simplified its training and use. [10].

As the name implies, RBM is a variation of the Boltzmann machine with a ban on connections between neurons of one layer, which converts the model into a two-particle graph in which one fraction corresponds to a layer of "visible" neurons v_i and the other to "hidden" h_i. The structure of the RBM is shown in Fig. 4.5.

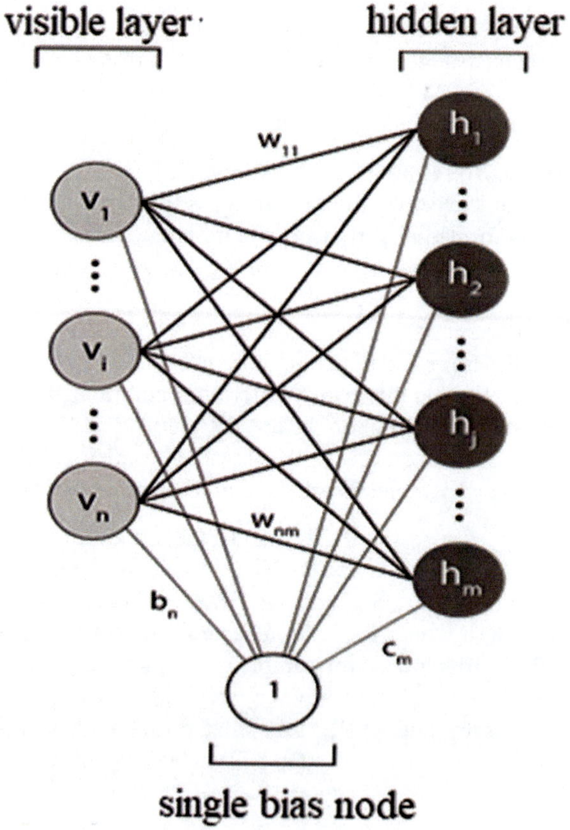

Fig. 4.5 Structure of RBM

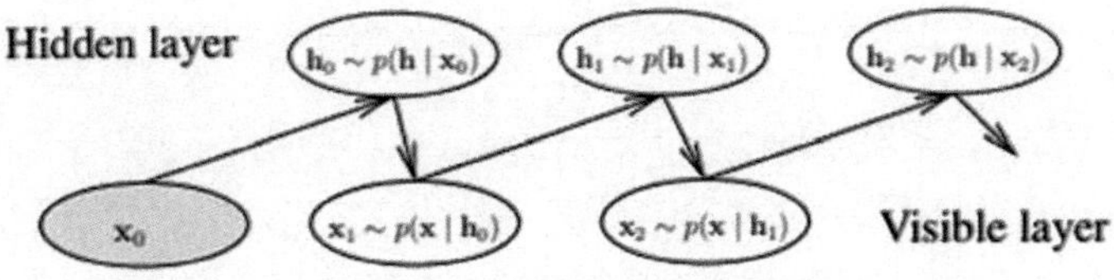

Fig. 4.6 Visualization of the idea of how the layer-wise Gibbs sampling in RBM

Restricted Boltzmann machines can also be used in depth learning networks. In particular, deep belief neural networks can be formed by "assembling" an RBM and, possibly, by fine-tuning the resulting DBN using gradient descent and back propagation algorithm.

Since RBM is a special case of Boltzmann machine (BM), it is possible to employ the same Gibbs sampling to learn. Thanks to its restricted structure, Gibbs sampling can be used more efficiently, as given one layer, either visible or hidden, the neurons in the other layer become mutually independent (Fig. 4.6). This possibility of the layer-wise sampling enables the full utilization of the modern parallelized computing environment.

However, as the number of neurons in RBM increases, a greater number of samples must be gathered by Gibbs sampling in order to properly explain the probability distribution represented by RBM. Moreover, due to the nature of Gibbs samplings, the samples might still miss some modes of the distribution.

4.7 Topology of RBM

As already noted, this approach is based on the representation of each layer of the neural network in the form of RBM. The restricted Boltzmann machines consists of two layers of stochastic binary neural elements, which are interconnected by bidirectional symmetric connections (Fig. 4.7). The input layer of neural elements is called visible (layer X), and the second layer is hidden (layer Y). A deep neural network can be represented as a collection of RBM. Restricted Boltzmann machines can approximate (generate) any discrete distribution if enough neurons of the hidden layer are used [14].

This network is a stochastic neural network in which the states of visible and hidden neurons change in accordance with the probabilistic version of the sigmoid activation function:

$$p(y_j|x) = \frac{1}{1+e^{-S_j}}, \quad S_j = \sum_i^n w_{ij}x_i + T_j, \tag{4.14}$$

$$p(x_j|y) = \frac{1}{1+e^{-S_i}}, \quad S_i = \sum_j^n w_{ij}y_j + T_i. \tag{4.15}$$

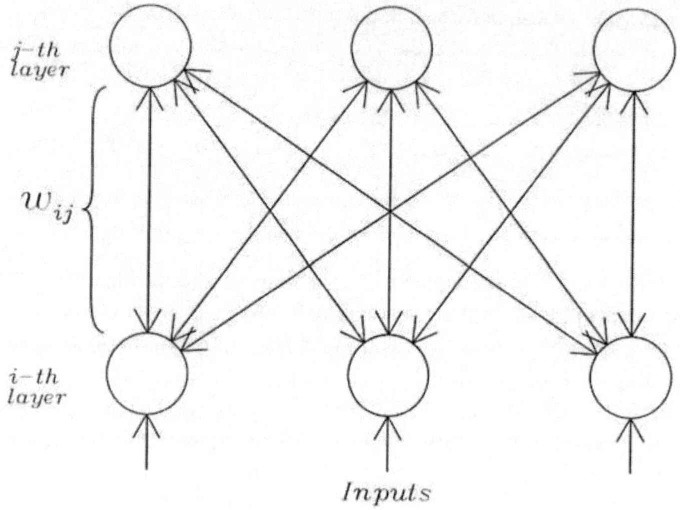

Fig. 4.7 Restricted Boltzmann machine

The states of visible and hidden neural elements are made independent:

$$P(x|y) = \prod_{i=1}^{n} P(x_i|y), \tag{4.16}$$

$$P(y|x) = \prod_{j=1}^{m} P(y_j|x). \tag{4.17}$$

Thus, the states of all the neural elements of RBM are determined through the probability distribution. In restricted Boltzmann machines, neurons of the hidden layer are feature detectors that determine the patterns of input data. The main task of training is to reproduce the distribution of input data based on the states of the neurons of the hidden layer as accurately as possible. This is equivalent to maximizing the likelihood function by modifying the synaptic connections of the neural network. Consider this in more detail.

The probability of finding the visible and hidden neuron in the state (x, y) is determined on the basis of the Gibbs distribution:

$$P(x, y) = \frac{e^{-E(x,y)}}{Z}, \tag{4.18}$$

where $E(x, y)$ is the energy of the system in the state (x, y); Z is the parameter of the probability normalization condition; the sum of the probabilities should be equal to one. This parameter is calculated as follows:

$$Z = \sum_{x,y} e^{-E(x,y)}. \tag{4.19}$$

The probability of finding visible neurons in a certain state is equal to the sum of the probabilities of the $P(x, y)$ configurations over the states of hidden neurons:

$$P(x) = \sum_{y} P(x, y) = \sum_{y} \frac{e^{-E(x,y)}}{Z} = \frac{\sum_{y} e^{-E(x,y)}}{\sum_{x,y} e^{-E(x,y)}}. \tag{4.20}$$

To find the rule for modifying synaptic connections, it is necessary to maximize the probability of reproducing the states of visible neurons $P(x)$ by the Boltzmann machine. In order to determine the maximum likelihood function of the distribution of data $P(x)$, we apply the method of gradient descent in the space of weights and network thresholds, where the gradient we take the log-likelihood function:

$$\frac{\partial \ln(P(x))}{\partial \omega_{ij}} = -\sum_{y} P(y|x) \frac{\partial E(x, y)}{\partial \omega_{ij}} + \sum_{x,y} P(y|x) \frac{\partial E(x, y)}{\partial \omega_{ij}}. \tag{4.21}$$

Then the gradient is

$$\frac{\partial \ln(P(x))}{\partial \omega_{ij}} = \frac{\partial}{\partial \omega_{ij}} \left(\sum_{y} e^{-E(x,y)} \right) - \frac{\partial}{\partial \omega_{ij}} \left(\sum_{x,y} e^{-E(x,y)} \right). \tag{4.22}$$

Transforming the last expression, we get

$$\frac{\partial \ln(P(x))}{\partial \omega_{ij}} = -\frac{1}{\sum_{y} e^{-E(x,y)}} \sum_{y} e^{-E(x,y)} \frac{\partial E(x, y)}{\partial \omega_{ij}} + \frac{1}{\sum_{x,y} e^{-E(x,y)}} \sum_{x,y} e^{-E(x,y)} \frac{\partial E(x, y)}{\partial \omega_{ij}}. \tag{4.23}$$

Since

$$P(x, y) = P(y|x) P(x). \tag{4.24}$$

Then

$$P(y|x) = \frac{P(x, y)}{P(x)} = \frac{\frac{1}{Z} e^{-E(x,y)}}{\frac{1}{Z} \sum_{y} e^{-E(x,y)}} = \frac{e^{-E(x,y)}}{\sum_{y} e^{-E(x,y)}}. \tag{4.25}$$

As a result, you can get the following expression:

$$\frac{\partial \ln(P(x))}{\partial \omega_{ij}} = -\sum_{y} P(y|x) \frac{\partial E(x, y)}{\partial \omega_{ij}} + \sum_{x,y} P(y|x) \frac{\partial E(x, y)}{\partial \omega_{ij}}. \tag{4.26}$$

In this expression, the first addendum determines the positive phase of the RBM when the network operates on the basis of images from the training set. The second term describes the negative phase of operation when the network operates in a free mode, regardless of the environment.

Consider the energy of the RBM network. From the point of view of the network energy, the task of training is to find the configuration of the output variables with the minimum energy on the basis of the input data. As a result, the network will have less energy in the training set compared to other states.

The energy function of the binary state (x, y) is determined similarly to the Hopfield network:

$$E(x, y) = -\sum_i x_i T_i - \sum_j y_j T_j - \sum_{i,j} x_i y_j w_{ij}. \quad (4.27)$$

In this case

$$\frac{\partial E(x, y)}{\partial \omega_{ij}} = -x_i y_j, \quad (4.28)$$

$$\frac{\partial \ln P(x)}{\partial \omega_{ij}} = \sum_y P(y|x) x_i y_j - \sum_{x,y} P(x, y) x_i y_j. \quad (4.29)$$

Since the expectation is calculated as

$$E(x) = \sum_i x_i P_i, \quad (4.30)$$

then

$$\frac{\partial \ln P(x)}{\partial \omega_{ij}} = E[x_i y_j]_{data} - E[x_i y_j]_{model}. \quad (4.30)$$

Similarly, gradients can be obtained for threshold values:

$$\frac{\partial \ln P(x)}{\partial \omega_{ij}} = E[x_i y_j]_{data} - E[x_i y_j]_{model}. \quad (4.31)$$

Since the calculation of the expectation based on the RBM network is very complex, J. Hinton proposed to use an approximation of the addendum data, which he called a contrast difference (contrastive divergence (CD)) [15]. More on this method and other training methods will be discussed in the following sections.

4.7.1 *Analysis of Training Algorithms*

In this section, algorithms for the training of RBM will be considered.

Markov chains play an important role in RBM training because they provide a method to draw samples from 'complex' probability distributions like the Gibbs distribution of a Markov Random Field (MRF). This section will serve as an introduction to some fundamental concepts of Markov chain theory. The section will describe Gibbs sampling as an Markov Chain Monte Carlo (MCMC) technique often used for MRF training and in particular for training RBMs.

Gibbs Sampling belongs to the class of Metropolis–Hastings algorithms [16]. It is a simple MCMC algorithm for producing samples from the joint probability distribution of multiple random variables. The basic idea is to update each variable subsequently based on its conditional distribution given the state of the others.

All common training algorithms for RBMs approximate the log-likelihood gradient given some data and perform gradient ascent on these approximations.

4.7.1.1 Contrastive Divergence

Obtaining unbiased estimates of log-likelihood gradient using MCMC methods typically requires many sampling steps. However, recently it was shown that estimates obtained after running the chain for just a few steps can be sufficient for model training [17]. This leads to CD learning, which has become a standard way to train RBMs [15, 17–20].

The idea of k-step contrastive divergence learning (CD-k) is quite simple: instead of approximating the second term in the log-likelihood gradient by a sample from the RBM-distribution (which would require running a Markov chain until the stationary distribution is reached), a Gibbs chain is run for only k steps (and usually $k = 1$). The Gibbs chain is initialized with a training example $v^{(0)}$ of the training set and yields the sample $v^{(k)}$ after k steps. Each step t consists of sampling $h^{(t)}$ from $p\left(h\middle|v^{(0)}\right)$ and subsequently sampling $v^{(t+1)}$ from $p\left(v\middle|h^{(t)}\right)$. The gradient, w.r.t. $\boldsymbol{\theta}$ of the log-likelihood for one training pattern $v^{(0)}$ is then approximated by

$$CD_k\left(\theta, v^{(0)}\right) = -\sum_h p\left(h|v^{(0)}\right)\frac{\partial E\left(v^{(0)}, h\right)}{\partial \theta} + \sum_h p\left(h|v^{(k)}\right)\frac{\partial E\left(v^{(k)}, h\right)}{\partial \theta}v. \quad (4.32)$$

This algorithm was invented by Professor Hinton [19], and is distinguished by its simplicity. The main idea is that mathematical expectations are replaced by quite definite values. This approximation is based on Gibbs sampling. The CD-k process looks like this:

- the state of visible neurons equates to the input image;
- the probabilities of states of the hidden layer are displayed;

- for each neuron of the latent layer, the condition "1" is brought in with the probability equal to its current state;
- the probabilities of the visible layer are derived based on the latent;
- if the current iteration is less than k, then return to step 2;
- the probabilities of states of the hidden layer are displayed.

$$\triangle w_{ij} = \eta\left(M[v_i h_j]^{(0)} - M[v_i h_j]^{(\infty)}\right),$$

i.e. the longer we do sampling, the more accurate our gradient will be. At the same time, the professor asserts that even for CD-1 (only one iteration of sampling) a quite good result is obtained. The first term is called the positive phase, and the second with the minus sign is called the negative phase.

In the Gibbs sampling, the first terms in the expressions for the gradient characterize the data distribution at time $t = 0$, and the second is the reconstructed or generated state of the state at time $t = k$. Proceeding from this, the CD-k procedure can be represented as follows:

$$x(0) \to y(0) \to x(1) \to y(1) \to \cdots \to x(k) \to y(k).$$

As a result, you can get the following rules for training the RBM network. In the case of the use of CD-1 $k = 1$ and taking into account that in accordance with the method of gradient descent

$$\omega_{ij}(t+1) = \omega_{ij}(t) + \alpha \frac{\partial \ln P(x)}{\partial \omega_{ij}(t)}, \tag{4.33}$$

for consistent training we have:

$$\begin{aligned} \omega_{ij}(t+1) &= \omega_{ij}(t) + \alpha(x_i(0)y_j(0) - x_i(1)y_j(1)), \\ T_j(t+1) &= T_i(t) + \alpha(x_i(0) - x_i(1)), \\ T_j(t+1) &= T_j(t) + \alpha\left(y_j(0) - y_j(1)\right). \end{aligned} \tag{4.34}$$

Similarly, for the CD-k algorithm

$$\begin{aligned} \omega_{ij}(t+1) &= \omega_{ij}(t) + \alpha(x_i(0)y_j(0) - x_i(k)y_j(k)), \\ T_j(t+1) &= T_i(t) + \alpha(x_i(0) - x_i(k)), \\ T_j(t+1) &= T_j(t) + \alpha\left(y_j(0) - y_j(k)\right). \end{aligned} \tag{4.35}$$

In the case of group learning and CD-k

$$\omega_{ij}(t+1) = \omega_{ij}(t) + \alpha \sum_{l=1}^{L} (x_i^l(0) y_j^l(0) - x_i^l(k) y_j^l(k)),$$

$$T_j(t+1) = T_j(t) + \alpha \sum_{l=1}^{L} (y_j^l(0) - y_j^l(k)),$$

$$T_i(t+1) = T_i(t) + \alpha \sum_{l=1}^{L} (x_i^l(0) - x_i^l(k)). \tag{4.36}$$

From the last expressions it can be seen that the rules of training RBM minimize the difference between the original data and the results generated by the model. The values generated by the model are obtained by Gibbs sampling.

4.7.1.2 Persistent Contrastive Divergence

The CD-1 is fast, has a low dispersion and is a reasonable approximation to the likelihood gradient, but it is still significantly different from the probability gradient when the mixing speed is low. This can be seen by drawing samples of the distribution that he studies. Generally speaking, CD-n for greater n is better than CD-1 if there is enough time to work. In [21] was suggested solutions for such situations. In the context of the IPC, the idea is as follows.

Although CD-1 is not a very good approximation to maximum likelihood learning, this does not seem to matter when an RBM is being learned in order to provide hidden features for training a higher-level RBM. CD-1 ensures that the hidden features retain most of the information in the data vector and it is not necessarily a good idea to use a form of CD that is a closer approximation to maximum likelihood but is worse at retaining the information in the data vector. If, however, the aim is to learn an RBM that is a good density or joint-density model, CD-1 is far from optimal.

At the beginning of learning, the weights are small and mixing is fast so CD-1 provides a good approximation to maximum likelihood. As the weights grow, the mixing gets worse and it makes sense to gradually increase the n in CD-n. When n is increased, the difference of pairwise statistics that is used for learning will increase so it may be necessary to reduce the learning rate.

Method, called Persistent Contrastive Divergence (PCD) solves the sampling with a related method, only that the negative particle is not sampled from the positive particle, but rather from the negative particle from the last data point [20].

The idea of persistent contrastive divergence (PCD) [22] is described in [23] for log-likelihood maximization of general MRFs and is applied to RBMs in [22].

The PCD approximation is obtained from the CD approximation (4.22) by replacing the sample $v^{(k)}$ by a sample from a Gibbs chain that is independent of the sample $v^{(0)}$ of the training distribution. The algorithm corresponds to standard

CD learning without reinitializing the visible units of the Markov chain with a training sample each time we want to draw a sample $v^{(k)}$ approximately from the RBM distribution. Instead one keeps "persistent" chains which are run for k Gibbs steps after each parameter update (i.e., the initial state of the current Gibbs chain is equal to $v^{(k)}$ from the previous update step). The fundamental idea underlying PCD is that one could assume that the chains stay close to the stationary distribution if the learning rate is sufficiently small and thus the model changes only slightly between parameter updates [22, 23]. The number of persistent chains used for sampling (or the number of samples used to approximate the second term of gradient) is a hyper parameter of the algorithm. In the canonical form, there exists one Markov chain per training example in a batch.

The PCD algorithm was further refined in a variant called fast persistent contrastive divergence (FPCD) [24]. Fast PCD tries to reach a faster mixing of the Gibbs chain by introducing additional parameters, $w_{ij}^f b_j^f,\ c_j^f$ (for $i = 1, \ldots, n$ and $ij = 1, \ldots, m$) referred to as the fast parameters. This new set of parameters is only used for sampling and not in the model itself. When calculating the conditional distributions for Gibbs sampling, the regular parameters are replaced by the sum of the regular and the fast parameters, i.e., Gibbs sampling is based on the probabilities

$$\tilde{p}(H_i = 1|v) = \mathrm{sig}\left(\sum_{j=1}^{m}\left(w_{ij} + w_{ij}^f\right)v_j + \left(c_i + c_i^f\right)\right) \tag{4.37}$$

and

$$\tilde{p}\left(V_j = 1|h\right) = \mathrm{sig}\left(\sum_{i=1}^{m}\left(w_{ij} + w_{ij}^f\right)h_i + \left(b_j + b_j^f\right)\right), \tag{4.38}$$

instead of the conditional probabilities given by (4.14) and (4.15). The learning update rule for the fast parameters is the same as the one for the regular parameters, but with an independent, large learning rate leading to faster changes as well as a large weight decay parameter. Weight decay can also be used for the regular parameters, but it has been suggested that regularizing just the fast weights is sufficient [24]. Neither PCD nor FPCD seem to increase the mixing rate (or decrease the bias of the approximation) sufficiently to avoid the divergence problem, as can be seen in the empirical analysis in [25].

4.8 Parallel Tempering

However, the study of contrast divergence is considered an effective way to study RBM, it has a drawback due to the biased approach in the learning curve. It is proposed using the advanced Monte Carlo method, which is called Parallel Tempering (PT), and experimentally shows that it works effectively.

The problem that was not solved by either the Gibbs sampling or the CD training is that the samples formed during the negative phase are not inclined to explain the whole state of the state. Thus, this section proposes to use another improved version of the Monte Carlo Markov Chain sampling method, called Parallel Tempering (PT).

The introduction of the PT sample occurs in [24] where introduced the Monte Carlo replica and applied it to the using model, which is equivalent to RBM with visible neurons. Simulation of the replica Monte Carlo suggested modeling several copies of particles (replicas) at different temperatures simultaneously, rather than simulating them consistently. Similarly, later introduced the use of a parallel MC sampling chain based on the mixing rate of samples through parallel chains to maximize the likelihood.

The basic idea of the PT sample is that the samples are collected from a plurality of Gibbs discretization chains with different temperatures from the highest temperature $T = 0$ to the current temperature $T = 1$. The term temperature in this context means the energy level of the overall system. The higher the temperature of the chain, the more likely it is that samples collected with Gibbs move freely.

For each pair of samples collected from two different chains, the probability of swapping is calculated, and the samples vary places according to the probability. The probability of a swap for a pair of samples is formulated in accordance with the Metropolis rule (see, for example, [26])

$$P_{swap}\left(x_{T_1}, x_{T_2}\right) = \min\left(1, \frac{P_{T_1}\left(x_{T_2}\right) P_{T_1}\left(x_{T_1}\right)}{P_{T_1}\left(x_{T_1}\right) P_{T_2}\left(x_{T_2}\right)}\right), \tag{4.39}$$

where T_1 and T_2 denote the temperatures of the two chains; x_{T_1} and x_{T_2} denote samples collected from two chains.

After each round of sampling and swapping, the sample at the true temperature $T = 1$ is gathered as the sample for the iteration. The samples come from the true distribution, $p(v,\ h \mid \theta)$ in case of RBMs, assuming that enough iterations are run to diminish the effect of the initialization.

It must be noted that the Gibbs sampling chain with the highest temperature ($T = 0$) is never multi-modal such that all the neurons are mutually independent and likely to be active with probability 1/2. So, the samples from the chain are less prone to missing some modes.

From the chain with the highest temperature to the lowest temperature, samples from each chain become more and more likely to follow the target model distribution. How PT sampling could be trapped into a single mode is illustrated in Fig. 4.8.

The red, purple, and blue curves and dots indicate distributions and the samples from the distributions with the high, medium, and cold temperatures, respectively. Each black line indicates a single sampling step.

This nature of swapping samples between the different temperatures enables better mixing of samples from different modes with much smaller number of samples than that would have been required if Gibbs sampling was used.

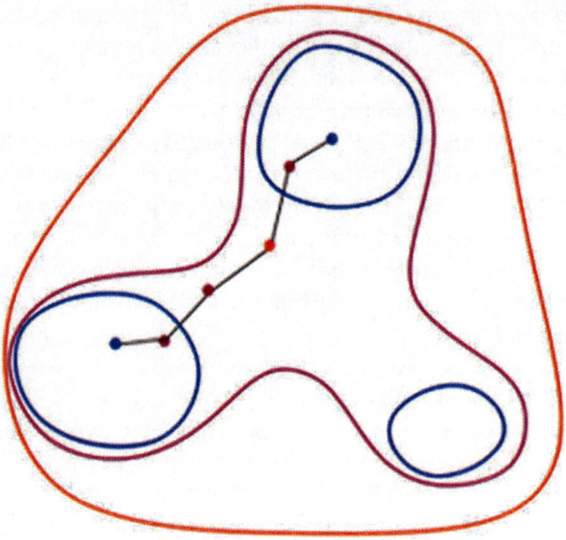

Fig. 4.8 Illustration of how PT sampling could avoid being trapped in a single mode

PT sampling in training RBMs can be simply uses as a replacement of Gibbs sampling in the negative phase. This method is, from now on, referred to as PT learning. Due to the previously mentioned characteristics, it is expected that the samples collected during the negative phase would explain the model distribution better, and that the learning process would be successful even with a smaller number of samples than those required if Gibbs sampling is used.

A brief description of how PT sampling can be carried out for RBMs is given in Algorithm 3. This is the procedure that is run between each parameter update during learning.

Algorithm 3 PT sampling steps for an RBM

Create a sequence of RBMs $(R_0, R_1, \cdots, R_K)$ such that parameters of R_k are $\boldsymbol{\theta}_k = (T_k\mathbf{W}, T_k\mathbf{b}, T_k\mathbf{c})$, where $0 \leq T_0 < T_1 < \cdots < T_K = 1$.
Create an empty set of samples $\boldsymbol{X} = \{\}$.
Set $\mathbf{x}_0 = (\mathbf{x}_{0,0}, \cdots, \mathbf{x}_{K,0})$ such that every $\mathbf{x}_{k,0}$ is a uniformly distributed random vector (or use old ones from the previous epoch).
for $m = 1 \cdots M$ **do**
  Sample $\mathbf{x}_m = (\mathbf{x}_{0,m}, \cdots, \mathbf{x}_{K,m})$ from the sequence of RBMs such that $\mathbf{x}_{k,m}$ is sampled by one-step Gibbs sampling starting from $\mathbf{x}_{k,m-1}$.
  for $j = 2 \cdots K$ **do**
    Swap $\mathbf{x}_{j,m}$ and $\mathbf{x}_{j-1,m}$ according to $P_{\text{swap}}(\mathbf{x}_{j,m}, \mathbf{x}_{j-1,m})$ computed using (3.1).
  end for
  Add $\mathbf{x}_{K,m}$ to $\boldsymbol{X}$.
end for
$\boldsymbol{X}$ is the set of samples collected by parallel tempering sampling.

Although, potentially, the PT sample algorithm for RBM has advantages over * CD family algorithms, but in the very nature of the PT sample, its major disadvantage is hidden, namely the necessity of large number of difficult computational operations for each sample obtaining, that CD algorithms successfully avoid.

Typically, different objects and different data sets require the individual models, both under structure and under the methods of its training and parameters. This approach is also applied to the RBM, for which it has been developed different learning algorithms that satisfy to different simulation conditions. To reduce the load on the expert that creates the model, search automating task of optimal algorithm and hyperparameters of training, which most closely corresponds to the training data and criteria of this model effectiveness. To solve the problem of automatic search of optimal algorithm of RBM learning can be used the search algorithm in the space of options, commonly used for neural network model determination, its structural synthesis, and a learning algorithm. It is appropriate to apply the same approach to RBM.

4.9 The Influence of the Choice of the Base Algorithm on the Boltzmann Constrained Machine Learning

Since three algorithms for its learning are common today, it is appropriate to use a grid search algorithm (due to the small capacity of the variant space), which implements the following steps:

1. formation of multiple algorithms $alg = \{CDk, PCDk, PT\}$;
2. generating a plurality of hyperparameters for each algorithm, for example, for CD-k and PCD-k is the Markov chain depth $k = \{x|1 \leq x \leq 100\}$. For the PT algorithm, this is the number of sampling steps k corresponding to different temperatures. The authors of the PT algorithm have shown that its optimal performance is achieved when k is close to the size of the training package of the input samples;
3. formation of the search grid by combining sets alg and k. For the current example, the grid size will be equal $|alg| \bullet |k|$;
4. start the model training process for each variant of the search grid;
5. validation of the trained models and selection of the best option according to the optimality criteria on the test data set.

Provided that sufficient computing resources are available, the lattice search can be performed in parallel, thus significantly reducing the process of finding and building the optimal model.

To evaluate the influence of a one or another learning algorithm on the convergence of RBM, a series of experiments were performed using a search by grid formed by a combination of the parameters given in the Table 4.1. The validated each variant was executed on a test data set in two modes: generation and classification. The structure of the test model is shown in Fig. 4.9.

Table 4.1 Variants of RBM learning hyperparameters for determination of search grid for conducting experiments

Algorithm	*k*
CD	[1, 2, 3, 5, 15, 100]
PCD	[1, 2, 3, 5, 15, 100]
PT	[10, 100]

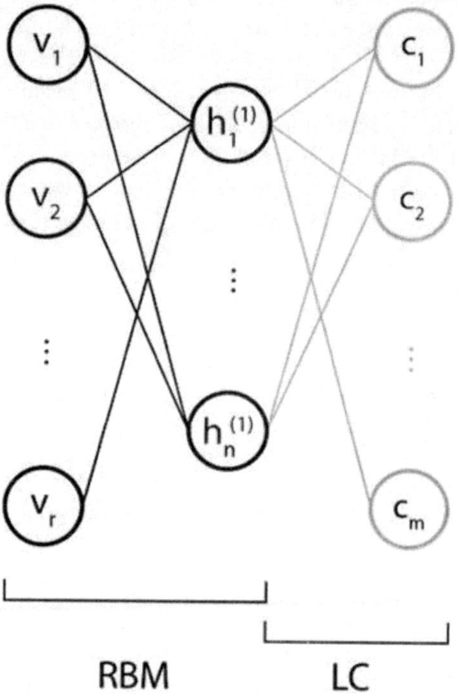

Fig. 4.9 Structure of the test model for assessing of generative and classification ability of RBM

where LC is a linear classifier; *r* is the size of the input layer (corresponds to the size of the input pattern); *n* is the size of the hidden layer; *m* is the size of the output layer is a linear classifier (corresponds to the number of classes in the test set).

In the generation mode, the validation was performed using with criterion of error of reproduction a test set sample on pure RBM. The following steps were additionally performed to validate RBM in classification mode:

1. the RBM parameters were fixed;
2. a linear classifier with random weights initialization was added to the RBM;
3. the weights of the linear classifier were additionally learned.

Restricted Boltzmann machines validation in the classification mode was performed using the classification accuracy criterion of the test set patterns.

Table 4.2 RBM experimental model parameters and training hyperparameters for determination the effect of a type of learning algorithm on learning results

Parameter	Value
Size of the visible layer	784
Size of the hidden layer	500
Size of the classifier layer	10
Training package size (batch size)	100
Number epochs of learning	50
Step learning	0.001
Number of learning epochs of the control classifier	30
RBM optimization algorithm	SGD
Size of the training data set	50,000
Size of the test data set	10,000

Table 4.3 The grid of optimal hyperparameters set search in experiments to determine the influence of the RBM learning algorithm

CD-1	CD-2	CD-3	CD-5	CD-15	CD-100
PCD-1	PCD-2	PCD-3	PCD-5	PCD-15	PCD-100
PT-10	PT-100	–	–	–	–

Other RBM training adjustments using the MNIST benchmark [27] are remained common to all options (Table 4.2).

For the PT algorithm, the use of small values k is doesn't make sense, since the mixing rate of the patterns falls significantly, in such manner, the algorithm degenerates into PCD. It makes sense for it to explore variance with $k \approx batch\ size$, and a control variant $k = 10$.

Thus, the search grid contains 14 variants, and the grid itself will look (Table 4.3).

Graphic results of experiments are reflected in the combined figures, which consist of 2 parts, reflecting the learning process in time (from above) and the diagram of the learning results after its finish (from bottom). The bottom part of the diagram columns displays the final learning error, and the top part of the diagram columns is showing the waste time on the full cycle of learning. Full column height reflects weighted learning marks: the higher the mark, the worse the result.

CD Algorithm

The results of the experiments are given in Table 4.4 and in Fig. 4.10.

It is interesting to study the effect of Markov chain length on the effectiveness of RBM when using the CD-k algorithm. Although, theoretically, an increase in this parameter improves chain convergence, studies confirm Carreira-Perpinan and Hinton's [18] finding that small values of k are sufficient for effective RBM training,

Table 4.4 The results of comparative experiments for the CD-k algorithm

k	Learning time, s	RBM learning error		Accuracy classification
		1 epoch	*50 epoches*	
1	59	0.062378	0.018724	0.7855
2	63	0.062007	0.018274	0.7993
3	64	0.062221	0.018197	0.8002
5	76	0.062358	0.018082	0.7787
15	117	0.062356	0.018077	0.7815
100	481	0.062376	0.018052	0.7984

since with increasing k the accuracy gain is negligible compared to the cost of computing resources.

As a result of the experiment, it is obvious that the use of $k > 5$ is not appropriate, and it is better to increase the number of training epochs, and hence the frequency of updating the model parameters.

PCD Algorithm

The results of the experiments are given in Table 4.5 and in Fig. 4.6, indicate that the value $k = 1$ for the PCD-k algorithm makes no sense, since in this version it becomes similar to the CD-k algorithm, and the effect from chain "storing" is offset by step-by-step update of network parameters. The algorithm achieves the highest efficiency with the value $k = 5$. The further increase of this parameter does not give its increase, but only increases the waste of time.

PT Algorithm

The results of the experiments are given in Table 4.6 and Fig. 4.11. The disadvantages of the PT algorithm compared with other RBM learning algorithms are immediately noticeable from the results of the experiments. For example, when the power of the temperatures set $k = 10$, the number of replication operations to obtain one pattern of the reproduced data is 10, wherein for each replica it is happened the model energy calculation, comparison obtained values for all adjacent replicas, calculation the probability of replica exchange, and replica exchange operation performance. This whole sequence of actions for obtaining only one pattern of data is much more expensive than a simple Gibbs sampling operation in *CD algorithms. Therefore, the use of the PT algorithm makes sense only in cases when the *CD algorithms do not provide the desired efficiency on specific data sets and under the presence of computing resources that allow to execute sampling operations in parallel. For example, for a training mini batch of 100, it is most efficient to use a computing cluster of 100 computers that generate data samples in parallel.

With a computer system that was used to perform the experiments the PT algorithm is inefficient. Further work on a similar system will be occured without the use of this algorithm.

The summary results of all experiments are shown in Fig. 4.12 (Fig. 4.13).

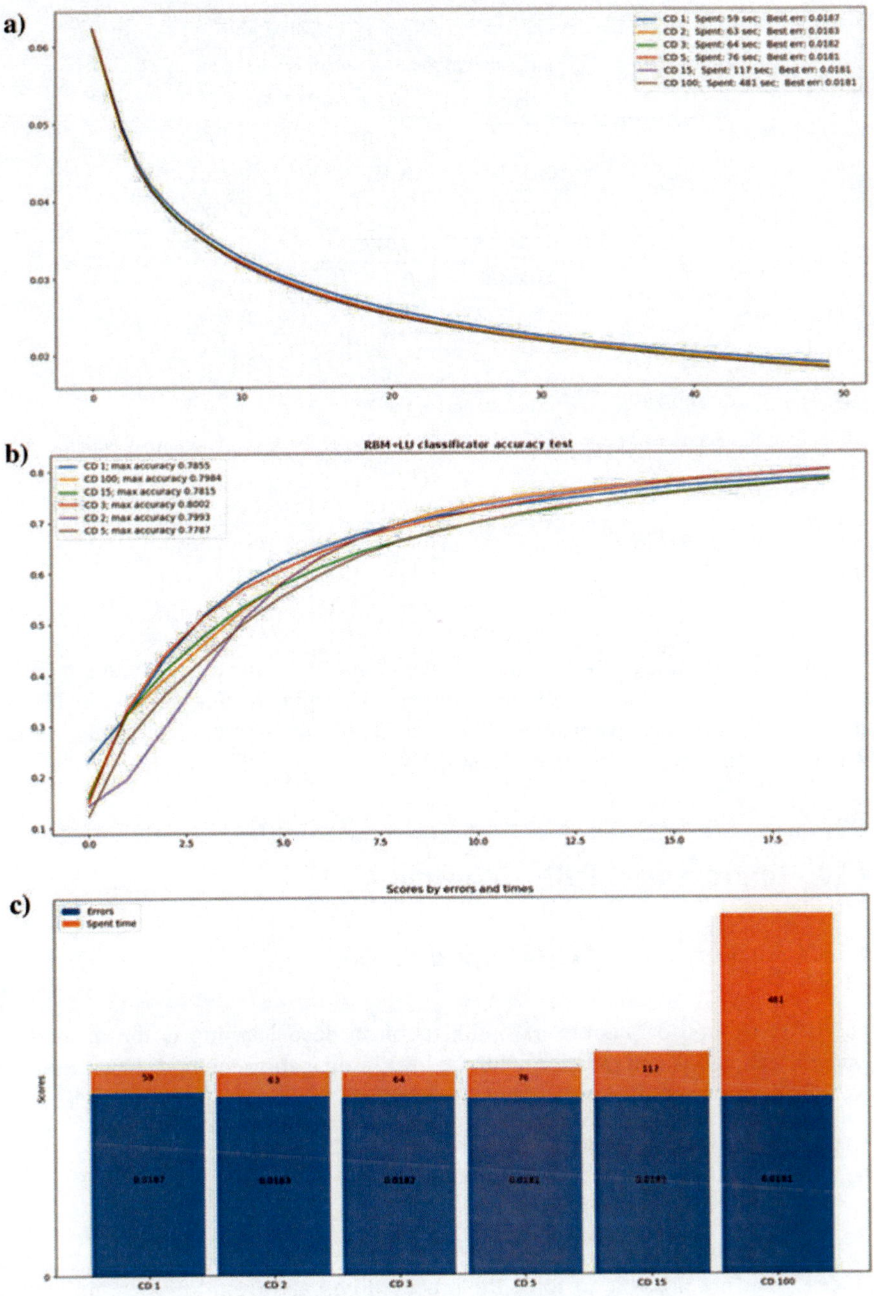

Fig. 4.10 Results for different values of k in the CD-k algorithm: **a** graph of generation error; **b** graph of classification accuracy; **c** diagram of learning results

Table 4.5 Results of comparative experiments for the PCD-*k* algorithm

k	Learning time, s	The value of the loss function		Accuracy classification
		1 epoch	*50 epoches*	
1	54	0.199377	0.020121	0.7675
2	63	0.188201	0.017520	0.7867
3	73	0.176479	0.017352	0.7882
5	76	0.154055	0.017339	0.8035
15	116	0.095482	0.017591	0.7884
100	463	0.073712	0.017661	0.7829

Table 4.6 Results of comparative experiments for the PCD-*k* algorithm

Number of temperatures *k*	Learning time, min	The value of the loss function		Accuracy classification
		1 epoch	*50 epoches*	
10	73	0.154912	0.063076	0.7887
100	803	0.155582	0.059686	0.7906

Since RBM training is based on the calculation and application of function gradients of input signal renewal error, improvement of learning results of individual RBM and DBN makes sense to consider the advanced of gradient optimization algorithms (Chap. 2), which was performed in part 4.10.

4.10 Improvement DBN Adjustment

4.10.1 Algorithms of RBM Optimization

The basic method of parameters optimization in deep learning is the method of gradient descent in combination with the back-propagation method. However, this combination has significant problems:

- high chance of falling into the local optimum;
- gradients attenuation with increasing network depth;
- "explosion" of parameters (weights).

Due to the constancy of the learning rate parameter λ during the model training process, it is not possible to force the process to escape from the local minimum if λ is sufficiently small, and vice versa, it is not possible to stop at an acceptable optimum if λ is sufficiently large.

Another problem is the fact that in a complex model, different parameters may vary in different ways: some parameters require a smaller learning step, while others,

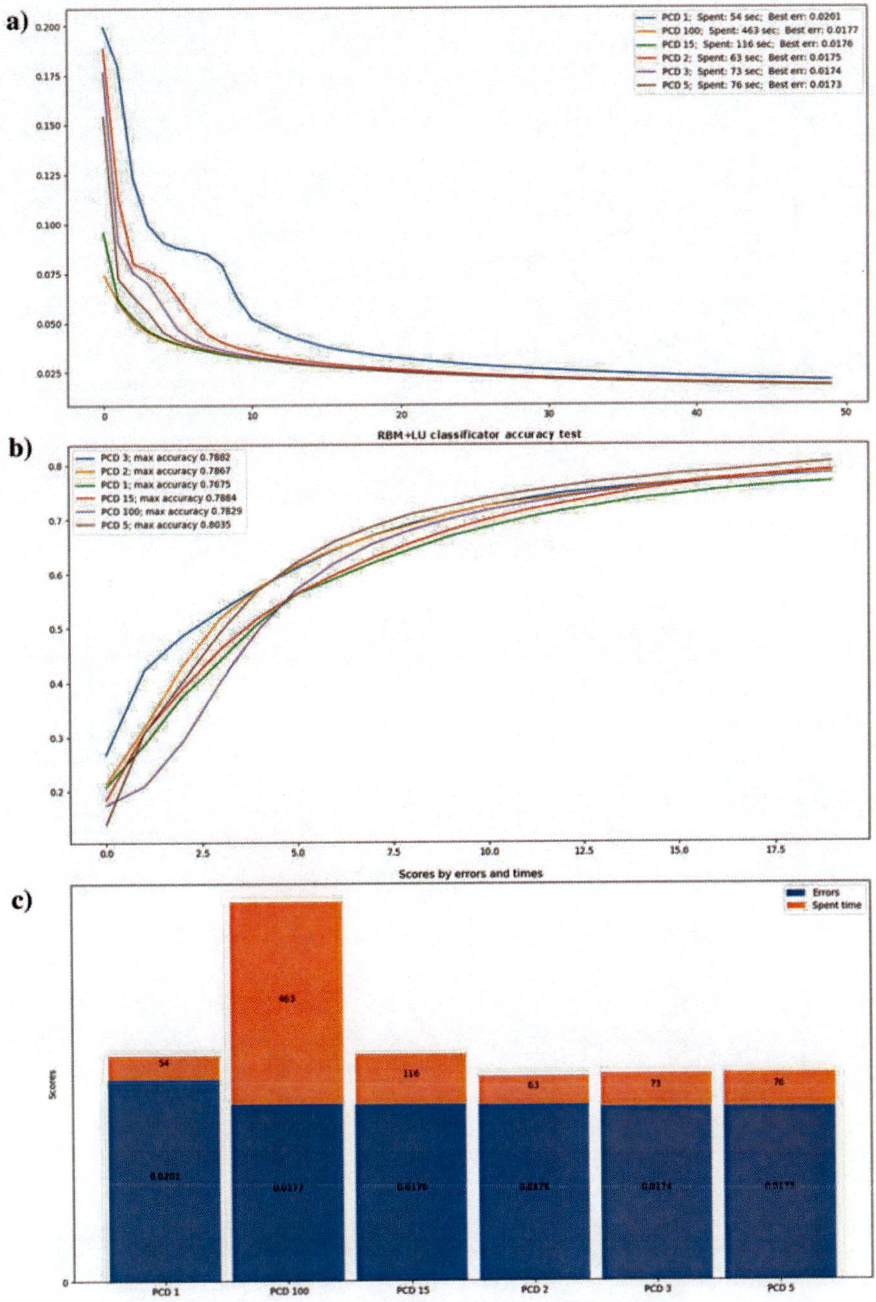

Fig. 4.11 Results for different values k in the PCD-k algorithm: **a** graph of generation error; **b** graph of classification accuracy; **c** diagram of learning results

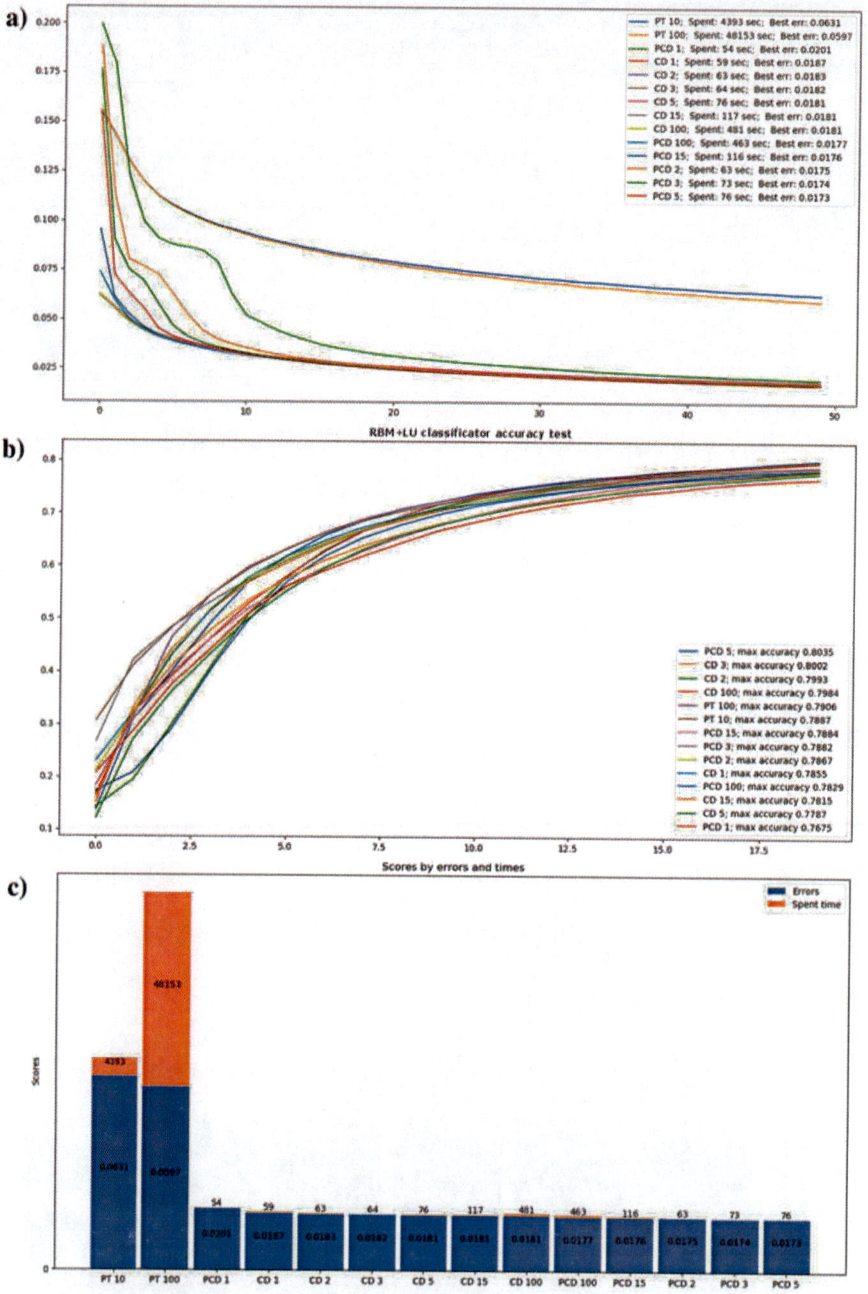

Fig. 4.12 Results for different values k in the PCD-k algorithm: **a** graph of generation error; **b** graph of classification accuracy; **c** diagram of learning outcomes

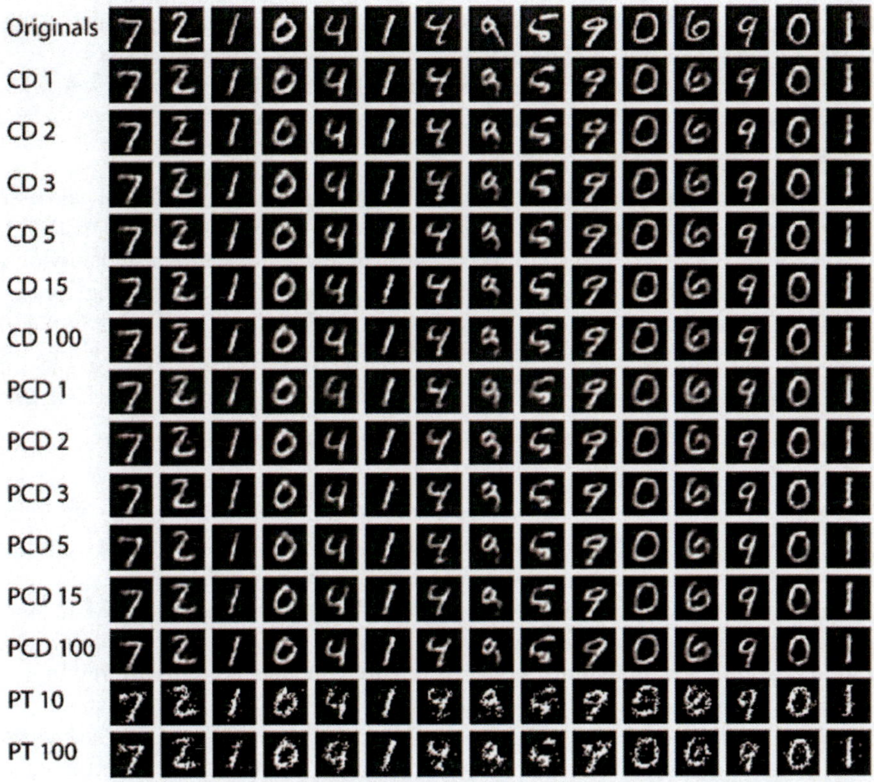

Fig. 4.13 The results of the MNIST test suite recovery models

on the contrary, more. This problem is especially relevant when input data are not normalized and unscaled.

These defects make it impossible to apply the pure gradient method for deep neural network models.

To avoid problems and improve the results of this method, a number of powerful upgraded methods-optimizes have been developed, among them: Stochastic Gradient Descent with Moment (SGDM), Nesterov accelerated gradient (NAG), Adaptive Gradient Algorithm (AdaGrad, AdaDelta), Root Mean Square Propagation (RMSProp), Adaptive Moment Assessment (AdaM), Nesterov Adaptive Moment (NAdaM) (see Sect. 2.2) [28–33].

For many particularly deep neural networks, the AdaM algorithm performs better than other optimizers, although for shallow networks it does not provide significant advantages over AdaGrad or NAG.

Advantages:

- the actual learning step in each epoch is closely related to the given step;

- the upgrade step is not directly dependent from the current gradient value, which avoids plateaus that is a problem for SGD, which is quickly getting stuck;
- AdaM was designed to combine the benefits of AdaGrad, which works well with sparse gradients, and RMSProp, which works well on small minibatch sizes. Also, this method is very easy to combine with NAG (NAdaM) and SGDM.

However, it has disadvantages. Despite its very rapid convergence [29], it has been experimentally found that AdaM sometimes converges to poorer optima than SGDM or NAG. Particularly on CIFAR family data set [34]. It was done the researches which shown the possibility of replacing AdaM after a certain amount of training epochs on SGDM, which has reached the better optimum [35].

4.10.2 The Influence of Optimizer Type for RBM Learning

Given the advancements in the development of optimization methods, the question arises of applying them to improve RBM training, with compared to the conventional gradient method. Although some of the disadvantages of the classical gradient methods with back propagation are offset by the stochastic nature of RBM, the application of momentum, grading estimation based on training history, control the training step in the training process merits research into the use of RBM training.

It is made a set of experiments with use the CIFAR-10 reference data set of the experimental model of RBM with parameters given in Table 4.7. The model structure for influence estimations of optimizer type for RBM training coincides with the experimental model structure for the previous experiments in Sect. 2.2. The experiments results validation process also coincides with the previous one, and the generative and classification ability of the RBM by the corresponding criteria.

Table 4.7 RBM experimental model parameters and optimization hyperparameters for the influence determination of optimizer type for the training process

Parameter	Value
Size of the visible layer	3072
Size of the hidden layer	1000
Size of the classifier layer	10
Size of the training package	100
Number epochs of learning	50
Number of learning epochs of the control classifier	30
Step learning	0.001
RBM algorithm	CD-1
Size of the training data set	50,000
Size of the test data set	10,000

Table 4.8 The results of comparative experiments for different moment values in the SGDM algorithm

Moment	Generation error		Classification accuracy	
	1 epoch	*The last epoches*	*1 epoch*	*The last epoch*
0.0 (SGD)	0.050145	0.027080	0.3432	0.4405
0.1	0.049340	0.026630	0.3477	0.4453
0.5	0.045789	0.024079	0.3397	0.4524
0.9	0.036009	0.015268	0.3473	0.4430
0.99	0.033236	0.013347	0.3421	0.4647

SGD and SGDM

For the SGDM algorithm, There were 4 experiments were performed with moment values $\mu = \{0.1,\ 0.5,\ 0.9,\ 0.99\}$. The results of the experiment are given in Table 4.8 and in Fig. 4.14.

The results of the experiments show the fundamental advantage of the SGDM algorithm over SGD, with a value of 0.99 the optimizer finds the best optimum of the loss function, moreover, much faster than other variants.

SGD and NAG

For the NAG algorithm, there were 4 experiments were also performed with moment values $\mu = \{0.1, 0.5, 0.9, 0.99\}$. The results of the experiment are presented in Table 4.9 and in Fig. 4.15.

The results of the experiments are very similar to the comparison results of SGDM algorithms over SGD. The NAG algorithm performs far better than SGD, and slightly better than SGDM at the same hyperparameters and learning step.

SGD and RMSProp

For the RMSProp algorithm, There were 4 experiments were also conducted with values of the coefficient σ_{ij}. The results of the experiment are given in Table 4.10 and in Fig. 4.16.

As a result of the experiments, the principal advantage of the RMSProp algorithm over the SGD can be noted, but the graph also shows volatility or the RMSProp wandering around the optima, indicating that there is too much learning for this algorithm with this model and data set.

SGD and AdaGrad

Two experiments were performed with the same initial settings of the common parameters. The results of the experiment are given in Table 4.11 and in Fig. 4.17.

As a result of the experiments, it is noticeable that the AdaGrad algorithm does not give a significant qualitative advantage over the SGD algorithm under use to training RBM on a given data set.

a)

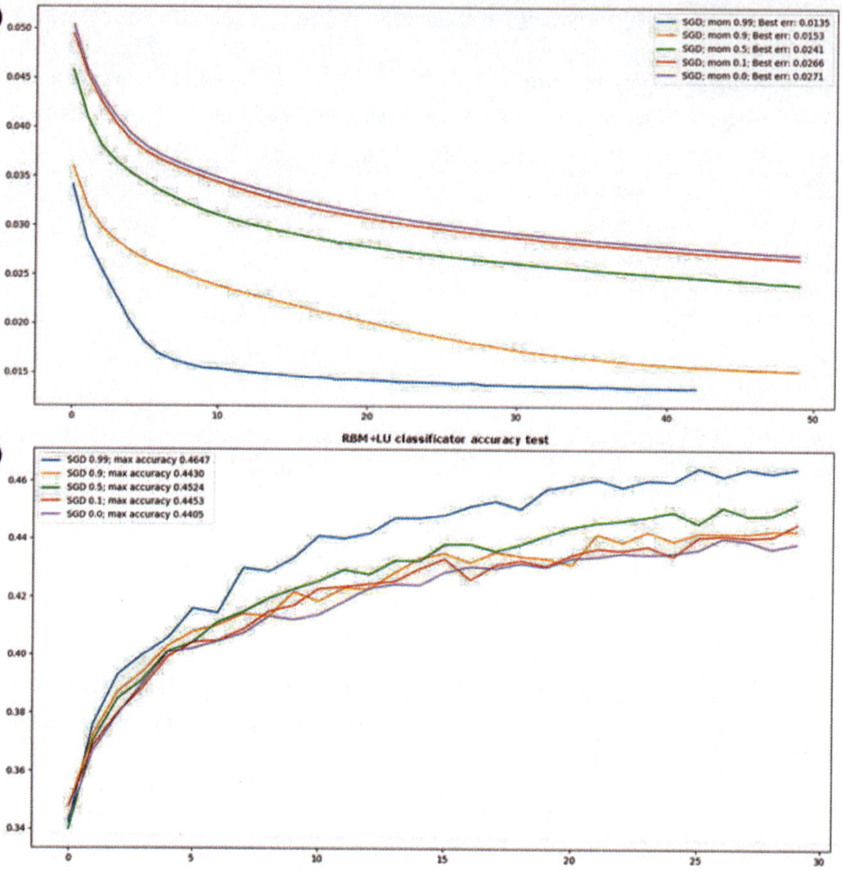

b)

Fig. 4.14 Results for different moments values in the SGDM algorithm. Under moment value $\mu = 0.0$ SGDM is converted to classic SGD: **a** graph of generation error; **b** graph of classification accuracy

Table 4.9 Results of comparative experiments for different moment values in NAG and SGD algorithms

Moment	Algorithm	Generation error		Classification accuracy	
		1 epoch	*The last epoch*	*1 epoch*	*The last epoch*
0.0	SGD	0.050145	0.027080	0.3432	0.4405
0.1	NAG	0.049338	0.026631	0.3375	0.4423
0.5	NAG	0.045778	0.024078	0.3482	0.4523
0.9	NAG	0.035837	0.015240	0.3438	0.4430
0.99	NAG	0.033236	0.013347	0.3300	0.4678

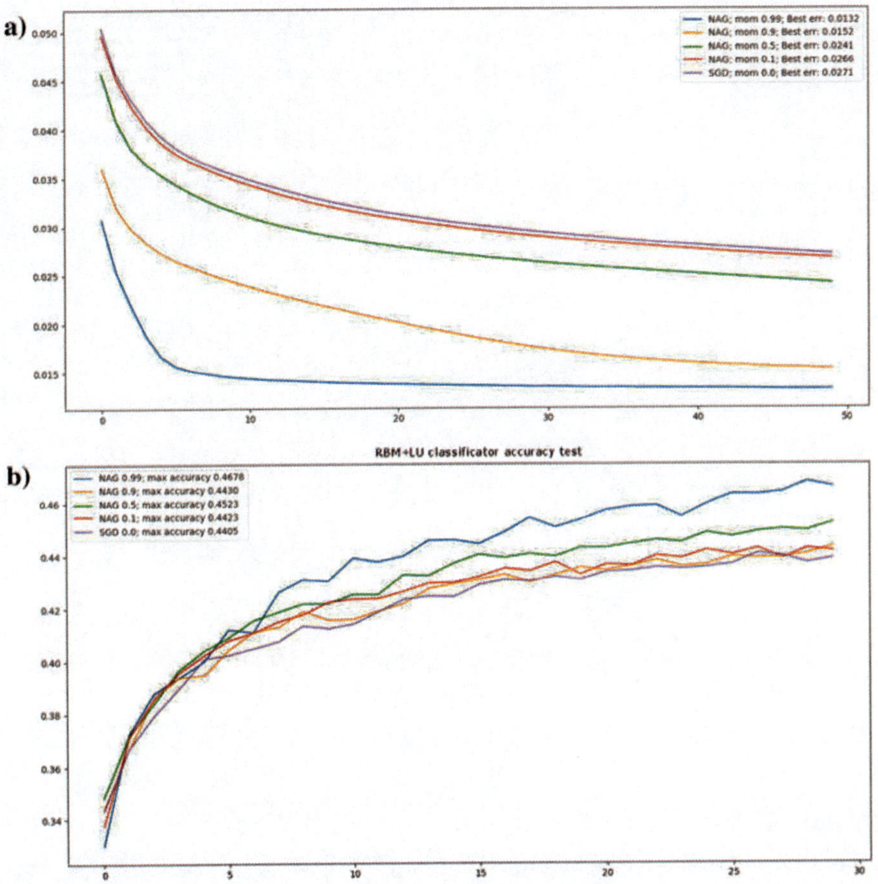

Fig. 4.15 Results for different moment values in NAG and SGD algorithms. **a** Graph of generation error; **b** graph of classification accuracy

Table 4.10 Results of comparative experiments for different moment values in Rmsprop and SGD algorithms

Moment	Algorithm	Generation error		Accuracy of classification	
		1 epoch	*The last epoch*	*1 epoch*	*The last epoch*
0.0	SGD	0.050145	0.027080	0.3432	0.4405
0.1	RMSProp	0.031899	0.018290	0.3486	0.4593
0.5	RMSProp	0.030798	0.016634	0.3517	0.4635
0.9	RMSProp	0.030436	0.015698	0.3585	0.4776
0.99	RMSProp	0.029422	0.015679	0.3522	0.4744

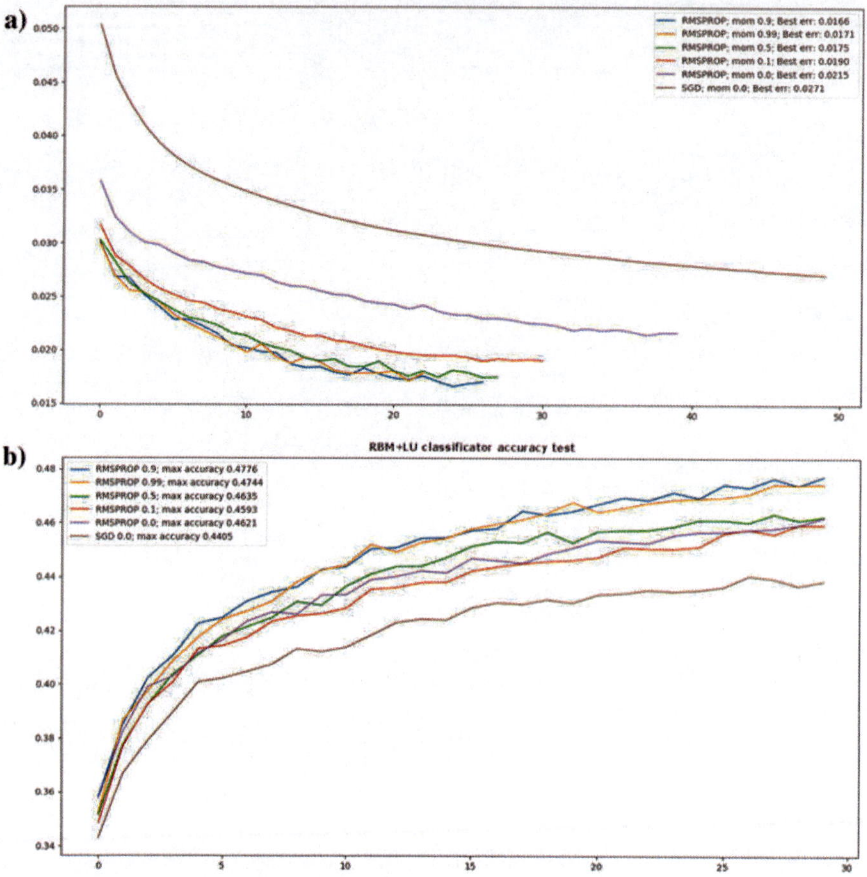

Fig. 4.16 The results for different values β in the RMSP algorithm are compared to SGD: **a** graph of generation error; **b** graph of classification accuracy

Table 4.11 Results of comparative experiments for SGD and Adagrad algorithms with the same hyperparameters

Algorithm	Generation error		Accuracy of classification	
	1 epoch	*The last epoch*	*1 epoch*	*The last epoch*
SGD	0.050145	0.027080	0.3432	0.4405
AdaGrad	0.037363	0.026286	0.3528	0.4534

SGD and AdaM

There were two experiments were performed with the same initial settings of the common parameters, and with the recommended parameters $\beta_1 = 0.9$ and $\beta_2 = 0.99$ of the AdaM algorithm [36]. The results of the experiment are presented in Table 4.12

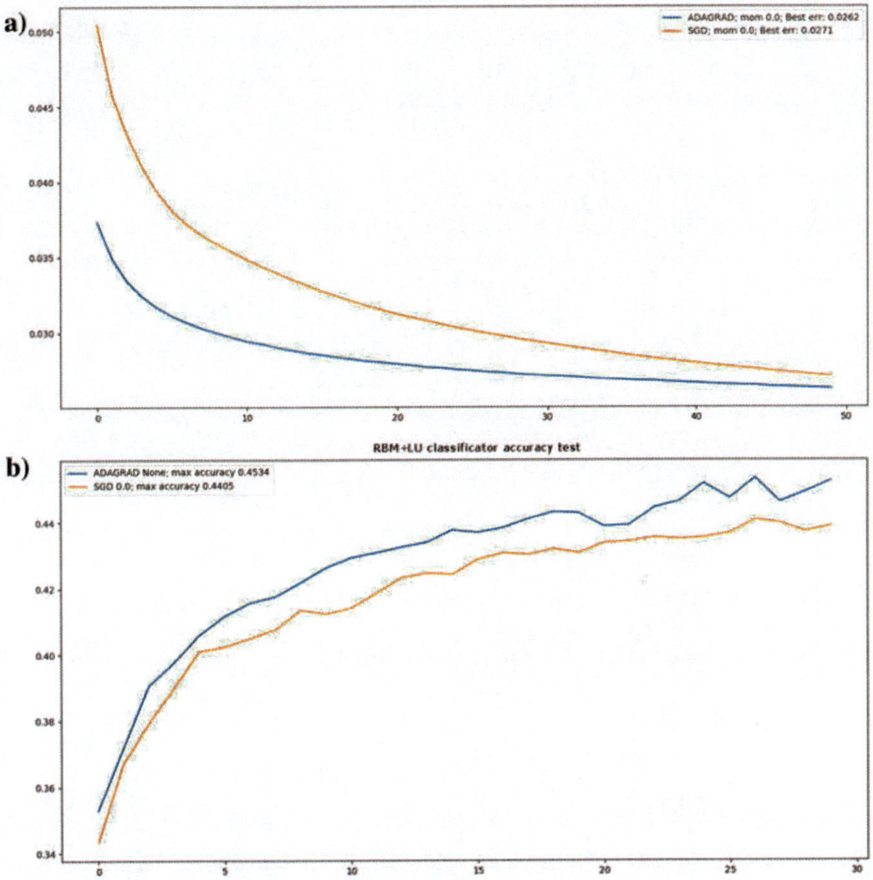

Fig. 4.17 Results of comparison of Adagrad and SGD algorithms: **a** graph of generation error; **b** graph of classification accuracy

Table 4.12 Comparison results for SGD and AdaM algorithms with similar settings

Algorithm	Generation error		Accuracy of classification	
	1 epoch	*The last epoch*	*1 epoch*	*The last epoch*
SGD	0.050145	0.027080	0.3432	0.4405
AdaM	0.026042	0.014054	0.3438	0.4503

and in Fig. 4.18. They testify to the principle advantage of using the AdaM algorithm over SGD, even without addition parameters adjusting of AdaM.

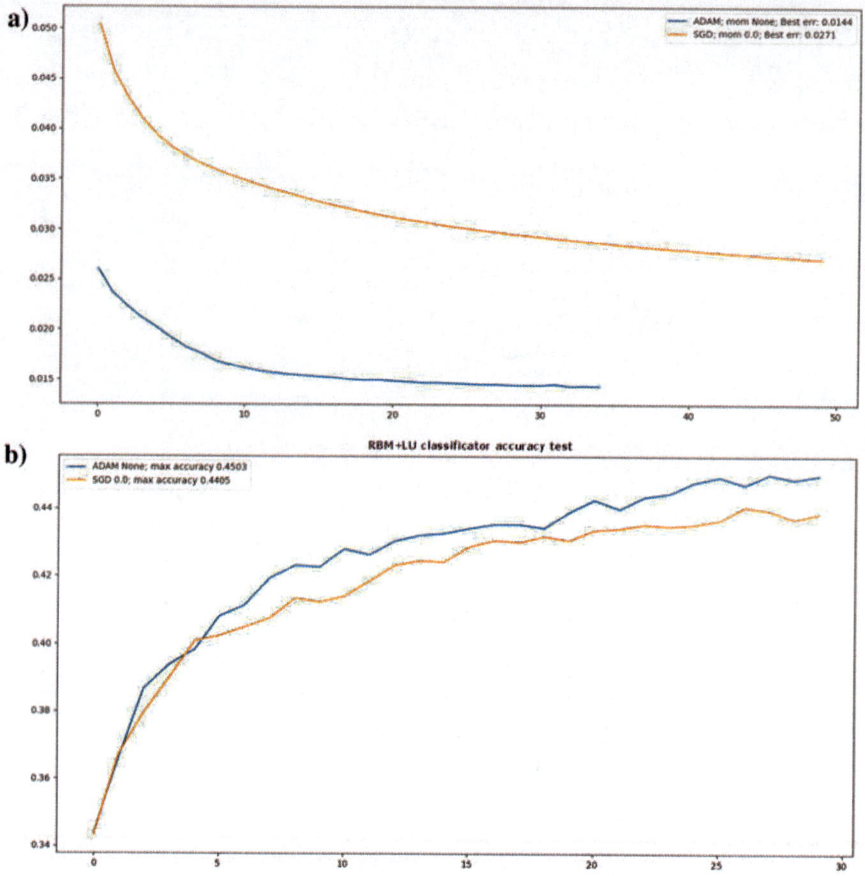

Fig. 4.18 Results of comparison of AdaM and SGD algorithms: **a** graph of generation error; **b** graph of classification accuracy

Figure 4.19 and Table 4.13 summary results of influence comparison of optimizer selection on the RBM convergence under common adjustments of learning step $\lambda = 0.001$, the number of training epoches equal to 50, and the RBM CD-1 training algorithm. For their formation, the best performing hyperparameters were taken for each optimization algorithm from the previous experiments.

The best optimization result has been achieved by the NAG and SGDM algorithms with moment value $\mu = 0.99$. The AdaM algorithm also showed a close-to-best result with the developers' default settings $\beta_1 = 0.9$ and $\beta_2 = 0.9$. The RMSProp algorithm demonstrates a walk around the optima of the objective function, that may indicate the need to find more acceptable values of the training step and the moment, but also gives much better results compared to standard gradient descent. Despite on researchers' positive evaluations of AdaGrad algorithm for deep neural networks, when used for

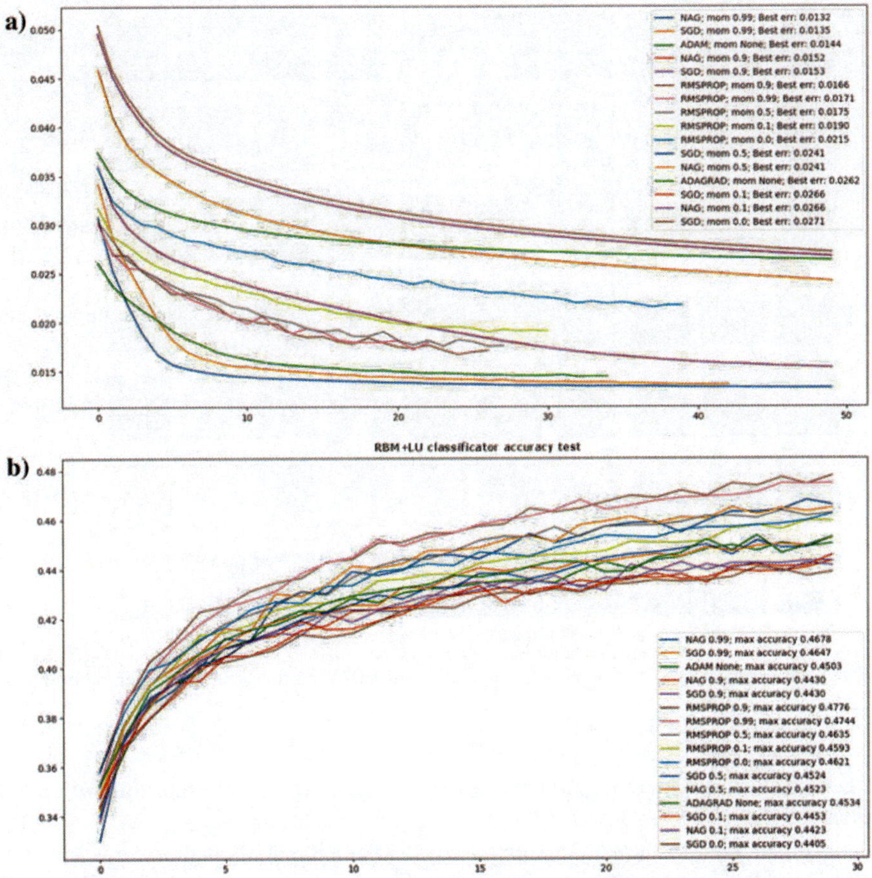

Fig. 4.19 Summary graph of optimization objective function parameters with using different optimization algorithms: **a** graph of generation error; **b** graph of classification accuracy

Table 4.13 Summarized results of comparative experiments for different algorithms of RBM optimization

Moment	Algorithm	Generation error		Accuracy of classification	
		1 epoch	*The last epoch*	*1 epoch*	*The last epoch*
0.0	SGD	0.050145	0.027080	0.3432	0.4405
0.99	SGDM	0.033237	0.013348	0.3421	0.4647
0.99	NAG	0.030135	0.013168	0.3300	0.4678
0.0	AdaGrad	0.037363	0.026286	0.3528	0.4534
0.99	RMSProp	0.029422	0.015679	0.3522	0.4744
0.0	AdaM	0.026042	0.014055	0.3438	0.4503

Fig. 4.20 Comparison of the generative ability of RBM trained with different optimization algorithms

RBM, it does not show significant advantages over the SGD algorithm without a moment.

In Fig. 4.20 it is shown the generative ability of RBM after 50 training epoches with the different use optimizers use. Data samples were taken from the CIFAR-10 non-training model data set. Which did not take part in model traning. Since the choice of optimization algorithm for the RBM training is obviously of fundamental importance, this process also would be including into search grid of model learning hyperparameters.

4.11 Method of Determining the Structure of the Neural Network Deep Learning

4.11.1 Problem Statement

A finite set $J = \{(R_j, Y_j)\}, \; j = 1, \ldots, P$ of the attribute-value pairs, where $R_j, \; Y_j$ are the input and output vectors of the NN, respectively.

It is necessary to synthesize such optimum topology of NN of deep learning, which would provide the most effective solution of the applied problem which is described and which is described by sample J [37]. The vector criterion is taken as the optimality criterion [38].

$$I = [I_1(x), I_2(x)] \to \text{opt}, \tag{4.40}$$

where $I_1(x) = E_{general}(x)$ is the generalization error, which determines the magnitude of the error of the solution of the problem; $I_2(x)$ is the complexity of the NN $x = (x_{1i}, x_2, x_3)^{\mathrm{T}}$, x_{1i} is the number of neurons in the ith hidden layer; $\overline{1, x_2}$, x_2 is the number of hidden layers (the number of restricted Boltzmann machines or auto-encoders for the NN of deep confidence); x_3 is the number of neural connections.

To learn and evaluate the accuracy of the constructed NN was used the CATS (for Competition on Artificial Time Series) (Fig. 4.21).

Optimization criteria:

- minimization of the mean square error of the network operation

$$I_1 = \sum_{i=1}^{n} (y_i - d_i)^2, \tag{4.41}$$

where y_i is the obtained NN output; d_i is the desired NN output; n is the number of sample items;

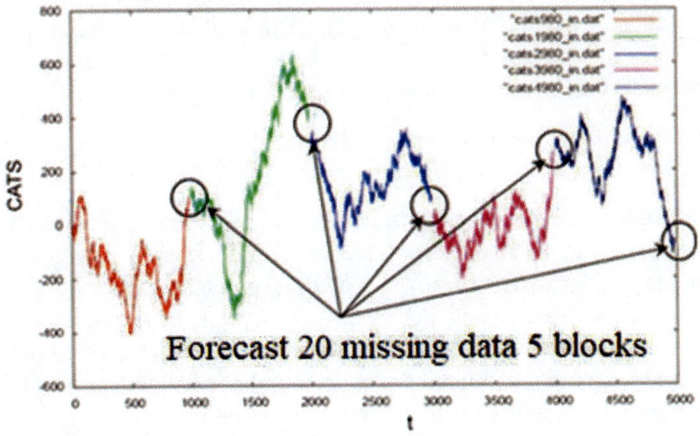

Fig. 4.21 Training and assessment of the accuracy of work built in NN when using CATS artificial sampling

$$i = 1, \ldots, P \quad I_1 = \sum_{i=1}^{k} \sum_{j=1}^{k} \sigma_{ij}, \tag{4.42}$$

where $\sigma_{ij} = 1$, if there is a connection between the ith and jth neurons; k is the number of neurons in the NN.

4.11.2 An Overview of Existing Methods

For the given problem solution can be used the evolutionary algorithms of optimization: swarm of particles [39, 40], genetic algorithms and hybrid algorithms consisting of them [41, 42] and others.

The swarm particle algorithm includes the following steps.

Step 1. We select the arbitrary size of the population (number of particles) P and set a limit on the number of searching iterations L and the desired error ε of neural network output.

Step 2. It is randomly constructed a set of vectors

$$\mathbf{Y_i}(y_1 y_2 \ldots y_n), \tag{4.43}$$

where y_1 is the number of neurons in the first layer; y_k is the number of neurons in the kth layer; y_n is the number of neurons in the output layer, $k = 1 \ldots n, \quad i = 1 \ldots P$.

Every such vector will correspond to one particle of the swarm.

The number of neurons in the first and the last layer can be determined by the task. In this case, only the other components of every vector are randomly generated.

Step 3. For each particle is randomly set its speed v_i, i is the particle number, $i = 1, \ldots, P$.

Step 4. For each particle it is built the corresponding it neural network. That is, the particle $\mathbf{Y_i}(y_1 y_2 \ldots y_n)$, will correspond to the neural network, in which there are y_1 neurons in the first layer, y_k neurons in the kth layer, etc., $\mathbf{Y}_i^{gbest}$.

We learn of each received network by any of the permissible methods of learning. For every neural network it is computed the criterion value, which is a convolution of $I_1(x)$ and $I_2(x)$ criteria. Computed for each neural network the criterion value will serve as a function of quality for the corresponding particle.

Every particle memorizes in its history the quality function value on each iteration.

Step 5. For every particle it is found the minimum value of quality function for its period of history (minimum value of general error, corresponding to neural network for the whole period of the algorithm operation). Designate the value of the vector Y_i (where i is the number of the particle), which corresponds to the minimum value of the quality function, Y_i^{lbest} is the best local solution for the corresponding particle.

Step 6. Among all the particles it is found the minimum value of the quality function for the whole swarm history (that is, the minimum criterion value for neural network for the whole period of algorithm functioning among all the generated configurations of neural networks). Then we designate the value of a vector **Y** which corresponds to the found value of the quality function as $\mathbf{Y}^{gbest}$ is the best global solution.

Step 7. We modify the speed v_i (i is the particle number) of each particle as follows:

$$v_{i,t+1} = v_{i,t} + \varphi_p r_p \left(\mathbf{Y}_i^{lbest} - Y_{i,t}\right) + \varphi_g r_g \left(\mathbf{Y}^{gbest} - Y_{i,t}\right), \tag{4.44}$$

where $v_{i,t}$ is the speed of the ith particle for the tth iteration of the algorithm; $\mathbf{Y_{i,t}}$ is the vector of the ith particle for the tth algorithm iteration; $\mathbf{Y}_i^{lbest}$ is the best local solution of the ith particle; $\mathbf{Y}^{gbest}$ is the best global solution found by all the particles; r_p, r_g are random numbers from the interval (0, 1); φ_p, φ_g are chosen arbitrarily weight coefficients are analogues of "learning speed" for NN.

Step 8. It is modified the value of the vector $\mathbf{Y_i}$ that corresponds to each particle as follows:

$$\mathbf{Y}_{i,t+1} = \text{int}\left[\mathbf{Y}_{i,t} + v_{i,t+1}\right], \tag{4.45}$$

where i is the number of particle; t is the number of iteration; int[·] is the whole particle.

Step 9. Stop criterion. If the value of the best global solution, found in step 6, provides the desired accuracy of $\mathbf{Y}^{gbest} < \varepsilon$, or if the limit of the iteration number $t \geq L$ is reached, then we will finish the search. Then the $\mathbf{Y}^{gbest}$ vector corresponds to the best NN configuration found, otherwise, we increase the iteration counter and go to *step 4*.

As a genetic algorithm it was used a hybrid genetic algorithm (see Sect. 2.6.8).

The results of modeling using swarm particle algorithm and genetic algorithm are given in Table 4.14.

Table 4.14 Simulation results using particle swarm and genetic algorithms

Algorithm	Algorithm parameters	Obtained ANN configuration	ANN error on the test sample	Configuration search iterations number
Particle Swarm algorithm	$1 < N < 20$, $P = 10$, $100 < L < 5000$, $E < 1250$	20–18–1	0.381	353
Genetic algorithm		20–13–1	0.463	387

N is the number of neurons in the layer; *P* is the population size; *L* is the number of iterations; *E* is the average quadratic error on the entire data array

Comparison of particle-swarm and genetic algorithms

The particle swarm and genetic algorithms have the same efficiency, but the optimum tuning algorithms differ for each task and with the growth of the problem dimension the range of algorithms efficiency values increases.

The particle swarm algorithm can achieve an acceptable effect for fewer iterations and/or less computational cost. In particular, the particle swarm algorithm requires a significantly less computational cost to solve the nonlinear of unconditional tasks.

The advantage of the particle swarm algorithm is the simplicity of its software implementation, higher intuitive clarity and visual geometric interpretation.

4.11.3 Combined Algorithm of Deep Learning Neural Network Structure Determination

The combined algorithm involves execution of the following steps.

The particle swarm algorithm provides high accuracy (due to particle memory) and quick acceptable solution. At the same time, the genetic algorithm is better suited for solving discrete problems and has better struggle mechanisms with local minimums (due to mutations and successful crossovers).

The given algorithm allows to combine the advantages of both algorithms and thereby to achieve a quick and exact solution of the task [41, 42].

The algorithm is based on the idea of consecutive execution of one search iteration by every base algorithm (swarm of particles and genetic algorithms), comparison of the obtained by each of the algorithms results and adding the best of the obtained solutions to each algorithm.

Step 1. Both algorithms (particle swarm and genetic) are started simultaneously in parallel mode for the same NN structure synthesis.

Step 2. It is execute done iteration of every algorithm according to the technique described in the preceding steps.

Step 3. After each iteration It is compared the results obtained by both algorithms and It is selected the best solution.

Let $Y_{pso}^{(i)}$ be the best solution obtained by swarm particle algorithm on ith iteration, and $W_{ga}^{(i)}$ is the best solution obtained by genetic algorithm.

If $I(Y_{pso}^{(i)} < I(W_{ga}^{(i)})$, that is, obtained with help of the swarm particle algorithm solution provides the less value of the quality (objective) function I, then go to *step 4a*. Otherwise, go to *step 4b*.

Step 4a. Replace the worst solution of genetic algorithm $W_{ga\_worst}^{(i)}$ by solution $Y_{pso}^{(i)} W_{ga\_worst}^{(i)} = Y_{pso}^{(i)}$ and go to *step 5*.

Table 4.15 Results of modeling with the use of swarm particles, genetic and combination algorithms

Algorithm	Algorithm parameters	Obtained ANN configuration	ANN error	Number of configuration search iterations
Swarm particle algorithm	$1 < N < 20$, $P = 10$, $100 < L < 5000$, $E < 1250$	20–18–1	0.381	353
Genetic algorithm		20–13–1	0.463	387
Combined algorithm		20–17–1	0.325	329

N is the number of neurons in the layer; *P* is the population size; *L* is the number of iterations; *E* is the average quadratic error on the entire data array

Step 4b. We replace the worst solution of the particle swarm algorithm $Y_{pso\_worst}^{(i)}$ by solution $W_{ga}^{(i)}$, $Y_{pso\_worst}^{(i)} = W_{ga}^{(i)}$ and go to *step 5*.

Step 5. If both algorithms continue their work (i.e., the stop criterion is not fulfilled for either of them), then go to *step 2*.

The described approach is repeated up to the stop of one of the algorithms. As a final solution, the best of the solutions obtained by both methods at the moment of stopping is taken.

The results of modeling with using of swarm particle, genetic and combined algorithms are given in Table 4.15.

4.11.4 Features of Deep Neural Networks Optimal Structures Calculation

Obtaining the optimal structure of a neural network is completely based on the hybrid algorithm, but for the solution coding and calculation of the fitness functions for every solution, specific features inherent to NN.

The chromosomes that encode each possible topology of neural network have the following form:

$$W_i(w_{11} \ldots w_{1k} w_{21} \ldots \; w_{2k} \; \ldots \; w_{n1} \ldots \; w_{nk}), \tag{4.46}$$

where $w_{11} \ldots w_{1k}$ encoded in binary format vector that describes the number of neurons in the first layer of the neural network; $w_{p1} \ldots w_{pk}$ encoded in binary format vector that describes the number of neurons in the layer of neural network; …; $w_{n1} \ldots w_{nk}$ is the binary encoded vector that describes the number of neurons in the last layer of the neural network; *n* is the number of layers; *k* is the number of bits used to encode each value (gene).

Each chromosome will correspond to a certain architecture of NN.

W_1	W_2	...	W_p	...	W_{n-1}	W_n

Fig. 4.22 Randomly generated the rest components of each vector

10	0	12	0	0	8	0

Fig. 4.23 Chromosome structure

The number of neurons in the first and last layer is usually determined by the task. In this case, it is randomly generated only the rest components of each vector (Fig. 4.22).

A similar structure of chromosomes allows the number of layers to be defined in addition to the number of neurons in layers.

Thus, the chromosome which is presented in Fig. 4.23, codes a deep belief network with three hidden layers, where the first hidden layer contains 10 neurons; the second hidden layer contains 12 neurons; the third hidden layer contains 8 neurons.

4.11.5 An Example of Deep Belief Network Optimal Structure Calculation

As a task to be solved using NN it is chosen the task of position prediction in the robot-manipulator space, based on the test data set k in 8 nm [41] was chosen. The chosen problem imposes such restrictions on NN parameters:

- number of NN-8 inputs;
- number of NN-1 outputs.

Additionally, we introduce the following limitations:

- number of hidden layers is the 2;
- the maximum number of neurons in the layer is a 16.

To demonstrate the functioning of the proposed algorithm, consider the simplified task of finding the optimal structure:

N (population size) = 5;
N_A (archive size) = 2;
T (maximum number of generations) = 5;
p_c (initial value of crossover probability) = 0.9;
p_m (initial value of mutation probability) = 0.05.

As a result of the fourth algorithm iteration execution the next population was obtained (Table 4.16).

For convenience, further when it is described the structure of NN, which is encoded by the individual, we will use the notation

Table 4.16 Fourth iteration chromosome set

Individual	Layer neurons number			
	Input layer	1st hidden layer	2nd hidden layer	Output layer
Individual 1	8	15	7	1
Individual 2	8	11	10	1
Individual 3	8	11	8	1
Individual 4	8	13	12	1
Individual 5	8	14	13	1

$$<N_{in} - N1 - N2 - N_{out}>,$$

where N_{in} is the number of neurons in the input layer; $N1$ is the number of neurons in the first hidden layer; $N2$ is the number of neurons in the second hidden layer; N_{out} is the number of neurons in the source layer.

Then, the individual 1 encodes structure <8–15–7–1>, individual 2—structure <8–11–10–1>, etc.

Step 1. Determination of fitness. For each chromosome from the population and the archive we build the corresponding neural network. That is, the individual 1 (<8–15–7–1>) corresponds to the NN in which there are 8 neurons in the first layer, 15 neurons in the second layer, etc.

We learn each of the obtained networks by any of the permissible learning methods. For each neural network it is computed the average quadratic error of its operation on the control data sample and the number of neurons in all layers S_n. The results are shown below in Table 4.17.

Based on the obtained values of accuracy and neural networks complexity we will compute the value of the fitness function F for each individual in P_t. Obtained results are given below in Table 4.18.

Step 2: Modernization of the archive. We copy all individuals whose solution vectors are not dominated in relation to P_t (i.e. whose fitness function values are less than 1) in the intermediate archive A_t^*.

As a result of this step, the individuals 2 and 3 will be copied to the intermediate archive A_t^*.

Table 4.17 The obtained results of neural networks training

Name	Structure	Average quadratic error	Complexity
Individual 1	<8–15–7–1>	0.06112	31
Individual 2	<8–11–10–1>	**0.05378**	30
Individual 3	<8–11–8–1>	0.05652	**28**
Individual 4	<8–13–12–1>	0.05875	34
Individual 5	<8–14–13–1>	0.06674	36

Table 4.18 Results of the F value function for each individual

Name	Mean square error	Complexity	Dominant individuals	The power of an individual	Rough value of adaptability
Individual 1	0.06112	31	5	1/6	1/2 + 1/2 = 1
Individual 2	**0.05378**	30	1, 4, 5	1/2	0
Individual 3	0.05652	**28**	1, 4, 5	1/2	0
Individual 4	0.05875	34	5	1/6	1/2 + 1/2 = 1
Individual 5	0.06674	36	–	0	1 − 6 + 1/2 + 1/2 + 1/6 = 8/6

Step 3. Ranking. The individuals in the population are sorted by descending of fitness functions values (from the best individual to the worst one). We get the following sequence of individuals:

- individual 2;
- individual 3;
- individual 1;
- individual 4;
- individual 5.

Step 4. Grouping. The individuals are divided into groups, each of which consists of two individuals. These two individuals are elected from the beginning of the sorted individuals list.

Obtained pairs:

- individual 2 and Individual 3;
- individual 1 and Individual 4;
- individual 5 does not participate in interbreeding because it does not have a pair.

Step 5. Crossover and mutation. Crossover and mutation are happening in each of the formed groups.

Let's be crossing the first two pairs of individuals and apply a mutation operation to the individual 5.

Crossing

First pair:

Parents: individual 2 (<8–11–10–1>) and individual 3 (<8–11–8–1>).

Subsidiaries individuals: individual 6 (<8–11–8–1>) and individual 7 (<8–11–10–1>).

Second pair:

Parents: individual 1 (<8–15–7–1>) and individual 4 (<8–13–12–1>).

Subsidiaries individuals: individual 8 (<8–15–12–1>) and individual 9 (<8–13–7–1>).

Mutation

Parental individual: individual 5 (<8–14–13–1>).

Subsidiary individual: individual 10 (<8–13–13–1>).

Adaptation of crossing and mutation probability

Step 6. All subsidiary individuals are merged into one group, which becomes the new population P_t:

- individual 6—<8–11–8–1>;
- individual 7—<8–11–10–1>;
- individual 8—<8–15–12–1>;
- individual 9—<8–13–7–1>;
- individual 10—<8–13–13–1>.

Step 7. End

Because a limit of iterations was reached $t \geq 5$, then A_t is the desired set of solutions:

- individual 2—<8–11–10–1>;
- individual 3—<8–11–8–1>.

4.12 Deep Neural Networks for Image Recognition and Classification Problems Solution

4.12.1 General Problem Statement of Pattern Recognition for the Detection of Unformalized Elements

Problem statement of classification problem. Let X be the set of objects descriptions, and Y be the set of classes numbers (or names). There is an unknown objective dependency—displaying $y^*: X \to Y$, the values of which are known only on objects of the finite training sample:

$$X^m = \{(x_1, y_1), \ldots, (x_m, y_m)\}, \tag{4.47}$$

where X^m is the training sample elements set with dimension m.

It is necessary to build an algorithm that can determine the membership arbitrary object $x \in X$ to the class $y \in Y$.

4.12.2 The Use of Deep Neural Networks to Solve an Unformalized Elements Images Recognition Problem

The use of multi-layered perseptron with traditional structure in solving the real problems of recognition and classification of images causes certain difficulties. Firstly, the images tend to have a large dimension, resulting in a growing number of neurons and synaptic connections in the network. In turn, this requires an increase of the training sample, resulting in increased time and computational complexity of the learning process. Secondly, the input data topology is ignored. Components of the input layer can be represented in any order, without taking into account the purpose of training. However, the images have a strict two-dimensional structure in which there is a dependency between space-neighbour pixels [43, 44].

From these disadvantages are free the so-called convolutional neural networks, which are a special class of multilayer perseptron, that were specially designed for recognition of two-dimensional surfaces with a high level of invariance to scaling, shift, rotation, change in angle and other spatial distortions, and deep NN are built on the basis of autopilots, the previous training is carried out on the basis of Restricted-machine Boltzmann [8–10, 45].

4.12.3 High-Performance Convolutional Neural Networks

4.12.3.1 Convolutional Network Organization Principle

The convolutional neural network (CNN) is a special architecture of the artificial neural network imitating the features of the visual area of the cerebral cortex [45].

In contrast to the multi-layered perceptron the CNN have the following distinctive features:

(1) Local fullness: according to the concept of receptive fields, it is used to use spatial locality by applying the local connectivity scheme between neurons in adjacent layers. Thus, this architecture provides the possibility of a well-trained "filters" (convolution kernels) to produce the strongest response for the spatial-Local input pattern. The structure of many such layers is equivalent to the use of the nonlinear filter and is sensitive for bigger area of pixel space. In this way, the network first generates image of small input details, and then creates the representation of bigger areas [45].

(2) Joint weights: in the CNN each filter is repeated throughout the visual field. These repeated nodes use the common parameterization (weights and bias vector) and form a map of features. This means that all neurons in a given convolutional layer react to the same feature within their receptive bounds. The

nodes repeating therefore permits to determine features regardless of their position in the visual field, ensuring that supplies the property of invariance offset [45].

Together, these properties allow CNN to achieve a better generalization on image recognition problems. Joint weights using sharply reduces the number of free parameters due to the network is learned by reducing the memory requirements for the network operation and allowing training of large, more powerful networks [45].

The convolution neural network is based on the basis of the convolution operation, which allows to learn the CNN on separate parts of the image, iteratively increasing the local training area of a separate convolution kernel.

Suppose $x(t)$—some function at $t \in \Re$ [45].

Then the convolution $x(t)$ with the kernel $k(t)$ is a function $S(t)$, which is defined as

$$S(t) = (x \cdot k)(t) \equiv \int_{-\infty}^{\infty} x(\tau)k(t-\tau)d\tau. \tag{4.48}$$

If the function is discrete, $t \in Z$ then

$$S(t) = (x \cdot k)(t) \equiv \int_{\tau=-\infty}^{\infty} x(\tau)k(t-\tau)d\tau. \tag{4.49}$$

If $I(i, j)$ an image, then the convolution of the image $I(i, j)$ with the kernel $K(t, s)$ will be written as

$$W = (W_p - F + P \cdot 2)/S + 1, \tag{4.50}$$

where m, n is the current position of the kernel relative to *the image* $I(i, j)$ with the dimension $i \times j$.

4.12.3.2 Classification of Convolutional Network Layers

The key point in understanding CNN is the concept of so-called common weights, that is, part of the neurons of some layer of the neural network can use the same weights coefficients. The neurons that use the same weights are combined into features maps (eng. feature maps), and every neuron of feature map is associated with part of neurons of the previous layer. Under the network calculating, every neuron realizes the convolution of some area of previous layer (which is determined by set of neurons associated with this neuron). The layers of a neural network which are constructed by the described manner are named by convolutional layers. Beside convolutional layers they can be aggregation layers (subdiscretization)in CNN, that realize the function of decreasing the feature map dimension, and fully connected layers (classifier, which is located at the network output). The convolutional and aggregation layers can be rotated, mostly the aggregation layers are located after convolution layers [21, 45].

The convolutional neural networks can have multidimensional layers (mainly is it used two-dimensional ones, for example, in networks that process the images, and three-dimensional ones, for example, for color image) of several types [21, 45].

With respect to the tasks of objects recognition in images, the input layer is most often represented as a three-dimensional grid, the size of which depends on the input image,

$$l_n = W \cdot H \cdot D, \tag{4.51}$$

where l_n is a dimension of the network input layer; W is the width of the input image; H is the height of the input image; D is the depth or number of color channels of the image.

A conditional view of CNN input layer the wrapping neural network is presented in Fig. 4.24, where an example of image representation in RGB format is given [21, 45].

The general view of CNN is shown in Fig. 4.25.

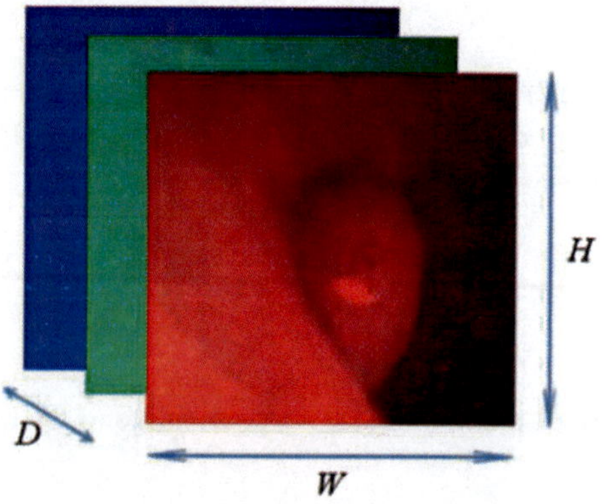

Fig. 4.24 Conditional view of input layer of CNN

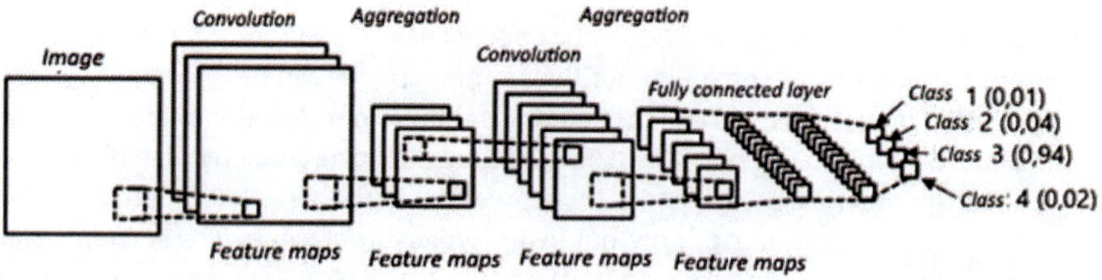

Fig. 4.25 General view of CNN

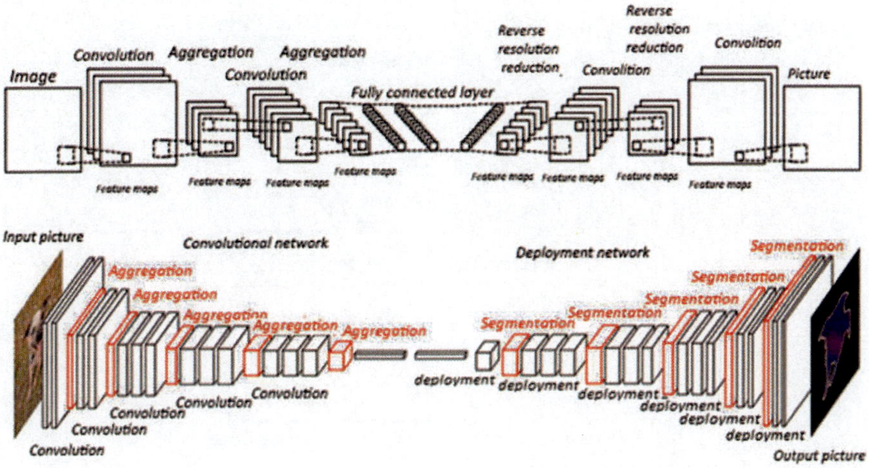

Fig. 4.26 Combined CNN

The combined network which is consisted from convolutional classifier and deconvolutional neural network, is shown in Fig. 4.26.

This architecture allows not only to recognize the elements of the image, but also to mark the recognized elements on it. A deconvolutional neural network is a mirror image of a convolutional neural network [21].

Consider the layers of CNN in more details.

4.12.3.3 Convolutional Layer

Convolutional layer is the main unit of CNN and is destined for image features separation and their transformation. The convolutional layer consists of a set of filters-multidimensional (usually two-dimensional or three-dimensional) matrices of the connection weights of previous layer neurons with a convolutional layer neurons. The convolutional layer includes its filter for every channel (in CNN, the values of filter kernel is called by convolutional layer neurons connections weights: $w_{00}, w_{01}, \ldots, w_{(F-1)(F-1)}$, where $F \times F$ is the filter kernel dimension), convolutional kernel which realizes a convolution operation for fragments of previous layer [45]. The process of convolutional layer neurons signals formation is shown in Fig. 4.27 [21].

The weight coefficients of the convolution kernel are unknown and determined in the learning process. A feature of the convolutional layer is a relatively small number of training parameters. For the color image of 100×100 pixels across three channels (this is 30,000 input neurons) under using 6 convolution kernels of 3×3 dimensions, is used only 9 parameters for every convolution kernel of each channel, that is, only $9 \times 6 \times 3 = 162$ parameters, which is much less than for fully connected layer [21].

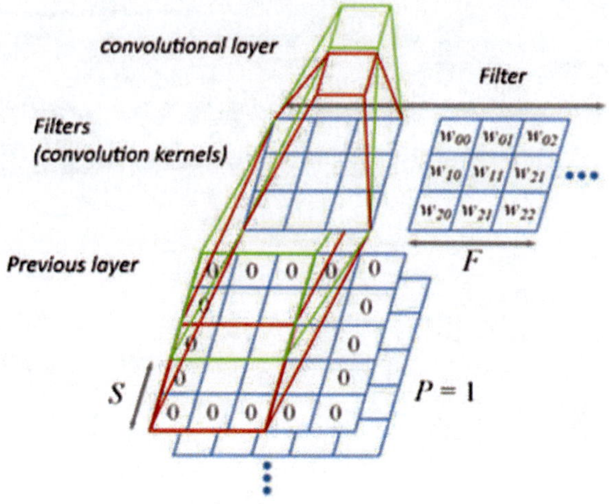

Fig. 4.27 Formation of a CNN feature map

The main parameters of the CNN are: the size of the convolution kernel, the number of convolution kernels (depends on the number of convolution layers), bias value under convolution kernel moving by image matrix (convolution kernel step), the parameters (by vertical and horizontal) of edge effects accounting (the initial position of the convolution kernel on the image matrix or features map before moving beginning by horizontal and by vertical in order to build features maps) (see Fig. 4.25), the initial filling of the kernel of a convolution [45].

The dimension of the convolution kernel determines how much the spatial variability of the image decreases after each convolution, and how much the image is decreased. The number of convolutions kernels determines the number of features maps that encode some feature of the image (for example, the presence of a straight or inclined line). The bias depends on how many steps the filter is shifted for the next convolution operation execution. Under large shift, some information can be lost. From the bias it is depending how much the image is decreased after each convolution operation. The edge effects are the parameters of convolution execution on the image edges. There are some types of edge effects, the most common of them are pruning and filling with zeros. Filling with zeros does not reduce the image matrix [21].

In order to reduce the computational costs, it is used pruning operation—this is the reduction of features maps after convolution operation. The size of features maps at pruning use is determined by the stride, the convolution kernel dimension and previous network layer sizes, according to the formulas:

$$W = (W_p - F + P \cdot 2)/S + 1, \tag{4.52}$$

$$H = (H_p - F + P \cdot 2)/S + 1, \tag{4.53}$$

where W and H are the width and height of the features maps, respectively; W_p and H_p are the width and height of the image matrix, respectively; F is the convolution kernel dimension of the convolutional network layer; P is the number of added zeros at the edges of the image matrix; S is the offset filter [45].

The mathematical model of the convolutional layer in a simplified form can be described by the following formula $x^l = f(x^{l-1} \cdot k^l + b^l)$, where x^l is the layer output; $f(\cdot)$ is the activation function; b^l is the shift factor, the symbol * denotes the convolution operation of the input x^{l-1} with kernel k^l. If we paint the formula in the standard form, we will get:

$$x_j^l = f\left(\sum_{i=0}^{n}\sum_{j=0}^{m} x_i^{l-1} * k_j^l + b_j^l\right), \tag{4.54}$$

where x_j^l is a features map j (layer output l; $f(\cdot)$ is the activation function; b_j^l is the shift factor for the features map j; k_j^l is the convolution kernel j; x_i^{l-1} is a features map of the previous layer; n, m is the image size. The scheme of the convolutional layer is shown in Fig. 4.28 [45].

An example of the convolution operation in CNN is shown in Fig. 4.29.

In a convolutional neural network, a set of weights is not one, but a whole gamma encoding image elements (for example, lines and arcs at different angles). Depending on the number of filters used, we get the same number of features maps.

Convolution kernels are initiated by random values to get started (see Sect. 4.12.3.6), and then are tuned by neural learning classical method of back propagation. The passage of each convolution kernel forms its own instance of features map that make the neural network by multichannel (many independent features maps on a single layer). It should also be noted that under sort through the layer by convolution kernel, it is usually moved not by a full step (determined by the dimension of

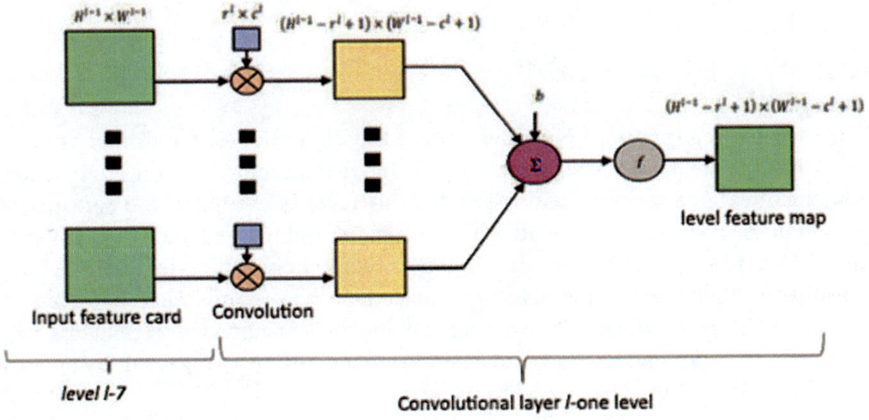

Fig. 4.28 Convolutional layer scheme

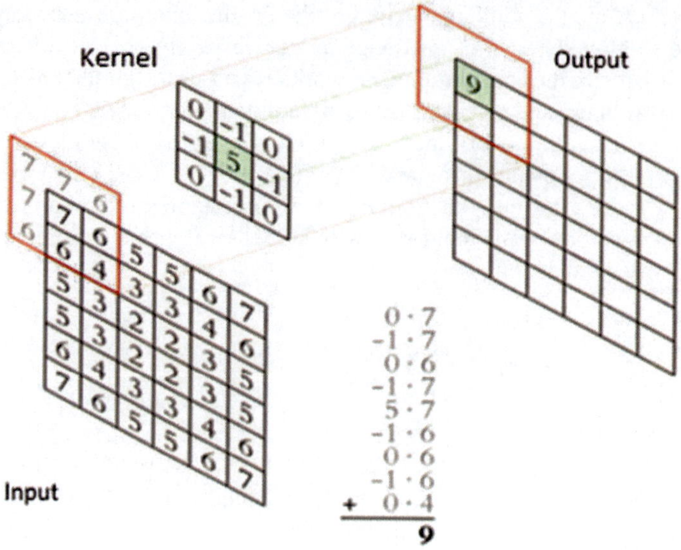

Fig. 4.29 An example of the operation of convolution in CNN

this matrix), but by several elements of the matrix (Fig. 4.30). This distance is called stride. So, for example, at the weights matrix dimension 5 × 5 it is shifted by one or two neurons (pixels) instead of five, so as not to "step over" the desired feature [45].

Each features map can recognize separate image details (such as vertical and horizontal lines) (Fig. 4.31). The layers alternating permits to compose the features maps from feature maps, the map decreased in dimension on each next layer, but it is increased the number of channels. In practice, this means the ability to recognize complex features hierarchies. Usually, after passing through several layers, the features map degenerates into a vector or even a scalar, but such feature maps become.

4.12.3.4 Aggregation Layer

The subsampling (pooling) is used to reduce the dimension of the formed features maps in order to reduce the probability of rapid retraining, as well as to reduce computational and memory costs. Usually, this layer is used after the convolution operation and converts the convolution layer signals, highlighting the most significant signals according to certain criteria. In this network architecture, it is believed that information about the desired feature presence fact is more important than accurate knowledge of its coordinates, so from some neighboring neurons of the features map, the maximum one is selected and taken as one neuron of the compacted feature map of smaller dimension [21].

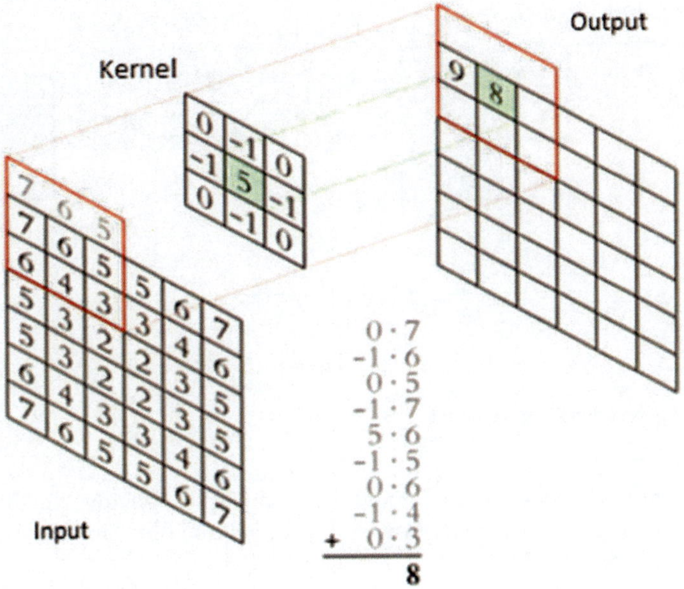

Fig. 4.30 An example of the Kernel Shift per unit

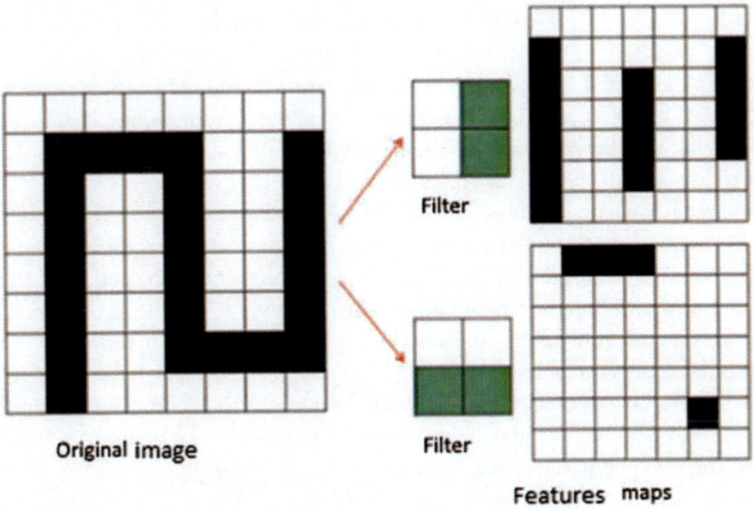

Fig. 4.31 Illustration of the reaction of each card to individual features

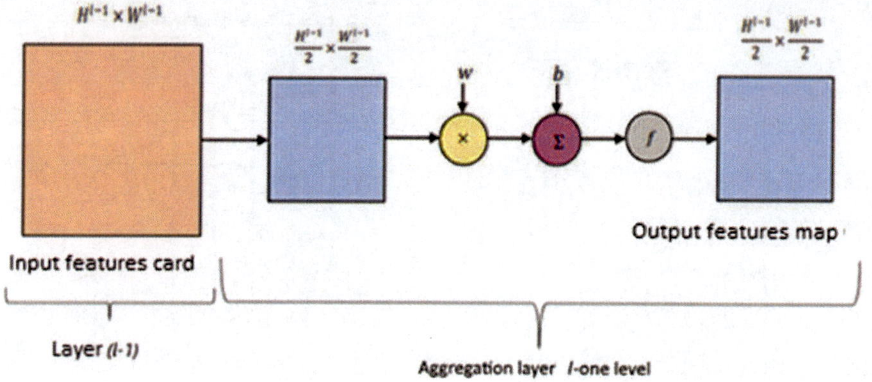

Fig. 4.32 Aggregation layer diagram l

The use of this layer allows to improve the examples recognition with the changed scale. The aggregation operation provides one more kind of invariance with respect to parallel transference [45].

The aggregation layer acts independently on each section of the input depth, and reduces its spatial size. Formally the layer can be described as follows:

$$x^l = f\left(a^l \cdot subsample\left(x^{l-1}\right) + b^l\right), \tag{4.55}$$

where x^l is the output layer l; $f(\cdot)$ is the activation function; a^l, b^l are coefficients; $subsample(\cdot)$ is the operation of sampling local maximum values. The scheme of the aggregation layer is shown in Fig. 4.32 [21].

There are such types of aggregation: maxpooling, average pooling, $L2$-norm pooling and stochastic pooling.

The most common type is the maxpooling layer with filters dimension 2×2, which used with step of 2, which reduces the sampling of each section of the input depth in 2 times, both in width and height, discarding 75% of the excitations. In this case, each operation of taking the maximum acts over the four numbers. The size after depth remains unchanged [45].

Historically, average pooling was used frequently, but lately less compared to maxpooling, whose performance in practice has proven to be better. An example of maxpooling and average pooling is shown in Fig. 4.33 [45].

At stochastic pooling the usual deterministic aggregation actions are replaced by a stochastic procedure in which the excitation within each aggregation area is chosen randomly, according to the polynomial distribution given by the excitations within the aggregation area. This approach is free of hyperparameters, and can be combined with other approaches of regularization such as dropout (see 4.12.3.7) and increase data [45].

An alternative view of stochastic pooling is that it is equivalent to standard maxpooling, but with many copies of the input image, each of which has some

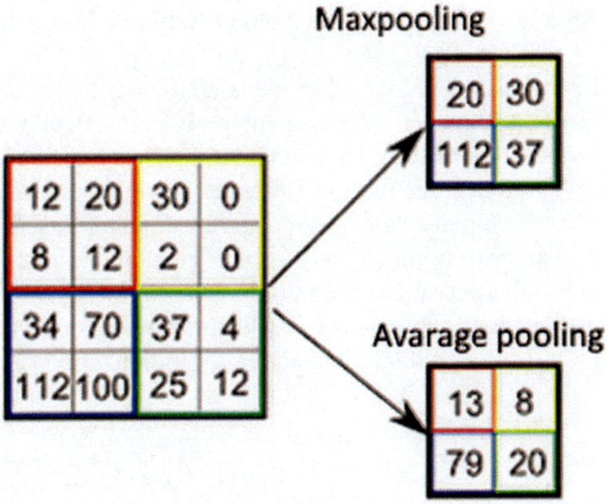

Fig. 4.33 Examples of aggregating functions

local deformations. This is similar to the explicit elastic deformations of the input images, which provide excellent performance in MNIST (see Sect. 4.12.3.8). The use of stochastic pooling in a multilayer model gives an exponential number of deformations, because the choice in the higher layers depend on the choice in the lower layers [21].

4.12.3.5 Fully Connected Layer

After several convolutional and maxpooling layers which are used to features highlighting a special area, a classifier is placed, that is realized by fully connected layers (eng. fully connected layers), and solves the classification tusk of the highlighted area. As a classifier in this work it is considered: autoencoder, Softmax, fuzzy classifier of deep learning. A fully connected layer can be described by the formula:

$$x_j^l = f\left(\sum_i^n x_j^{l-1} w_{ij}^{l-1} + b_j^{l-1}\right), \tag{4.56}$$

where x_j^l is the jth layer output l; $f(\cdot)$ is the activation function; b_j^{l-1} is the jth bias of lth layer; w_{ij}^{l-1} is the ith and jth element of the weight coefficient matrix $l-1$ layer; n-number of neurons in a fully connected layer.

4.12.3.6 Convolutional Neural Network Learning

Learning of CNN can be divided into four stages: initialization of weight coefficients, direct passage of reference input signals, calculation of the error function, back propagation of the error and updating of weight coefficients [45].

The first stage of CNN learning is weight coefficients initialization. In general, if CNN contains convolution aggregating and fully connected layers, then it will be necessary to initialize only kernels convolution of convolution and fully connected layers. If a network contains the convolutional layers too then it is necessary to initiate their weight coefficients. For CNN learning in this work it is used the normalize initialization, which is called Glorot initialization [21]

$$W \cong U\left[-\frac{\sqrt{6}}{\sqrt{n_j + n_j}}, \frac{\sqrt{6}}{\sqrt{n_j + n_{j+1}}}\right], \tag{4.57}$$

where U is the uniform distribution on the segment; n_j is the number of neurons on the current network layer; n_{j+1} is the neurons number on the next network layer. The use of normalized initialization leads to a neuron saturation decrease and error signal is propagated much better.

The method of weight coefficient normalized initialization has been adapted for the ReLU activation function when initial weight coefficient $W$are initialized as follows:

$$W \cong U\left[-\frac{-2}{n_j}, \frac{2}{n_j}\right], \tag{4.58}$$

where U is the uniform distribution on the segment; n_j is the neurons number n the current network layer. The method of normalized initialization of weight coefficients allowed to achieve high-quality learning of deep neural networks without a necessity of preliminary unsupervised learning used.

For CNN learning it is used back propagation method [45]. For the output layer (MLP), the error is calculated as follows:

$$\delta = (T - Y) \cdot f'(u)$$

where T is the expected output; Y is a real output; $f'(u)$ is the derivative of the activation function by its argument. For hidden layers the error looks like:

$$\delta^{l-1} = (W^l)^T \cdot \delta^l \cdot f'(u^{l-1}), \tag{4.59}$$

where δ^l is the error of the layer l; $f'(u)$ is the derivative of activation function; u^l is the state (not activated) of l layer neurons; W^l is the weight coefficients matrix of l layer.

The error at the output of the convolutional layer is formed by simply increasing the error matrix dimension of the next aggregated layer

$$\delta^{l-1} = subsample(\delta^{l}) \cdot f'(u^{l-1})$$

where δ^l is the error of the layer l; $f'(u^l)$ is the derivative of activation function; u^l is the state (not activated) of neurons l; $subsample(\cdot)$ is the operation which is inverse to aggregation.

The aggregation layer output error is calculated by features map inverse convolution executing of the next after him convolutional layer, that is, above features map is executed the convolution with the corresponding "inverted" kernel, wherein the dimension of output matrices are increased due to boundary effects. Then, under the obtained maps, partial sums are obtained by convolution kernel number, according to neighbouring matrix of the aggregating and convolution layers [45]

$$\delta^{l-1} = (W^{l})^{T} \cdot \delta^{l} \cdot f'(u^{l-1}), \tag{4.60}$$

where δ^l is the error of the layer l; $f'(u^l)$ is the derivative of activation function; u^l is the state (not activated) of neurons l; k is the convolution kernel; rot180(k) is the rotation of convolution kernel on 180°.

A reverse convolution is the same convolution, only with convolution kernel rotation 180° and changed edge effects. That is, it is necessary to take the convolution kernel, rotate it on 180° and make an ordinary convolution with help of calculated before features maps errors, but so that scan window goes beyond the borders map. The result of inverse convolution operation is shown in Fig. 4.34 [45].

For minimization of considered above error function, it is necessary to calculate its gradient, that requires the knowledge of partial derivatives by the base network parameters [10].

The gradient for the convolution kernel can be calculated as the convolution of convolution layer input matrix with the "inverted" error matrix for the selected kernel

$$\Delta k_j^l = \mathrm{rot110}\left(x^{x-1}\mathrm{rot180}(\delta_j^l)\right), \tag{4.61}$$

where δ^l is the error of the layer l; x^l is the layer input; k is the convolution kernel; rot180(k) is the convolution kernal rotation on 180°.

The shift gradient for the convolutional layer is calculated as the sum of values of the corresponding error matrix $\Delta\ b^l = \sum \delta_j^j$, where δ_j^j is the error of the convolutional layer l.

The gradient for the aggregation layer coefficients is calculated as follows:

$$\Delta a^l = \sum \delta_j^l subsample\left(x^{l-1}\right), \tag{4.62}$$

where δ_j^j is the error of the layer l; x^l is the layer input l; $subsample(\bullet)$ is the sample operation of the local maximum values.

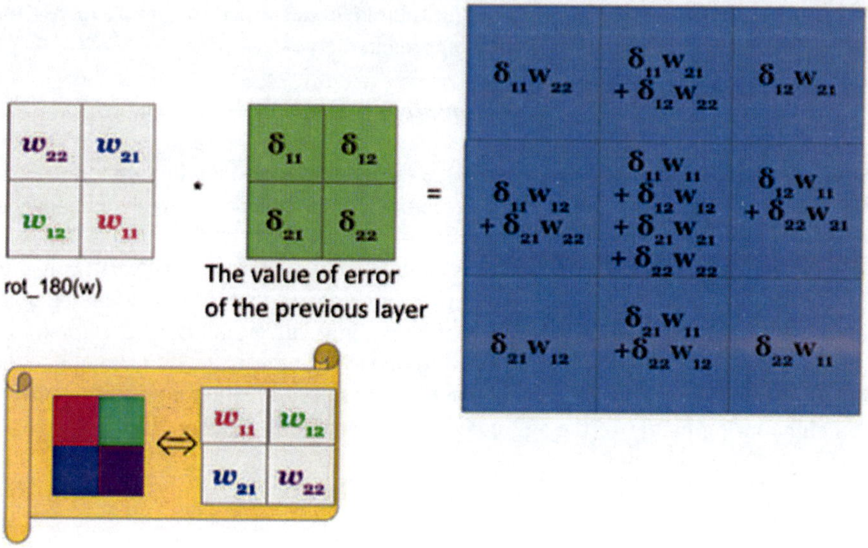

Fig. 4.34 Reverse convolution

The gradient for the bias coefficient for the aggregation layer is calculated as the sum of the values of the corresponding error matrix $\triangle b^{pl} = \sum \delta_j^p$, where δ_j^p is the error of the aggregating layer p.

The local gradient for fully connected layer weights coefficients tuning is calculated as follows:

$$\Delta W^l = (\delta^l)^T \cdot x^{l-1}, \tag{4.63}$$

where δ^l is the error of the layer l; x^l is the layer input l; W^l is the weight coefficients matrix of the layer l.

4.12.3.7 Methods to Prevent Overfitting

One of the series learning problems of CNN is the problem of overfitting, when explains the model only patterns from the learning sample, adapting to the training patterns, instead of learning to classify the patterns that did not participate in the learning (losing the ability to generalize). There are the next approaches to avoid such situation: artificial data, early stop, limiting the number of parameters, methods of thinning or dropout, dropout of connections, weakening of weights, hierarchical coordinate grids [10].

In this work, dropout is used to prevent overfitting. There are the following modifications of this method use: Dropout of individual neurons, Dropout of connections, Inverted Dropout [45].

In the case of the Dropout of individual neurons at each stage of training, the individual neurons are either "drop outed" from the network with probability p, or remained with probability $q = 1 - p$, hence a reduced network remains; the connections of the Dropout ed neurons are also eliminated. For each neuron, separately the probability of Dropout is calculated separately. At the next stage, only reduced network is trained on the data. After that, the eliminated nodes are repeatedly included into the network with their primary weights [10].

The mathematical model of this process at the stage of learning applied to the I-th neuron can be presented as follows [46]:

$$O_i = \frac{1}{q} X_i a\left(\sum_{k=1}^{d_i} w_k x_k + b\right) = \begin{cases} a\left(\sum_{k=1}^{d_i} w_k x_k + b\right), & \text{if } \mathrm{X}_i = 1, \\ 0, & \text{if } X_i = 0, \end{cases} \tag{4.64}$$

where $h(x) = xw + b$ is the linear projection of input d_i vector $\mathbf{x}$ on d_k dimensional space of input values; d_i-dimensional vector of $\mathbf{x}$ on d_h-dimensional space of initial values; $a(h)$ is the activation function, where $X = (X_1, \ldots, X_{dh})$ is the d_h-dimensional vector of random values X^i, distributed by Bernoulli's law:

$$f(k; p) = \begin{cases} p, & \text{if } k = 1, \\ 1 - p, & \text{if } k = 0. \end{cases} \tag{4.65}$$

$f(k; p)$ is the probability distribution, where k are all possible initial values. This random value corresponds to the Dropout procedure applied to a single neuron (the neuron is Dropped out with probability $p = P(k = 1)$, and if not, then leave included) [10].

At the testing stage, it is necessary to emulate the neural networks ensemble behavior. Using under learning period, it is necessary on testing stage to multiply the activation function by the coefficient q at the testing stage. Thus, a mathematical model of this process at the testing stage applied to the Ith neuron can be presented as follows [46]

$$O_i = qa\left(\sum_{k=1}^{d_i} w_k x_k + b\right). \tag{4.66}$$

The disadvantage of direct Dropout is the necessity to change the NN for testing, because without multiplication by q, the neuron will return a value greater than those expected by the next neurons; therefore, the reverse Dropout has more perspective [45].

In this case, we multiply the activation function by the coefficient not during the test stage, but during training.

The coefficient is equal to the inverse value of probability that the neuron will remain in the network: $O_i = 1/1 - p = 1/q$, thus, mathematical models of

Inverted Dropout at the stages of training and testing applied to the Ith neuron can be represented, respectively, in the following form [46]:

$$O_i = \frac{1}{q} X_i a\left(\sum_{k=1}^{d_i} w_k x_k + b\right). \quad (4.67)$$

The advantage of inverted Dropout is consisted of the necessity only one time to describe the model once, under its use and then start training and testing on this model, changing only the parameter (Dropout coefficient).

4.12.3.8 Determination the Most Significant CNN Parameters from the Point of View Effectiveness

The determination of the most significant from the point of view CNN parameters effectiveness is executed under experiment on a combined network which consisted of a CNN, a classifier, and the convolution NN [47].

A database of patients with thyroid disease was used for the experiment.

Design of experiment its necessary follow to the next:

(1) It is analyzed the number of convolution layers in the range from 2 to 4.
(2) The aggregation layers will be placed after the convolutional layers and before the deconvolutional layers, one aggregation layer is used for one, two or three convolutional layers. In each aggregation layer it will be used the kernel dimension 2 × 2 as optimal. Underlarge dimension of aggregation kernel it is lost too much information. The aggregation will the maxpolling function because it is simple and reliable.
(3) The deconvolution layers will be used in the same number as the convolution layers in an amount of 2–4.
(4) The dimension of the convolution kernel should be odd and not too large, so it is possible dimensions are 3 × 3, 5 × 5, 7 × 7.
(5) The number of features maps can be from 4 to 64, it is also desirable to use an even number, and preferably the number is a power will be the degree of two to simplify the calculations.

The offset will be used the same for horizontal and vertical, always unit for simplifying the network architecture. Next, 2 × 2 or 3 × 3 offset for the 5 × 5 and 7 × 7 convolution kernel will be considered.

For the image matrix convolution at the edges of the matrix, it is used the filling with zeros of the missing image matrix fields without features maps decreasing.

As an activation function it is used ReLU function and as the classifier—autoencoder.

Considering the above, the experiment was performed by changing the following parameters:

1. The number of convolution layers from 2 to 4.
2. The number of aggregation layers from 1 to 4.
3. It is considered three variants: one aggregate layer after one convolutional layer, one aggregate layer after two convolutional layers, one aggregation layer after three convolutional layers.
4. Convolution layers (on each layer separately):
   - dimension of convolution kernals: 3×3, 5×5, 7×7;
   - number of features maps 8, 16 and 32;
   - the offset value is a fixed parameter (Table 4.19);
   - used boundary effect—fixed parameter (Table 4.19).
5. Aggregation layers:
   - aggregation kernel dimension—fixed parameter (Table 4.19);
   - the function of the aggregation kernel is a fixed parameter (Table 4.19).
6. Classifier:
   - type of classifier—fixed parameter (Table 4.19);
   - the number of layers from 2 to 4;
   - the dimension of each layer is 64, 128, 256 and 512.
7. Dropout operations for each layer: percentage of dropout—fixed parameter (Table 4.19).

Fixed neural network parameters:

1. Convolution layers:
   - offset value 1 for each layer;
   - at the edges of the image matrix for the convolution it is used zeros filling, the dimension of the feature maps is equal to the dimension of the image matrix;
   - the activation function is ReLU.
2. Aggregation layers:
   - dimension of aggregation kernel 2×2;
   - as an aggregation function it is used maxpulling.
3. Only fully connected layers are used, activation function is ReLU.
4. Dropoutis not used.
5. Initialization functions—normalization initialization (Glorot).
6. Error functions—MSE.
7. The number of eras of learning 20.

Table 4.19 The results of the experiment

№	The first convolution layer	The second convolution layer	The size of the classifier	Accuracy	Training time
0	16 features maps with kernel dimension 3 × 3 and offset—1. Maxpolling layer with kernel dimension 2 × 2	16 features maps with kernel dimension 3 × 3 and offset—1. Maxpolling layer with kernel dimension 2 × 2	512–128–512	0.001742294	391.118135
1	8 features maps with kernel dimension 3 × 3 and offset—1. Maxpolling layer with kernel dimension 2 × 2	8 features maps with kernel dimension 3 × 3 and offset—1. Maxpolling layer with kernel dimension 2 × 2	256–128–256	0.001828256	315.2135794
2	8 features maps with kernel dimension 3 × 3 and offset—1. Maxpolling layer with kernel dimension 2 × 2	16 features maps with kernel dimension 3 × 3 and offset—1. Maxpolling layer with kernel dimension 2 × 2	512–128–512	0.001830939	344.9284253
3	8 features maps with kernel dimension 5 × 5 and offset—1. Maxpolling layer with kernel dimension 2 × 2	16 features maps with kernel dimension 3 × 3 and offset—1. Maxpolling layer with kernel dimension 2 × 2	512–256–512	0.001847606	400.4870558
4	16 features maps with kernel dimension 5 × 5 and offset—1. Maxpolling layer with kernel dimension 2 × 2	16 features maps with kernel dimension 3 × 3 and offset—1. Maxpolling layer with kernel dimension 2 × 2	512–256–512	0.001856388	432.3665767
5	8 features maps with kernel dimension 3 × 3 and offset—1. Maxpolling layer with kernel dimension 2 × 2	16 features maps with kernel dimension 3 × 3 and offset—1. Maxpolling layer with kernel dimension 2 × 2	256–256–256	0.001867345	323.8970006
6	8 features maps with kernel dimension 5 × 5 and offset—1. Maxpolling layer with kernel dimension 2 × 2	16 features maps with kernel dimension 3 × 3 and offset—1. Maxpolling layer with kernel dimension 2 × 2	512–64–512	0.001872922	395.7997081
7	8 features maps with kernel dimension 5 × 5 and offset—1. Maxpolling layer with kernel dimension 2–2	16 features maps with kernel dimension 3 × 3 and offset—1. Maxpolling layer with kernel dimension 2 × 2	256–128–256	0.001883072	375.089128

(continued)

Table 4.19 (continued)

№	The first convolution layer	The second convolution layer	The size of the classifier	Accuracy	Training time
8	8 features maps with kernel dimension 5 × 5 and offset—1. Maxpolling layer with kernel dimension 2 × 2	16 features maps with kernel dimension 3 × 3 and offset—1. Maxpolling layer with kernel dimension 2 × 2	512–128–512	0.001907312	396.17189
9	16 features maps with kernel dimension 3 × 3 and offset—1. Maxpolling layer with kernel dimension 2 × 2	16 features maps with kernel dimension 3 × 3 and offset—1. Maxpolling layer with kernel dimension 2 × 2	512–64–512	0.001910896	391.7317238

The results of the experiment are given in Table 4.19.

The dependence of time and accuracy from the size of the fully connected layers for the two-layered CNN is shown in Table 4.20.

The dependence of accuracy from the size of the fully connected layer for the two-layered CNN is shown in Table 4.21.

The dependence of the accuracy from the size of the fully connected layer for the two-layered CNN is shown in Table 4.22.

On the basis of the results analysis of the combined CNN testing on the testing sample presented in Tables 4.19, 4.20, 4.21 and 4.22 it is possible to define the significant parameters of the CNN:

Table 4.20 Dependence of time and accuracy from the size of fully connected layers

The size of the classifier	Average learning time	The average error
128–128–128	359.03918	0.002336
128–64–128	365.41794	0.002331
256–128–256	386.7935667	0.002081833333
256–256–256	377.0971714	0.002196285714
256–64–256	372.98418	0.00224
512–128–512	408.1395429	0.002040428571
512–256–512	402.3346875	0.00207175
512–64–512	407.6629857	0.002037285714
In general	**387.538866**	**0.00215036**

Table 4.21 Dependency of accuracy from fully connected layer size for a two-layered CNN

The size of the classifier	Average learning time	The average error
128–128–128	577.138937	0.003233888889
128–64–128	585.9992231	0.003113
256–128–256	579.5261	0.002881666667
256–256–256	579.3374333	0.002944111111
256–64–256	577.4629667	0.003002851852
299–203–10	926.1383	0.005402
400–200–400	470.1731	0.0026045
512–128–512	584.2283185	0.002777777778
512–256–512	585.362237	0.002738296296
512–64–512	583.114363	0.002727
607–270–378	227.3493	0.003207
849–504–823	207.6274	0.007604
958–458–858	1046.459	0.00434
In general	**580.8619891**	**0.002963579186**

Table 4.22 Dependence of accuracy from the size of the fully connected layer for two-layer CNN

The size of the classifier	Average learning time	The average error
128–128–128	783.8963694	0.004702472222
128–64–128	784.9493389	0.00488425
256–128–256	789.4073444	0.004520722222
256–256–256	799.4197829	0.004283371429
256–64–256	787.7435611	0.004556388889
400–200–400	501.48954	0.0037154
512–128–512	787.0071278	0.004362583333
512–256–512	789.0946806	0.004270833333
512–64–512	792.0630917	0.0043455
In general	**784.2361459**	**0.00447819863**

1. The number of convolution layers.
2. The number of aggregation layers.
3. Mutual arrangement of convolution layers and aggregating layers.
4. Convolution layers (on each layer separately):

   - the size of the convolution kernel (on each layer separately);
   - the number of features maps (on each layer separately);
   - the value offset (on each layer separately);
   - edge effect parameter.

5. Aggregation layers (on each layer separately):

   - the size of the aggregation kernel;
   - aggregation kernel function.

6. Fully connected layers (each layer separately):

   - number of fully connected layers;
   - the size of each layer;
   - classifier type: autoencoder.

7. Presence of Dropout operation for each layer: Percentage of Dropout and random function.

For CNN structure and parameter optimization it is used the genetic algorithm (see Sect. 2.2).

4.12.3.9 Structure and Parameters Optimization of CNN

Chromosome description

In our algorithm, every individual is corresponded 114 bits. They can be represented as 6 genes for the number of convolution and classifier layers. In the first three genes(convolution genes) it can be marked 7 fields:

1. The presence of a layer—1 bit (at 0 in the first gene the individual enters into bioreactor).
2. The number of features maps—7 bits, from 4 to 256.
3. Kernel size—3 bits, the kernel has a size from 3 to 8 inclusive.
4. Offset—2 bits, from 1 to 4 inclusive, when offset more than half plus $0.5(n/2 + 0.5)$, the individual enters the bioreactor. That is, the most possible offset 4 when the size of the kernel 7 or 8.
5. Dropout. The gene has a size of 5 bits, possible Dropout values from 0 to 30%.
6. Edge convolution effects, 1 bit—determine image size decreasing under convolution operation performing
7. Aggregation layer (pooling)—3 bits. There are five possible variances: presence of aggregation layer aggregation dimension 2×2 or 3×3, and aggregation function—maxpooling and averaging.

In other three genes it can be marked 3 fields:

1. The presence of a layer—1 bit (at 0 in the first gene the individual enters into bioreactor).
2. The fully connected layer is 11 bits—the layer is between 0 and 1024 inclusive.
3. Dropout. The gene has a size of 5 bits, the possible values of Dropout from 0 to 30%.

Schematic representation of one chromosome is shown in Table 4.23.

The results of the genetic algorithm for CNN optimization are given in Table 4.24.

All the best results have 2 convolution layers, so this table does not list networks with 3 or 4 convolution layers. In each network, the convolutional layer is followed by an aggregating layer.

4.12.4 Building a Training Sample for Deep Neural Networks Processing Unstructured Images

The problem with training deep NN image processing used in medical diagnostics is the difficulty of obtaining a training sample. The training sample is based on the small number of examples available by turning them into small angles in different planes, with further processing of the images in another way [44].

The problem of image recognition in such a formulation is a rather complex problem with a number of limitations. The paper proposes a systematic approach to image processing based on structural decomposition (Fig. 4.35) [48].

Table 4.23 Scaled representation of the chromosome, totaling 30 genes

Presence 1 bit	Presence 1 bit	Presence 1 bit	Presence	Presence	Presence
The number of features maps—7 bits, from 4 to 256	The number of features maps—7 bits, from 4 to 256	The number of features maps—7 bits, from 4 to 256	Fully connected layer size—11 bits, from 0 to 1024 inclusive	Fully connected layer size—11 bits, from 0 to 1024 inclusive	Fully connected layer size—11 bits, from 0 to 1024 inclusive
Kernel size—3 bits, from 3 to 8 inclusive	Kernel size—3 bits, from 3 to 8 inclusive	Kernel size—3 bits, from 3 to 8 inclusive			
Offset, 2 bits	Offset, 2 bits	Offset, 2 bits			
Dropout, 5 bits, possible values of Dropout from 0 to 30%	Dropout, 5 bits, possible values of Dropout from 0 to 30%	Dropout, 5 bits, possible values of Dropout from 0 to 30%			
			Dropout, 5 bits, possible values of Dropout from 0 to 30%	Dropout, 5 bits, possible values of Dropout from 0 to 30%	Dropout, 5 bits, possible values of Dropout from 0 to 30%
Edge effects	Edge effects	Edge effects			
Aggregation (pooling) layer, 3 bits	Aggregation (pooling) layer, 3 bits	Aggregation (pooling) layer, 3 bits			

Table 4.24 10 best results

0	The first convolution layer	The second convolution layer	No of chromosome	Parents	Accuracy	Learning time, p	Network parameters
1	12 features maps with kernel dimension 3 × 3 and offset—1. Maxpolling layer with kernel dimension 2 × 2	72 features maps with kernel dimension 3 × 3 and offset—1. Maxpolling layer with kernel dimension 2 × 2	46	M = 10, F = 5	0.0008146313731	788.423846	{'FCL': [256], Bias: [1, 1], EE: ['0', '0'], 'P': 'M = 10, F = 5', 'ID': 46, 'PS': [2, 2], 'FCLD': [0], 'CLD': [0, 0], 'CKD': [3, 3], 'NCK': [12, 72]}
2	16 features maps with kernel dimension 3 × 3 and offset—1. Maxpolling layer with kernel dimension 2 × 2	16 features maps with kernel dimension 3 × 3 and offset—1. Maxpolling layer with kernel dimension 2 × 2	551	M = 370, F = 22	0.0009253731321	813.7425597	{'FCL': [256], Bias: [1, 1], EE: ['0', '0'], 'Anc': 'M = 370, F = 22', 'ID': 551, 'PS': [2, 2], 'FCLD': [0], 'CLD': [0, 0], 'CKD': [3, 3], 'NCK': [136, 16]}
3	12 features maps with kernel dimension 3 × 3 and offset—1. Maxpolling layer with kernel dimension 2 × 2	16 features maps with kernel dimension 3 × 3 and offset—1. Maxpolling layer with kernel dimension 102 × 102	460	M = 271, F = 20	0.0009349162266	401.5859036	{'FCL': [768], Bias: [1, 1], EE: ['0', '0'], 'P': 'M = 271, F = 20', 'ID': 460, 'PS': [2, 102], 'FCLD': [0], 'CLD': [0, 0], 'CKD': [3, 3], 'NCK': [12, 16]}
4	12 features maps with kernel dimension 3 × 3 and offset—1. Maxpolling layer with kernel dimension 2 × 2	200 features maps with kernel dimension 3 × 3 and offset—1. Maxpolling layer with kernel dimension 2 × 2	167	M = 37, F = 44	0.0009524901024	2961.137158	{'FCL': [768], Bias: [1, 1], EE: ['0', '0'], 'P': 'M = 37, F = 44', 'ID': 167, 'PS': [2, 2], 'FCLD': [0], 'CLD': [0, 0], 'CKD': [3, 3], 'NCK': [12, 200]}
5	12 features maps with kernel dimension 3 × 3 and offset—1. Maxpolling layer with kernel dimension 2 × 2	136 features maps with kernel dimension 3 × 3 and offset—1. Maxpolling layer with kernel dimension 2 × 2	2	M = 37, F = 44	0.0009657892627		{'FCL': [256], 'id_net': Bias: [1, 1], EE: ['0', '0'], 'P': '-', 'ID': 2, 'PS': [2, 2], 'FCLD': [0], 'CLD': [0, 0], 'CKD': [3, 3], 'NCK': [12, 136]}

(continued)

Table 4.24 (continued)

0	The first convolution layer	The second convolution layer	No of chromosome	Parents	Accuracy	Learning time, p	Network parameters
6	12 features maps with kernel dimension 3 × 3 and offset—1. Maxpolling layer with kernel dimension 2 × 2	200 features maps with kernel dimension 3 × 3 and offset—1. Maxpolling layer with kernel dimension 2 × 2	530	M = 23, F = 12	0.00100064638	2045.003203	{'FCL': [256], Bias: [1, 1], EE: ['0', '0'], 'P': 'M = 23, F = 12', 'ID': 530, 'PS': [2, 2], 'FCLD': [0], 'CLD': [0, 0.0], 'CKD': [3, 3], 'NCK': [12, 200]}
7	140 features maps with kernel dimension 3 × 3 and offset—1. Maxpolling layer with kernel dimension 2 × 2	16 features maps with kernel dimension 4 × 4 and offset—1. Maxpolling layer with kern el dimension 2 × 2	259	M = 16, F = 15	0.001004185923	993.5887313	{'FCL': [256], Bias: [1, 1], EE: ['0', '0'], 'P': 'M = 16, F = 15', 'ID': 259, 'PS': [2, 2], 'FCLD': [0], 'CLD': [0, 0], 'CKD': [3, 3], 'NCK': [140, 16]}
8	140 features maps with kernel dimension 3 × 3 and offset—1. Maxpolling layer with kernel dimension 2 × 2	200 features maps with kernel dimension 3 × 3 and offset—1. Maxpolling layer with kernel dimension 2 × 2	218	M = 17, F = 2	0.001006102076	2822.352327	{'FCL': [256], Bias: [1, 1], EE: ['0', '0'], 'P': 'M = 17, F = 2', 'ID': 218, 'PS': [2, 2], 'FCLD': [0], 'CLD': [0, 0], 'CKD': [3, 3], 'NCK': [140, 200]}
9	140 features maps with kernel dimension 5 × 5 and offset—1. Maxpolling layer with kernel dimension 2 × 2	72 features maps with kernel dimension 3 × 3 and offset—1. Maxpolling layer with kernel dimension 2 × 2	7	M = 17, F = 2	0.001033267347		{'FCL': [256], 'id_net': Bias: [1, 1], EE: ['0', '0'], 'P': '-', 'ID': 7, 'PS': [2, 2], 'FCLD': [0], 'CLD': [0, 0], 'CKD': [3, 5], 'NCK': [140, 72]}
1	88 features maps with kernel dimension 5 × 5 and offset—1. Maxpolling layer with kernel dimension 2 × 2	200 features maps with kernel dimension 4 × 4 and offset—1. Maxpolling layer with kernel dimension 2 × 2	204	M = 22, F = 77	0.001038146274	5195.876984	{'FCL': [1155], Bias: [1, 1], EE: ['0', '0'], 'P': 'M = 22, F = 77', 'ID': 204, 'PS': [2], 'FCLD': [0], 'CLD': [0.0, 0], 'CKD': [4, 5], 'NCK': [88, 200]}

*FCL is the fully connected layer; EE are edge effects; P are parents; M is the mother; F is the father; ID is the identificator; PS is the pooling size (aggregation); FCLD is the fully connected layer dropout; CLD is the convolutional layer Dropout; CKD is the convolution kernels dimension; NCK is the number of convolution kernels

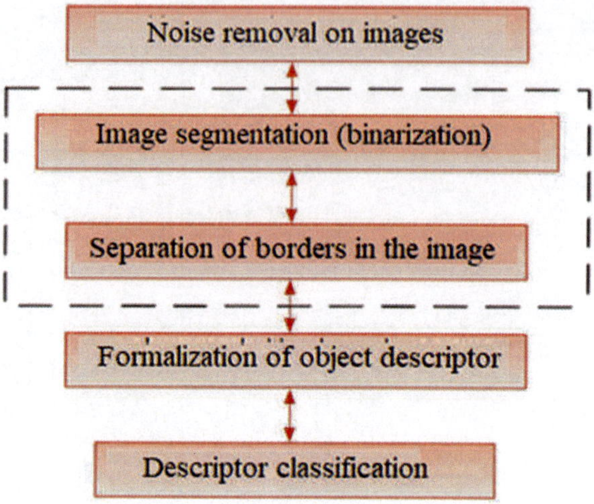

Fig. 4.35 Structural diagram of image classification algorithm on noisy image

Object descriptor is required for pattern recognition. It can be written effectively enough by analyzing the contour of an object. Therefore, one of the sub-tasks will be to define the boundaries of the objects in the image, formalize their contours and further classify the suspicious contours.

To analyze the contour of the object, it is necessary to correctly select the boundaries of the object in the image [43, 49]. But the selection of object boundaries has a major drawback—there are often errors in noisy, non-segmented images. Therefore, before delimiting, the image must be segmented.

Segmentation is the process of dividing a digital image into several segments (many pixels, also called superpixels) [49, 50]. The result of segmentation can be considered a binary image.

A binary image is a type of digital bitmap where each pixel can represent only one of two colors [48].

Since the image input is noisy, before any further transformations, it must be cleared of noise.

The algorithms for solving the problems indicated in Fig. 4.35 are given in Annex 3.

The results of the algorithm of image classification [51] on the noisy image to create a training sample and convolutional NN, are shown respectively in Figs. 4.36 and 4.37.

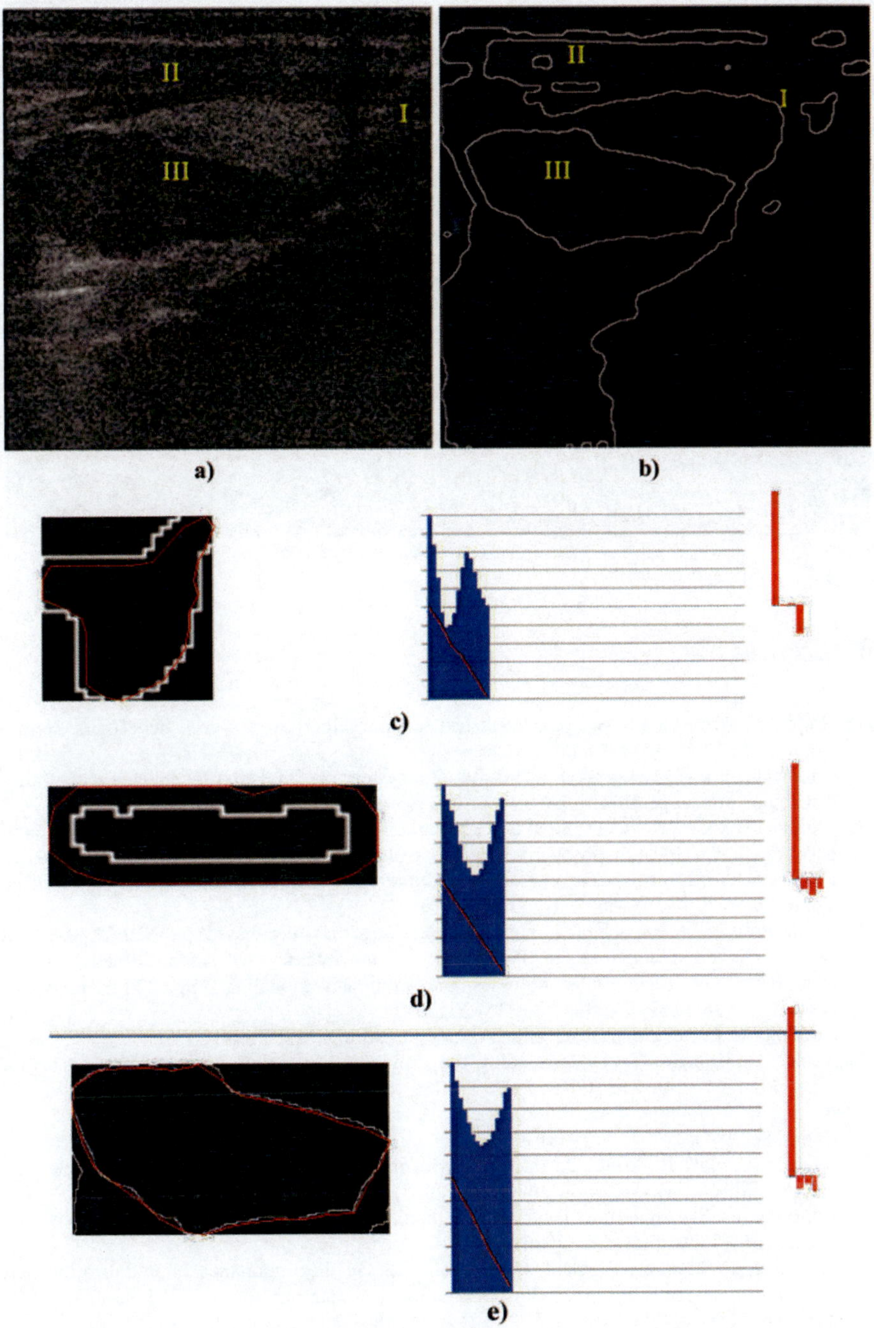

Fig. 4.36 Calculation of the contour descriptor for the image: **a** is the original image; **b** is the image with the outlined contours; **c** is object I and its ACF; **d** is the object II and its ACF; **e** is the object III and its ACF

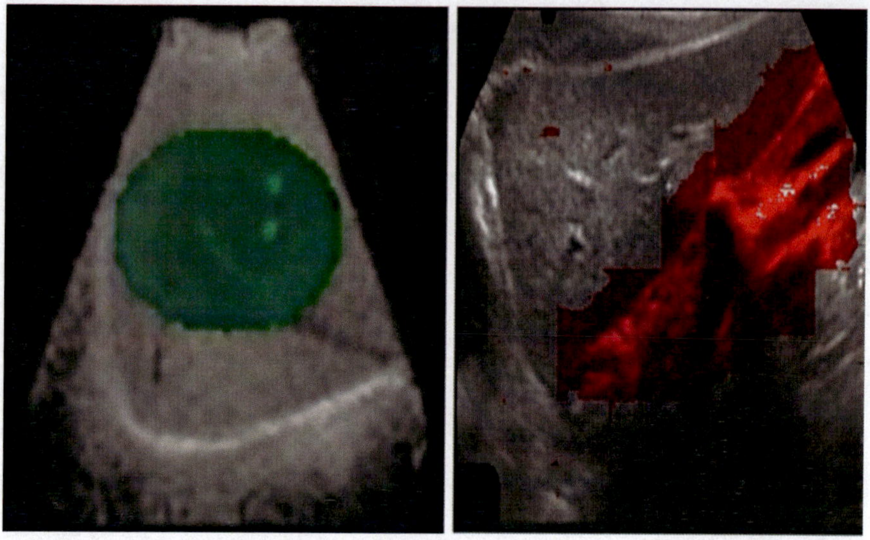

Fig. 4.37 Results of the CNN

References

1. Hinton, G.E., Osindero, S., Teh, Y.: A fast learning algorithm for deep belief nets. Neural Comput. **18**, 1527–1554 (2006)
2. Chumachenko, H., Luzhetskyi, A.: Building a system of simulation modeling for spatially-distributed processes. Electron. Control Syst. **1**(39), 108–113. Kyiv, NAU (2014)
3. Chumachenko, H., Koval, D., Sipakov, G., Shevchuk, D.: Using ANFIS and NEFCLASS neural networks in classification problems. Electron. Control Syst. **1**(43), 93–98. Kyiv, NAU (2015)
4. Chumachenko, H.: Deep learning classifier based on NEFPROX neural network. Electron. Control Syst. **4**(50), 63–66. Kyiv, NAU (2016)
5. Sineglazov, V., Chumachenko, H.: Deep learning classifier based on NEFCLASS and NEFPROX neural networks. In: Proceedings of the International Scientific and Practical Conference "Information Technologies and Computer Modeling", pp. 278–281, Ivano-Frankivsk, Yaremche, Ukraine, May 15–20, 2017
6. Goodfellow, I., Bengio, Y., Courville, A.: Deep Learning. MIT Press (2016)
7. Xavier, G., Bengio, Y.: Understanding the difficulty of training deep feedforward neural networks. J. Machine Learn. Res. Proc. Track **9**, 249–256 (2010)
8. Ackley, D.H., Hinton, G.E., Sejnowski, T.: A learning algorithm for Boltzmann machines. Cogni. Sci. **9**(1), 147–169 (1985)
9. Buades, A., Coll, B., Morel, J.M.: A review of image denoising algorithms with a new one. SIAM Multiscale Model. Sim. **4**, 490–530 (2005)
10. Hinton, G.E., Salakhutdinov, R.: Reducing the dimensionality of data with neural networks. Science **313**(5786), 504–507 (2006)
11. Sutskever, I., Martens, J., Dahl, G., Hinton, G.: On the importance of initialization and momentum in deep learning. In: 30th International Conference on Machine Learning, ICML 2013, pp. 1139–1147
12. Sutskever, I., Hinton, G.E.: Learning multilevel distributed representations for high-dimensional sequences. In: AI and Statistics. Puerto Rico (2007)
13. Chapelle, O., Scholkopf, B., Zien, A.: Semi-Supervised Learning. MIT Press, Cambridge, MA (2006). ISBN 978-0-262-03358-9

14. Zhu, X.: Semi-supervised learning. In: Sammut, C., Webb, G.I. (eds.) Encyclopedia of Machine Learning and Data Mining. Springer, Boston, MA (2017)
15. Lu, Z., Pu, H., Wang, F., Hu, Z., Wang, L.: The expressive power of neural networks: A view from the width. In: Neural Information Processing Systems, pp. 6231–6239 (2017)
16. Liu, C., Zhang, Z., Wang, D.: Pruning deep neural networks by optimal brain damage. In: Proceedings of the Annual Conference of the International Speech Communication Association, INTERSPEECH, pp. 1092–1095 (2014)
17. Hinton, G.E.: A practical guide to training restricted Boltzmann machines. Department of Computer Science, University of Toronto. UTML TR 2010–003 (2010)
18. Carreira-Perpinan, M., Hinton, G.: On contrastive divergence learning. In: Artificial Intelligence and Statistics (2005)
19. Hinton, G.E.: Training products of experts by minimizing contrastive divergence. Neural Comput. **14**(8), 1771–1800 (2002)
20. Tieleman, T.: Training restricted Boltzmann machines using approximations to the likelihood gradient. In: Proceedings of the 25th International Conference on Machine Learning, pp. 1064–1071 (2008). https://doi.org/10.1145/1390156.1390290
21. Lee, H., Grosse, R., Ranganath, R., Ng, A.Y.: Convolutional deep belief networks for scalable unsupervised learning of hierarchical representations. In: ICML '2009 (2009)
22. Tieleman, T., Hinton, G.: Using fast weights to improve persistent contrastive divergence. In: Proceedings of the 26th International Conference on Machine Learning. In ICML' 2009, p. 130. https://doi.org/10.1145/1553374.1553506
23. Cho, K., Raiko, T., Ilin, A.: Parallel tempering is efficient for learning restricted Boltzmann machines. In: Proceedings of the International Joint Conference on Neural Networks (IJCNN 2010), pp. 1–8. https://doi.org/10.1109/IJCNN.2010.5596837
24. Swendsen, R.H., Wang, J.S.: Replica Monte Carlo simulation of spin glasses. Phys. Rev. Lett. **57**, 2607–2609
25. Guillaume, D., Aaron, C., Bengio, Y., Pascal, V., Olivier, D.: Parallel tempering for training of restricted Boltzmann machines, p. 9 (2010)
26. Bottou, L.: Online algorithms and stochastic approximations. In: Online Learning and Neural Networks. Cambridge University Press (1998). ISBN 978-0-521-65263-6
27. LeCun, Y., Cortes, C., Burges, C.J.C.: The MNIST database of handwritten digits (1998) [Electronic resource]. Access mode http://yann.lecun.com/exdb/mnist
28. John, D., Elad, H., Yoram, S.: Adaptive subgradient methods for online learning and stochastic optimization. J. Machine Learn. Res. **12**, 2121–2159 (2011)
29. Hinton, G., Tieleman, T.: Lecture 6.5—RMSProp: divide the gradient by a running average of its recent magnitude. COURSERA: Neural Netw. Machine Learn. **4**(2), 26–31 (2012)
30. Kingma, D., Ba, J.: Adam: a method for stochastic optimization. In: International Conference on Learning Representations (2015) [Electronic resource]. Access mode https://arxiv.org/pdf/1412.6980
31. Ning, Q.: On the momentum term in gradient descent learning algorithms. Neural Netw. Official J. Int. Neural Netw. Soc. **12**(1), 145–151 (1999)
32. Reddi, S., Kale, S., Kumar, S.: On the convergence of Adam and beyond (2018) [Electronic resource]. Access mode https://arxiv.org/pdf/1904.09237
33. Nesterov, Y.: Method for minimizing convex functions with a convergence rate of O ($1/k^2$). In: Report of the Academy of Sciences of the USSR, vol. 3, pp. 543–547, 269 edn. (1983)
34. Krizhevsky, A., Nair, V., Hinton, G.: The CIFAR-10 dataset [Electronic resource]. Access mode https://www.cs.toronto.edu/~kriz/cifar.html
35. Muresan, H., Oltean, M.: Fruits-360: a dataset of images containing fruits and vegetables. Fruit recognition from images using deep learning. Acta Univ. Sapientiae, Informatica **10**(1), 26–42
36. Ruder, S.: An overview of gradient descent optimization algorithms (2016) [Electronic resource]. Access mode https://ruder.io/optimizing-gradient-descent/
37. Bottou, L., Bousquet, O.: The tradeoffs of large scale learning. In: Advances in Neural Information Processing Systems, vol. 20, pp. 161–168 (2008)

38. Chumachenko, H., Zakharov, S.: The study of the structure of the neural network in the diagnostic task. In: X International Scientific and Technical Conference "Avia-2011", pp. 22.40–22.43, April 15–21 (2011) (K.: – 2011)
39. Chumachenko, H., Levitskiy, O.: The algorithm training radial-basis networks based on particle swarm algorithm. In: Intellectual System for Decision Making and Problems of Computational Intelligence. ISDMC-2012 Congrece Proceeding, May 27–31, pp. 425–427 (2012). Yevpatoria, Ukraine
40. Sineglazov, V., Chumachenko, H., Levitsky, O.: Algorithm for training radial basis networks based on the particle swarm algorithm. Cybernet. Comput. (167), 25–32 (2012)
41. Chumachenko, H., Koval, D.: A hybrid evolutionary algorithm for forming a deep neural network topology. In: Proceedings of the International Scientific and Practical Conference "Information Technologies and Computer Modeling", Ivano-Frankivsk, Yaremche, Ukraine, May 23–28, pp. 20–22 (2016)
42. Sineglazov, B.V., Chumachenko, H., Koval, D.: Improvement of hybrid genetic algorithm for synthesis of deep neural networks. In: IV International Scientific-Practical Conference "Computational Intelligence", pp. 142–143, Kyiv, Ukraine, May 16–18, 2017
43. Al-Marzouqi, H.: Data clustering using a modified Kuwahara filter, neural networks. In: International Joint Conference, pp. 128–132 (2009)
44. Chumachenko, H., Levitsky, O.: Development of an image processing algorithm for diagnostic tasks. Electron. Control Syst. **1**(27), 57–65. K., NAU (2011)
45. LeCun, Y., Boser, B., Denker, J.S., Henderson, J.S., Howard, R.E., Hubbard, W., Jackel, L.D.: Backpropagation applied to handwritten zip code recognition. Neural Comput. 541–551 (1989)
46. Srivastava, N., Hinton, G., Krizhevsky, A., Sutskever, I., Salakhudinov, R.: Dropout: a simple way to prevent neural networks from overfitting. J. Machine Learn. Res. **15**, 1929–1958 (2014)
47. Chumachenko, H., Zakharov, S.: Algorithmic support of distributed databases. Artifi. Intell. **3**, 37–42 (2012)
48. Katkovnik, V., Egiazarian, K., Astola, J.: Adaptive window size image denoising based on intersection of confidence intervals (ICI) rule. J. Math. Imag. Vis. **16**(3), 223–235 (2002)
49. Gonzalez, R., Woods, R.: Digital Image Processing, p. 635. M., Technosphere (2005)
50. Fisenko, V., Fisenko, T.: Computer processing and image recognition: a training manual, p. 192. St. Petersburg State University of Information Technologies, Mechanics and Optics, St. Petersburg (2008)
51. He, K., Zhang, X., Ren, S., Sun, J.: Delving deep into rectifiers: surpassing human-level performance on imagenet classification. In: IEEE International Conference on Computer Vision (ICCV 2015), p. 1502. https://doi.org/10.1109/iccv.2015.123

Chapter 5
Intelligence Methods of Forecasting

5.1 The Solution to the Forecasting Problem Based on the Use of "Intelligent" Methods

Since in most different statements of forecasting problems it is believed that the initial data contains some random component that can degrade the quality of the constructed predictive model, it is desirable to carry out some pre-processing data in order to reduce this component. On the other hand, it is not known in advance whether there is some emission in the source data as a result of the influence of a random component, or if this emission is informative, and its smoothing can again worsen what the model predicts. Thus, the algorithm for solving the forecasting problem consists of two parts.

1. Pre-processing of data.
2. Actually building a forecasting model.

Data pre-processing is considered in Chap. 6.

There are two main approaches to building a forecasting model:

1. not intelligence (regression analysis [1], ARIMA, group method of data handling (GMDH));
2. intelligence.

In this work, the forecast problem is considered based on the use of neural networks.

M. Zgurovsky et al., *Artificial Intelligence Systems Based on Hybrid Neural Networks*, Studies in Computational Intelligence 904,
https://doi.org/10.1007/978-3-030-48453-8_5

5.2 Formulation of the Problem of Short-Term Forecasting of Nonlinear Nonstationary Processes

For a class of nonlinear nonstationary processes:

$$\{X(k, \varepsilon)\} \in \Re^N,$$

where $E[X(k)] \neq \text{const}$, $\text{var}[X(k)] \neq \text{const}$, build forecasting models (neural networks) $x(k) = f[\theta, z(k), \varepsilon(k)]$, where θ vector of model parameters; $k = 0, 1, 2,$ is the discrete time; $z(k)$ are independent explanatory variables; $\varepsilon(k)$ is the random disturbing process with arbitrary distribution.

Based on the model (in the form of a neural network) to create the forecast functions as follows

$$\hat{x}(k+s) = E_k\big[x(k+s)|x(k), x(k-1), \ldots, x(0), \hat{\varepsilon}(k), \ldots, \hat{\varepsilon}(0)\big],$$

where $x(k+s)$ is a function that allows you to calculate future values of a main (dependent) variable based on known historical data $\{x(k), x(k-1), \ldots, x(0)\}$ and estimates of random process values $\{\hat{\varepsilon}(k), \hat{\varepsilon}(k-1), \ldots, \hat{\varepsilon}(0)\}$; s is the number of forecasting steps.

The quality analysis of forecast estimates is executed by statistical criteria set: mean square error (MSE) and mean absolute percentage error(MAPE):

$$\text{MAPE} = \frac{1}{s}\sum_{i=1}^{s}\frac{|x_f(k+i) - \hat{x}(k+i)|}{|x_f(k+i)|}100\%,$$

where $x_f(k+i)$ the actual values of the main variable pertaining to the test sample; $\hat{x}(k+i)$ are forecasting estimates.

5.3 Forecasting Time Series Using Neural Networks

5.3.1 Combining ANN Approaches and Group Method of Data Handling

After analyzing the various currently available methods for constructing a forecast model, the most flexible methods can be clearly called the ANN and the Group Method of Data Handling (GMDH). Flexibility refers to the lack of the need for an expert to define a clear forecasting model, such as in linear regression or Box-Jenkins methodology. However, when using ANN, the question of choosing the optimal network architecture raises, and when using GMDH- the optimal support functions.

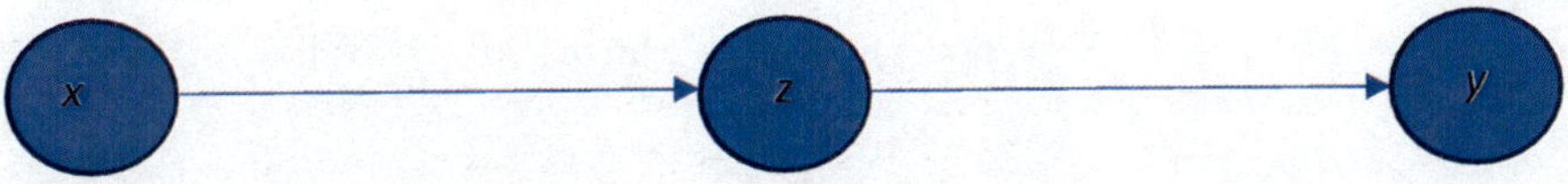

Fig. 5.1 The simplest case of a neural network that implements the GMDH method

In the works of the author [2–6] an algorithm that combines both approaches is used, where the ANN is used as a support function of the GMDH.

Due to this, this algorithm combines the advantages of both ANN and GMDH: a gradual increase in the complexity of the GMDH model and the opportunity to learn ANN.

Advantages and disadvantages of the proposed algorithm

The advantage of this algorithm in comparison with GMDH is that it is not necessary to explicitly specify the type of support functions, the necessary dependence will be found by neural networks, which are known to do very well with this task.

This algorithm is also this algorithm is deprived of the known disadvantage of ANN: when building a network, its optimal complexity is not known in advance, and a too simple neural network may model the process badly, and too complex networks are prone to so-called "overfitting", or retraining, which causes the network begins to model noise, presented in the training sample and as a result shows poor results in the validation sample. The proposed algorithm for each iteration uses simple networks that are not prone to retraining, but due to cascading complication is able to forecast very complex processes.

Despite the computational efficiency of the proposed approach, it is not possible to determine the optimal values of the weight coefficients. Let us prove this fact by example.

Let's take the simplest case (Fig. 5.1).

$z = a_0 + a_1 x + a_2 x^2$, $y = b_0 + b_1 z + b_2 z^2$, or if you rewrite y as a function of x:

$$\begin{aligned} y =& b_0 + b_1(a_0 + a_1 x + a_2 x^2) + b_2(a_0 + a_1 + a_2 x^2)^2 = (b_0 + a_0 b_1 + a_0^2 b_2) + \\ &+ x(a_1 b_1 + 2a_0 a_1 b_2) + x^2(a_2 b_1 + 2a_0 a_2 b_2 + a_1 b_2) + x^3(2a_1 a_2 b_2) + x^4(a_2^2 b_2). \end{aligned}$$

According to the multilevel algorithm of GMDH, the vector of parameters $\vec{a} = [a_0, a_1, a_2]^{\mathrm{T}}$ is as follows:

$$\vec{a} = (X'^T X')^{-1} X'^T y, X' = \begin{bmatrix} 1 & x_1 & x_1^2 \\ \ldots & \ldots & \ldots \\ 1 & x_n & x_n^2 \end{bmatrix},$$

where $\mathbf{X} - 3 \times n$ is the input matrix, $\mathbf{y}$ is the output column vector.

Going to the next level where the vector of parameters will be calculated $\vec{b} = [b_0, b_1, b_2]^{\mathrm{T}}$, vector of values of new input variable $\vec{z}$ is determined by the formula

$\vec{z} = X'\vec{a} = X'(X'^T X')^{-1} X'^T \vec{y}$, and a new input matrix **Z** has the following form:

$$Z' = \begin{bmatrix} 1 & z_1 & z_1^2 \\ \dots & \dots & \dots \\ 1 & z_n & z_n^2 \end{bmatrix}.$$

Parameters vector $\vec{b}$ then calculated by the formula $\vec{b} = (Z'^T Z')^{-1} Z'^T \vec{y}$. But since the function $y(x)$ is just the 4th degree polynomial $y(x) = c_0 + c_1 x + c_2 x^2 + c_3 x^3 + c_4 x^4$, then the optimal coefficients vector $\vec{c}$ should be calculated as follows:

$$\vec{c} = (X''^T X'')^{-1} X''^T \vec{y},\ X'' = \begin{bmatrix} 1 & x_1 & x_1^2 & x_1^3 & x_1^4 \\ \dots & \dots & \dots & \dots & \dots \\ 1 & x_n & x_n^2 & x_n^3 & x_n^4 \end{bmatrix}.$$

Obviously, for example, the coefficient $\tilde{n}_4$ will not be equal $a_2^2 b_2$, namely such factor at x^4 will be obtained using multilevel GMDH. As an example, consider the following: we define a true function $y(x)$:

$$y(x) = 2 + 0,2x - 0,8x^2 + 0,1x^3 - 0,2x^4.$$

Next, we generate a training sample. To do this, we first set the vector of input values $\vec{x}$:

$$\vec{x} = [-5, -4.9, -4.8, \dots, 4.9, 5]^T,$$

then calculate the vector of the output values for learning as the output of the true function plus some random noise:

$$\vec{y}_i = y(x_i) + N(0, 1), i = 1, \dots, 101.$$

The resulting training sample can be visualized as a graph shown in Fig. 5.2.

Next, let's evaluate the model parameters $\hat{y}(x) = \hat{\tilde{n}}_0 + \hat{\tilde{n}}_1 x + \hat{\tilde{n}}_2 x^2 + \hat{\tilde{n}}_3 x^3 + \hat{\tilde{n}}_4 x^4$. by least square method (LSM), which introduces additional variables $x^{(2)} = x^2$, $x^{(3)} = x^3$, $x^{(4)} = x^4$, then the model $\hat{y}(x)$ can be described as a linear model for variables x, $x^{(2)}$, $x^{(3)}$, $x^{(4)}$:

$$\hat{y}(x, x^{(2)}, x^{(3)}, x^{(4)}) = \hat{c}_0 + \hat{c}_1 x + \hat{c}_2 x^{(2)} + \hat{c}_3 x^{(3)} + \hat{c}_4 x^{(4)}.$$

We obtain the following estimates of the coefficients:

$$\hat{\tilde{n}}_0 = 2.0486,\ \hat{\tilde{n}}_1 = 0.1565,\ \hat{\tilde{n}}_2 = -0.8121,$$
$$\hat{\tilde{n}}_3 = 0.1043,\ \hat{\tilde{n}}_4 = -0.1990.$$

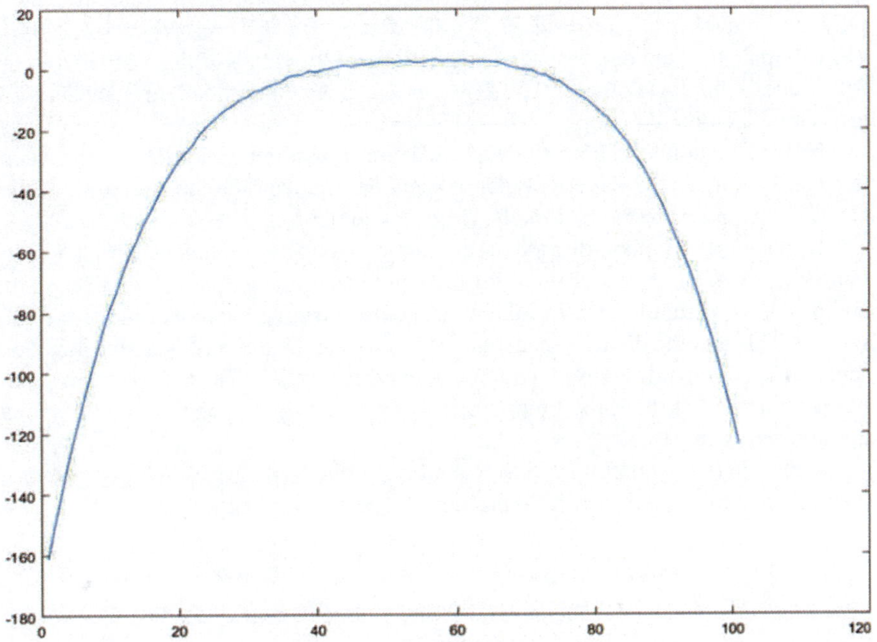

Fig. 5.2 Graph of the training sample

The normalized mean square error of the model with these parameters is $4,8641 \times 10^{-4}$, the total absolute deviation of the parameter estimates from the true values is equal 0.1095.

Now let us learn the two-layer polynomial network with quadratic neurons shown in Fig. 5.3, according to the algorithm GMDH—that is, first we find the value of the parameters vector $\vec{a} = [a_0, a_1, a_2]^{\mathrm{T}}$, at which the minimum mean square error (MSE) of the model $z(x) = a_0 + a_1 x + a_2 x^2$ is reached, then we fix this parameters vector and find the value of the parameters vector $\vec{b} = [b_0, b_1, b_2]^{\mathrm{T}}$, at which the minimum MSE of the model $y_{gmdh}(x) = b_0 + b_1 z(x) + b_2 z(x)^2$ is reached.

The parameters of both models will be found according to LSM. We obtain the following values of vectors: $\vec{a} = [13.1323, \quad 1.7523, \quad -5.1606]^{\mathrm{T}}$, $\vec{b} = [-1.36, \quad 0.3708, \quad -0.0072]^{\mathrm{T}}$. The error of such a network in the training sample is equal 0.0016—approximately 3.5 times bigger than the error of the polynomial whose parameters were evaluated directly. Having vectors $\vec{a}$, $\vec{b}$ you can calculate the corresponding parameters vector $\vec{c}_{gmdh}$ of the 4th degree polynomial, which given by this

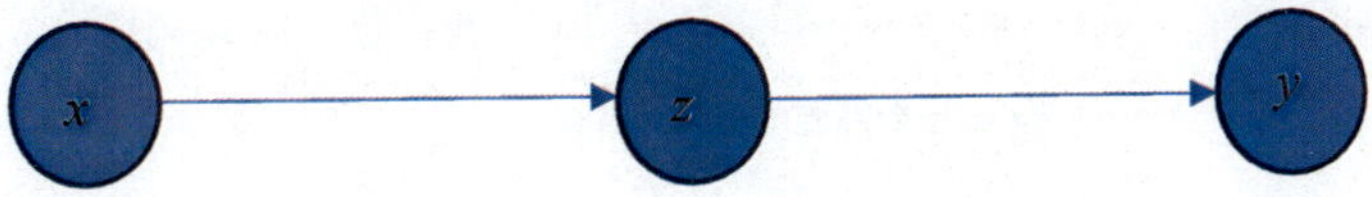

Fig. 5.3 A two-layer polynomial network with quadratic neurons

polynomial network $c_{gmdh} = [2.2743, \ 0.3201, \ -0.9467, \ 0.1296, \ -0.1908]^T$, and calculate the sum of the absolute deviations of the values of this vector from the true values, which is equal to 0.5979—about 5 times bigger than the same sum for estimates calculated directly.

Here are the graphs of both models (4th degree polynomial with parameters calculated directly and a two-layer polynomial network trained according to the algorithm GMDH) compared to the real model (Figs. 5.4 and 5.5).

This example confirms that the partial model parameters obtained at each level of multilevel GMDH are not optimal for the whole "multilayer" model obtained at the end of the algorithm in terms of the mean square error of the model in the training sample. This is quite obvious since the parameters of each level-layer are calculated separately, with fixed values of parameters in other levels—that is, all parameters are never optimized at the same time, which is what should be done to find the global minimum of a function.

To eliminate the identified disadvantage in this work, we developed a hybrid method of solving the forecasting problem based on deep learning.

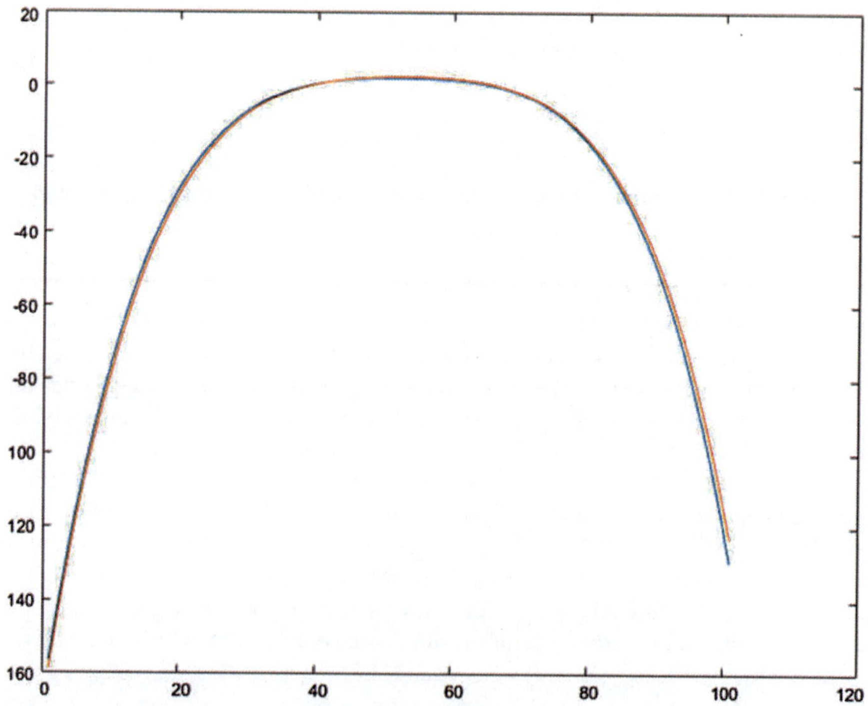

Fig. 5.4 Graph of the model obtained as a result of tuning the parameters of the two-layer polynomial network according to the GMDH algorithm

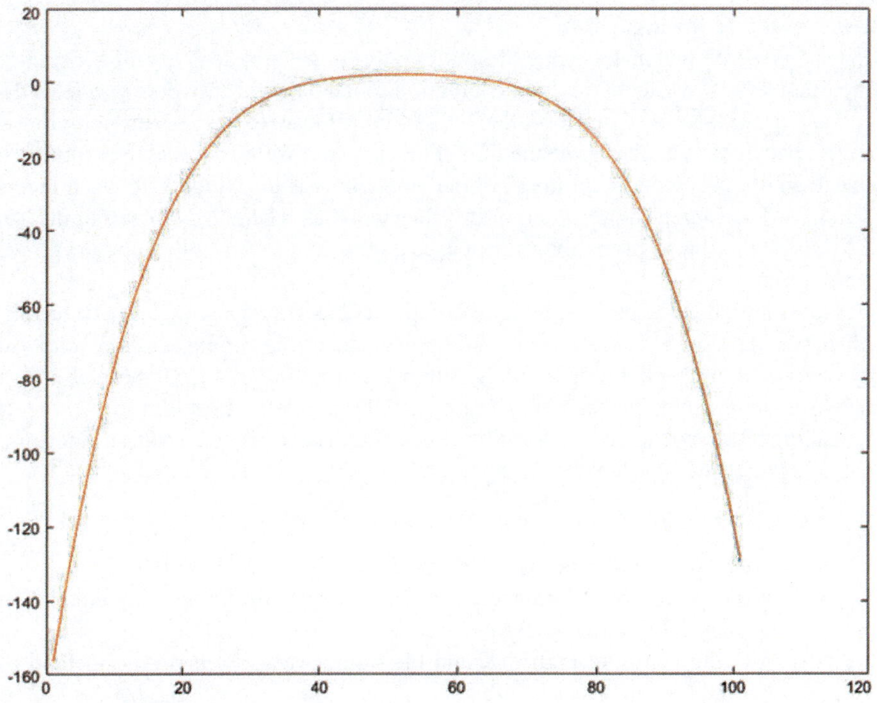

Fig. 5.5 Graph of the real model

5.3.2 *Hybrid Method of Solving the Forecasting Problem Based on Deep Learning*

As proved in Chap. 3, the most perspective direction of ANN study at the moment is deep learning and deep neural networks. In the works [7, 8] first proposed the solution of this problem: initially the network grew layer *by layer, with the initial weights between the two layers were using the algorithm of a restricted* Boltzmann machine, and then the resulting network was "adjusted" with the use of the same algorithm of error back propagation is the initial weights already provided "reasonable" behavior of the network, and the problem of fading or over-growing gradients has disappeared. Thus, there are two main stages of deep learning neural networks.

1. "Advanced" training of deep learning network, the essence of which is to add new layers one by one, wherein the weights between the two layers is learnt separately—most often with the use of the Boltzmann restricted machine learning algorithm.
2. "Retraining" of the obtained network structure with the use of the error back propagation algorithm (or some of its modifications), sometimes—with the use of regularization methods (the most popular regularization method for deep-network training for this moment is the dropout algorithm [9]).

Description of the algorithm

In [8] it is mentioned that, despite the comparative novelty of deep learning methods, the first method to effectively train (polynomial) deep neural networks was GMDH. Indeed, a multi-level algorithm GMDH can be represented as a deep polynomial neural network. In addition, the method of this network weights learning is similar to the stage of "preliminary learning" of deep networks is the weights between layers i and $i + 1$ are learnt independently of other weights. The stage "of retraining" in GMDH methodology probably was not used due to the insufficient computers power for that time.

Nowadays, the computing power of computers has grown significantly to accomplish this stage in a reasonable time. The essence of the proposed approach is to use the Resilient backpropagation algorithm for "retraining" of polynomial neural network weights obtained as a result of a GMDH multilevel algorithm.

Taking into account the foregoing, the deep neural network constructing algorithm canb be presented as follows.

1. Forming a sample from the initial time series $\{x_n\}$:

   (1) *obtaining* a difference time series: $d_i = x_{i+1} - x_i$;
   (2) *normalizing* the difference time series to zero mean and single standard deviation: $d_i = (d_i - \mu)/\sigma$;
   (3) *forming* the input matrix **X** and the output vector y using the method of embedding time series with some dimension of embedding k:

$$\mathrm{X} = \begin{pmatrix} d_1 & \cdots & d_k \\ \vdots & \ddots & \vdots \\ d_{N-k} & \cdots & d_{N-1} \end{pmatrix}, \vec{y} = \begin{bmatrix} d_{k+1} \\ \vdots \\ d_N \end{bmatrix}$$

2. The "pre-training" stage, that uses a multi-level GMDH algorithm to obtain the initial structure and weights of the polynomial neural network:

   (1) *the whole sample* is randomly divided into training and validation in the ratio 0,7–0,3 (that is, 70% of the patternss fall into the training sample, 30%—into the validation sample);
   (2) *learnt* C_k^2 models

$$f(x_i, x_j) = a_{i1}x_i + a_{j1}x_j + a_{ij}x_ix_j + a_{i2}x_i^2 + a_{j2}x_j^2,$$

   using a linear regression on the training sample;
   (3) *for each model* f its error is calculated on the validation sample $E(f) = \sum_{(\vec{x},y)\in(X_v,\vec{y}_v)} (f(\vec{x}) - y)^2$;
   (4) *the models* with the least error are selected s_l, where l is the row number (often, for ease of algorithm implementation it is chosen $s_l = k$);
   (5) *the outputs* of these models for each pattern of training sample form a new matrix of inputs $\mathbf{X}^{(1)}$ for the next row (layer) of models:

$$X^{(1)} = \mathrm{X} = \begin{pmatrix} f_1(\vec{d}_1) & \cdots & f_{s_l}(\vec{d}_1) \\ \vdots & \ddots & \vdots \\ f_1(\vec{d}_{N-k}) & \cdots & f_{s_l}(\vec{d}_{N-k}) \end{pmatrix},$$

where $\vec{d}_p = [d_p, d_{p+1}, \ldots, d_{p+k-1}]^{\mathrm{T}}$, but a record $f_m(\vec{d}_p)$ means that from the vector $\vec{d}_m$ will be taken only elements below the indexes i and j which corresponding to the model f_m.

(6) *among the errors* on the validation sample of all models of the current row, the minimum one is selected, and if it is less than the minimum error of the models from the previous row (or if it is the first row), it is executed the transition to the next row (beginning from step 2), otherwise, the algorithm stops, and the best model of the previous row is selected as "final"; thus, the stop criterion of the GMDH algorithm can be written as:

$$\begin{cases} l > 1, \\ \min\limits_{f \in F_l} E(f) \geq \min\limits_{f \in F_{l-1}} E(f), \end{cases}$$

where l is the number of the current row; F_l is the set of all models of the current row;

(7) *at the output* of this step we obtain a polynomial neural network of the following form (Fig. 5.6).

3. Stage of "retraining", on which the weights of the obtained network are learnt with the use of the algorithm of Resilient backpropagation error:

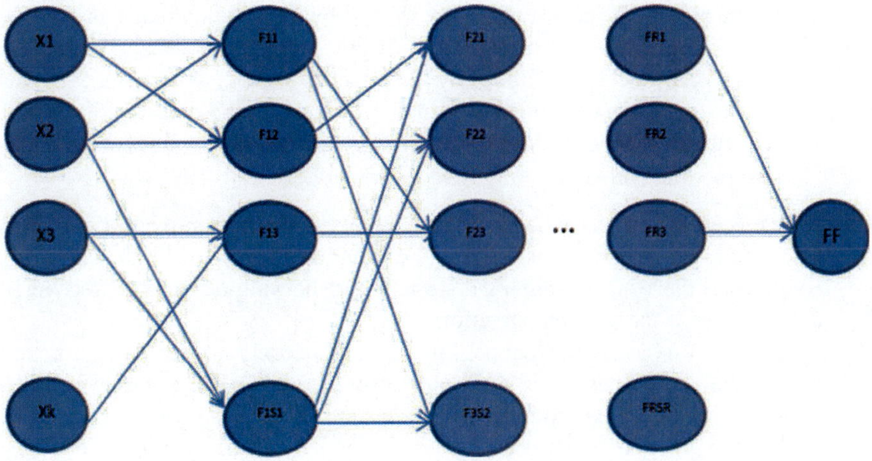

Fig. 5.6 Polynomial neural network parameters of the two-layer polynomial network according to the algorithm GMDH

(1) *for each pair* < input vector, output value > from the training sample two so-called passes are executed:
"straight pass" is the vector of input values is entered the first layer of the network, and the outputs of each polynomial "neuron" are calculated up to the last one, "output" neuron;
"back pass" are calculated the derivatives of error function by each weight using the following formulas:

$$\frac{\partial f_{l,m}}{\partial \vec{a}} = [x_i,\ x_j,\ x_i \cdot x_j,\ x_i^2,\ x_j^2]^{\mathrm{T}},$$

$$\frac{\partial f_{l+1,h}}{\partial f_{l,m}} = \begin{cases} a_{i1} + a_{ji}x_j + 2a_{i2}x_i, & \text{if } f_{l,m} \text{ input } x_i \text{ for } f_{l+1,h}, \\ a_{j1} + a_{ji}x_i + 2a_{j2}x_j, & \text{if } f_{l,m} \text{ input } x_j \text{ for } f_{l+1,h}, \\ 0, \text{ if } f_{l,m} \text{ not input for } f_{l+1,h}, \end{cases}$$

(2) *all the calculated* derivatives are summed up by all the patterns;
(3) *each weight* is updated according to the following rule:

$$\Delta_{ij}^{(t)} = \begin{cases} a^+ \cdot \Delta_{ij}^{(t-1)}, & \frac{\partial E^{(t)}}{\partial w_{ij}} \cdot \frac{\partial E^{(t-1)}}{\partial w_{ij}} > 0, \\ a^- \cdot \Delta_{ij}^{(t-1)}, & \frac{\partial E^{(t)}}{\partial w_{ij}} \cdot \frac{\partial E^{(t-1)}}{\partial w_{ij}} < 0, \end{cases}$$

$$w_{ij}^{(t)} = w_{ij}^{(t-1)} - sign\left(\frac{\partial E^{(t)}}{\partial w_{ij}}\right) \cdot \Delta_{ij}^{(t)}.$$

If the sign of the error function derivative by the weight coincides with the sign of the derivative on the previous iteration is the correction value for this weight will be multiplied by some factor $a^+ > 1$. Otherwise, the value of the update is multiplied by a factor $a^- < 1$. At the first iteration, some constant correction value is used $\Delta_{ij}^0 = c$. Usually such values of constants are recommended: $a^+ = 1.2,\ a^- = 0.5,\ c = 0.1$.

(4) *After all the weights* have been updated, the network error on the test sample is calculated—if the error is less than on the previous iteration one—the training is continued, otherwise the training is stopped and the weight values are recovered to the values at the previous iteration.

4. As a result, we have a conventional polynomial neural network, the prediction on new data is done as usual: the input vector $\vec{\mathrm{X}}$ is entered the first layer of the network, and then the outputs of all neurons are calculated layer by layer, until the last layer with one neuron, the output of which will be the prediction.

5.3.3 *The Use of the GMDH Multilevel Algorithm and Single-Layer Networks with Sigm_Piecewise Neurons*

In some practical cases of the prediction task a number of input variables can be large enough and reach thousands or even tens of thousands of variables. Usually it is the tasks in which, in addition to the main time series that is predicted there are exogenous (external) time series that potentially affect to he main one. Under solving the problems with so many input variables the following problems occur:

- due to the increasing number of parameters that need to be tuned, the learning process can last quite a long time—it makes it practically impossible to apply the approach of brute force of neurons number in a hidden layer to find the optimal network structure, since the process of complete brute force may demand too much time;
- it is unknown in advance which input variables actually carry information useful for predicting the objective variable, and which can be ignored—and the number of "unnecessary" variables can be very large;
- in certain tasks you need to get a model with a limited number of parameters, since there is some limit either on the "size" of the model in the memory of the forecasting device, or on the time spent to obtain the forecast in the trained model—it is obvious that when the number of model parameters increases it is increased both the number of memory units required to reproduce this model and the time need to obtain a prediction when using this model.

To solve all these problems, it is proposed to use the approach of the GMDH multilevel algorithm [10], which will allow:

- much faster than under full brute force use, to find in a certain sense the optimal network structure, with given restrictions on the number of network parameters;
- automatically filter out "uninformative" variables.

The algorithm for automatic synthesis of optimal network structure with sigm_piecewise neurons based on the GMDH multilevel algorithm consists of the following steps:

1. Input sample $< X, y : X \to R >$, where the set x is a finite subset of space R^n, $n > 2$ is in some way divided into two samples: training R^n, $n > 2$ and validational—$< X^{(V)}, y^{(V)} : X^{(V)} \to R >$, $X^{(V)} \subset X, X^{(V)} \cap X^{(T)} = \emptyset, X^{(T)} \cup X^{(V)} = X$. Usually, about 70% of the patterns are selected in the training sample, and the simplest variant, which works well in practice, is random selection.
2. Independently one from one C_n^k neurons of the sigm_piecewise type are learing, where each neuron has inputs—x_i, x_i; $i, j = 1, \ldots, n$, $i < j$—that is, all possible pairs of inputs are sorted out, and a separate neuron is trained for each pair. For learning it is used training sample, the criterion according to which the parameters of each neuron are tuned is the following:

$$E_{i,j}(\vec{w}_+, \vec{w}_-, \vec{h}) = \sum_{\vec{x} \in X^{(T)}} (y^{(T)}(\vec{x}) - \text{sigm\_piecewise}([x_i, x_j]^T; \vec{w}_+, \vec{w}_-, \vec{h}))^2.$$

1. *The value of the "external" criterion* is calculated for each neuron. The most common criterion is the MSE model on the validation sample:

$$C_{i,j}(\vec{w}_+, \vec{w}_-, \vec{h}) = \sum_{\vec{x} \in X^{(V)}} (y^{(V)}(\vec{x}) - \text{sigm\_piecewise}([x_i, x_j]^T; \vec{w}_+, \vec{w}_-, \vec{h}))^2.$$

2. $\left\lfloor \alpha \cdot \tilde{N}_k^n \right\rfloor$, $a \in (0, 1)$ *the neurons* with the worst external criterion value are rejected. Parameter value α is selected depending on the restriction of the number of network parameters, or according to some heuristic, if such restriction is absent. For example, if between input variables there are variables that are completely unrelated with the predicted variable, then the external criterion value for the neurons that used these variables as inputs is usually much worse than the similar value for the neurons that used the "informative" variables (Fig. 5.7).
   It is clearly seen here that the neurons that used the input variables x_7, x_8, x_9 have a much larger error in the validation sample, so it is logical to get rid of these neurons. The structure of the network at the output of this step is shown in Fig. 5.8.
   An output neuron with a linear activation function is added to the network, as result we obtain a network with the structure shown in Fig. 5.9.
3. *At the same time*, both adjusting the parameters of the output linear neuron and adjusting the parameters of the neurons in the hidden layer are performed according to the algorithm of error backpropagation. The following formulas of derivatives are used:

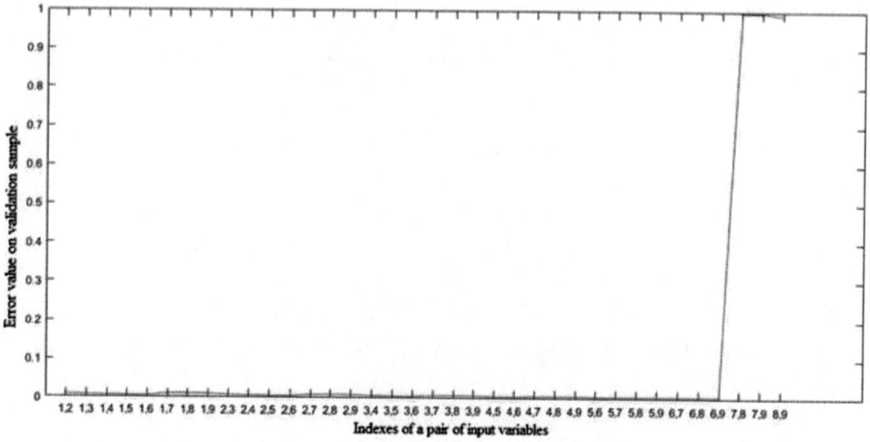

Fig. 5.7 An example of "model error external criterion values on validation sample" of different neurons with the presence of "uninformative" variables

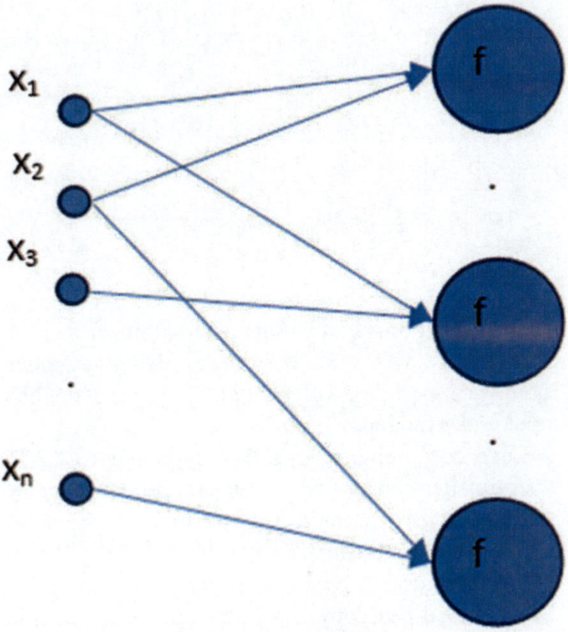

Fig. 5.8 Structure of the reduced neural network

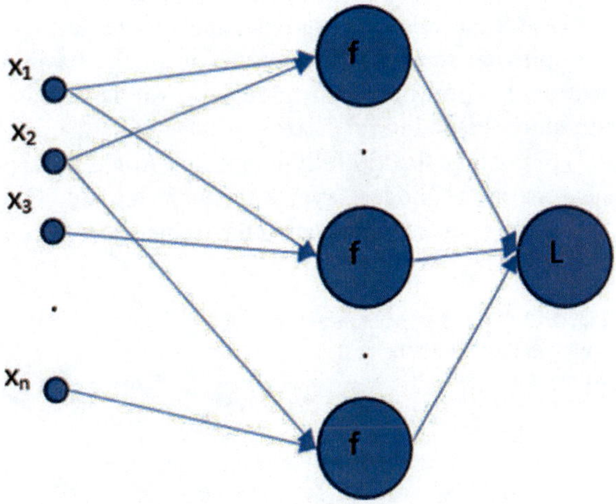

Fig. 5.9 Network structure with the addition of an output neuron with a linear activation function

$$\frac{\partial E}{\partial v_{i^*}} = \sum_{\vec{x}} \left(\sum_i f_i(\vec{x}) v_i - y(\vec{x}) \right) f_{i^*}(\vec{x}),$$

$$\frac{\partial E}{\partial w_{+q,i^*}} = \frac{\partial E}{\partial f_{i^*}} \frac{\partial f_{i^*}}{\partial w_{+q,i^*}} = \sum_{\vec{x}} \left(\sum_i f_i(\vec{x}) v_i - y(\vec{x}) \right) v_{i^*} \cdot \left(\frac{x_q}{1 + e^{-k\vec{h}^T \vec{x}}} \right),$$

$$\frac{\partial E}{\partial w_{-q,i^*}} = \frac{\partial E}{\partial f_{i^*}} \frac{\partial f_{i^*}}{\partial w_{-q,i^*}} = \sum_{\vec{x}} \left(\sum_i f_i(\vec{x}) v_i - y(\vec{x}) \right) v_{i^*} \cdot \left(\frac{x_q}{1 + e^{k\vec{h}^T \vec{x}}} \right),$$

$$\frac{\partial E}{\partial w_{q,i^*}} = \frac{\partial E}{\partial f_{i^*}} \frac{\partial f_{i^*}}{\partial w_{q,i^*}} = \sum_{\vec{x}} \left(\sum_i f_i(\vec{x}) v_i - y(\vec{x}) \right) v_{i^*} \cdot k x_q \left(\frac{\vec{w}_+^{\mathrm{T}} \vec{x} - \vec{w}_-^{\mathrm{T}} \vec{x}}{2 + e^{k\vec{h}^T \vec{x}} + e^{-k\vec{h}^T \vec{x}}} \right),$$

where v_{i^*} is the weights of the original linear neuron; $f_i(\vec{x})$ is the activation function of sigm_piecewise neuron, in the hidden layer under the index i, w_{+q,i^*}, w_{-q,i^*}, h_{q,i^*} is the corresponding weights of the neuron in the hidden layer under the index i^* for input under the index q.

Let's consider applying this approach to CATS time series forecasting, and wherein after converting the initial time series according to the procedure described above, we will add to each vector of input variables $\vec{x}$ several randomly generated numbers that imitate extra variables.

So, let's turn the initial time series into a sample $< X, y : X \rightarrow R >$ using the described procedure with attachment dimension of 6 and prediction horizon of 3, and add 3 randomly generated numbers to each vector $\vec{x}$. It will be chosen 60% of patterns from obtained sample we will randomly select for the training sample, 20%—for the validation sample, and another 20%—for the test sample.

Applying the described approach on the basis of GMDH and leaving only 3 neurons with the best error value in the validation sample, we will obtain the network structure shown in Fig. 5.10.

For comparison, 20 fully-connected single-layer perceptrons with the number of neurons in the hidden layer 2–21 were trained. The best value of the error in the test sample was obtained using the trained perceptron with 10 neurons in the hidden

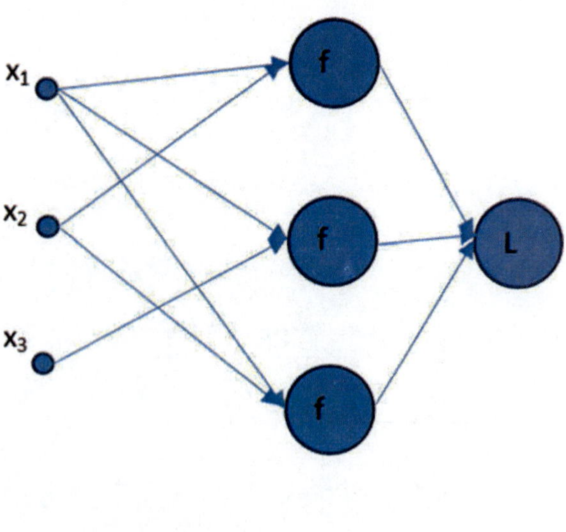

Fig. 5.10 Network structure with 3 neurons with the best error value

layer, and this error was 0.021—about 9% better than in the network built using the proposed approach. But this perceptron consists of $10 \times 10 + 11 = 111$ parameters, while the first network has only $3 \times 3 + 4 = 13$ parameters.

5.3.4 Testing the Performance of Sigm_Piecewise Neuron Networks for the Time Series Forecasting Task

In Sect. 3.4, NN with sigm_piecewise neurons was proposed. Let's test the effectiveness of these networks in real-time forecasting tasks.

The first is a sample of monthly electricity consumption in southeastern part of Brazil from January 1976 to December 2000 (Fig. 5.11).

Let's compare the efficiency of using the following networks to predict this sample: single-layer perceptron with sigm_piecewise neurons in the hidden layer, single-layer perceptron with neurons of tansig type (activation fucntion $\tan\mathrm{sig}(x) = \frac{e^x - e^{-x}}{e^x + e^{-x}}$) in a hidden layer and a polynomial neural network using the GMDH algorithm to construct and train it. The method of comparison is as follows:

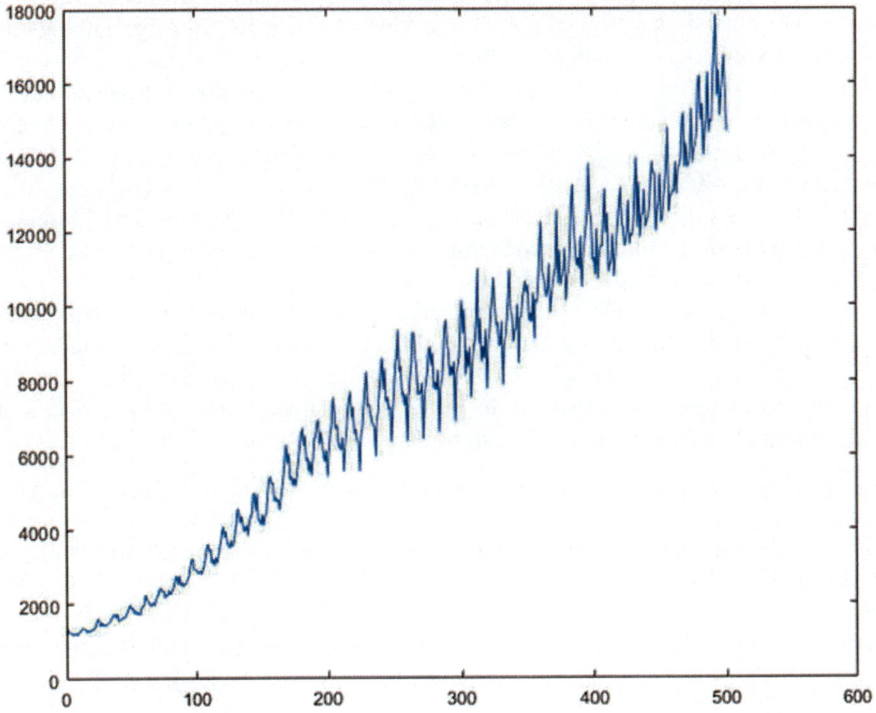

Fig. 5.11 monthly consumption of electiricity sample

(1) The input time series is converted to a sample $< X, y : X \to \Re >$ by using a time series embedding method with an embedding dimension of 6 and a forecast horizon of 3—that is, to predict the value of g_{i+3} it was used the values $g_{i-5}, g_{i-4}, g_{i-3}, g_{i-2}, g_{i-1}, g_i$.

(2) The sample was transformed as follows:

   2.1 $\forall \vec{x} \in X : y(\vec{x}) := y(\vec{x}) - x_n, \vec{x} \in R^n$—that is, instead of forecast value of the time series itself g_{i+3} it is predicted the deviation of this value from the last known one—$g_{i+3} - g_i$.

   2.2 $\forall \vec{x} \in X : \vec{x} := [x_1 - x_n, \ldots, x_{n-1} - x_n]^{\mathrm{T}}$—that is, to predict the deviation from the last known time series value g_i, the following deviations $g_{i-5} - g_i$, $g_{i-4} - g_i$, $g_{i-3} - g_i$, $g_{i-2} - g_i$, $g_{i-1} - g_i$ are used.

(3) 70% of the patterns were randomly chosen into the training sample $< X^{(Train)}, y^{(Train)} : X^{(Train)} \to \Re >$

(4) As the training criterion it is used the mean square error MSE on the training sample. As a criterion for comparison of obtained by different methods of forecasting models it is used MAPE as the result of models functioning on the test sample.

(5) The MAPE value of a naive model that "believes" that $g_{i+3} = g_i$, equals $MAPE_{naive} = 10.23\%$. By sorting different numbers of neurons in a hidden layer, a network of 15 sigm piecewise neurons was selected. After training this network in the test sample $MAPE_{sigm\_piecewise} = 3.12\%$.

(6) After using the GMDH algorithm, a polynomial network of 6 layers and 5 neurons was obtained in each layer except the original one—that is, this architecture was optimal in terms of the external criterion. The error of such a polynomial network was $MAPE_{GMDH} = 5.33\%$.

(7) A network of 36 tansig-type neurons was selected by brute force different numbers of neurons in the hidden layer. After training the prediction criterion value of this network was $MAPE_{\mathrm{tansig}} = 4.21\%$.

(8) Therefore, among all types of networks tested, the network with sigm piecewise neurons had the smallest error in the test sample, wherein the prediction criterion value of approximately 25% less than that of the tansig network and approximately 40% less than the polynomial network one, that was built and learnt according to the GMDH algorithm.

Visually, the predictions of the three networks are as follows (Figs. 5.12, 5.13 and 5.14).

Thus, this sample obtained the following results of comparison of several methods (Table 5.1).

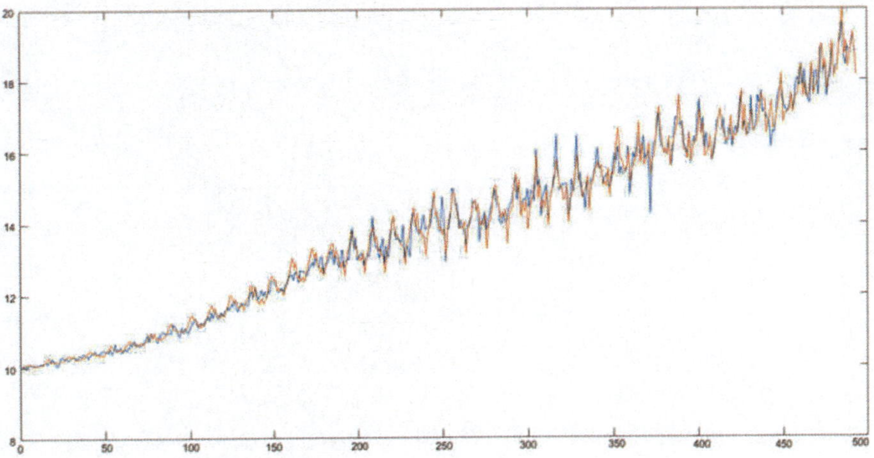

Fig. 5.12 Network prediction with sigm_piecewise neurons

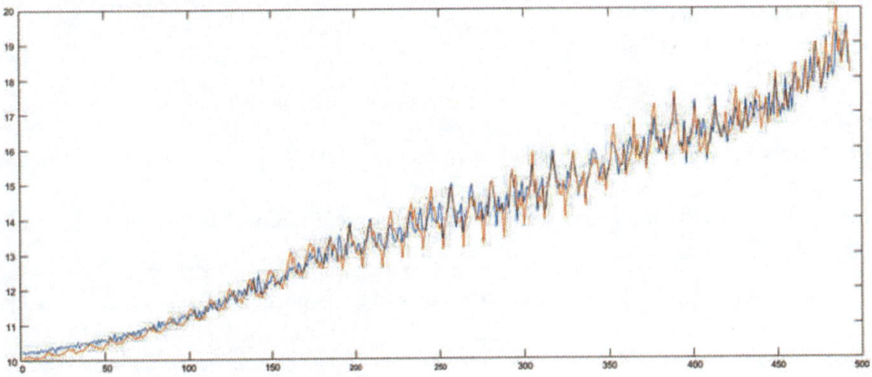

Fig. 5.13 Network forecast with tansig neurons

5.3.5 Integration of Multiple ANN Topologies

To improve the quality of the prediction, you can use a integration of estimates obtained from different models. Integration means the weighted sum of estimates obtained with the generated set of models. Weight coefficients are determined by the external model optimality criterion—variance in the examination sample. Many models are obtained by brute froce the variants of splitting the original sample into sub-samples and brute force of different prediction methods [11]. So, having k_1 subdivision for sub-samples and k_2 prediction methods, we receive $k_1 \cdot k_2$ different models. For a final prediction, having an input vector $\vec{\mathbf{x}}$, it is needed to:

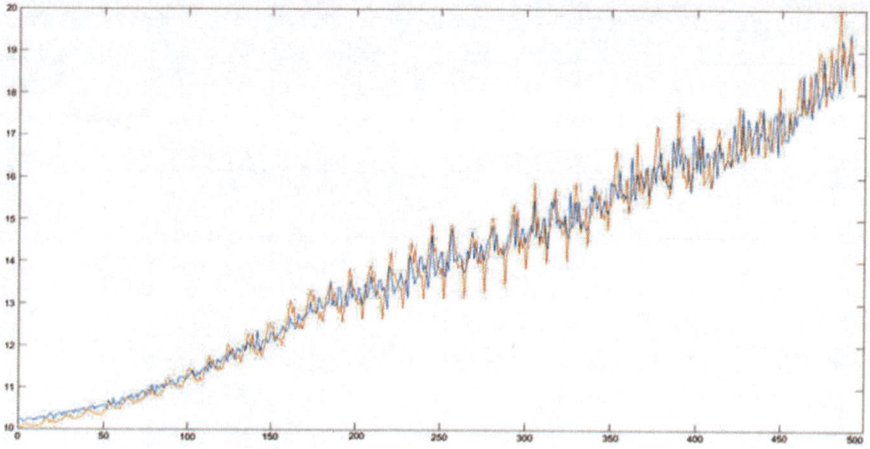

Fig. 5.14 Polynomial network forecast

Table 5.1 MAPE of the methods on the test sample

Sigm piecewise	Tansig	Multilevel GMDH
3.12	4.21	5.33

(1) apply this vector to the input of each model, thus obtaining a vector of estimates $\hat{y} = \hat{y}_1[\hat{y}_1, \ldots, \hat{y}_{k_1 k_2}]$;

(2) get final prediction $\hat{y}_f$ as a weighted sum of the elements of the vector of estimates $\hat{y}$.

These steps require the determination of weights coefficients $a_i, i = 1 \ldots k_1 \cdot k_2$.

5.3.5.1 Division of the Original Sample into Sub-Samples

By order. To the training subsample the first $C_1 \cdot N$ points are selected, to the verification one—the follow $C_2 \cdot N$ points and into testing—points, that are left, that is $(1 - C_1 - C_2)N$ (where N is the total points, $C_1 + C_2 < 1$; $C_1, C_2 > 0$ are coefficients, usually choose $C_1 = 0.6$; $C_2 = 0.2$).

Randomly. Similarly to the previous method, only the points are selected not by order but randomly, but the proportions between the subsamples are preserved.

Every ith. To the verification sample every ith point is selected, from the rest every jth point is selected to the testing sample, all other points are selected in the training sample (usually choose $i = 3$, $j = 4$).

By variance. All points are ranked by variance (one point means one pattern) and then are chosen into sample similarly to the first method.

5.3.5.2 Obtaining Models Through a Enumeration of Methods

For each of the chosen division methods for samples and forecasting methods, a forecast process model is constructed. In [11], the following forecasting methods are used: ANN, GMDH, a combination of GMDH and ANN. Thus, with four division methods for samples and three forecasting methods, we obtain $4 \times 3 = 12$ different models.

5.3.5.3 Integration of Obtained Results

To obtain a real forecast, it is used the integration of forecasts of all models obtained in the previous stages [12, 13]. That is, having estimates of the predicted value $\hat{y} = \hat{y}_1[\hat{y}_1, \ldots, \hat{y}_{k_1 k_2}]$, the prediction of the value itself will be equal to $\hat{y}_f = \sum_i a_i \hat{y}_i$. The coefficients a_i are determined using external criterion—forecast variance of the ith model on the testing sample: $a_i = \frac{1}{\sigma_i}$, where σ_i is the mean square deviation of predictions of the ith model on the testing sample from the real forecasted process' values. After calculation of all a_i they have to be normalized using the following formula $a_i^n = \frac{a_i}{\sum_i a_i}$.

In order to evaluate the suitability of integration to improve the quality of the prediction, we compare the mean square deviation of the prediction (on some artificial data) obtained with help of the prediction algorithm proposed in this work and with help of forecast integration of several methods (including the proposed one). We generate an artificial time series with the following true model:

$$f(t) = 0.1t^3 - 10t^2 + 6 + 300\sin(t),\ t = 1 \ldots 100.$$

Let's add to the mathematical model the normal noise $\varepsilon(t)$ with the following characteristics: $E(\varepsilon) = 1000$, $\sigma(\varepsilon) = 500$ (Fig. 5.15).

After that, we build a predictive model using the algorithm GMDH + ANN and with the use of complex estimation of several methods [7]. The mean square deviation of the prediction obtained by the algorithm GMDH + ANN, is equal to 0.0038 (Fig. 5.16).

The mean square deviation of the prediction obtained by integration several algorithms (GMDH + ANN, GMDH, ANN), is equal to 0.0020 (Fig. 5.17).

As you can clearly see from the figures, integration really greatly improves the quality of the prediction.

The proposed method of constructing the predictive model was tested on several datasets available in the open access on the Internet, and the following results were obtained (Table 5.2).

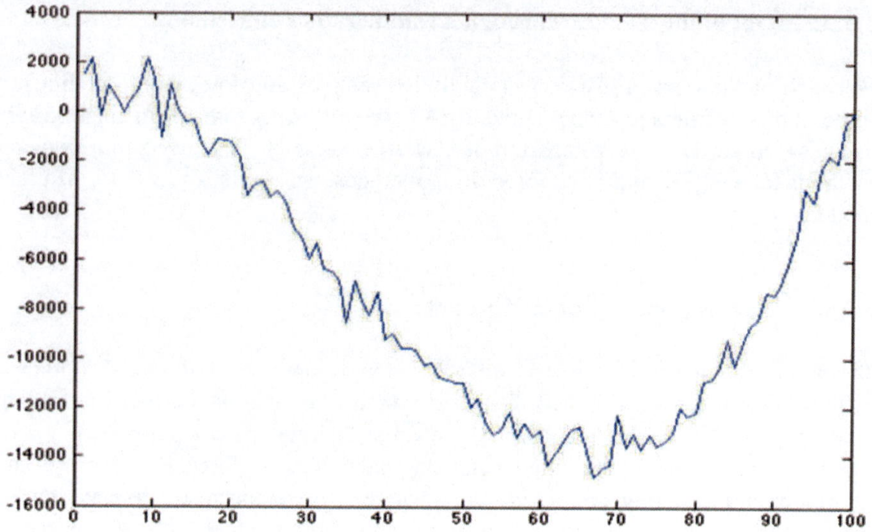

Fig. 5.15 Artificial time series

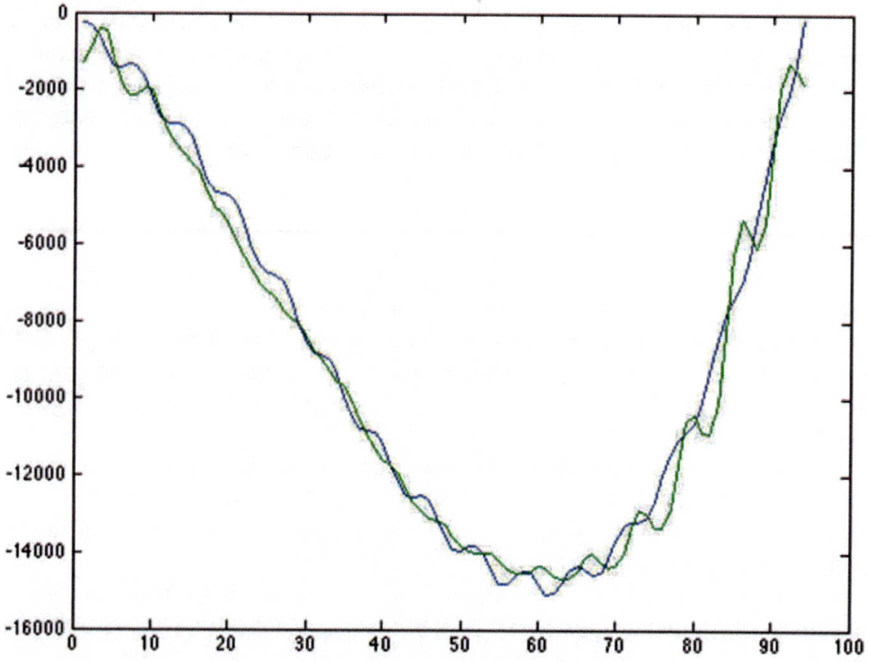

Fig. 5.16 Prediction obtained using algorithms GMDH + ANN

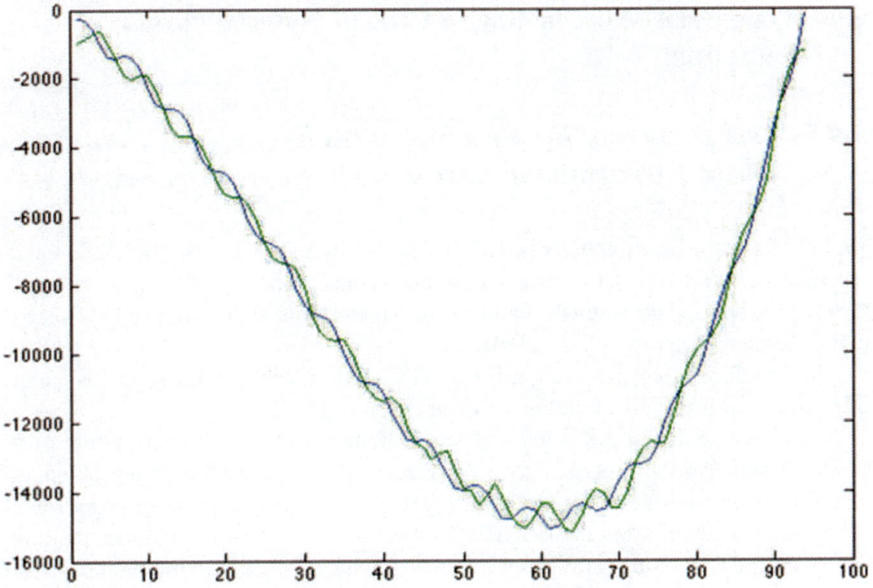

Fig. 5.17 The prediction obtained by integration several algorithms

Table 5.2 Normalized root mean square error

Sample name	ANN	GMDH	Proposed method
Power generation in Australia	0.017662	0.019721	0.012685
Test of CATS	0.002894	0.002696	0.002901
Dollar to euro exchange rate	0.063802	0.05511	0.062086
The dollar to pound exchange rate	0.055874	0.050277	0.058154
CPI index	5.50E-05	0.007696	2.22E-05
Electricity consumption in Spain	0.019363	0.024104	0.017655
Average interest rates in Spain	0.055512	0.048002	0.053009
Stock exchange index in Spain	0.002652	0.005721	0.002495
Number of spots on the sun	0.5811	0.45099	0.17865
Aircraft production in the USA	0.20121	0.15927	0.13734
NAO winter index	1.0566	0.98757	1.0009
Total error	2.0567	1.8112	1.5259

5.4 Time Series Forecasting in Case of Heterogeneous Sampling

5.4.1 Justification of Problem Statement Necessity of Time Series Forecasting in Case of a Heterogeneous Sample

Under the sample heterogeneity is understood the situation when the probability characteristics of the process, the sample of which is observed, "significantly" changes in the observed period, or when the patterns from the sample belong to several "clusters" that differ significantly in characteristics.

The problem of constructing a forecast model under the heterogeneous sample often arises in many different areas of life.

Most of the existing predicted problem statements assume that the nature of the predicted process does not change in the observed period, and thus, the problem can be solved by finding the desired model and estimation of its parameters using all available data. However, often the behavior of the predicted process can change significantly several times during the observed period, which makes such problem statement incorrect—a one single model will not be able to describe several different states of the process. Consider the statement of the predictive problem, which explicitly takes into account the problem of the inconsistent nature of the forecasted process.

Let there be:

(a) sample of patterns $S = \{(\vec{x}_i, y_i) : i = 1 \ldots n\}$, where every pattern is $(\vec{x}_i, y_i)$ consists of a vector of values of inputs and the corresponding output value;
(b) model f, which depends on some vector parameter $\vec{\theta} = [\theta_1, \ldots, \theta_m]^T$ and in the most general way it describes the predicted process—keep in mind that, depending on the parameter values, the model is able to describe all the possible states of the projected object.

Need to find:

(a) set of parameter vectors $\Theta = \{\vec{\theta}_1, \ldots, \vec{\theta}_k\}$, such that the vector $\vec{\theta}_j$ is optimal according to the criterion L_j on some subset of the output data S_j: $\Theta = \left\{\vec{\theta}_j : \vec{\theta}_j = \arg\min_{\theta}\left(L_j[f, \vec{\theta}, S_j]\right), S_j \subseteq S, j = 1 \ldots k\right\}$. Each parameter vector will thus describe one state of the predicted process on some subset of the initial sample. This problem statement assumes that the available data describe several different states of the process, and each such state can be described by a specific vector of parameters. However, the number of such states is unknown—this is the task of an appropriating method of prediction;
(b) the function of determining the finite vector of parameters θ_s for new input $\vec{x}_s$ based on the found set of parameters vectors Θ and the initial data itself: $\vec{\theta}_s = F(\vec{x}_s, \Theta, S)$, such that minimize some estimation of the prediction error $\hat{e} = e(F, \Theta, S)$.

For further consideration of approaches for the solution of formally given problem, firstly it is necessary to consider certain typical classes of heterogeneous samples—since certain approaches are oriented towards specific types of heterogeneous samples. Depending on which exactly probability characteristics of the process change significantly in the observed sample, distinguish the following typical types of heterogeneous samples:

- different states distinguish only in different mathematical expectations (Fig. 5.18);
- different states distinguish only by variance (Fig. 5.19);
- different states distinguish only in the values of the autocorrelation function (Fig. 5.20);
- different states distinguish only in probability distribution (Fig. 5.21);
- different states distinguish by combination previous characteristics, such as mathematical expectation and variance, distribution, etc.

When constructing a forecast model it is necessary taken into account in some way the probability that the training sample is heterogeneous—that is, the internal state of the predicted process changed several times during the observation. Ideally, having a training sample $S = \{(\vec{x}_i, y_i) : i = 1 \ldots n\}$, need to find:

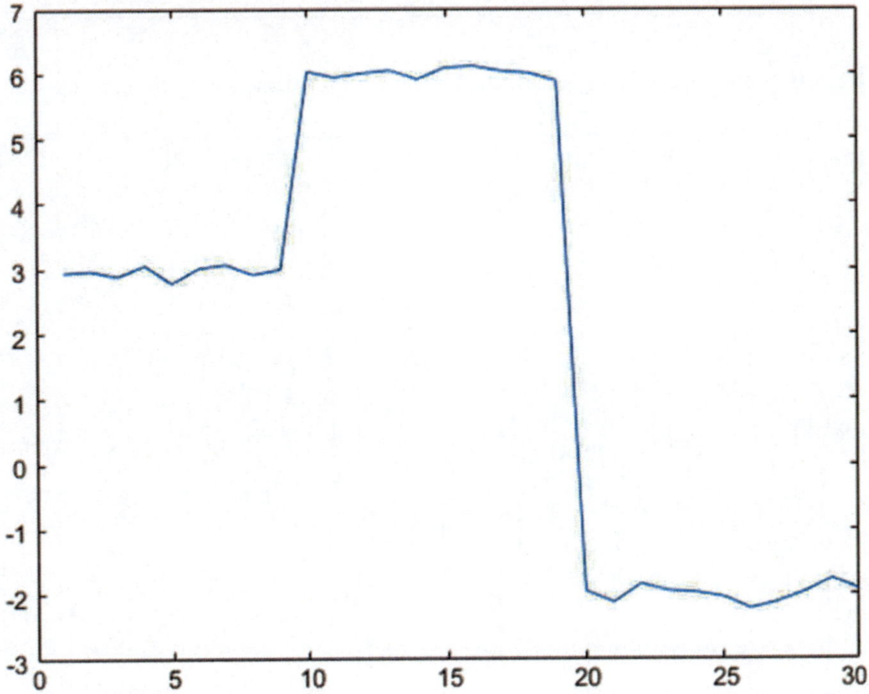

Fig. 5.18 Pattern of the heterogeneous sample with different mathematical expectations of states

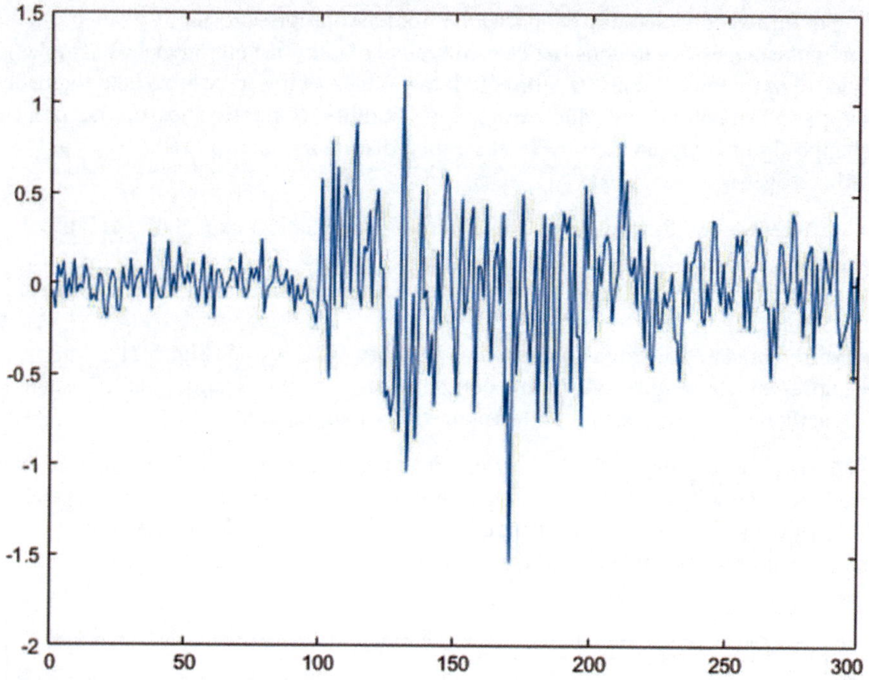

Fig. 5.19 Pattern of the heterogeneous sample with different variances for each of the states

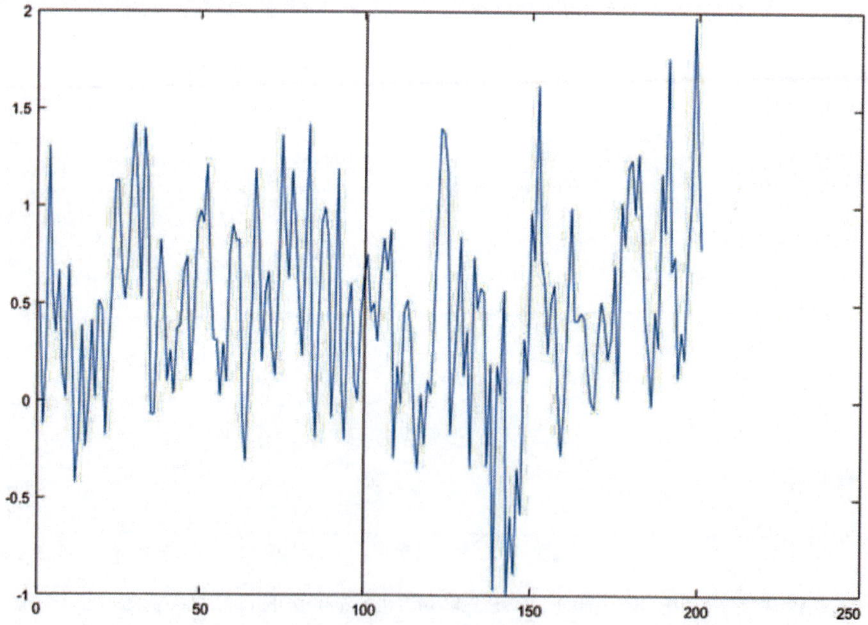

Fig. 5.20 Pattern of the heterogeneous sample with a different autocorrelation function

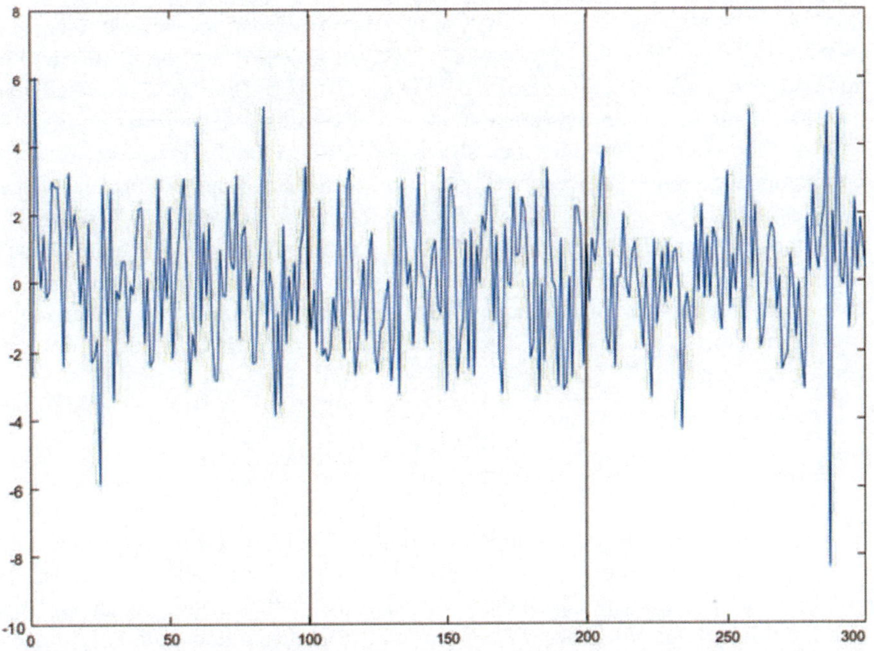

Fig. 5.21 Pattern of the heterogeneous sample with a different probability distribution for each state (but with the same mathematical expectation and variance): the first state is the normal distribution, the second state is the uniform distribution, the third estate is the lognormal distribution

1. The number of states K with significantly different behavior of the predicted process.
2. For each state $k \in \{1, 2, \ldots, K\}$ find the probability distribution of the input vectors $p(\vec{x}/k)$ and the a priori probability of this state $p(k)$.

Then, for each pattern of learning sample $\vec{x}_i$, $i \in \{1, 2, \ldots, n\}$ it is possible to calculate the probability of appurtenance of this example to each state k as:

$$p(k/\vec{x}_i) = \frac{p(\vec{x}_i/k) * p(k)}{\sum_{k' \in \{1,2,\ldots,K\}} p(\vec{x}_i/k') * p(k')}, \ \forall k \in \{1, 2, \ldots, K\},$$

and take these probabilities into account under construction of a foreacast model.

There are several common approaches for the solution of this problem: the use of ARCH (GARCH) models and the approach of inhomogeneous samples forecast based on the use of clustering and ANN methods [8], proposed by the author.

In [14], the model of autoregressive conditional heteroscedasticity (ARCH) was first proposed. This model is a statistical model for the variance description of the current values of a random process as an autoregressive (AR) function from the true observed values of past errors of the forecast model. Formally, this can be written as follows: denote a random value that describes the error of some forecast model at time t as e_t, then $\{e_t\}$, $t = 1, 2, \ldots$ will be an appropriate random process. According to the ARCH model, $e_t = \sigma_t z_t$, where $z_t \sim N(0, 1)$—then the variance e_t will be equal to σ_t^2 it is assumed that $\sigma_t^2 = a_0 + \sum_{i=1}^{q} a_i e_{t-i}^2$, $a_i > 0,\ a_i \geq 0,\ i > 0$, that is, the current variance is an AR function of the squares of previous errors e_{t-1}^2.

Using the assumptions described above, the ARCH model allows us to describe non-stationary random processes in which the variance of current random values changes over time, including processes that "exhibit" "inhomogeneous" behavior—when there are periods of relative "rest" (i.e., with a small variance of random values) and periods of high volatility. A large number of modifications to the ARCH model have been proposed in the following—[15, 16], of which the most popular model is the generalized ARCH (GARCH) [15], which uses the ARMA model to describe the current variance:

$$\sigma_t^2 = \omega + \sum_{i=l}^{q} a_i e_{t-i}^2 + \sum_{i=l}^{p} \beta_i \sigma_{t-i}^2.$$

The main disadvantage of ARCH-based models is that these models work well only when their assumptions are fulfilled—thus, these models allow us to describe nonstationary processes in which the nonstationary source is a nonconstant variance of random values at different points in time. That is, if in a non-stationary process the joint distribution of random values is compatible $p(y_t, y_{t-l}, \ldots, y_{t-k})$ at different points in time t changes "substantially" and in such a way that it cannot be described only by a non-constant variance of values y_t—ARCH "family" models will not be able to describe this process.

5.4.2 *Clustering and Model Building for Each Cluster/Segment*

In the case of the task of time series forecasting, the input values to the forecast model are usually past values of the time series. Then it is reasonable to assume that in the significantly different states of the predicted process, the probabilistic distribution of process values will differ substantially—and therefore the distribution of input vectors $\vec{x}$ will be significantly different. That is, to search for different states of the process, it is necessary to group patterns from the training sample on the principle of similar probabilistic distribution—for this purpose it is possible to use different clustering algorithms [17], that is, grouping patterns with similar characteristics into clusters. By applying clustering and obtaining clusters as an estimate of the probabilistic distributions of patterns in different states, it is possible to build separate models for each cluster, using only the patterns from that cluster for training, so that each model will be "specialized" for the patterns of the specific cluster, and if the clusters are indeed relevant different states—then the model will describe the corresponding state. This approach has some disadvantages, namely:

- the "forecasting quality" of the models is highly dependent on the quality of the clustering—if the clusters obtained do not correspond to the actual states, then the models studied on these clusters will not describe the corresponding states;
- each cluster must have a sufficient number of patterns to avoid retraining—that is, more patterns are needed than when building a single model on all available data without clustering.

The work discusses a new approach to clustering.

5.4.2.1 Problem Statement of Clustering

Having some sequence of values is written as a vector $\vec{y} = [y_1, \ldots, y_n]^{\mathrm{T}}$ is necessary for each value under the number $i \in \{1, 2, \ldots, n\}$ put the number in line to cluster $c_i^* \in N$, thus that the obtained vector of clusters numbers $\vec{c}^* = [c_1^*, \ldots, c_n^*]^{\mathrm{T}}$ minimizes a certain loss function $L(\vec{y}, \vec{c})$, which depends on the input vector $\vec{y}$ and vector of cluster numbers $\vec{c}$:

$$\vec{c}^* = \underset{\vec{c}}{\operatorname{argmin}}\{L(\vec{y}, \vec{c})\}.$$

It is important to emphasize that the number of clusters is not predefined.

It is assumed that all values from a cluster of number i have a normal distribution with unknown mathematical expectation μ_i and variance i and are independent of each other.

In most cases, the loss function is given informally, and the task of formalizing it is much more complicated than the problem of finding a value, which minimizes a formally given function—so many papers devoted to the problem of clustering inhomogeneous samples are concentrated only on the formalization of the loss function for the problem of clustering heterogeneous samples.

For example, the informal loss function can be specified as "all values in a particular cluster should have the most similar probabilistic distribution, and values from different clusters should be as different as possible". Obviously, such an informal loss function is rather fuzzy, and can be formalized in various ways, depending on the assumptions that will be allowed under the formalization. It is also assumed that clustering with fewer number of clusters is "better" than clustering with more clusters number. Then for each clustering it is possible to count its "quality" as the sum of the probability of the respective clustering and the value of a certain strictly decreasing function of the number of clusters. It is necessary to find the highest quality clustering. In fact, in this informal description of the loss function, we only need to formalize the clustering quality function from the number of clusters—for simplicity, we choose it as $Q(N) = N$, where N is the number of clusters. Then the formal loss function will look like:

$$L(\vec{y},\vec{c}) = \sum_{i=1}^{N}\sum_{j=1}^{n_i}\left(y_{ij} - \frac{\sum_{j=1}^{n_i} y_{ij}}{n_i}\right)^2 + N,$$

where y_{ij} is the jth element from ith cluster ($i \in \{1, 2, \ldots, N\}, j \in \{1, 2, \ldots, n_i\}$); N is the number of clusters; n_i is the number of elements in ith cluster; $\sum_{j=1}^{n_i} y_{ij}/n_i = \bar{y}_{ij}$ is the center of ith cluster.

That is, it is necessary to find such clustering that the values from each cluster are close to the arithmetic mean of all values of the cluster and the number of clusters is not too large, otherwise clustering "by cluster for each value" would be optimal.

Similar to the loss function are the "classic" clustering methods that find the vector of cluster numbers $\vec{c}$ with some fixed predetermined number of clusters $N : 2 \le N \le n-1$. Such methods usually have the following loss function:

$$L(\vec{y},\vec{c}) = \sum_{i=1}^{N}\sum_{j=1}^{n_i} D_i(y_{ij}, \bar{y}_i),$$

where $D_i(y_{ij}, \bar{y}_i)$ is the function of distance between jth element ith cluster and center ith cluster.

5.4.2.2 Soft Clustering Algorithm Based on Separating Hypersurfaces

As is well known, most clustering algorithms solve a particular problem of discrete optimization, which in the general case can be described as follows.

Having many examples $X = (\vec{x}_1, \ldots, \vec{x}_n)$ where each pattern is a vector in space R^d, each cluster number must be matched to each pattern $k \le n$ so that the vector of cluster numbers is obtained $\vec{k} = [k_1, \ldots, k_N]^T$ minimized a certain criterion $C\left(\vec{k}, X\right)$:

$$\vec{k}^* = \arg\left\{\min_{\vec{k}} C\left(\vec{k}, X\right)\right\}.$$

It is also known that even for very simple criterion views and small dimensions d this task is very difficult.

Probably the most common criterion is the total mean distance between points in one cluster:

$$CR\left(\vec{k}, X\right) = \sum_{i}^{K} \sum_{\vec{x}_j : k_j = i} \vec{x}_j - \vec{\mu}_i^2.$$

Which can also be represented in the following form:

$$CR\left(\vec{k}, X\right) = \sum_{i}^{K} \frac{1}{|C_i|} \sum_{\vec{x}_j, \vec{x}_l : k_j = i, k_l = i} \vec{x}_j - \vec{x}_l^2.$$

If you enter a function $k(\vec{x})$, that for each pattern assigns a cluster number, then you can also write this criterion so:

$$CR(k, X) = \sum\nolimits_i^K \frac{1}{\sum_{\vec{x} \in X} 1(k(\vec{x}) = i)} \sum\nolimits_{\vec{x}_j, \vec{x}_l} 1\left(k\left(\vec{x}_j\right) = i\right) \times 1(k(\vec{x}_l) = i) \times \left\|\vec{x}_j - \vec{x}_l^2\right\|$$

where $1\ (\text{condition}) = \begin{cases} 1, & \text{if the condition is done,} \\ 0, & \text{if the condition is not done.} \end{cases}$

Consider the simplest case when we only have 2 clusters, that is $k(\vec{x}) \in \{0, 1\}$. Perform the "softening" of the original problem—let us function $k(\vec{x})$ take values across the range [0; 1], that is, for pattern $\vec{x}$ function value $k(\vec{x})$ will set something similar to the probability of membership of this pattern to the cluster 1; respectively, the value $1 - k(\vec{x})$ specifies a certain analog of probability of pattern membership to the cluster 0. In this case, the "softened" version of the criterion will look like this:

$$\begin{aligned} CR(k, X) = & \frac{1}{\sum_{\vec{x} \in X} k(\vec{x})} \sum\nolimits_{\vec{x}_j, \vec{x}_l} \left\{k(\vec{x}_j) k(\vec{x}_l) \left\|\vec{x}_j - \vec{x}_l^2\right\|\right\} \\ & + \frac{1}{\sum_{\vec{x} \in X} [1 - k(\vec{x})]} \sum\nolimits_{\vec{x}_j, \vec{x}_l} \left\{[1 - k(\vec{x}_j)][1 - k(\vec{x}_l)] \left\|\vec{x}_j - \vec{x}_l\right\|^2\right\}', \end{aligned}$$

where the first component determines the contribution of the cluster with number 1 and the second one- the cluster with number 0.

If we have a certain "cluster-splitting surface model" as a function $k(\vec{x}; \vec{w}) \in [0, 1]$, depending on a certain parameters vector $\vec{w}$ and is differentiated by these parameters—the criterion $C(k, X)$ will also be a differentiated function from the vector $\vec{w}$, and therefore, to minimize it, you can already use the whole apparatus for minimizing continuous nonlinear differentiated functions, which has been developing very rapidly lately. Therefore, the "softened" version of the 2-cluster clustering problem can be solved as a continuous nonlinear optimization problem, for example by using some modification of the gradient descent algorithm.

To solve the softened variant of the clustering problem into K clusters, we can apply the "one against all"—we first divide all the examples into 2 clusters, then select a cluster with a greater average distance between its points, and divide it into 2 clusters, and so on further until we get the right number of clusters.

Perhaps the simplest model of a separating surface clusters is a logistical sigmoid: $k(\vec{x}; \vec{w}) = \frac{1}{1+e^{-\vec{w}^T \vec{x}}}$. In essence, such a model will be a certain approximation of the linear separating hypersurface given by the parameter vector $\vec{w}$—one side, for example $\vec{x} : \vec{w}^T\vec{x} > 0$ the model value will be approximately equal to 1 on the other side, for examples $\vec{x} : \vec{w}^T\vec{x} < 0$—approximately equal to 0. That is, using such a model, while minimizing the criterion, we will try to divide all the patterns of "almost linear" hypersurface into 2 clusters so that the total average distance of these clusters is minimal.

Obviously, this model is very simple and will not work well if the clusters in the existing set of examples are not linearly separated. In this case, more complex models of the separating hypersurface are needed, namely neural networks. The only limitation is that the network output should be in the range [0,1], but to do this, skip the network output through the logistic sigmoid mentioned above.

Considering the advantages of networks with one hidden layer of "sigm_piecewise" type neurons given in the chapter one, it is proposed to use these networks as models of cluster separating surfaces. Whatever model you choose, the model-based soft clustering algorithm itself can be completely abstracted from it by presenting the model as a function $k(\vec{x}; \vec{w})$. Then we get the following general description of this algorithm:

The inputs of the algorithm:

- set of patterns $X = (\vec{x}_1, \ldots, \vec{x}_n), \vec{x}_i \in R^d, i = 1, \ldots, n$;
- the number of K clusters to which all patterns should be divided;
- some model of the separating surface clusters $k(\vec{x}; \vec{w})$, which depends on the parameter vector $\vec{w}$ and is differentiated by it.

The steps of the algorithm

1. The whole set of patterns is divided into 2 clusters:

   1.1 The It is initialized randomly the initial vector of the parameters $\vec{w}_0$.

   1.2 Some gradient descent modification is performed to minimize the criterion value $C(\vec{w})$. For this purpose it is used the following formulas for derivatives:

$$
\begin{aligned}
\frac{\partial CR}{\partial w_t} &= \frac{1}{[\sum_{\vec{x}} k(\vec{x}, \vec{w})]^2} \cdot \left(\sum_{\vec{x}} \frac{\partial k(\vec{x}, \vec{w})}{\partial w_t}\right) \cdot \sum_{\vec{x}_j, \vec{x}_l} \left\{k(\vec{x}_j, \vec{w}) k(\vec{x}_l, \vec{w}) \|\vec{x}_j - \vec{x}_l\|^2\right\} \\
&+ \frac{1}{[\sum_{\vec{x}} k(\vec{x}, \vec{w})]^2} \sum_{\vec{x}_j, \vec{x}_l} \left\{\|\vec{x}_j - \vec{x}_l\|^2 \left[\frac{\partial k(\vec{x}_j, \vec{w})}{\partial w_t} k(\vec{x}_l, \vec{w}) + \frac{\partial k(\vec{x}_l, \vec{w})}{\partial w_t} k(\vec{x}_j, \vec{w})\right]\right\} \\
&+ \frac{1}{[n - \sum_{\vec{x}} k(\vec{x}, \vec{w})]^2} \cdot \left(\sum_{\vec{x}} \frac{\partial k(\vec{x}, \vec{w})}{\partial w_t}\right) \cdot \sum_{\vec{x}_j, \vec{x}_l} \left\{[1 - k(\vec{x}_j, \vec{w})][1 - k(\vec{x}_l, \vec{w})] \|\vec{x}_j - \vec{x}_l\|^2\right\} \\
&+ \frac{1}{\sum_{\vec{x}} [1 - k(\vec{x}, \vec{w})]} \sum_{\vec{x}_j, \vec{x}_l} \left\{\|\vec{x}_j - \vec{x}_l\|^2 \left(\frac{\partial k(\vec{x}, \vec{w})}{\partial w_t} [k(\vec{x}_l, \vec{w}) - 1] + \frac{\partial k(\vec{x}, \vec{w})}{\partial w_t} [k(\vec{x}_j, \vec{w}) - 1]\right)\right\}.
\end{aligned}
$$

   1.3 As a result, we get tuned parameter vector $\vec{w}_f$.

   1.4 All patterns are divided into 2 clusters—those for which the model values $f(\vec{x}; \vec{w}_f) < 0.5$ are selected in cluster 0, all other patterns (i.e. those for which $f(\vec{x}; \vec{w}_f) \geq 0.5$)—into cluster 1.

2. If the current number of clusters < K then for both clusters their average distance between points is calculated by the formula:

$$MD(C) = \frac{1}{2|C|}\sum\nolimits_{\vec{x}_j,\vec{x}_l \in C} \vec{x}_j - \vec{x}_l^2,$$

and a cluster whose value $MD(C)$ is bigger it is selected as a new set of patterns for further division into clusters, and then proceeds to the first step of the algorithm. Otherwise, obtained K clusters are the result of the algorithm work.

Consider the operation of the proposed algorithm on artificially generated data.

To do this, we generate 3 clusters with a normal distribution of patterns in each cluster (Fig. 5.22).

As a model of a separating clusters surface, we use a logistic sigmoid, which approximately corresponds to a certain line in space R^2, which attributes all the patterns from one side to one cluster and all other patterns to the other.

After adjusting the parameters of the selected model according to the described algorithm, we obtain approximately the following separating clusters hyperplane (Fig. 5.23).

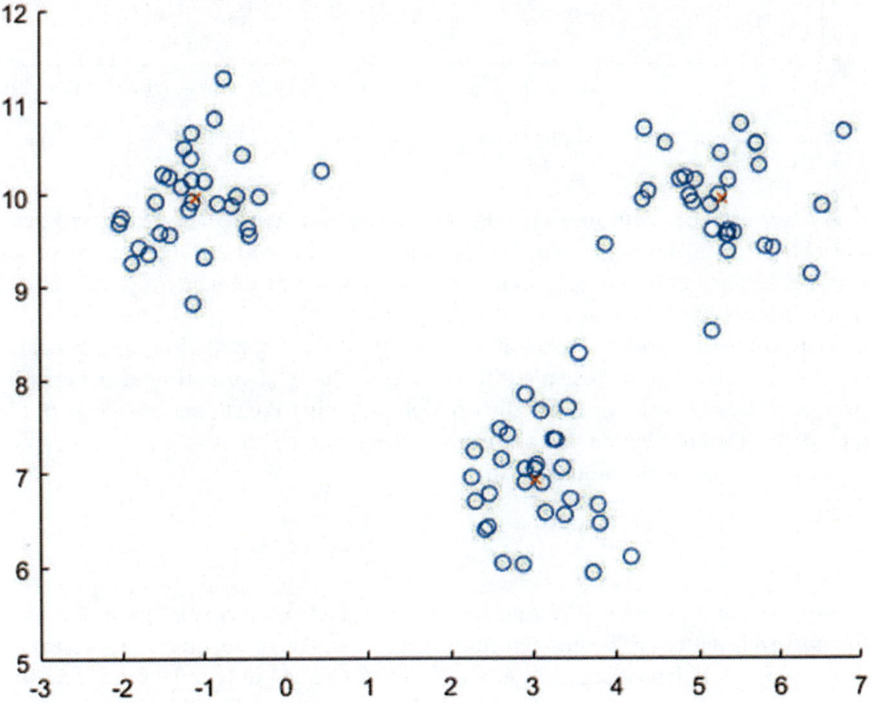

Fig. 5.22 Generated 3 clusters with a normal patterns distribution in each cluster

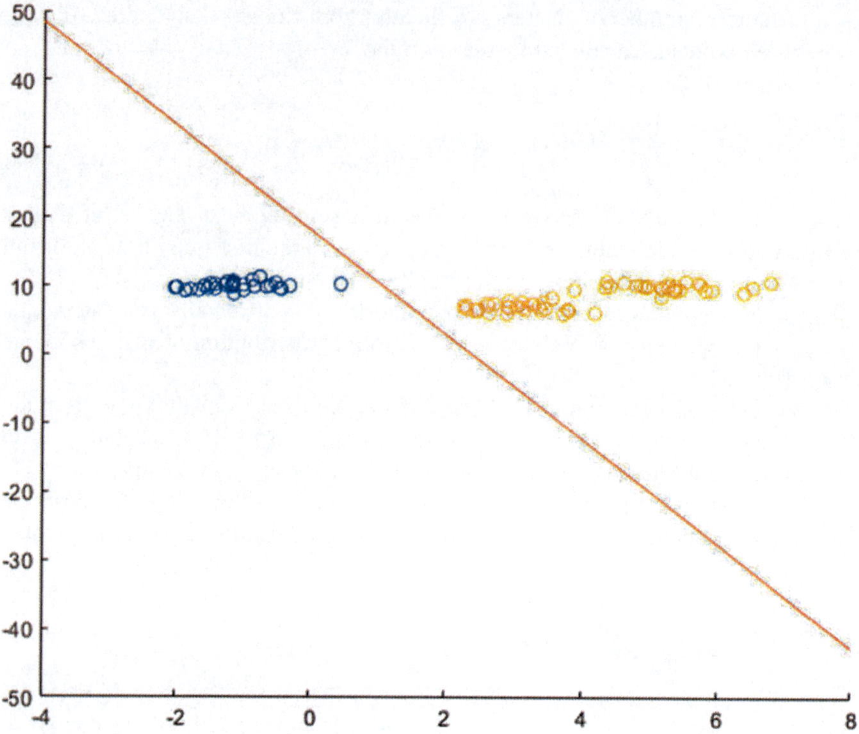

Fig. 5.23 Obtained separating clusters hyperplane

As you can see, with these parameters, a separate hyperplane really separates one cluster from the other two. Next, select a cluster with a larger value of the average distance between points, and repeat the same procedure for it. We obtain approximately the following separating hyperplane (Fig. 5.24).

The proposed algorithm has some similarity to the fuzzy clustering algorithms [18, 19]. Similarly with these algorithms instead "hard" distribution of patterns by clasters it is used "soft" or fuzzy distribution—that is for each pattern $\vec{x}^{(i)}$ is put in accordance the distribution vector of this pattern memberships to each claster (i.e., the ideas of fuzzy logic are used [20]:

$$\vec{\mu}^{(i)} \in R^K, \sum_j \vec{\mu}_j^{(i)} = 1, \vec{\mu}_j^{(i)} \geq 0.$$

However, unlike fuzzy clustering algorithms, which directly seek optimal vectors $\vec{\mu}^{(i)}$ for each pattern $\vec{x}^{(i)}$, this algorithm uses a model of separated hypersurface $\vec{\mu}^{(i)} = f\left(\vec{x}^{(i)}; \vec{w}\right)$, depending on a certain parameter vector and for pattern $\vec{x}^{(i)}$ returns the corresponding vector of membership $\vec{\mu}^{(i)}$, and is tried to find the optimal parameter vector for the model. Thus, the algorithm uses a number of parameters that is independent of the size of the training sample—and for large samples this can

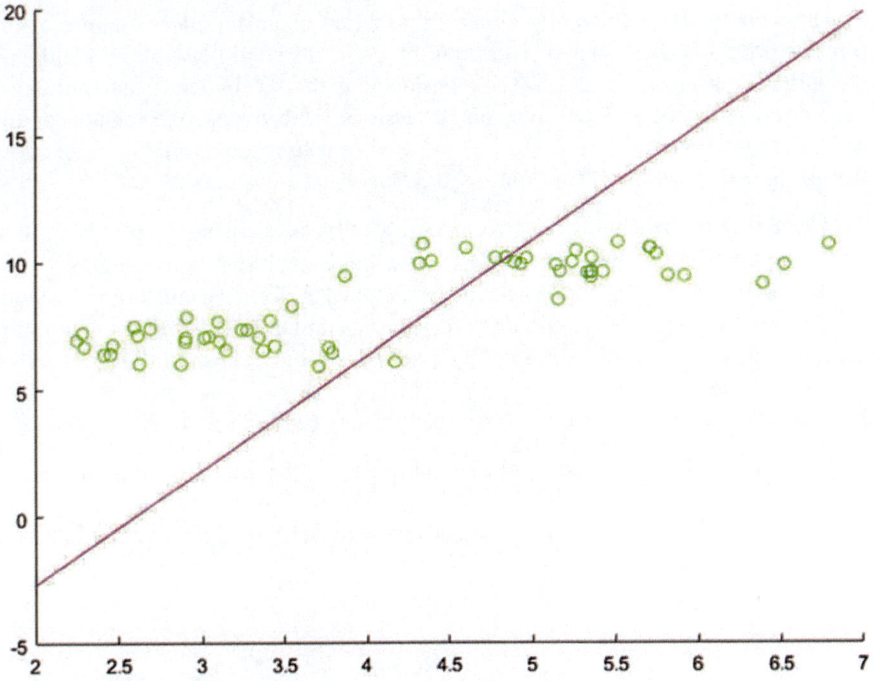

Fig. 5.24 Obtained separating hyperplane

significantly reduce the number of parameters that need to be tuned. In addition, the proposed algorithm does not require finding clusters centers—thus, it allows you to potentially find clusters that are not "centered" with respect to some point-center.

5.5 General Approach for Forecasting Inhomogeneous Samples Based on the Use of Clustering and ANN Methods

This algorithm consists of the following steps

1. A direct propagation neural network is trained using the full set of available examples $S = \{(\vec{x}_i, y_i) : i = 1 \ldots n\}$. The standard error of the network is the standard root mean square error:

$$E = \sum_{i=1}^{n} \left[\text{net}(\vec{x}_i) - y_i\right]^2.$$

The weights $\vec{\theta}^g$ of this network are information about the global nature of the forecast process. The issues of choosing the optimal neural network structure and the optimal learning algorithm are not considered in this thesis—the usual multilayer perceptron with one hidden layer of 10 neurons, with hyperbolic tangent as the activation function of the hidden layer, and the linear function as an output function; Rprop algorithm was used for training [8].

2. Clustering of the sample is performed S. Ideally, each cluster C_j should contain examples that are best described by some single set of network weights $\vec{\theta}_j, j = 1 \ldots k$, where k is the number of clusters, that is, performing clustering, assumes that similar examples should be described by similar model weights. Naturally, the quality of the forecasting in this case will strongly depend on the clustering method used.
3. Neural networks k are trained—one per cluster. The training is as follows:
   (a) the weights of each network are initialized by the previously found weights $\vec{\theta}^g$;
   (b) network net_j learns only from cluster examples C_j using the following error function:

$$E_j = E_j = \gamma \sum_{\vec{x}_i \in C_j} (\mathrm{net}_j(\vec{x}) - y_i)^2 + \sum_{\theta_l \in \vec{\theta}_j} (\theta_l - \theta_l^g)^2, \sum_{\vec{x}_i \in C_j} (\mathrm{net}_j(\vec{x}) - y_i)^2$$
$$+ \sum_{\theta_l \in \vec{\theta}_j} (\theta_l - \theta_l^g)^2,$$

where γ is the regularization parameter.

Thus, when learning network weights will strive to reduce the network error in the examples from their cluster, and not be too far from the original weights $\vec{\theta}^g$—that is, no loss of information about the global nature of the process [8].

4. After learning local networks, the prediction for new examples is as follows: for the input vector, the cluster corresponding to it C_h, is found by finding the closest center of the cluster, and the network trained in the examples of this cluster is used for prediction.

5.6 Forecast Algorithm Based on General Approach

The general approach described in the previous section does not identify the specific clustering algorithm and the specific neural network (i.e., its structure and topology) that should be used to construct the final predictive model. Consider one of the possible concretizations of this approach, namely, the use of a soft clustering algorithm based on single-layer perceptron separators with sigm_piecewise neurons as a model of separating hypersurfaces, and the use of a similar single-layer perceptron with signeuropeans in sigm_speechwise.

The inputs of the algorithm

- vector of time series observations $\vec{y} = [y_1, \ldots, y_n]^T$;
- the value of the forecast horizon h;
- the dimension of the attachment of the time series d;
- expected number of conditions K.

Algorithm

1. The nesting of a time series with parameters h and d is performed, resulting in a matrix of input examples $X \in R^{m \times d}$ and the target variables vector $\vec{o} \in R^m$ where $m = n - d - h + 1$.

2. The algorithm of greedy retrain of networks with one hidden layer of neurons sigm_piecewise is applied to automatically determine the optimal number of neurons in the hidden layer, after which the obtained network is trained using the algorithm Rprop. According to the general approach, standard standard error is used in both stages of training:

$$E = \sum_{i=1}^{n} [\text{net}(\vec{x}_i) - y_i]^2.$$

3. Clustering of vector vectors from matrix **X** to K clusters is performed using a soft clustering algorithm based on separating hypersurfaces, where the network obtained in the previous step is used as a model of separating hypersurface, adding the final neuron with logistic function. As a result, for each vector of the example $\vec{x}_i, i = 1, \ldots, m$ we get the corresponding cluster number $k_i \in \{1, \ldots, K\}, i = 1, \ldots, m$.
4. There are local networks for each cluster according to the general approach—that is, the weights of the network for the cluster $k_i \in \{1, \ldots, K\}$ are initialized by the weights of the "global" network trained in the second step, and the following error function is used to train it:

$$E_j = \gamma \sum_{\vec{x}_i \in S_k} (net_k(\vec{x}_j) - y_j)^2 + \sum_{\theta_l \in \vec{\theta}_k} (\theta_l - \theta_l^g)^2.$$

5. The use of trained LANs to obtain a forecast for new data is performed according to the general approach—for the input vector is the corresponding cluster k (using the trained model of the separating cluster of the hyperplane), and for the forecast uses the network trained in the examples of this cluster.

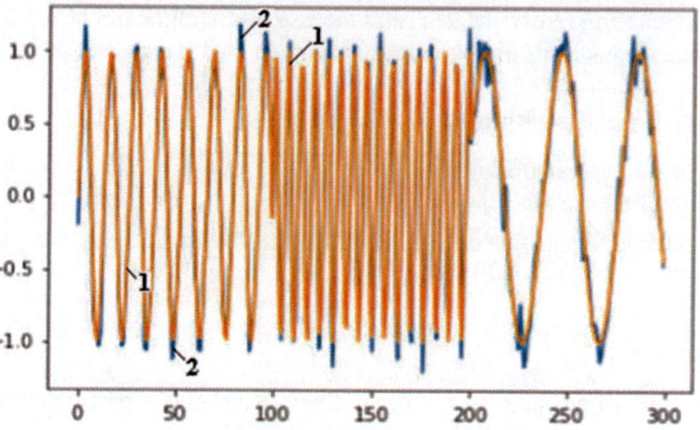

Fig. 5.25 Graph consisting of 3 Different sine wave with added noise with normal distribution: orange chart (1) "pure" signal; blue (2)—noisy

The proposed methods and approaches have been tested both on artificially generated data and on real samples. A sample of the following form was generated to test for artificial data:

$$y_t = \sin(w_t * t) + N(0, \sigma^2),\ t = 1, \ldots, 300,$$

where $w_{1:100} = \frac{3}{2\pi}$, $w_{101:200} = \frac{6}{2\pi}$, $w_{201:300} = \frac{1}{2\pi}$, $\sigma^2 = 0, 01$, that is, we actually have a graph consisting of three different sinusoids with the addition of noise with normal distribution (Fig. 5.25).

Obviously, this time series is heterogeneous. Next, the time series nesting method was usedя $d = 5$ and a two-step forecast, resulting in a sample $< X, y >$, which was randomly separated into an educational one—$< X\_train,\ y\_train >$ and test—$< X\_test,\ y\_test >$. After training on a training sample of a simple two-layer perceptron with 12 ReLU-type neurons in both hidden layers and a linear output neuron, a root mean square error of 0.0984 was obtained, a mean absolute error—0,2006 (Fig. 5.26).

Next, the forecast model was studied according to the method of mixture of experts with three experts, where single-layer perceptrons with 12 ReLU-type neurons in the hidden layer were used as expert models and gating models. After training, the mean squared error of 0.0825 was reached, the mean absolute error—0.1745 (Fig. 5.27).

As can be seen from the graph—the model based on the method of mixture of experts has much better predictions for the 3—"state" of the process, that is, for the last sine wave. Last but not least, a predictive model was built and trained based on the method of regularized mixture of experts with the same values of the hyperparameters that were used in the construction of the model based on the classic method of mixture of experts. After training, the root mean square error of 0.0797 was reached, the mean absolute error—0.1627 (Fig. 5.28).

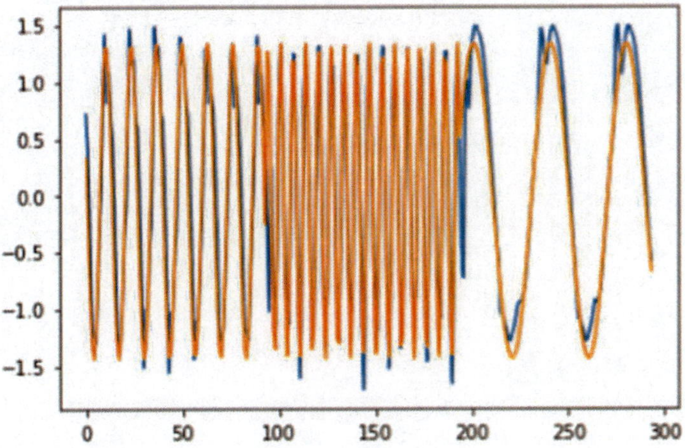

Fig. 5.26 The orange graph is the time series, the blue is the perceptron forecast

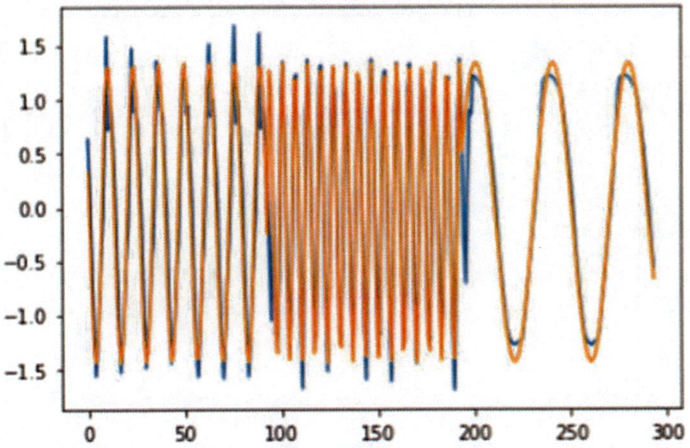

Fig. 5.27 The orange graph is the time series, the blue is the forecast of the expert mix method

As can be seen from the graph—regularization allowed to obtain a model with more accurate forecasts in all 3 "states" of the process.

The data provided by the Federal Reserve Bank of St. Louis, namely the index sample, was used to verify the actual sample T10Y2Y (Fig. 5.29).

For this sample, a time series nesting method with a $d = 10$ nesting dimension and a 2-step forecast was applied, followed by repeating all steps performed for artificial data with the same hyperparameters. The following model error values were obtained.

As can be seen from the Table 5.3, the application of the expert mix method to this time series has led to a worsening of the model's prediction accuracy over the use of a single network—in practice, this problem occurs quite often and can be caused by

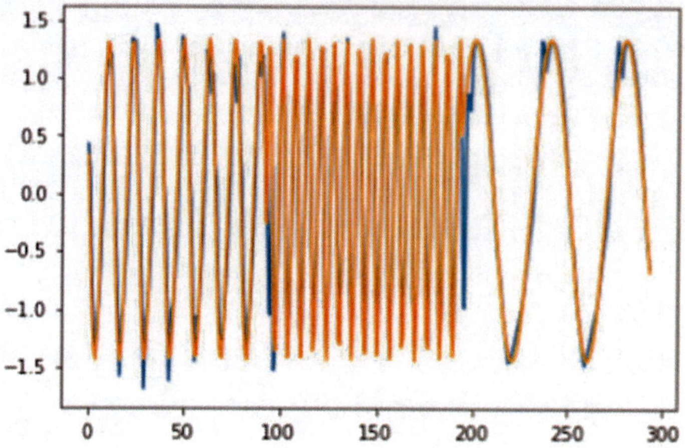

Fig. 5.28 The orange graph is the time series, the blue one is the forecast of the regular mix of experts

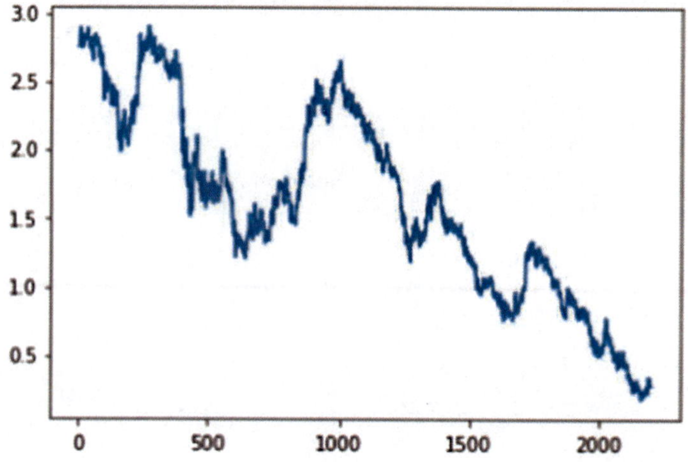

Fig. 5.29 Sample under index T10Y2Y, accessed online by the federal reserve bank of St. Louis

Table 5.3 Errors of different models in the test sample

Model	MSE error	Average absolute error
Unified neural network (two-layer perceptron)	0.0057	0.0571
Method of mixture of experts	0.0064	0.0625
The method of a regularizable mixture of experts	0.0052	0.0551

many different factors, one of which is the retraining of local experts. In this case, the use of a regular mix of experts allowed us to "fix" this problem and get the best model in terms of selected criteria.

The application of the proposed approaches to solve the application problems is discussed in the 5.7.

5.7 Use the Suggested Approaches to Solve Application Problems

Consider the practical application of the above methods in the specific examples.

5.7.1 Forecasting Sales of Aviation Equipment

Demand forecasting is one of the most important inventory management issues. Forecasting forms the basis for inventory planning and is probably the most significant problem in the repair industry. One common problem facing airlines around the world is the need to know the short-term forecast of demand for aircraft with the highest possible degree of accuracy. The high cost of modern aircraft and spare parts to be repaired, such as aircraft engines and avionics, make up the majority of the total investment of many air carriers. Production of an insufficient number of aircraft can lead to large financial losses, and overproduction, in turn, leads to downtime.

Forecasting techniques have evolved over the years and are widely used in aviation today. Due to the sporadic nature of the demand for aircraft, its forecasting is an important task requiring modern and accurate forecasting methods.

Next, we consider an example of forecasting the demand for aircraft. The data for this example were taken from a publicly available data archive using the link [3]. They represent the number of aircraft sold annually from 1947 to 2011 (65 points) (Fig. 5.30).

As can be clearly seen in the data visualization, the predicted "process" is not cyclical and has no visible outliers.

For forecast we use an algorithm based on ANN and GMDH. We set the algorithm parameters as follows.

(1) The size of the variable window to obtain a matrix of training examples $k = 5$.
(2) The ratio of the sizes of the training sample and the test sample is—0.7 m/0.3 m where m is the initial number of examples. The first 0.7 m were used to build a training sample.
(3) Support functions of the form: $f_l = f(x_i, x_j, x_i x_j, x_i^2, x_j^2)$, $1 = 1 \ldots C_5^2$ $i = 1 \ldots 5, j = 1 \ldots 5, i \neq j$.
(4) Neural network structure: one output neuron, one hidden layer with three neurons, five input neurons.

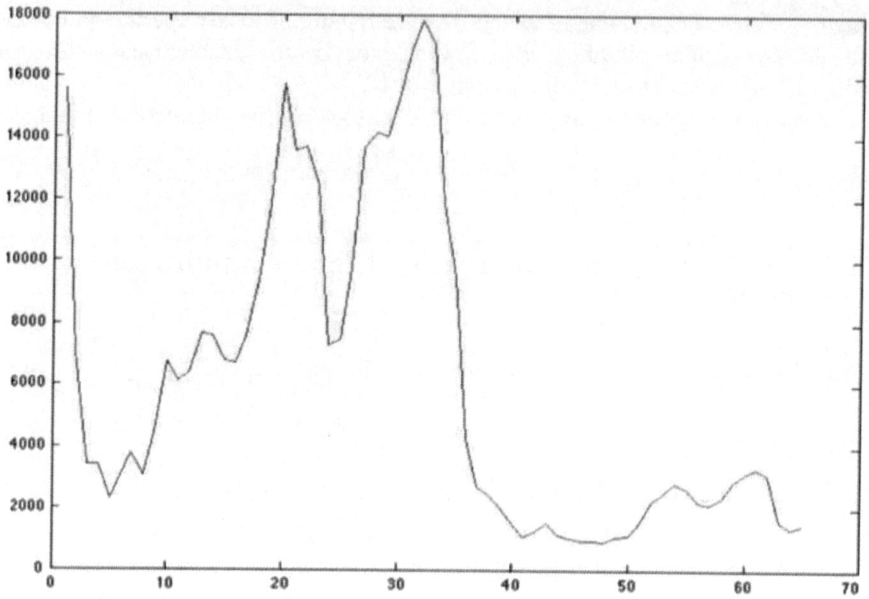

Fig. 5.30 Output data

(5) The MSE error for the selection of networks is calculated on all output examples (both on examples from the training sample and on examples from the test sample).
(6) When constructing the initial matrix for the next iteration, the outputs of the three networks with the best error were selected as the new variables, as well as the two output variables, which were the inputs of the network with the smallest MSE.

A multilayer perceptron with one output neuron, one hidden layer with 8 neurons in it and ten inputs was selected as the forecast method used to compare the results of the application of the described algorithm. FM, like the algorithm, studied only on a training sample. The Levenberg-Marquardt algorithm was chosen as the learning algorithm. Because ANNs are known to be prone to "jamming" at local minima of the error function, about ten contenders were constructed and trained, and a multilayer perceptron with a minimum standard error across the entire sample was selected.

Comparison of the results obtained using the ANN and the proposed algorithm is shown in Fig. 5.31.

5.7.2 Forecasting Meteorological Quantities

Briefly describe the subject area of meteorological forecasting.

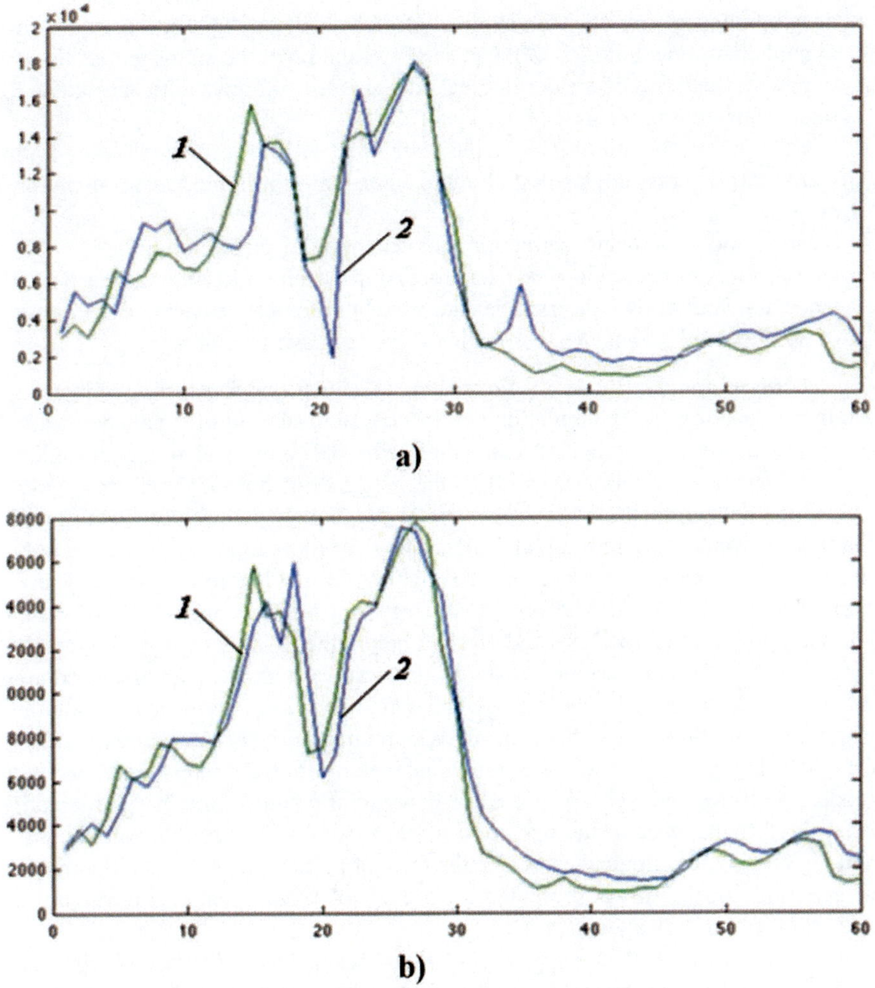

Fig. 5.31 Comparison of prediction results obtained using ANN **a** and proposed algorithm **b** 1 is the initial data; 2 is the forecast

By the end of the twentieth century, the global meteorological community had made significant progress in short- and medium-term weather forecasting. These successes include:

- scientific advances in understanding global atmospheric processes and atmospheric dynamics, in the mathematical description of radiation from the Sun, transfer, reflection, absorption of short-wave and long-wave radiation, condensation and evaporation processes, melting/freezing of precipitation, mixing mechanisms of air masses, including convection and turbulence, processes of interaction with land and ocean;

- development in a number of countries of global, regional and mesoscale hydrodynamic numerical models of the general circulation of the atmosphere, allowing to predict the fields of meteorological elements for 5–7 days with acceptable for many consumers accuracy;
- creation in large meteorological centers equipped with powerful computing equipment unique technologies that allow to implement these models in operational practice;
- creation and organization of continuous operation of global international observation, telecommunication and data processing systems allowing weather surveillance, transmission of observation data to meteorological centers and distribution of products to National Meteorological Service forecast centers.

The international nature of the forecasting system. Rapid forecasting of hydrodynamic models of general circulation requires constant support of expensive systems of observations and automated technologies for collecting and processing global meteorological information, as well as the presence of powerful scientific potential for the development and improvement of the models themselves. Therefore, in the field of weather monitoring and forecasting for more than 130 years, there has been close international cooperation through the World Meteorological Organization (WMO). The World Meteorological Organization is a comprehensive system consisting of national facilities and services belonging to individual WMO member countries. WMO members undertake, in accordance with their ability to commit themselves, to an agreed scheme so that all countries can benefit from the combined efforts. Within the framework of the WMO, an international forecasting industry is established, consisting of world and regional meteorological centers, equipped with modern facilities and technologies at the expense of countries that have voluntarily committed themselves to the operation of such centers. The production of global and regional meteorological centers in the form of numerical analyzes and forecasts of meteorological fields is presented for use by all WMO members through their national meteorological centers.

Changes in weather forecasting technology by operational meteorologist. As a result of scientific advances, the weather forecasting technology of a meteorologist at a particular point or area has changed dramatically compared to previous years. Successes in the development of numerical modeling of the atmosphere have led to the centralization and even globalization of the main stage of the forecast—the forecast of fields of meteorological values, based on which the operative meteorologist makes the prediction of the elements and phenomena of the weather for a particular point, area or territory. However, the crucial role of the forecaster in the interpretation of the output of numerical models and the use in the preparation of objective methods of forecasting meteorological values and weather phenomena, as well as operational data of different surveillance systems. This role is especially important when forecasting hazardous weather events.

Weather Predictability Restriction. The operational models used in large meteorological centers have a predictability limit of 5–7 days and differ in their characteristics, numerical procedures, processing technology and power of computing facilities. Therefore, the forecast values of meteorological values may be different, although the values can be compared. It is also important to note that major advances in numerical modeling of the atmosphere are mainly related to large-scale weather systems. Small-scale formations of several tens and even hundreds of kilometers, with which dangerous hydro-meteorological phenomena are connected, cannot yet be predicted by numerical models.

The forecast of such entities is made by a specialist forecaster based on the interpretation of the production of numerical models and the use of additional information that reflects the development of mesoscale processes (radar observations, satellite data, etc.). Therefore, despite the development of mesoscale numerical models and automated surveillance tools, local weather forecasts will always be associated with some uncertainty about the specific location, time, and intensity of meteorological phenomena. This is especially true of extreme phenomena, which occur rarely and suddenly, exist for a short time and which can often be predicted with only a small (1–3 h) time.

Struggle with undetermined. Strictly speaking, uncertainty is inherent not only in weather forecasts but also in the degree of assessment of the current state of the atmosphere. If quantifiable inherent uncertainty could be expressed, the value of forecasts to decision makers would be greatly increased. The solution to this problem is to use a prediction group (ensemble) under a number of initial conditions that differ for a single model or a group of numerical prediction models with different but equally possible approximations. The forecasting ensemble covers a number of possible outcomes, providing a range of data where uncertainties can grow. As a result, the forecasting ensemble can automatically obtain information about the capabilities that are relevant to customer requirements.

Long-term weather forecasts. Detailed forecasts of meteorological values and weather phenomena or sequences of meteorological systems by month, season and beyond are unreliable. The chaotic nature of atmospheric movements is determined by the basic predictability limit of 10 days for such deterministic forecasts. However, some predictability of average temperature and precipitation anomalies persists over a longer period due mainly to the interaction between the atmosphere and the ocean, as well as the land and ice surface. However, compared to the atmosphere, the ocean has been poorly studied, and therefore further progress in long-term weather forecasting is impossible without intensifying research into regional and global ocean processes.

Climate forecast. In climate forecasting, the most important inputs of the models are future changes in greenhouse gases and other radiation-active substances. They change the radiation impact on the planet and cause climate change on a very long-time scale. Therefore, when modeling the possible future climate, the term "prospective estimate" should be used, not "forecast" or "prediction".

Physical processes that are not important in the prognostic models of the general circulation for a period of 5–7 days, and even in the long-term forecasting, are decisive in climate modeling. This is especially true of the dynamics of oceanic circulation, the changing landscape of the underlying surface, and the evolution of the snow-ice cover. Studying these processes takes considerable effort before the opportunity to reproduce many aspects of the climate is realistic. However, despite the complexity of the physical processes, there is some confidence that existing climate models provide a useful perspective on how to change it.

Already, many models now allow satisfactory climate modeling. Moreover, the simulation is quite able to reproduce the observed large-scale changes that occurred in ground-level air temperature over the twentieth century. This large-scale consistency between simulation results and observations provides confidence in estimates of the next century's warming rate. The observed natural variability modeling (e.g., the El Niño phenomenon, monsoon circulation, north Atlantic fluctuations) has also improved.

On the other hand, the systematic errors are still too large. One of the factors limiting confidence in the prospective assessment of climate change is the uncertainty of the external impact (for example, the future concentration of atmospheric carbon dioxide and other greenhouse gases and aerosol loads).

As with mid-term and long-term forecasts, prospective ensemble climate estimates are also extremely important. Ensembles allow for a more clearly statistically significant signal of climate change.

Despite the fact that, on an industrial scale, predetermined meteorological values use predetermined physical models, intellectual forecasting methods can still produce good results, which will be shown below.

For example, we take the data from the site [21], namely the average annual temperature of the Indian region from 1901 to 2002 (Fig. 5.32). As the graphical representation clearly shows, the projected process is cyclical and contains a small

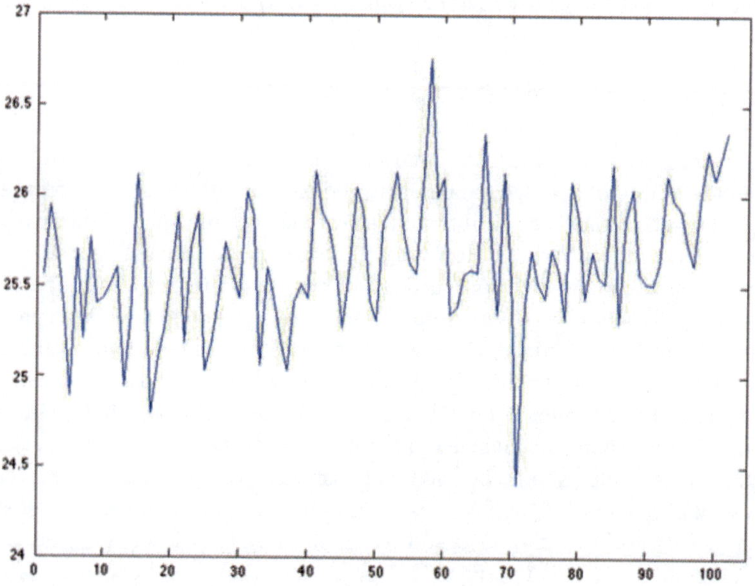

Fig. 5.32 Baseline—the average annual temperature of the area of India from 1901 to 2002

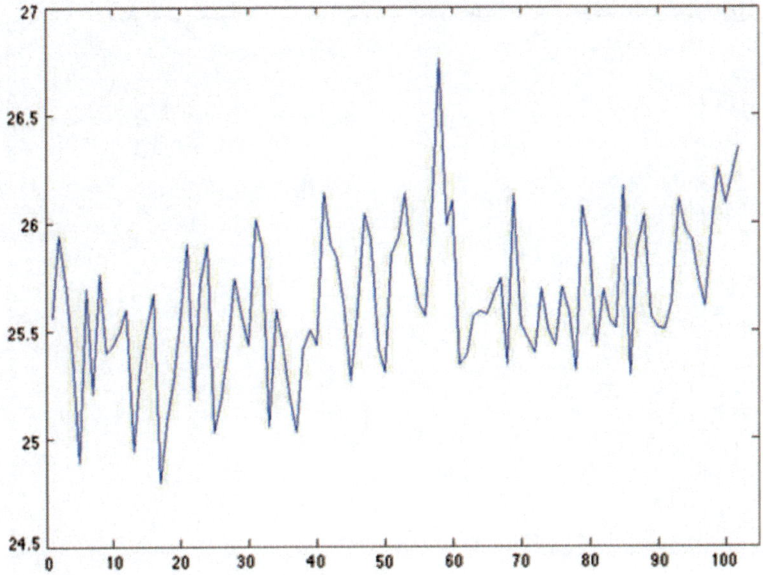

Fig. 5.33 Data processed by Tukey 53H algorithm

amount of emissions. Therefore, the data should be eliminated from emissions, for example, using the Tukey 53H algorithm (Fig. 5.33).

Forecasting will be made one year in advance, case studies for all methods will be made on a continuous basis window with a window size $k = 5$.

So, let's start with the group argument method. We use standard polynomial functions from two arguments as reference functions

$$y_b = a_o + a_1 x_1 + a_2 x_2.$$

We choose polynomial neural networks as the GMDH algorithm. We will use the adjusted Akaike criterion for the model search. The resulting model will have a UPC on test sample 1.4429e-04 (Fig. 5.34).

The next model is a multilayer perceptron. Set the following parameters:

- learning algorithm is the Levenberg–Marquardt;
- number of hidden layers—one;
- number of neurons in the hidden layer—eight;
- activation functions of neurons in the hidden layer—sigmoid, activation function of the original—linear.

The resulting model will have MSE on the sample sample 2,4392e-04 (Fig. 5.35).
Next is an algorithm based on ANN and GMDH. Algorithm parameters:

- topology of neural networks used as reference functions—multilayer perceptron;
- type of support functions—$f(x_1, x_2, x_1 x_2, x_1^2, x_2^2)$;

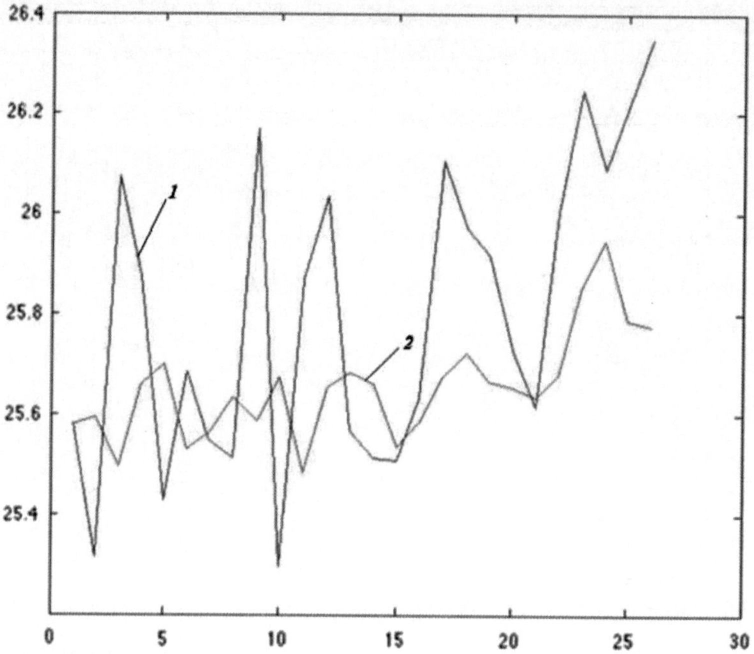

Fig. 5.34 The forecast obtained with the help of GMDH: 1 is the initial data; 2 is the forecast

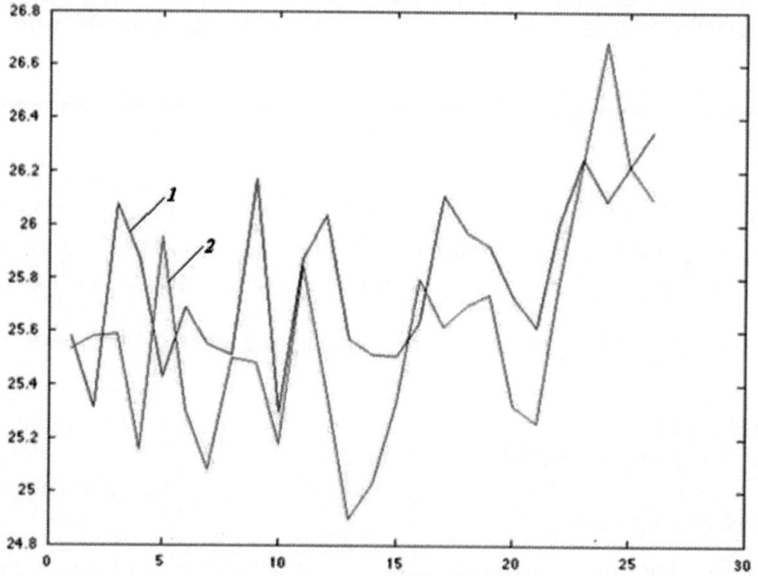

Fig. 5.35 The forecast obtained with the help of ANN: 1 is the initial data; 2 is the forecast

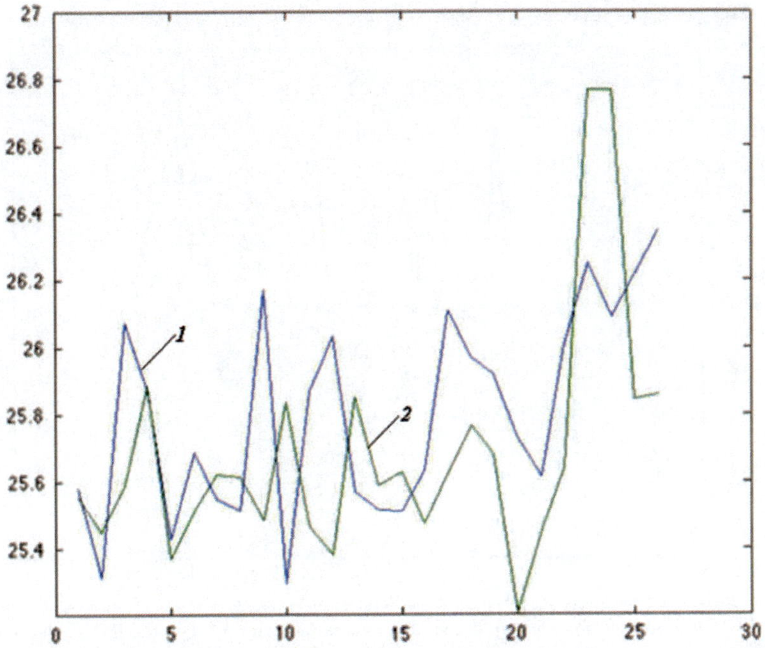

Fig. 5.36 The forecast obtained with the combination of GMDH and ANN: 1 is the initial data; 2 is the forecast

- the number of neurons in the only hidden layer used FM-3;
- other FM parameters—as in the previous algorithm.

The resulting model will have MSE on the sample sample 2,1200e-04 (Figs. 5.36, 5.37).

Finally, we use a combination of all three methods already used. Method parameters will not be changed. The resulting model will have MSE on test sample 1.3824e-04 (Fig. 5.37).

In summary

- combining all three algorithms, as expected, gave the best result;
- in second place—GMDH, in the third—algorithm based on GMDH and ANN, in the last—ANN;
- the gap between the complexation of algorithms and GMDH is very small, and given the complexity of compiling three algorithms in this case, GMDH should be preferred;
- from the graphs shown in Figs. 5.36 and 5.37. It can be seen that ANN-based algorithms behave "unstable" on this task—their outputs far exceed the real values, while the forecasts obtained by pure GMDH and the complexization of the three algorithms behave relatively "calmly". As a possible way to improve the quality

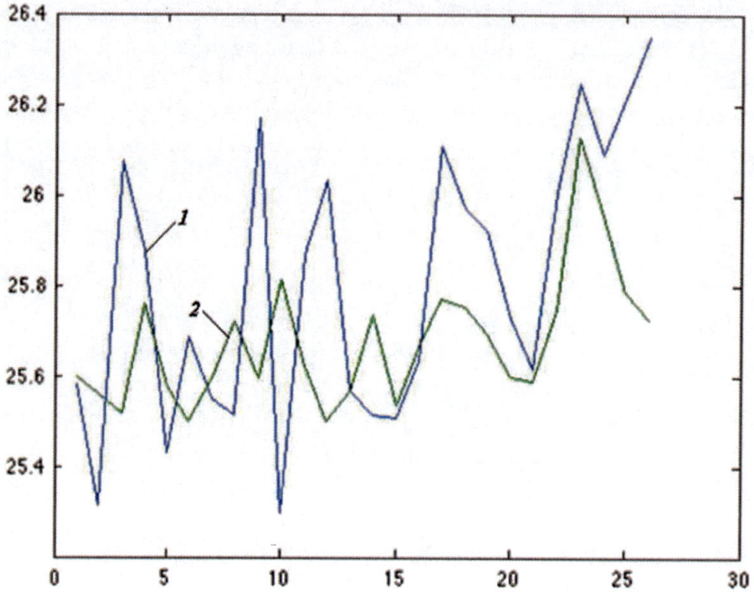

Fig. 5.37 The Forecast obtained by combining several algorithms: 1 is the initial data; 2 is the forecast

of prediction of ANN-based algorithms, a more thorough pre-processing of data should be performed and models should be retrained.

References

1. Radchenko, S.: Methodology of regression analysis: monograph. In: Radchenko, S. (ed.) K.: Korniychuk, p. 376 (2011)
2. Chumachenko, H.: Forecasting the demand for UAV using different neural networks topology. In: Chumachenko, H., Gorbatiuk, V. (eds.) The 2nd International Conference, Actual Problems of Unmanned Air Vehicles Development Proceedings, Kyiv, Ukraine, pp. 62–64 15–17 Oct 2013
3. Chumachenko, H.: Algorithms for solving the forecasting problem. In: Chumachenko, H., Gorbatiuk, V. (eds.) Intellectual System for Decision Making and Problems of Computational Intelligence. ISDMC-2012 Congrece proceeding, Yevpatoria, Ukraine, pp. 423–425 27–31 May 2012
4. Sineglazov, V.: An algorithm for solving the problem of forecasting. In: Sineglazov, V., Chumachenko, H., Gorbatiuk, V. (eds.) Aviation. Latvia, vol. 17(1), pp. 9–13 (2013)
5. Sineglazov, V.: Applying different neural network's topologies to the forecasting task. In: Sineglazov, V., Chumachenko, H., Gorbatiuk, V. (eds.) 4th International Conference in Inductive Modelling ICIM, pp. 217–220 (2013)
6. Sineglazov, V.: One approach for the forecasting task. In: Sineglazov, V., Chumachenko, H., Gorbatiuk, V. (eds.) Proceedings, the 5th World Congress "Aviation in the XXI-st Century",

Safety in Aviation and Space Technologies, vol. 2, pp. 3.5.49–3.5.53. Kyiv, Ukraine, 25–27 Sept 2012
7. Chumachenko, H.: Algorithms for solving the forecasting problem. In: Chumachenko, H., Gorbatiuk, V. (eds.) Artificial Intelligence, vol. 2, pp. 23–31 (2012)
8. Sineglazov, V.: A method for building a forecasting model with dynamic weights In: Sineglazov, V., Chumachenko, H., Gorbatiuk, V. (eds.) Vostochno-Evropejskij zhurnal peredovyh tehnologij, vol. 2(4), pp. 4–8
9. Chumachenko, E.: Development of an image processing algorithm for diagnostic tasks. In: Chumachenko, H., Levitsky, O. (eds.) Electronics and Control Systems. NAU, vol. 1(27), – pp. 57–65 (2011)
10. Chumachenko, H.: Method of encouraging a predictive model with dynamic parameters. In: Sineglazov, V., Chumachenko, H., Gorbatiuk, V. (eds.) Materials of the International Science-Practical Conference "Information Technology and Computer Simulation", Ivano-Frankivsk—Yaremche, Ukraine, pp. 23–26 23–28 May 2016
11. Sineglazov, V.: The method of solving the forecasting problem based on the aggregation of estimates. In: Sineglazov, V., Chumachenko, H., Gorbatiuk, V. (eds.) Inductive Modeling of Folding Systems: Science and Technology Book.—K.: IESC of ITS of the NASU and the MESU, vol. 4, pp. 214–223 (2012)
12. Chumachenko, H.: Complexation of several algorithms in solving the prediction problem. In: Chumachenko, H., Gorbatiuk, V. (eds.) Bulletin of the Zhytomyr State Technical University. Series: Technical Sciences, vol. 1(76), pp. 101–107 (2016)
13. Chumachenko, H.: Complex decision-making algorithm for the most varied forecast tasks. In: Chumachenko, H., Gorbatiuk, V. (eds) Abstracts of the VIII International Science and Technology Conference "Information and Computer Technology—2016", Zhytomyr, Bulletin of the Zhytomyr State Technical University, Ukraine, pp. 95–96 22–23 April 2016
14. Engle, R.F.: Autoregressive conditional heteroskedastisity with estimates of the variance of U.K. Inflation Econometrica, vol. 50, pp. 987–1007 (1982)
15. Bollerslev, T.: Generalized autoregressive conditional heteroscedastisity. J. Econometrics **31**, 307–327 (1986)
16. Engle, R.F.: ARCH: selected readings. Oxford University Press, Oxford (1995)
17. Al-Marzouqi H. Data Clustering Using a Modified Kuwahara Filter, Neural Networks// International Joint Conference, pp. 128–132 (2009)
18. Bezdek, J.C.: Pattern Recognition with Fuzzy Objective Function Algorithms. Plenum Press, N.Y. (1981)
19. Gath, I.: Unsupervised optimal fuzzy clustering. In: Gath, I., Geva, A. (eds.) Pattern Analysis and Machine Intelligence, vol. 2(7), pp. 773–787 (1989)
20. Zade, L.A.: The concept of a linguistic variable and its application to making approximate decisions.—M.: Mir (1976)
21. Sineglazov, V.: Intelligence method of forecasting.In: Sineglazov, V., Chumachenko, H., Gorbatiuk, V. (eds.) K.: Osvita Ukraine, p. 220 (2013)

Chapter 6
Intelligent System of Thyroid Pathology Diagnostics

6.1 Classification of Thyroid Cancer, Clinical Picture

Thyroid cancer is a tumor that progresses from the cells of the epithelium of the thyroid gland. In the absence of radiation influence, the risk of thyroid cancer formation increases with age. If it is rare happened with children, then, in people over 60 years, in half of the cases can be found nodal forms of cancer. The increase of morbidity risk occurs in the 4th decade in both men and women, but the ratio of morbidity risk remains 1:3, respectively. The occurrence of thyroid cancer in all countries has two zones: a smaller one—in the age range of 7–20 years, and a larger one—50–65 years. Thyroid cancer accounts for 0.4–2% of all malignancies in the structure of malignancy.

The detailed study of patients of thyroid cancer diseases makes it possible to make certain judgments about of thyroid cancer causes.

It is decisively shown by experiments that the increased content of thyroid-stimulating pituitary hormone (TSH) in the blood is an important etymological and pathogenetic factor of thyroid tumors progression. However, the suppression of TSH secretion by thyroid hormones causes a treatment effect in differentiated thyroid cancer. It was noted that the baseline of TSH under thyroid cancer is significantly higher than under the pathology absence in the organ. Recently, more and more observations indicate that one of the reasons of thyroid cancer progression is ionizing radiation.

In 1978, the research showed that it was detected the thyroid cancer in 19.6% of people who had been exposed to x-ray in the head and neck in childhood. Among the Japanese who were exposed to the atomic bombs in Hiroshima and Nagasaki, thyroid cancer is 10 times more common than the rest of Japan's population.

Attention should be paid to the increase in the incidence of thyroid cancer in people exposed to ionizing radiation after the Chernobyl accident. The level of incidence of thyroid cancer in children aged 5–9 years after the accident increased by 4.6–15.7 times compared with average values [1].

M. Zgurovsky et al., *Artificial Intelligence Systems Based on Hybrid Neural Networks*, Studies in Computational Intelligence 904,
https://doi.org/10.1007/978-3-030-48453-8_6

It is noted that in patients with papillary and follicular thyroid cancer, favorable factors were found in 84–86% of cases, with the combination of several factors occurring in most patients (60.5%). The question of the relationship between cancer and "background" processes is one of the main ones in oncology, as it relates to cause and effect relationships. It is established that iodine deficiency in the body is the main cause of thyroid hyperplasia. Usually, such hyperplasia is compensatory, but sometimes it becomes irreversible. This process can be promoted also the factors that block the synthesis of thyroid hormones. Thus, the progress of malignant tumors in the thyroid gland is often preceded the nodular goiter, diffuse and nodular hyperplasia, benign tumors (adenoma). The highest percentage of early cancer detection against the background of adenoma and thyroid adenomatosis, but hyperplastic diseases can be the background for the development of thyroid cancer in 23.6% of cases. This proves once again the need for morphological verification of any nodal formation in the thyroid gland [2].

With the account of the available data on thyroid cancer, it should be noted that the at-risk group should include:

- women who have long-standing inflammatory or tumoral diseases of the genitals and mammary glands;
- persons who have a hereditary inclination to tumors and glandular dysfunction;
- patients suffering from thyroid adenoma or adenomatosis;
- recurrent euthyroid goiter in epidemic areas;
- persons who have received general or local effects on the head and neck area by ionizing radiation, especially in childhood.

The greatest widespread obtained the malignant tumors distribution by 4 stages, each of which is characterized by the degree of the primary tumor spread, presence of regional and distant metastases [3].

Distribution of thyroid cancer by clinical signs (stages)

Stage I is a single tumor in the thyroid gland without deformation and germination of the capsule of the gland under the regional and distant metastases present.

Stage II

- single or multiple tumors in the thyroid gland that cause it to deform, but without germination of the gland capsule and without limitation of displacement, in the absence of regional and distant metastases;
- single or multiple tumors in the thyroid gland that cause or do not cause deformity, without gland capsulegermination and without displacement limitation, but in the presence of regional metastases displaced on the affected side of the neck and in the absence of distant metastases.

Stage III

- the tumor has spread beyond the capsule of the thyroid gland and is associated with the surrounding tissues or squeezes the adjacent organs (compression of the trachea, esophagus, etc.) with limited displacement of the gland, but under absence of regional and distant metastases;

- tumor of the thyroid gland stages I, II and III, but in the presence of bilateral displacements of metastases on the neck, or metastases on the neck on the side opposite to the defeat of the thyroid gland.

Stage IV is the tumor sprouts into the surrounding structures and organs, the thyroid gland does not move.

Malignancies classification by TNM system

- T—primary tumor.
- T_X—not enough data to evaluate the primary tumor.
- T_0—primary tumor is undetermined.
- T_1—tumor up to 2 cm in the largest dimension, limited by thyroid tissue.
- T_2—tumor up to 4 cm in the largest dimension, limited by the tissue of the thyroid gland.
- T_3—tumor larger than 4 cm in the largest dimension, limited by thyroid tissue.
- T_4—a tumor of any size extending beyond the thyroid capsule or any tumor with minimal spread beyond the thyroid capsule.
- T_{4a}—a tumor sprouts a capsule of the thyroid gland and extends to any of the following structures: subcutaneous soft tissue, larynx, trachea, esophagus, nerve.
- T_{4b}—the tumor extends to the pre-vertebral fascia of the vessel or the carotid sheath.
- T_{4a*}—is only an undifferentiated (anaplastic carcinoma) tumor (of any size), limited by the thyroid gland **.
- T_{4b*}—only undifferentiated (anaplastic carcinoma) tumor (of any size), extends beyond the thyroid capsule **.

* All undifferentiated (anaplastic) carcinomas are classified T4.

** An undifferentiated (anaplastic) tumor that sprouts a capsule is considered nonresectable.

Regional lymph nodes

- N_x—insufficient data to assess regional lymph nodes.
- N_0—no signs of metastatic damage of regional lymph nodes.
- N—is a damage of regional lymph nodes by metastases.
- N_{la}—metastases at V1 level lymph nodes, including the peritoneal and lymph nodes.
- N_{ib}—metastases affected by other cervical lymph nodes on one side, or on both sides, or on the opposite side of upper anterior mediastinal.
- M—distant metastases.
- M_x—insufficient data to determine distant metastases.
- M_0—no signs of distant metastases.
- M_1—there are distant metastases.

Category M, depending on the localization of metastases can be supplemented by the following symbols:

- lungs—PUL;
- bone—OSS;
- liver—HEP;
- brain—BRA;
- leather—SKI.

Histological types

There are four common histopathological types [24].

1. Papillary carcinoma (including follicular focus).
   The papillary thyroid cancer represents for about 70% of all thyroid cancer cases. The patients with papillary type of thyroid cancer are 35–45 years old. This type of thyroid cancer is characterized by asymptomatic growth, which is why in 20% cases of its diagnostics it is diagnosed the metastases to other organs or lymph nodes. In childhood, papillary thyroid cancer is extremely rare and has a more aggressive course of the disease (Figs. 6.1 and 6.2).
2. Follicular carcinoma (including the so-called Hurt-le cell) carcinoma.
   The follicular thyroid cancer ranks second in frequency and about 20% of all diagnosed thyroid cancers. People suffering from thyroid follicular cancer are predominantly people over 50 years old. This type of thyroid cancer is differed by a more aggressive course and a faster process of metastasis (Figs. 6.3 and 6.4).
3. Medullary carcinoma.
   The thyroid medullary cancer, as a rule, is quite rare - about 5% of all cancers diagnosed. The medullary thyroid cancer can develop without previous pathological changes in the thyroid gland (Figs. 6.5 and 6.6).
4. Undifferentiated (anaplastic) carcinoma.

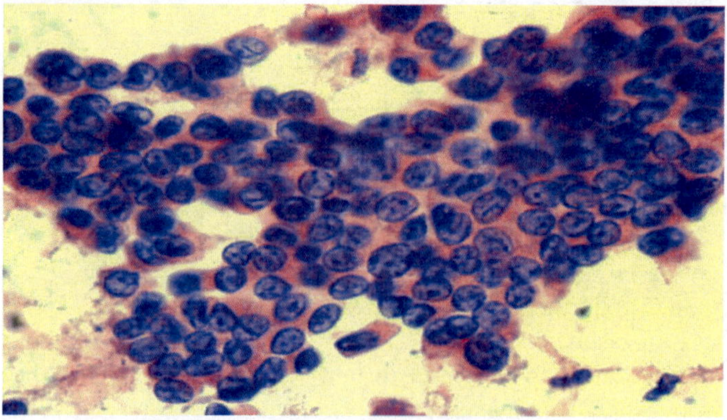

Fig. 6.1 Macrophotography of papillary thyroid carcinoma

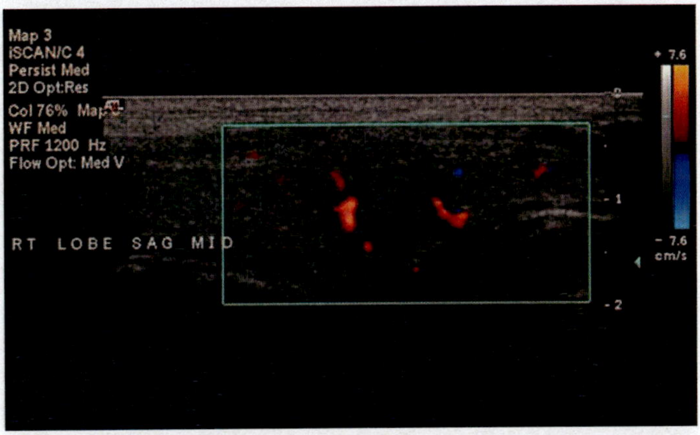

Fig. 6.2 Ultrasound image of papillary thyroid cancer

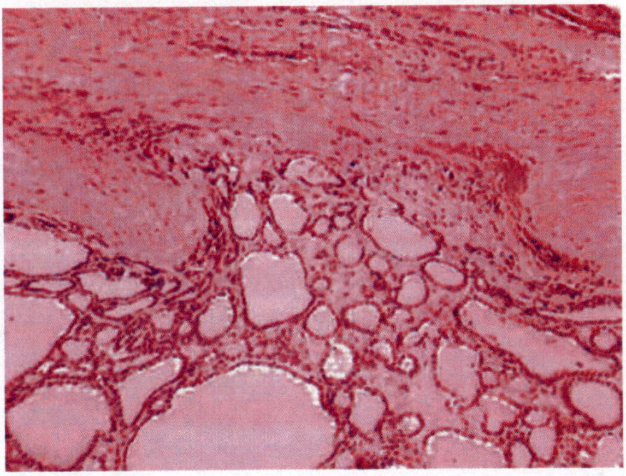

Fig. 6.3 Macrophotography of the follicular thyroid carcinoma

The anaplastic thyroid cancer is considered to be an aggressive form of thyroid cancer due to its low-differentiated structure. It is found in 3% of all diagnosed thyroid cancers. People over 50 are at risk for this disease. The risk factor for the development of anaplastic thyroid cancer is a long-term nodular goiter. The anaplastic thyroid cancer is characterized by an aggressive course, frequent metastasis to the brain, and, unfortunately, an unfavorable prognosis - the average life expectancy is only 6 months (Figs. 6.7 and 6.8).

After determining the T, N, M or/and RT, pN, RM categories can be grouped by stages.

The established degree of spread of the cancer process in system TNM or in stages should be remain unchanged in medical documentation.

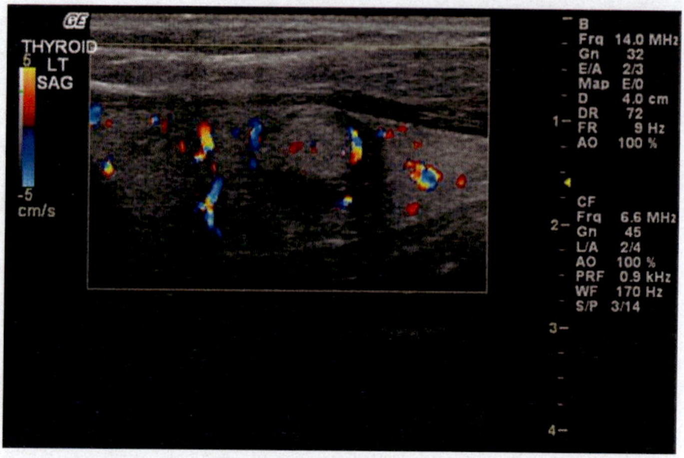

Fig. 6.4 Ultrasound image of thyroid follicular cancer

Fig. 6.5 Macrophotography of medullary thyroid carcinoma

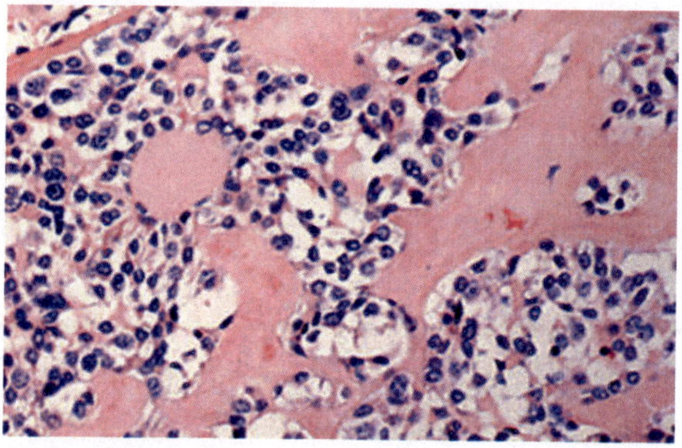

The differentiated thyroid cancer for a long time is not accompanied by any common disorders: there is neither pain nor fever, does not suffer the general condition of the body and working capacity. The patient cares only with the fact of the node existence in the projection of the thyroid gland. Often, the patient does not suspect that he has cancer, which is an unexpected find during a medical examination Thus, in the initial stages, the clinical manifestations of differentiated carcinomas have much in common with the symptoms of nodal non-toxic goiter. An exception is the metastatic variant of differentiated cancer. Thyroid carcinomas, even with large tumor sizes, are not accompanied by clinical signs of hypothyroidism. A typical complaint of patients with differentiated carcinomas is the presence of a tumor node in the neck. At small sizes the node is more often located in the region of one of the poles of the particle, fit snugly against the surface of the trachea. Flat, very dense (woody), adjacent to the trachea node—a typical sign of thyroid cancer.

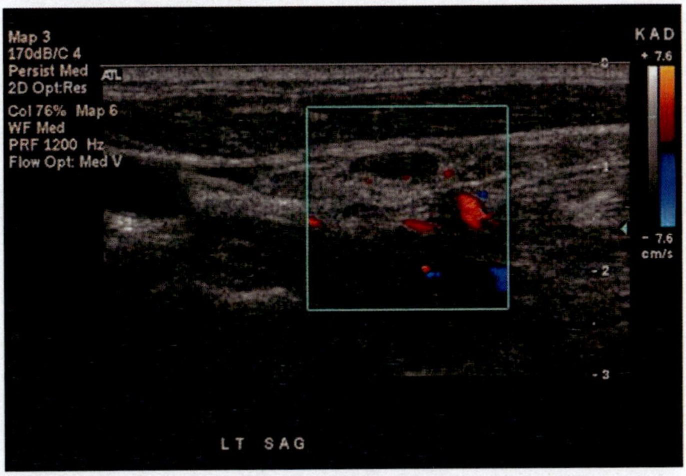

Fig. 6.6 Ultrasound image of medullary thyroid cancer

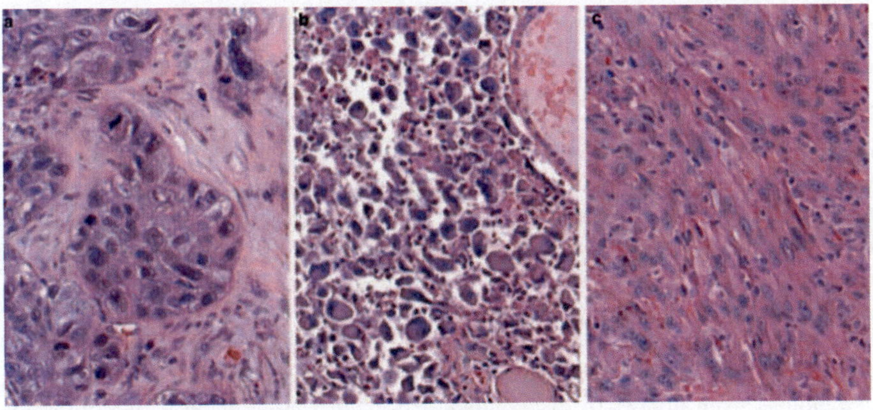

Fig. 6.7 Macrophotography of anaplastic thyroid carcinoma

One of the characteristic symptoms of thyroid cancer is considered the compression of the trachea and esophagus. These symptoms are very characteristic for low-differentiated thyroid carcinomas, which, having rapid infiltrating growth, reach considerable sizes, circularly cover the trachea and esophagus, causing stenosis of these organs.

Differentiated thyroid carcinomas, in particular papillary cancers, grow in the form of a single node, which, reaching certain sizes, can cause a tracheal shift. Clinical signs of tracheal stenosis in differentiated thyroid cancer occur at tumor localization when the lower pole of the tumor descends beyond the sternum.

In other cases, the trachea is shifted towards the soft tissue of the neck. One of the first clinical symptoms of differentiated thyroid cancer may be hoarseness. Unlike

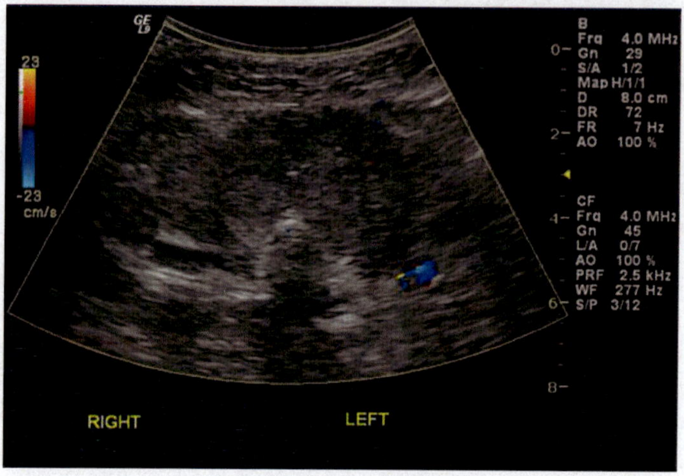

Fig. 6.8 Ultrasound image of anaplastic thyroid cancer

other thyroid tumors, metastases in regional lymph nodes often play a leading role in the clinical picture of differentiated carcinomas. Regional metastases can appear very early when the primary tumor is so small that it is not clinically determined. In such cases, regional metastases are the first and often the only clinical manifestation of the disease. Similar tumors of the thyroid gland were called "hidden" cancer.

To the "hidden" cancer can be attributed the tumors that are clinically undetermined (from microscopic to 1.5 cm) and only diagnosed by metastases, mainly regional, or randomly detected by histological examination of the thyroid gland, distant about the alleged benign growth. More successful is the term "small" cancer.

The "small" carcinomas of the thyroid gland are also called "minimal" or "small" cancer, depending on the size of the primary thyroid tumor. Extremely rapid rate and diffusion, infiltrative growth character is the main clinical differences of anaplastic carcinomas from tumors of differentiated structure. These features determine the nature of the patient's complaints and the objective manifestations of the disease.

By the time of hospitalization, all patients suffering from anaplastic thyroid cancer had tumors that cause neck deformity. Rapid growth of a tumor is accompanied by its inevitable necrosis, and absorption of decay products causes the phenomena of general intoxication: fever, weakness, anemia. These kinds of common manifestations of the disease were observed in 1/3 patients with anaplastic cancer, and as noted, are completely absent in differentiated thyroid carcinomas.

Locally undifferentiated thyroid cancer is defined as a dense, hilly tumor, which often occupies all parts of the gland and has the character of infiltrate. Regional metastases appear early, but, unlike carcinoma differentiated metastases, they are represented by conglomerates of soldered nodes that merge with the primary tumor into a single tumor infiltrate that occupies the entire anterior surface of the neck. In some cases, regional metastases are clearly defined. A rapidly growing tumor quickly extends beyond the thyroid gland and extends to adjacent anatomical formations, which are manifested by relevant clinical symptoms.

First of all, when germinating into the surrounding pretrachial muscles, the tumor is fixed and does not shift when swallowed. Soon enough skin is involved in the process, the latter one becomes hyperemic and infiltrated, there is a risk of bleeding. The clinical picture is growing so rapidly that progressive hyperemia is joined by fever, which gives the false impression of a possible inflammatory nature of the process.

Thyroid cancer is characterized by two ways of metastasis: lymphogenic and hematogenous. The main pathways of lymphogenous metastasis are the lymph nodes of the following groups: deep jugular, lateral triangle of the neck, including the accessory zone, peritoneal trachea, including the area of the anterior-upper mediastinum and lobes. According to oncology clinics, more than 40–60% of patients with thyroid cancer come to treatment with regional metastases of the vascular-nerve bundle of the neck and peritoneal trachea. As noted above, metastatic lesions of the lymph nodes may be the first clinical symptom of this disease.

The favorite areas of hematogenous thyroid cancer metastasis are the lungs—from 4.4 to 14% of cases, and the bones—from 1 to 8% of observations. More rarely, mainly under the undifferentiated forms of the disease, there are metastases in the liver, brain and other organs.

6.2 Analysis of the Diagnostic Importance of Examination Types for the Thyroid Pathologies Detection

Clinical methods of diagnosis are important in the detection of nodular growth in the neck: in the thyroid and regional areas. Properly collected anamnesis should be aimed at establishing the order and terms of nodal growth, their relationship with the symptoms of hyper and hypothyroidism, the rate of progression, methods and results of previous treatment. If the patient was previously operated on, it is necessary to specify the volume of the performed intervention, the data of histological examination of the removed drug.

The examination should begin with a careful examination of the neck, which should pay attention to the presence of deformation, especially in the area of the location of the organ. It is important to identify changes in hormonal status, noting the clinical manifestation of hypo or hyperthyroidism. Palpation of the thyroid gland is advisable to hold in a standing position and lying down. In a standing position, the doctor, while behind the patient's back, 2–4 fingers of both hands examines the thyroid gland, pressing its lobes during swallowing movements to the trachea. In the presence of nodes in the thyroid gland and regional areas, it is mandatory to research the otolaryngological organs to eliminate their tumor damage and establish the mobility of the vocal folds.

In the early stages of development, differential diagnosis of these diseases without the use of special methods of examination is almost impossible. At present, ultrasound is the main method of diagnosis, which is obligatory in case of suspected tumor lesions of the thyroid gland. The use of modern ultrasonic devices with special sensors 7.5 and 5 MHz allows to detect tumor growth cells up to 0.2–0.5 cm in the largest dimension. This provides an opportunity to identify additional, not stipulated by other methods of pre-operative examination, nodal growth in every second patient who has received surgical treatment. The sensitivity of this method in detecting non-palpable focus of tumor growth reaches 91%. In addition, some ultrasound symptoms are significantly more common in malignant tumors of the thyroid gland. Such criteria of malignancy, in addition to the well-known features (the exit of the tumor process for the capsule of the thyroid gland and the presence of metastatic nodes), should include the inequality of the outline of the tumor focus, the absence of a rim around it and the uneven structure of the node with predominance of hypoechoic solid areas.

Important is the ultrasound method for the detection of palpatory indeterminate lesions of regional lymph nodes, especially in paratracheal areas. Mandatory is an ultrasound examination of the abdominal cavity for the detection of primary multiple tumors in the medullary form of thyroid cancer.

Cytological examination of points from the focus of tumor growth is determining for establishing the correct diagnosis and choosing the optimal treatment option. This method is the most important for the differential diagnosis of benign and malignant lesions in the thyroid gland. It is advisable to obtain material for cytological examination from all nodes in the thyroid gland, the morphological nature of which may affect the choice of treatment method and volume of surgery. This method allows to accurately differentiate metastatic lesions of lymph nodes from other ones, primary-multiple tumors. Upon receipt of sufficient material for cytological conclusion of the material in most patients (60%), it is possible to establish the morphological form of the malignant tumor: papillary, follicular, medullary, undifferentiated cancer or sarcoma. However, according to the puncture of the metastatic node, it is possible to determine the localization of a clinically indeterminate primary tumor.

In the absence of morphological verification of the diagnosis before the operation the method of urgent cytological examination of scrapes or prints from a distant tumor node is high efficient. It must be used in case of preservation of the part of the infected part of the organ or the detection of lymph nodes suspected of metastasis. For the same purpose, it is possible to use urgent histological examination of frozen sections of the tumor. However, the difficulties of interpreting morphological changes in highly differentiated tumors and high frequency of hyper and hypodiagnostic errors require the participation of an experienced morphologist.

The most difficult is the pre-operative differential diagnosis of the initial stages of thyroid cancer and benign nodular formations of this organ: adenoma, nodular goiter, chronic thyroiditis. The most effective for this purpose is the cytological method with preoperative examination of the puncture from the node and intraoperative—scraping from the tumor. Use of biopsy (using special needles) and urgent histological examination is possible. Also important is the evaluation of the above criteria for malignancy in the ultrasound method.

For differential diagnosis of "hidden thyroid cancer" with tumors of the lymph nodes, which has other nature, and cysts of the neck is mainly used ultrasound. The detection of a hidden tumor in the thyroid gland and its cytological verification allows to establish the correct diagnosis. Cytological examination of the punctate from the nodes on the neck, also in most patients makes it possible to determine the nature of the detected changes.

There are some diagnostic difficulties with "hidden thyroid cancer", which is manifested by distant metastases. Pulmonary metastases are differentiated with miliary tuberculosis for which it is less characteristic: lesion of predominantly lower lungs, absence of general response of the organism, including temperature reaction, inefficiency of specific anti-tuberculosis treatment. Bone metastases from benign cystic lesions and primary bone tumors are differentiated by a characteristic predominantly osteolytic and multiple lesion pattern. The accurate diagnosis requires morphological verification of the detected changes, which is possible using trepanbiopsy. For any lesion of the lungs and bones, suspected of distant metastasis, it is advisable ultrasound examination of the thyroid gland with morphological verifications revealed changes in it.

6.3 Intelligent System for Diagnosis of Thyroid Pathology

The diagnosis of malignant tumors is performed on the basis of the following types of research [5, 7, 8, 11, 19, 21, 23]: doctor's consultation, radiological methods, ultrasound, computed tomography, magnetic resonance imaging, radioisotope diagnostics, radio immunoscintigraphy, thermography [22].

Currently, for the diagnosis of thyroid disease, doctors conduct researches according to a certain scheme, which consists of a mandatory diagnostic minimum (MDM) and a set of additional examinations. The signs that determine the type of thyroid pathology by ultrasound [10, 29] are shown in Fig. 6.9.

The values of the factors that are determined by the results of the ultrasound and are used as inputs of the intelligent diagnosis system are [11]: the type of rim of the tumor, the structure of the tumor, the echogenicity of the tumor, the size of the tumor. Additional factors used as inputs neural network are: general blood analysis, cancer-embryonic antigen (CEA), thyroid-stimulating hormonelevel (TSH) level, T4 level, T3 level, cervical lymphadenopathy, solid node consistency, hoarseness and somnolence, neck and head irradiation anamnesis, age, gender. The general structure of the diagnostic system is shown in Fig. 6.10.

Types (methods) of examinations with MDM differ in varying degrees of importance.

For their analysis, it is suggested to estimate the degree of importance using the rank coefficient of importance (RI) obtained expert way. Under the rank coefficient of importance is understood the row number in the list of methods ranked (sorted) in order of decreasing of their importance. For the main MDM methods, the RI values are presented in Table 6.1.

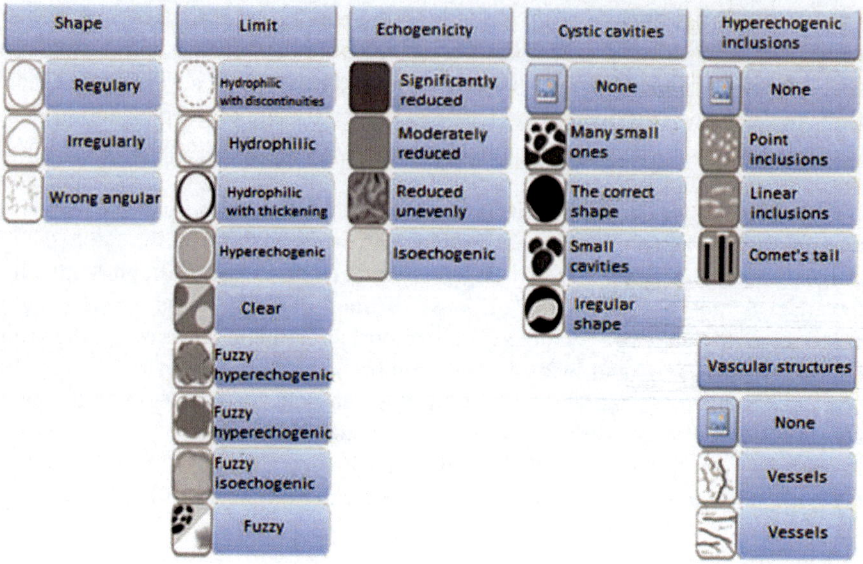

Fig. 6.9 Signs that determine the type of pathology of the thyroid glandon the results of ultrasound

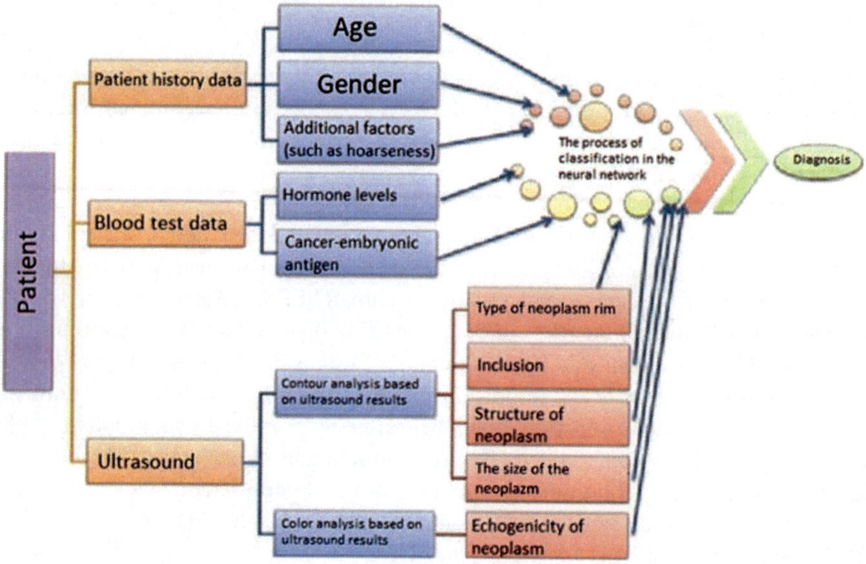

Fig. 6.10 General structure of the diagnostic system

Table 6.1 Types (methods) of examination from the diagnostic minimum

Method number (i)	The name of the method	Description of the method	RI	DSI
1	Surveys and medical examination	Recording of the main parameters of the body, complaints, description of visible parts of organs and systems (hoarseness and hoarseness of the voice)	2	13%
2	Palpation	Probing the thyroid gland to determine lymphadenopathy	4	3.4
3	General blood test	Determination of ESR (erythrocyte sedimentation rate), hemoglobin content, blood cells	3	9
4	Biochemical analysis of blood	Determination of the content of a number of chemicals	3	9
5	Cancer markers	Determination of Cancer-Embryonic Antigen (CEA)	2	11
6	Blood test for hormones	Determination of TSH, T4, T3, thyrocalcitonin levels	1	14
7	USE	Research of the thyroid gland	1	18
8	ECG	Detection of cardiovascular system (CVS) disorders	4	1.7

On the base of RI, as well as the relative number of obtained diagnostic parameters n_1 and their total number *N*, it is possible to determine the Diagnostic Significance Index (DSI) for these research methods using the following formula

$$DSI_i = \frac{\mathrm{RImax} - RI_i + 1}{RI\mathrm{max}} \cdot \frac{n_i}{N} \cdot 100\% \tag{6.1}$$

where *i* is the number of the diagnostic minimum method; $i \in \overline{1, M} RI_i$ is the rank coefficient of importance of *i*th method; $R\mathrm{Imax}$ is the maximum rank coefficient of importance; n_i is the number of diagnostic parameters that are obtained by *i* method; *M* is the total number of diagnostic minimum methods; *N* is the total number of parameters obtained in MDM $N = \sum_{i-1}^{M} n_i$.

An ideal diagnostic test based on formula (6.1) will have DSI = 100%, if $RI = 1$, $n = N$, which corresponds to the absence of examination results by other methods ($M = 1$). Really it is impossible to achieve such a DSI. However, the larger the total DSI of the examination types, the more accurate the diagnosis based on it.

The values of DSI and RI for the main methods of MDM are given in Table 6.1.

It is advisable to split the examination set into 3 groups:

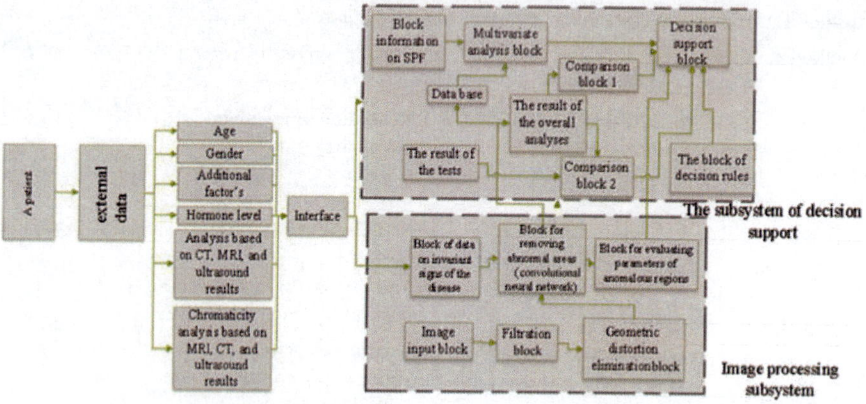

Fig. 6.11 Block diagram of intelligent diagnostics

(1) strong significance (types 6 i 7, for which DSI $\geq$ 14%);
(2) average significance (types 1, 3, 4 i 5 for which 9% < DSI $\leq$ 13%);
(3) low significance (types 2 i 8 for which DSI < 3.5%).

In this regard, it is necessary to conduct examinations of the 3rd group only if it is impossible to conduct examinations of the 1st and 2nd groups.

The block diagram of the intelligent diagnostic system (DSI) is shown in Fig. 6.11.

The information from the patient enters the system through the interface and includes data according to the Table 6.1. Video images in digital form enter the blocks of filtering and elimination of geometric distortions to eliminate the influence of noise. After that, the video image sent the anomalous region selection block, which is implemented on the basis of the CNN (see Chap. 4). Evaluation of signs that determine the type of thyroid pathology by the results of Ultrasound Examinations is executed in anomalous region estimation block.

After that, the video image is sent to the anomalous region selection block, which is implemented on the base of the CNN (see Sect. 6.4.2.3).

Then the parameters of the signs obtained from the image analysis are compared with the normal state of the organ being examined and with the pathological changes that are in the database [2]. It stores survey and sample data, pre-generated using the statistical processing block, and signs of diseases. This block interacts with the decision support block. Based on the data obtained from the information block on the set of pathogenetic factors (SPF), the multi-factor analysis block allows you to create regression equations for them [27]. Then, using the decision rules from the block of the same name, the recommended solution is formed.

The decision support block generates and displays special data on the monitor in a convenient form for the doctor, as well as evaluates their information reliability and makes recommendations to the doctor for making a diagnosis, taking into account the factors that affect the disease.

6.4 Ultrasound Video Image Processing Subsystem

In accordance with the structural diagram of the intelligent diagnostic system, it is necessary to carry out primary processing of the images. In order to analyze existing methods, it is necessary to consider the types of noise in the image.

6.4.1 *Noise Types of Image*

A fundamental problem with image processing is the effective removal of noise, with preserving important for the next semantic descriptions/recognition of image details. The complexity of the solution of this problem depends essentially on the noise model under consideration [5].

Considering existing types of noise.

White noise is a stationary noise which spectral components are uniformly distributed over the entire frequency range involved. As white noises are any noises which have the same (or slightly different) spectral density in the frequency range under consideration.

White noise characteristics:

$$\mu_w(t) = \mathrm{E}\{\mathbf{w}(t)\} = 0, \qquad \mathbf{R}_{ww}(t_1, t_2) = \mathrm{E}\{\mathbf{w}(t_1)\mathbf{w}(t_2)\} = \sigma^2\delta(t_1 - t_2), \tag{6.2}$$

where $\mathbf{w}$ is a vector of random numbers; $\mathbf{R}$ is a covariance matrix; $\mu_w(t)$ is the mathematical expectation of the vector $\mathbf{w}$; σ^2 is the variance; $\delta(t)$ is the delta-function.

The spectrum of this signal (Fig. 6.12) has the following form.

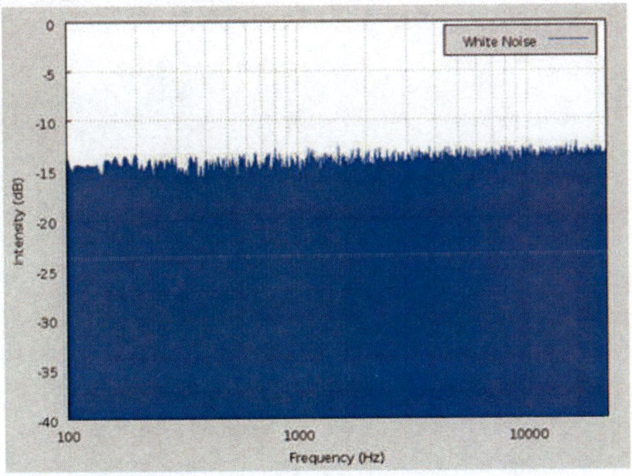

Fig. 6.12 White noise spectrum

Gaussian noise is a kind of white noise that typically occurs at the stage of digital image formation. It is possible to characterize this noise by adding to each pixel of image the values of the Gaussian distribution with zero mathematical expectation

$$\mathbf{N_{i,j}} = \frac{1}{2\pi\sigma^2} e^{-(i^2+j^2)/(2\sigma^2)}, \quad M[\mathbf{N_{i,j}}] = 0, \tag{6.3}$$

where $\mathbf{N_{i,j}}$ is the noise characteristic matrix; σ is the standard deviation of the Gaussian distribution.

Gaussian noise has the same characteristics as white noise.

Color noise is a signal whose spectral density has its color definition in accordance to the analogy between the spectra of the visible light color signals. Most often it is happened noises: pink, red, blue, purple and gray. Characteristics of the noise data:

$$\mu_w(t) = \mathrm{E}\{\mathbf{w}(t)\} = 0, \qquad \mathbf{R}_{ww}(t_1, t_2) = \mathrm{E}\{\mathbf{w}(t_1)\mathbf{w}(t_2)\}, \tag{6.4}$$

where $\mathbf{w}$ is a vector of random numbers; $\mathbf{R}$ is a covariance matrix; $\mu_w(t)$ is the mathematical expectation of the vector $\mathbf{w}$.

Pink noise is a signal which spectral density is inversely proportional to its frequency. In other words, this signal decreases uniformly on a logarithmic frequency scale. In some cases, pink noise is called any noise in which the spectral density decreases with increasing frequency (Fig. 6.13).

Brownian (red) noise is noise that has a large power reserve at low frequencies compared to high ones. Comparing the red noise with the mentioned above, the white noise will be more muted (Fig. 6.14).

Blue noise is a signal which spectral density is increased with frequency increasing. Blue noise is obtained by differentiating pink noise—their spectra are mirror (Fig. 6.15).

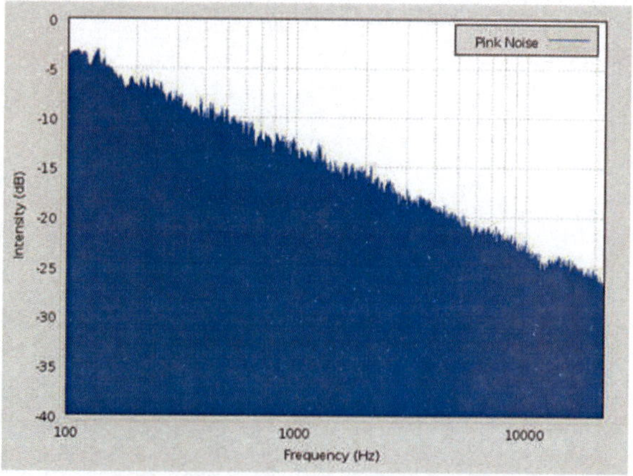

Fig. 6.13 Pink noise spectrum

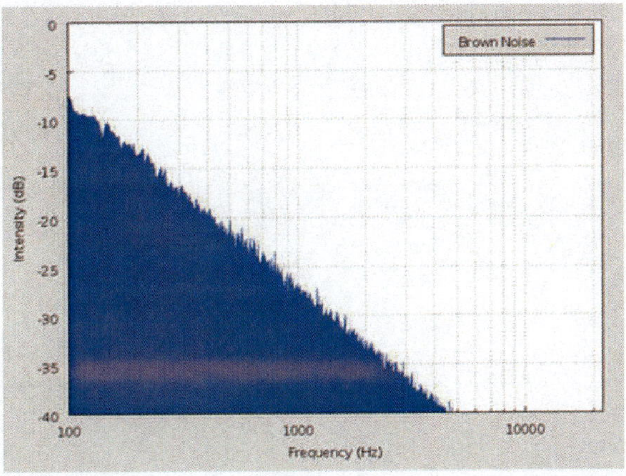

Fig. 6.14 Red noise spectrum

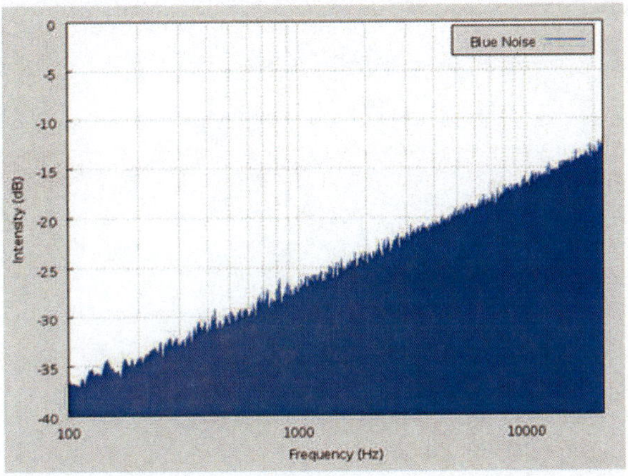

Fig. 6.15 Blue noise spectrum

Purple noise is a signal which spectral density is proportional to the square of the frequency and, similarly to white noise, in practice it must be limited in frequency. Purple noise is obtained by differentiating the white noise. The spectrum of purple noise is mirror opposite to red one (Fig. 6.16).

Gray noise is a signal that has the same density at all frequencies. The gray noise spectrum is obtained by adding the spectra of Brownian and purple noise. The spcctrum of this noise contains a large “dip” at medium frequencies (Fig. 6.17).

Impulse noise. Impulse signal means the distortion of the signal by impulses, i.e. emissions with very large positive and negative values and small duration. Pulse noise is characterized by the change of a portion of the pixels for fixed or a random values in the image. This model of noise is related, for example, to errors in image transmission [4].

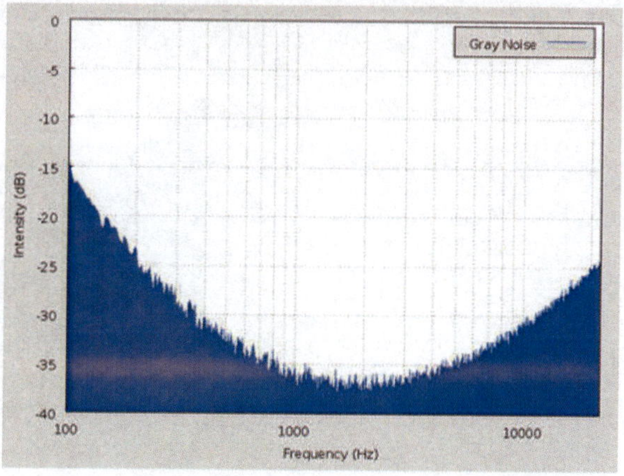

Fig. 6.16 Purple noise spectrum

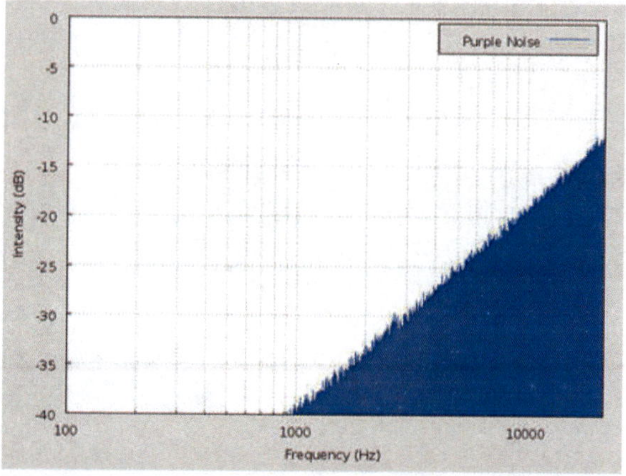

Fig. 6.17 Gray noise spectrum

During the image processing, impulse noise occurs, for example, as a result of decoding error that result in black and white dots appearing on the image. Therefore, it is often called point noise. This noise can be characterized by two different models.

Pulse noise

Model 1. The occurrence of noise emission at each point (i, j) of the image has a probability p and does not depend on the presence of noise at other points of the image or on the origin image. The law of distribution of this noise can be represented as follows:

$$\mathbf{N_{i,j}} = \begin{Bmatrix} d \text{ with probability } p \\ 0 \text{ with probability } (1-p) \end{Bmatrix} \quad M[\mathbf{N_{i,j}}] = d, \tag{6.5}$$

where $N_{i,j}$ is the noise characteristic matrix.

Since the noise emission does not depend from the presence of noise at other points in the image, the covariance matrix for a given distribution law will look like as a unit matrix.

Model 2. This model differs from model 1 only in that the distorted points acquire random, not fixed, values of $Z_{i,j}$. It is assumed that they are independent random variables. The noise matrix can be represented as follows:

$$\mathbf{N_{i,j}} = \left\{ \begin{matrix} d \text{ with probability } p \\ 0 \text{ with probability } (1-p) \end{matrix} \right\} \quad M[\mathbf{N_{i,j}}] = d/2, \tag{6.6}$$

$$Z_{i,j} = \frac{1}{2\pi\sigma^2} e^{-(i^2+j^2)/(2\sigma^2)}, \quad M[Z_{i,j}] = 0, \tag{6.7}$$

where $\mathbf{N_{i,j}}$ is the noise characteristic matrix; σ is the standard deviation of the Gaussian distribution.

Since the noise emission does not depend on the presence of noise at other points in the image, the covariance matrix for a given distribution law will look like a unit matrix.

In the real image it is possible to meet both additive and impulse noise, such noise is called **combined**. Under combined noise it is usually understood signal distortion by Gaussian pulse noise.

$$G_{m_1,m_2} = F_{m_1,m_2} + \mathbf{N1}_{m_1,m_2} + \mathbf{N2}, m_1 = \overline{0, M_1 - 1}, m_2 = \overline{0, M_2 - 1}, \tag{6.8}$$

where **N1** and **N2** are the Gaussian impulse noises characteristics matrices, respectively.

Gaussian, impulsed, and combined noise models are the most adequate in terms of practical applications.

6.4.1.1 Problem Statement of Noise Filtering on the Image

Consider a noisy model:

$$\mathbf{G}_{m_1,m_2} = \mathbf{F}_{m_1,m_2} + \mathbf{N}_{m_1,m_2}, \quad m_1 = \overline{0, M_1 - 1}, \quad m_2 = \overline{0, M_2 - 1}, \tag{6.9}$$

where $\mathbf{G}_{m_1,m_2}$ is the noisy image matrix; $\mathbf{F}_{m_1,m_2}$ matrix of the original image; $\mathbf{N}_{m_1,m_2}$ noise characteristic matrix.

The task is to recover $\mathbf{F}_{m_1,m_2}$ from $\mathbf{G}_{m_1,m_2}$ (Fig. 6.18).

6.4.1.2 Overview of Methods of Image Noise Filtering Problem Solution

Noise suppression algorithms typically specialize in suppressing a particular type of noise. There are no universal filters yet that detect and suppress all kinds of noises.

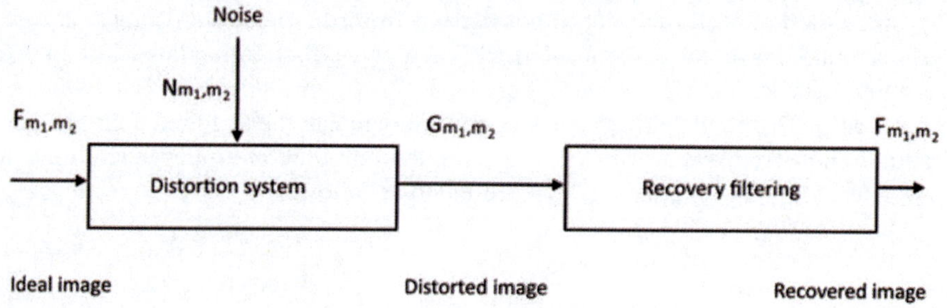

Fig. 6.18 Structural scheme of the image recovery process

However, numerous noises can be approximated quite well by the model of white Gaussian noise, so most algorithms are focused on suppressing this kind of noise. Here is a classification of filters (Fig. 6.19).

Below we consider the following filters [6, 17]:

- Linear filter.
- The Wiener filter.
- The Gaussian filter.
- Median filter.
- Extreme filter.
- Kuwahara Filter.

Algorithms of discrete convolution and aperture adaptation

The following concepts are used for discrete convolution and adaptation of the aperture (Figs. 6.20, 6.21 and 6.22):

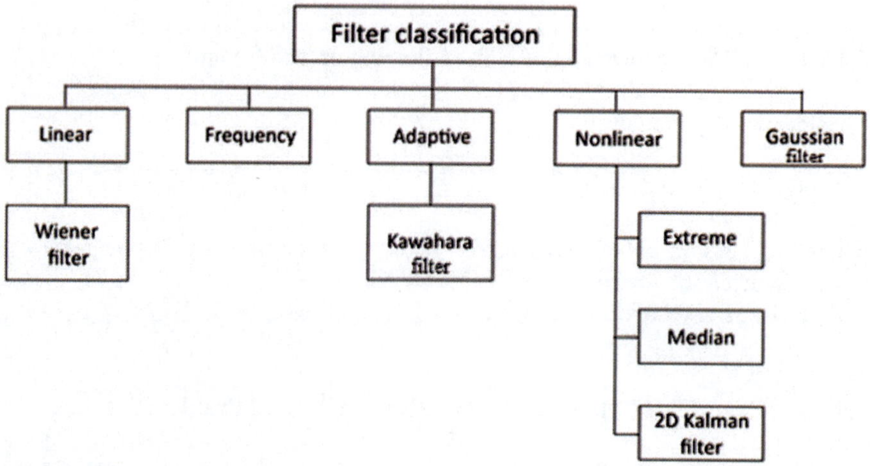

Fig. 6.19 Filter classification

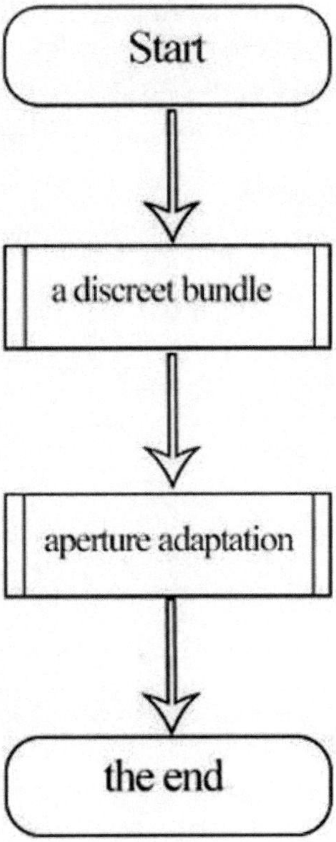

Fig. 6.20 Discrete convolution and aperture adaptation

- the original image $\mathbf{G}_{m_1,m_2}$;
- the resulting image $\mathbf{F}_{m_1,m_2}$;
- x, y are coordinates of the current pixel in the image;
- $\mathbf{C}_{i,j}$ is the convolution matrix;
- I, J is the dimension of the convolution;
- $r = \pm(0, 1, \ldots, R), q = \pm(0, 1, \ldots, Q)$;
- $\Delta_{i,j}$ is the weighted difference;
- $\mathbf{v}$ is the binary matrix of elements membership $G_{i,j}$;
- P_Δ is the threshold whose value depends on aperture size and number of brightness quantization levels;
- $h_{i,j}$ is the outline and image noise $G_{i,j}$.

Linear image filtering algorithm

Used to suppress Gaussian noise.

$$\mathbf{F}_{m_1 m_2} = \sum_i \sum_j \mathbf{G}_{ij} \mathbf{G}_{x+i,y+j}. \tag{6.10}$$

Fig. 6.21 Discrete convolution algorithm

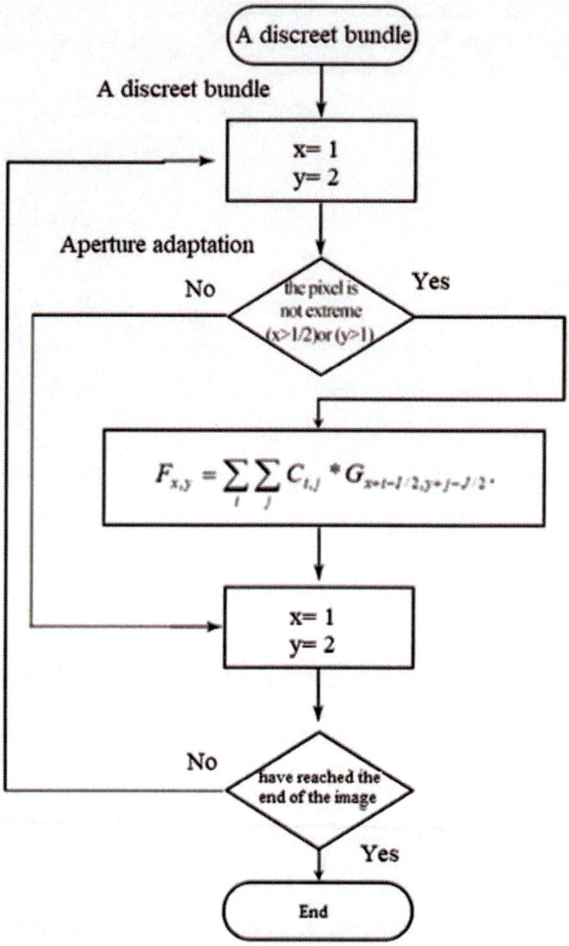

In the first case, the intensity value at the central point is assigned to the mean of the neighbors' intensities. In other cases, the weighted average according to the coefficients [29]. An example of the algorithm (Fig. 6.23).

The *Wiener filter algorithm.* Used to suppress Gaussian noise.

$$\hat{F}(u,v) = \left(\frac{1}{H(u,v)} \frac{|H(u,v)|^2}{|H(u,v)|^2 + S_\eta(u,v)/S_f(u,v)}\right) G(u,v), \qquad (6.11)$$

where $H(u, v)$, $F(u, v)$ is the Fourier images of the distorting function and the function of the initial image; S is the noise energy spectra.

$$G(u,v), G(u,v), = H(u,v)F(u,v) + N(u,v), \qquad (6.12)$$

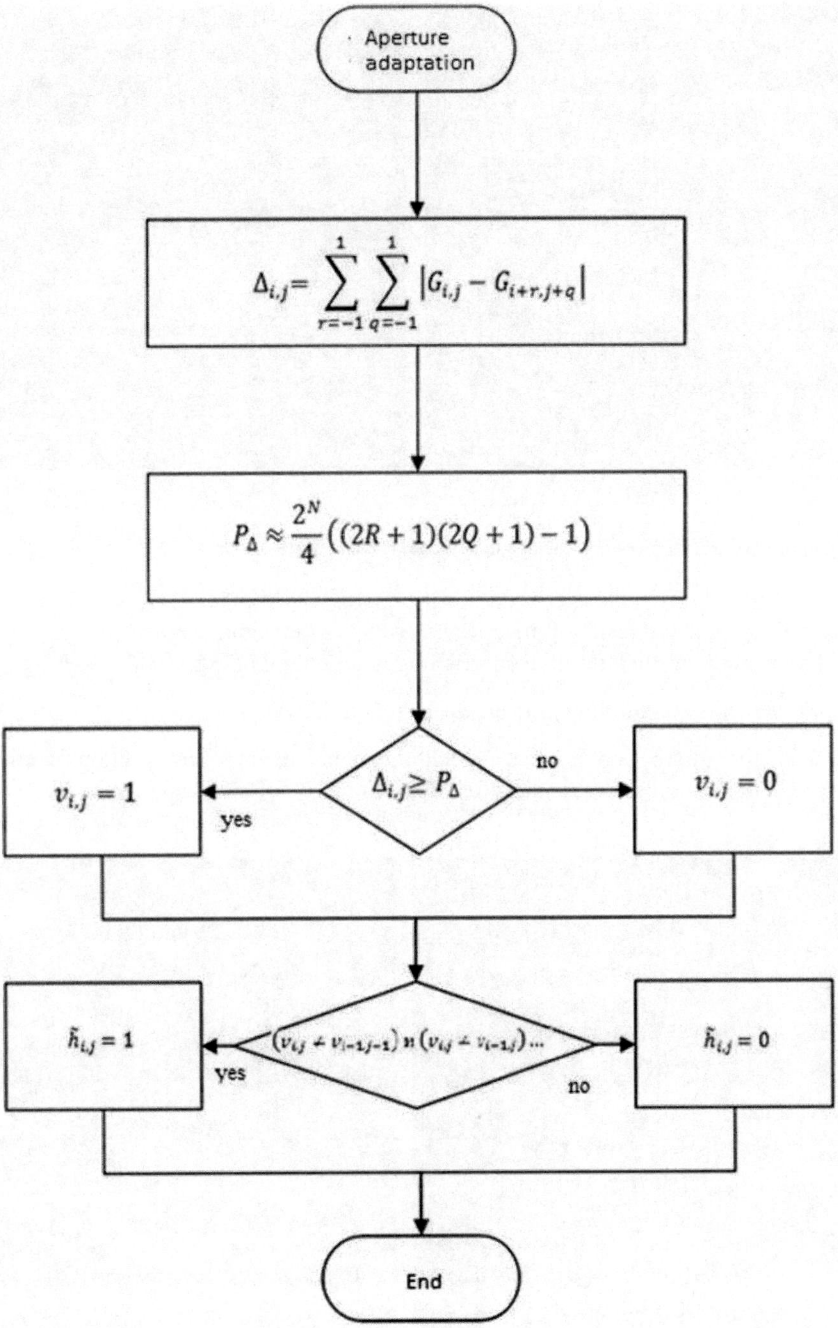

Fig. 6.22 Aperture adaptation

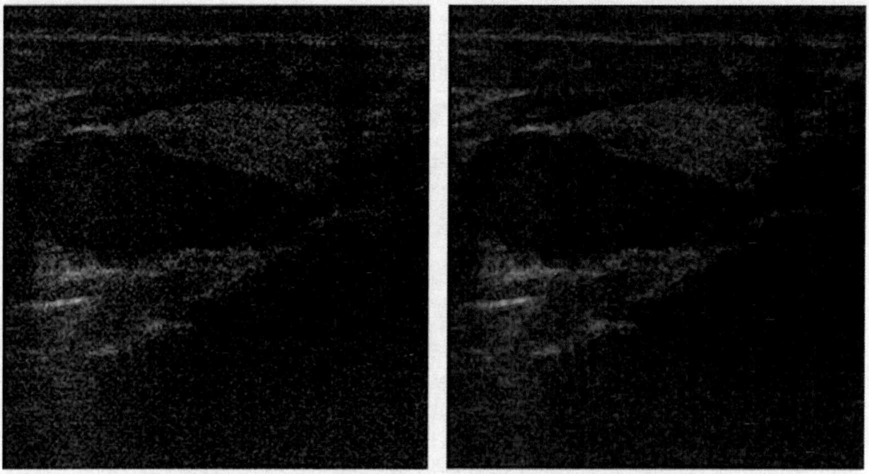

(a) (b)

Fig. 6.23 Linear filtration: **a** is the original image; **b** is the processed image

where $G(u, v)$ is the result of distortion; $N(u, v)$ is the additive noise.

An example of the algorithm functioning is presented in Fig. 6.24.

An algorithm for smoothing filtering according to Gauss

It is used for suppression of pulse and additive noise. Its impulse transition function is the Gaussian function. It is used for suppression of pulse and additive noise.

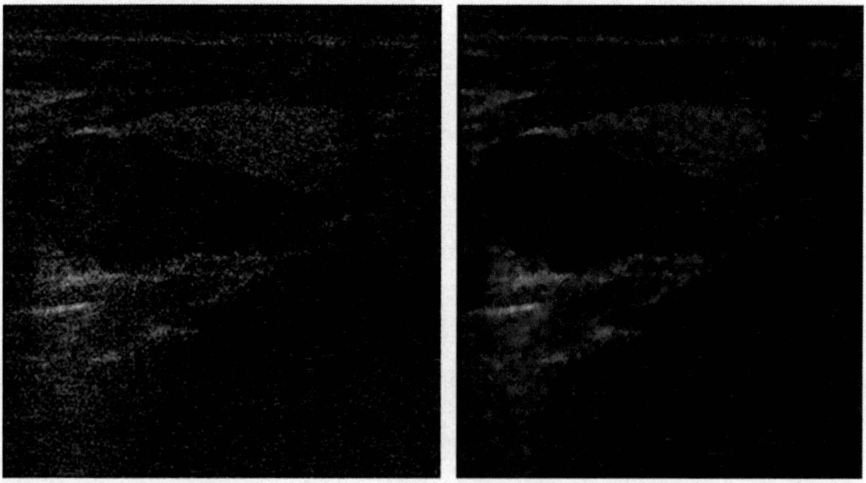

(a) (b)

Fig. 6.24 Filtration using the wiener filter: **a** is the original image; **b** is the processed image

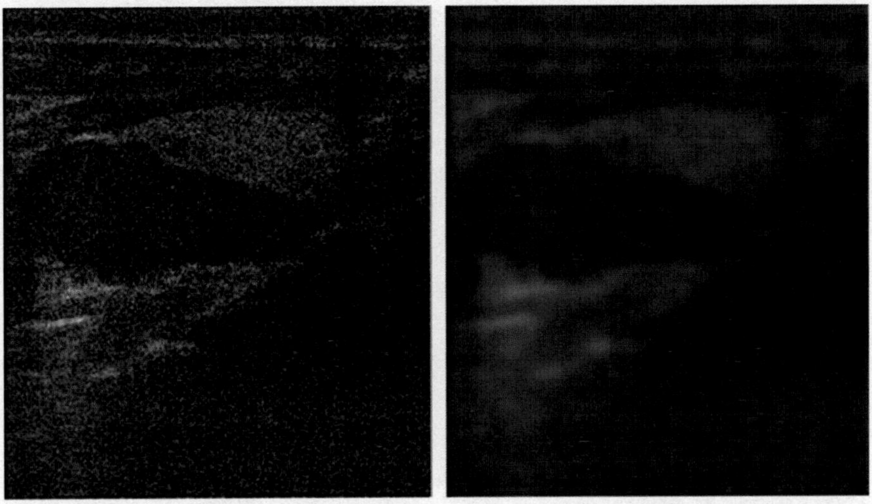

Fig. 6.25 Filtration using a gaussian filter (10% blur): **a** is the original image; **b** is the processed image

It uses a convolution operation with the kernel of the species

$$C_{i,j} = \frac{1}{2\pi\sigma^2} e^{-(i^2+j^2)/(2\sigma^2)}, \tag{6.13}$$

where is the blur radius $r^2 = u^2 + v^2$; σ is the standard deviation of Gaussian distribution.

In cases of two dimensions, this formula gives a surface that looks like concentric circles with a Gaussian distribution from the center point.

An example of this filter functioning is presented in Fig. 6.25.

Median filtering algorithm

It is used for suppression of pulse and additive noise.

The median filter, like any convolutional filter, has the form of a window sliding across the field window W, covering an odd number of elements. The center count is changed by the median of all the image elements that hit the window. The median of discrete sequence $x_1, x_2, \ldots, x_L$ for the odd L called the element for which it is existed $(L-1)/2$ elements smaller or equal to it by size, and $(L-1)/2$ elements greater or equal to it by the value.

$$B(x, y) = median\{N(x, y)\}, \tag{6.14}$$

where $B(x, y)$ is the point of the image being processed; $N(x, y)$ is the point of the original image.

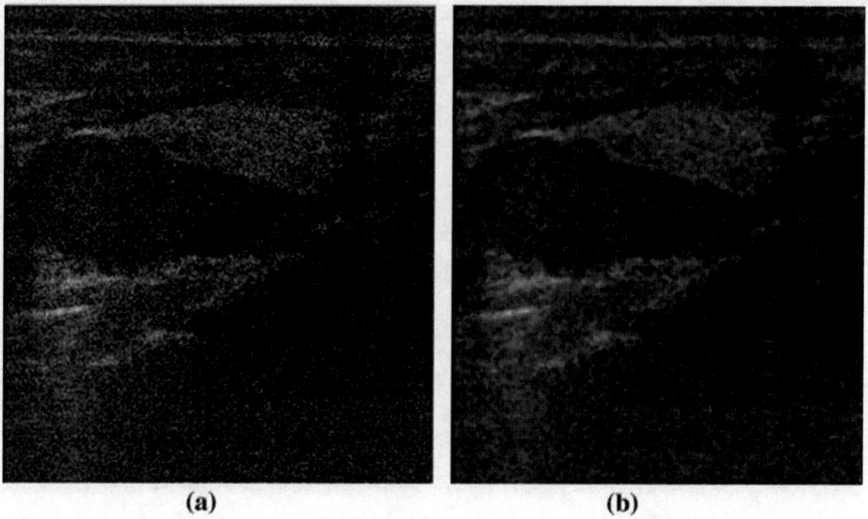

Fig. 6.26 Median filtering: **a** is the original image; **b** is the processed image

Example of filter functioning is presented in Fig. 6.26.

Extreme filtration algorithm

It is used for suppression of pulse and additive noises.

$$B_{\min}(x, y) = \min\{N(x, y)\},$$
$$B_{\max}(x, y) = \max\{N(x, y)\}. \tag{6.15}$$

Smoothing filtering algorithm Kuwahara

It is used to suppress impulse noise, Kuwahara filter kernel is presented in Fig. 6.27.

The algorithm consists of such steps:

Step 1. Calculation of arithmetic mean m dispersions D for areas **abcd** given the dimension extension

Fig. 6.27 Kuwahara filter kernel of dimension 5 × 5

a	*a*	*a/b*	*b*	*b*
a	*a*	*a/b*	*b*	*b*
a/c	*a/c*	*a/b/c/d*	*b/d*	*b/d*
c	*c*	*c/d*	*d*	*d*
c	*c*	*c/d*	*d*	*d*

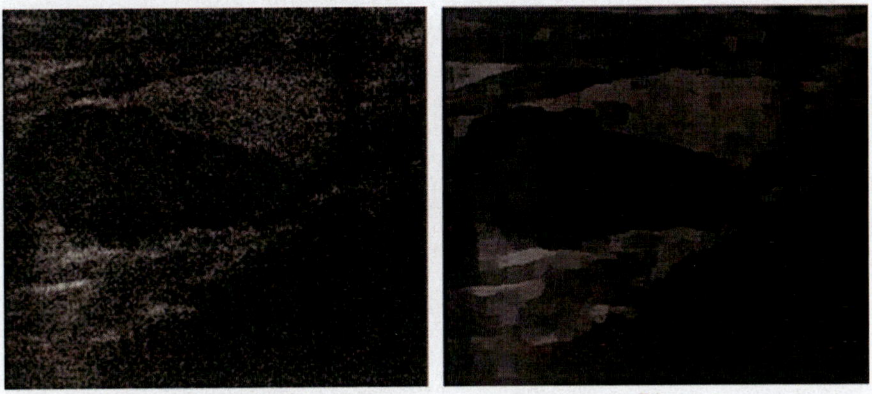

(a) (b)

Fig. 6.28 Filtering with a Kuwahara filter (4% Blur): **a** is the original image; **b** is the processed image

$$m_s = \frac{1}{m \times n} \sum_{k=0}^{m} \sum_{i=0}^{n} \int (x_k, y_i),$$

$$D_s = \frac{1}{m \times n} \sum_{k=0}^{m} \sum_{i=0}^{n} \left[\int (x_k, y_i) - m_s \right]^2, \tag{6.16}$$

where S is determined the number of areas on which the filter window should be divided; $\int (x_k, y_i)$ is the value of the pixel at coordinates (x_k, y_i); $m \times n$ is the number of pixels in the current area; f is the source of image function.

Step 2. Identify the area with the least dispersion

$$\Phi(x, y) = \begin{cases} m_1(x, y), \sigma_1(x, y) = \min_i \sigma_i(x, y), \\ m_2(x, y), \sigma_2(x, y) = \min_i \sigma_i(x, y), \\ m_3(x, y), \sigma_3(x, y) = \min_i \sigma_i(x, y), \\ m_4(x, y), \sigma_4(x, y) = \min_i \sigma_i(x, y), \end{cases} \tag{6.17}$$

Step 3. Assign a response as the lowest average of the area with the least variance. Example of filter functioning is presented in Fig. 6.28.

6.4.1.3 Comparative Analysis of Image Filtering Algorithms

For the determination the area of filter application for image filtration it is necessary to execute the comparative analysis of filtering algorithms. The comparative analysis of filtering algorithms is presented in Table 6.2.

Table 6.2 Comparative analysis for filtering algorithms

Algorithm	Type	Benefits	Disadvantages
Median	Smoothing	The filter does not change the step and dust functions; the filter well suppresses single impulse interference and random noise emissions	Not effective enough against additive noise
Extreme		The filter allows not only compensate noise, but also eliminates (partially) the effects of blurring the image	Ineffective against impulse noise
Wiener	Frequency	Has the advantage of wide adaptation; for some problems the transfer function of the filter can be obtained exactly and, accordingly, using the components of the simple physical configuration of the Wiener filter network	The image is not considered as a function of a pixel matrix
Linear	Convolutional	Effective against impulse noise	Not effective enough against additive noise
Gauss		Effectively removes noise	Loses boundary information
Kuwahara		Effectively removes noise without losing image boundaries	It loses some of the irrelevant information

6.4.2 Algorithms of Diseases Diagnostic Significant Signs Determination Based on Ultrasound Image Processing

6.4.2.1 Algorithms of Image Processing Subsystem Parameters Adjustment

When use in the composition of intelligent diagnostic system (DSI) the Ultrasound devices with different characteristics, there is a need to adjust some parameters of the processing function. The algorithm of adjusting the parameters of the Image processing subsystem is shown in Fig. 6.29 [30].

To adjust the filtering parameters of the corresponding function in the technical modeling environment Matlab it is possible the changes the maximum values of $S_{\max}$, which limit the sizes S of sliding window within of which the mean and standard deviation of brightness values are estimated. One more parameter *noise* sets the Gaussian noise power which the image is damaged. This parameter is determined from the a priori information about image. If this is not possible, then under execution of filtration function *wiener* 2 the parameter *noise* can't be considered. In this case, the noise power is estimated automatically.

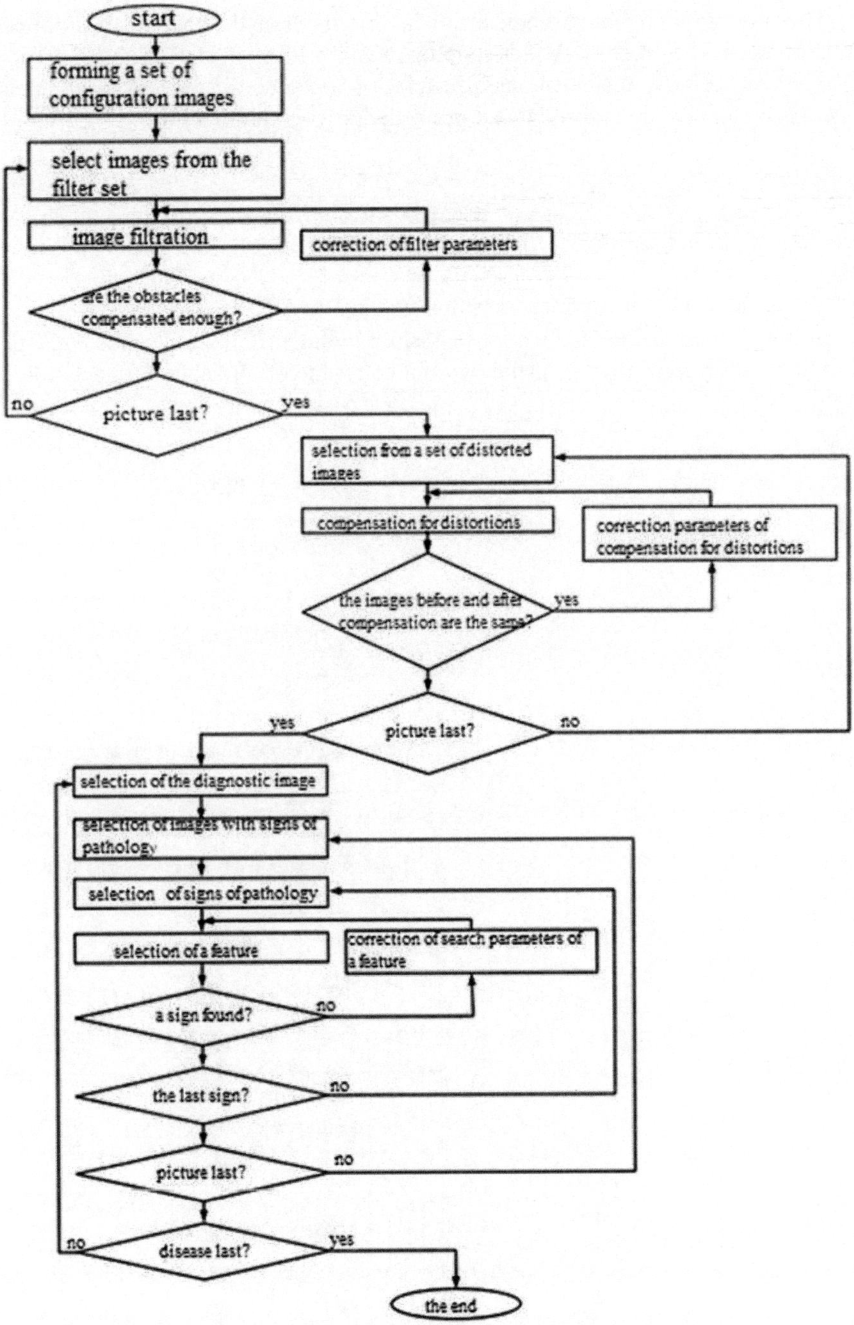

Fig. 6.29 Algorithm of video processing subsystem parameters adjustment

The correction of geometric distortion is carried out on the basis of the nonlinear image predistortion method, which supplies the compensation of expected distortions [15, 28]. For this the image is transformed into a polar coordinate system, and radial inverse distortions are realised by a third-order polynomial function:

$$S = r + ar^3,$$

where r is the radius-vector of polar coordinate system (θ, r); a is the conversion amplitude.

The error correction algorithm is presented in Fig. 6.30.

In accordance to Fig. 6.9 we will consider the algorithms for signs evaluation that mine the type of thyroid pathology on the results of ultrasound. In accordance

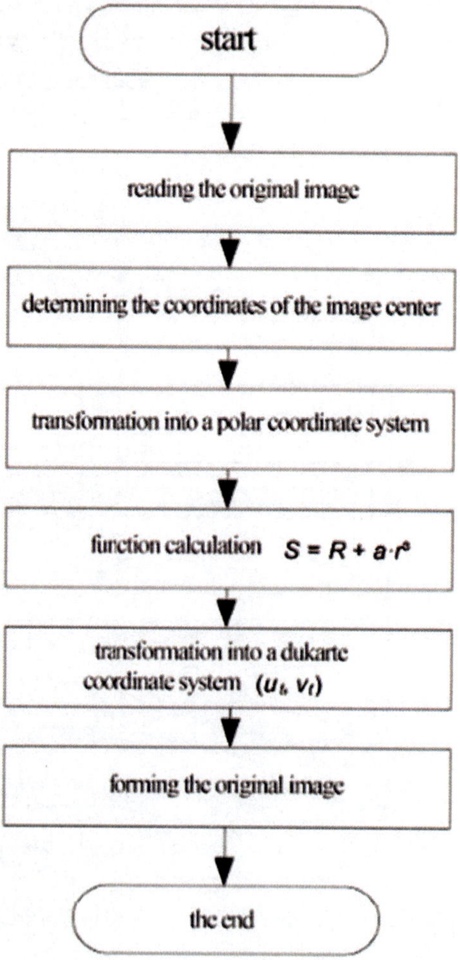

Fig. 6.30 Algorithm of geometric distortions correction

to Fig. 6.11 of the Intelligent Diagnostic Decision Support System (DDSS), we consider the algorithms of its adjustment and functioning. The algorithm of the DDSS adjustment is shown in Fig. 6.31, where it is considered the determination of input and output factors values ranges.

The pair correlation coefficients $C_{i,j}$ are then calculated to reflect the relationships between the factors [9]. This process can be described by the algorithm shown in Fig. 6.32.

Initially, it is determined the columns of the experimental data table for which the coupling coefficient will be calculated. Then the number of possible factors values in

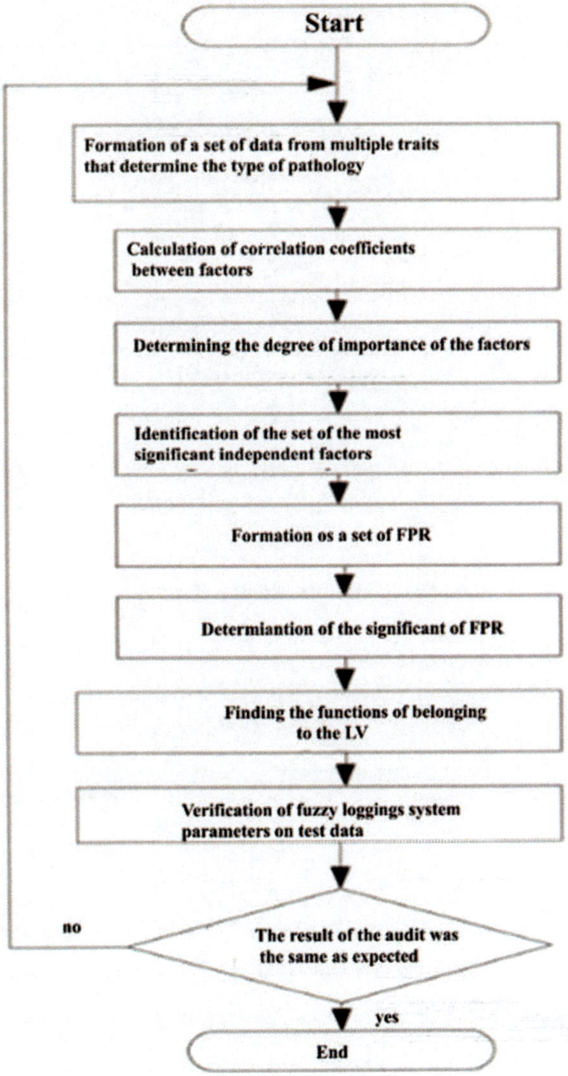

Fig. 6.31 Algorithm of DDSS adjustment: FPR is the fuzzy products rule; LV is a linguistic variable

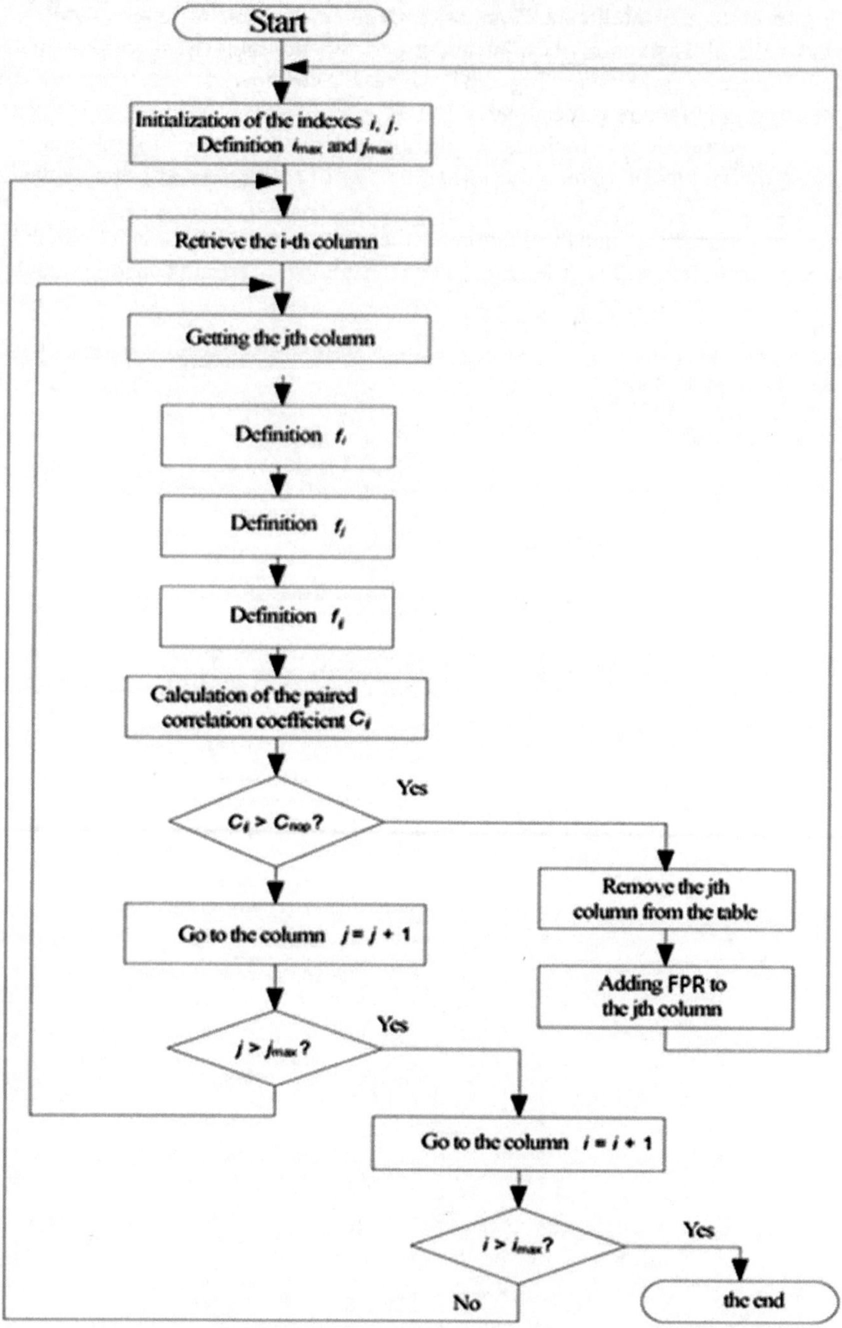

Fig. 6.32 An algorithm for calculating the coefficients of the relationship between factors

each column, and the number of pairs that can form the given values are determined. A correlation coefficient is calculated that reflects the relationship between these factors [24].

The resulting value is then compared to a threshold that is determined based on expert opinions. If the threshold is exceeded, it is made a conclusion about strong relationship between the *i*th and *j*th factors. As a result, the *j*th column is removed from the experimental data table.

After that, the following fuzzy products rule (FPR) is added: "IF Inp LV_j is the LV_k, then Inp LV_i is approximately LV_p", where LV is the value of linguistic significance; LD is the value of linguistic validity; "Approximately" denotes a standard modifier of a fuzzy set of degrees $\alpha = 0.5$, k and p are numbers of values of *i*th and *j*th Inp LV, for which correspondence is established based on the simultaneous appearance in the output table, which is taken into account in the calculation of f_{ij}.

6.4.2.2 Developing a Video Stream Storage and Locking System

When processing images, the doctor does not record all the videos from the ultrasound device, and manually makes several video snapshots that he finds most informative. There are several options for receiving videos from the ultrasound device:

- using a video capture card;
- ultrasound as a source that transmits video via HDMI/VGA;
- a flash card on which the video was pre-recorded;
- via Ethernet cable;
- an HDD that also includes pre-recorded video.

There is a problem with the video format when using different ultrasound devices so it is necessary to convert the video to a single format before storage. The size of the file and its quality varies depending on the format. To store video files and images, you must create directories with previous names "/train,/test, /images, /labels, /videos, /app, /trash, /save". Each of the directories is responsible for the corresponding data "/train" for the training sample "/test" for test and for further image to be processed by NN. "/labels" training sample image masks. "/videos" for videos created by the ultrasound device. "/app" the main software folder. "/trash" no video or just an image. "/Save" in this directory stores the weight of the NN after training.

In order to prepare images that will be segmented by the NN and processed by the identification system, it is necessary to convert the video image into a set of simple images.

Today, in one second, the video is from 15 to 30 frames, i.e. 2-minute video will consist of 1800–3600 images, which significantly increases the load on the system. It should be noted that not all images will be different. There is also a problem with the frames created by the ultrasound apparatus (Fig. 6.33). To identify unique images and to reduce the load on the pre-processing system, you need to get rid of these frames. First, using the OpenCV library, the video stream becomes a set of images (Fig. 6.34).

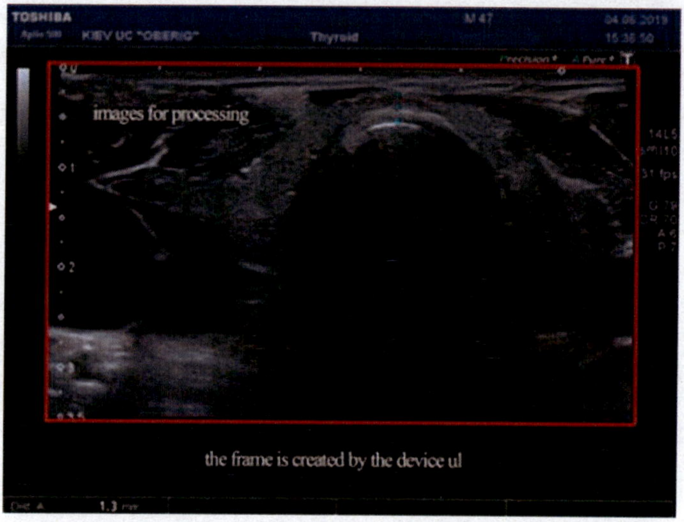

Fig. 6.33 Image taken from ultrasound device

Fig. 6.34 A set of images created from a single video file

OpenCV is a licensed open source library that includes several hundred algorithms. The main advantage is that this library is created in C++, the fastest language performance in the world. Therefore, implementing a video conversion on this library is the most optimal implementation of this system.

Depending on the configuration of the program, you can manually enter the boundaries. If it is not possible to specify precisely the boundaries, the program is configured to automatically define the frame. Automatic detection will not take much time as the algorithm will determine coordinates only once and then take it as a template.

An algorithm for converting a video image into a set of images

1. Upload a video.
2. Divide 1 s video image into frames.
3. Saving frames (15–30 pieces).
4. Move to the next time interval.
5. Items 2, 3, 4 are repeated until the video is complete.

Algorithm and definition of picture frames

1. Upload an image.
2. Determining the image size.
3. Determining the center of each side.
4. Determination of the first horizontal component.
5. Determination of the first vertical component.
6. Definition of the second horizontal component.
7. Definition of the second vertical component.
8. Determination of initial and final coordinates.
9. Delete frames in other images.

For correct processing it is necessary to implement a queue. This is explained by the fact that processing data by multiple streams, incorrect data distribution is possible and this will lead to image distortion Also, to avoid distortions and problems with further image processing, you must convert to another extension (uniform for the whole system).

To find the frames that need to be cut, the algorithm takes the image input as an array of values, not as a file. In this array, the vertical components of the previous one are compared with the next one (Fig. 6.35). If the pixel color is RGB (0,0,0) or GS (0) formats, and the next component is different from the previous one, it means that the previous values are 0, that is, black, then the coordinates are marked as initial (Fig. 6.36).

That is, from the middle of each side the arrays 1×20 are compared the previous and next values so the M1, M2, M3, M4 are determined. Knowing these points, it is simple to find the coordinates X1, Y1, X2, Y2. First, the coordinates of the centers are located, then two extreme coordinates, both in height and in width. This determines

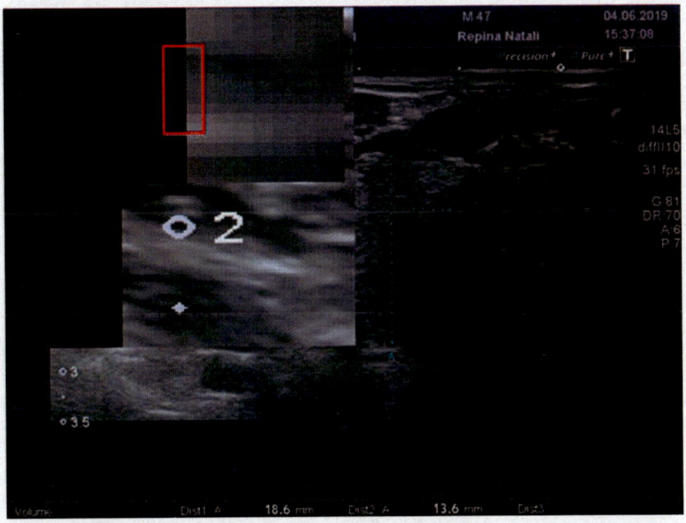

Fig. 6.35 Visualize a sharp changing values

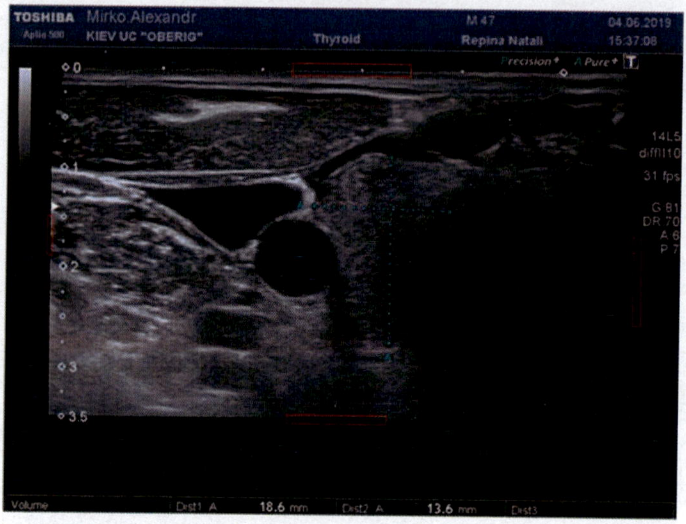

Fig. 6.36 The meaning of boundaries to define a frame

the coordinates of the beginning of the desired image, which is converted and stored without frames for further processing.

To identify images that are repeated several times or similar to the previous ones, it is necessary to compare the previous image with the next one (Fig. 6.37). It's also a very resource-intensive process, so it needs to be feasible for processing through multi-core processors.

The algorithm for finding the "personality" coefficient is as follows:

1. Uploading the 1st and 2nd images.
2. Convert the above images into arrays.
3. Determination of the array components size.
4. Elemental comparison of array values.
5. Number of identical values determinations.
6. If the number of identical values in the array is greater than the set factor, the image is deleted.
7. Repeat paragraphs 1–6 until there is an image.

Fig. 6.37 An example of the output of a personality determination program

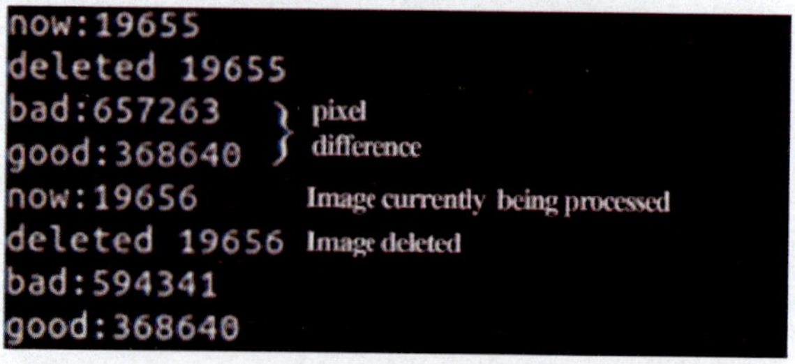

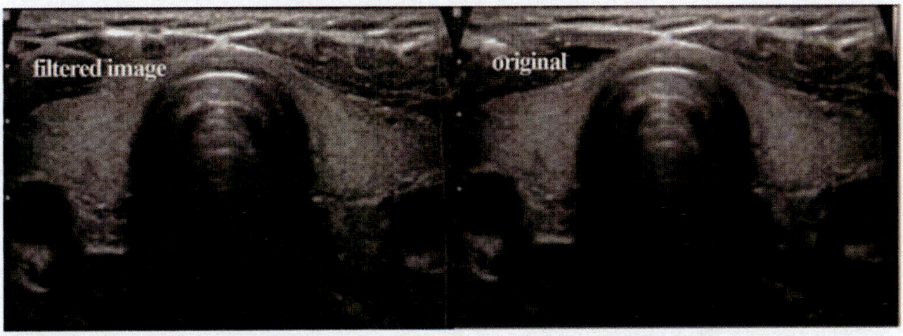

Fig. 6.38 An example of an image filtering system

Depending on the quality of the video and its volume, the "personality" coefficient is chosen in a practical way. Due to the coefficient being too large, a small number of images will be saved. In this work, the coefficient is equal to 0.35, since the obtained images are different from each other and a large number for further processing.

Algorithm for image clarity is developed using the PIL library to generate conclusions (Fig. 6.38).

This library supports many file formats and has powerful image processing capabilities.

The main advantage is that this library is designed for quick access and image processing, which is critically important to the algorithm created.

When using a different set of filters, you can get different variations of the images, but this system is configured to increase the clarity of images. By increasing clarity, the accuracy of the boundaries of the thyroid gland and the boundaries of the cystic cavities increases. The image becomes not so blurry and has clear boundaries, which positively affects the result after image segmentation.

6.4.2.3 Algorithms for Determining Diagnostically Significant Signs of Diseases

The main features of thyroid disease are the shape, boundary, echogenicity, the presence of cystic cavity and hyperechogenic inclusions. The shape of the thyroid gland can be divided into 2 types of correct and wrong (Fig. 6.39).

This feature is processed after the segmentation of the image of the NN (Fig. 6.40).

Edge detection is implemented using the Canni algorithm. This is an edge detection algorithm that uses a multi-step approach to detect a wide range of image boundaries. The process of edge detection algorithm is followed:

- apply a Gaussian filter for image smoothing and noise removal;
- image Intensity Gradient Search;
- applying maximal values suppression to eliminate the false reaction to the detection of object boundaries;

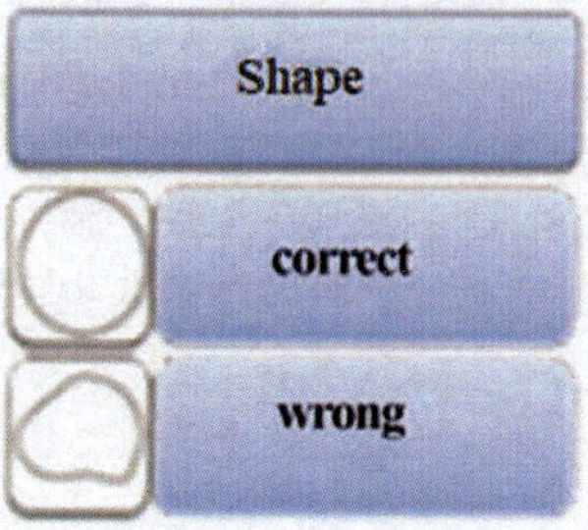

Fig. 6.39 Shape types

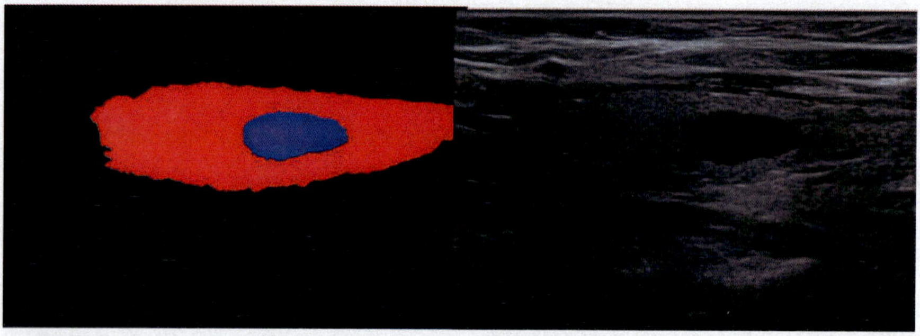

Fig. 6.40 NN image segmentation

- applying a double threshold to determine the potential boundaries of an object;
- determination of the object boundaries using hysteresis. Ending the processing by suppressing all fuzzy edges that are not associated with clear boundaries.

Because all edge detection results are easily affected to the image noises, then the erroneous detection of edges in the image is happened, so it is necessary to filter this noise, for example with help of Gaussian filter. This step slightly smoothest the image to reduce the effect of explicit noise on the edge detector. The equation for a Gaussian filter kernel of size $(2k + 1) \times (2k + 1)$ is given by:

$$H_{ij} = \frac{1}{2\pi\sigma^2}\exp\left(-\frac{(i-(k+1))^2 + (j-(k+1))^2}{2\sigma^2}\right), \quad 1 \le i,\ j \le (2k+1), \tag{6.18}$$

It is important to understand that choosing a Gaussian kernal size will affect the performance of the detector. The larger the size, the lower the sensitivity of the detector to noise. The localization error for edge detection will increase slightly with the increase of the Gaussian filter kernel size. The kernel size 5×5 is the optimal size for most cases.

Edges in the image can be in different directions, so the Canni algorithm uses four filters to determine the horizontal, vertical and diagonal edges of the blurred image. The edge detection operator returns the value for the first derivative in the horizontal direction (G_x) and vertical direction (G_y). From this you can determine the gradient and direction of the edge:

$$\begin{aligned} G &= \sqrt{G_x^2 + G_y^2}, \\ \Theta &= a\tan(G_y, G)_x, \end{aligned} \tag{6.19}$$

where G can be calculated using the hypothetical function and atan 2 is the two-argument arctangent function. The edge angle is rounded to one of four angles representing vertical, horizontal and two diagonals (0°, 45°, 90° and 135°). Hypot is a mathematical function defined to calculate the length of a hypotenuse of a right triangle. It was designed to avoid errors caused by calculations on computers with limited accuracy.

Not the maximum suppression is used to determine the "largest" edge. After applying the gradient calculation, the edge remains sufficiently blurred. Thus, not maximal suppression can help suppress all gradient values setting them to 0, except for local maxima, which indicate the places with the sharpest change in intensity value. The algorithm for each pixel of the gradient image is as follow:

1. Comparison of the edge values of the current pixel with the value of the edge of the pixel in the positive and negative directions of the gradient.
2. If the edge value of the current pixel is the greatest than other pixels in the mask with the same direction (for example, a pixel pointing in the **y** direction will be compared to a pixel above and below it in the vertical axis), the value will be saved. Otherwise, the value will be suppressed.

In some implementations, the algorithm classifies continuous gradient directions into a small set of discrete directions, and then moves the 3 × 3 filter over the output of the previous step. On each pixel, it suppresses the value of the center pixel (by setting its value to 0) if its value does not exceed the magnitude of the two neighbors in the gradient direction.

The application of nonmaximum suppression provides a more accurate representation of the real edges of the image. However, some erroneous edge pixels remain, which are caused by noise and discoloration. To take into account these errors, it is necessary to filter the edge pixels with a weak gradient value and store the pixels of the edges with a high gradient value. This is achieved by high and low thresholds selection. If the boundary pixel gradient value is higher than the high threshold, it is denoted as a strong boundary pixel. If the boundary pixel gradient value is less than the high threshold and greater than the low threshold, it is denoted as a weak boundary pixel. If the pixel value of the edge is less than the low threshold, it will be suppressed. Two limit values are determined empirically, and their determination will depend on the content of the given input image (Fig. 6.41).

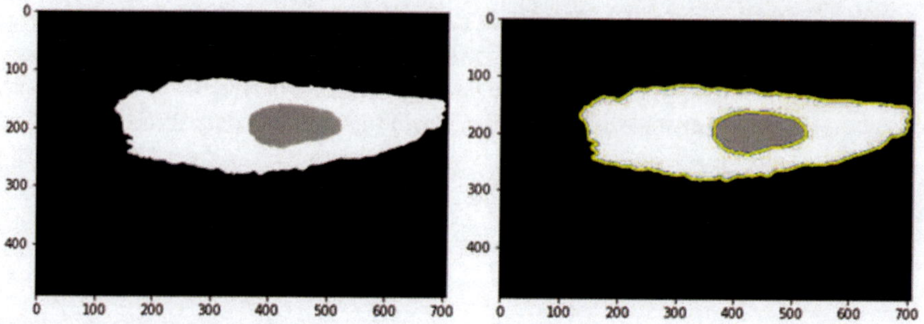

Fig. 6.41 Determined edge of thyroid

If the derivative of given function changes the sign where in on a specific diapason that it's changed the sign very often it means that the shape is wrong (Figs. 6.42 and 6.43).

Cystic cavities may be absent, small, irregular in shape, regular in shape and set of small ones. To determine the presence of cystic cavities (Fig. 6.44), similar to the previous algorithm, it is not determined the difference between the no background and the object but the difference between the object and the selected area inside the object (marked in blue).

An algorithm for determining cystic cavities is follows [16]:

1. Convert a segmented image into array.
2. Selection of coordinates obtained by the previous algorithm.
3. Compare the vertical components and determine the coordinates of the difference in the values of the array (color difference) in the middle of the object from step 2.

Fig. 6.42 The example of the wrong shape

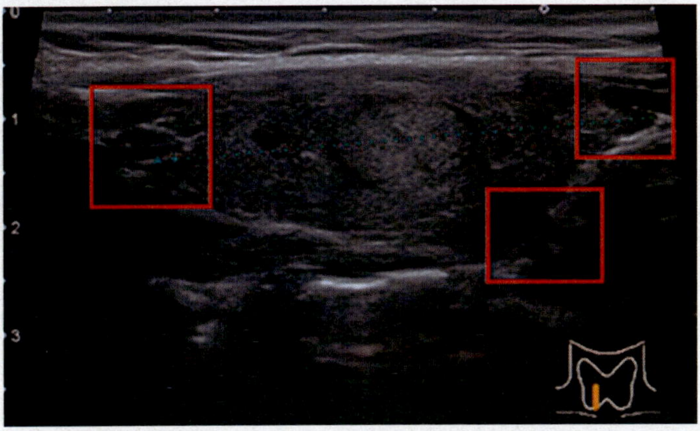

Fig. 6.43 An irregular example

Fig. 6.44 The types of cystic caves

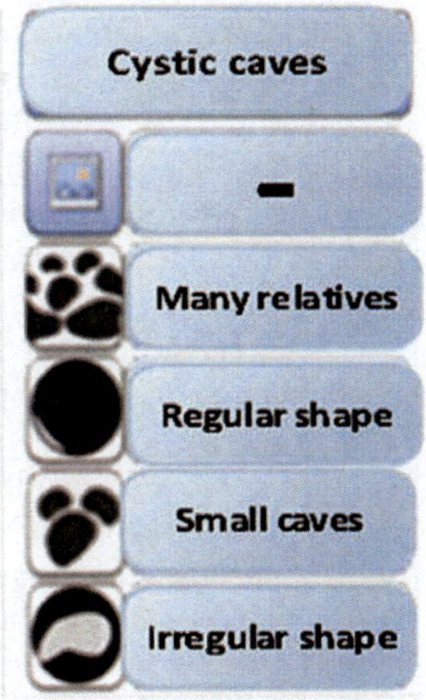

4. If all values in the middle of the coordinates from step 2 are identical then the cystic cavities are absent.
5. If under comparing vertical components it is determined from 5 or more objects that have between themselves a certain distance (50–150 pixels—up to about 3 cm), then it is a set of small cystic cavities.
6. If there are up from 5 to 2 objects that have a certain distance between themselves, then they are small cavities.
7. If 1 object is determined, then its shape is determined.
8. Approximation of obtained values is from step 7.
9. Determination of the sign of the derivative at each point of the function.
10. Determination of a number of sign change on 50 pixels of the image.

Fig. 6.45 The example of determination of the center of an object

11. If the sign has changed more than 3 times the definition of the shape is as incorrect.

The system creates an array of coordinates and determines the shape of the cystic cavity. To determine the size the algorithm takes two opposite coordinates and calculates the distance between them. The size is determined by the maximum length.

To determine the opposite coordinates by Bertram Brockhouse method, the cystic cavity is considered as an object of square or rectangular shape (Fig. 6.45), the points belonging to the boundary of the region and being to the left and above the others create one angle and similarly, the opposite points belonging to the boundary of the region create another angle. This allows you to determine the center of the area, and through the center and the extreme left or upper points to draw straight lines and find the coordinates of opposite points, which makes it possible to determine the size of the cystic cavity.

If there are several cavities, each is processed separately and the shortest distance between them is additionally found. Between the coordinates of one of the cavities and with the other, the distance is determined element-wise.

When a new smaller distance is found, the points of the boundary are fixed as well as the distance between them. This happens until a minimum distance is found. This procedure is necessary to clarify the sign of "multiple small cystic caves". All distances that are small and the number of multiple cavities that will be considered as a set are preset by the doctor in the graphic interface (3 cm). In the absence of caves, the algorithm will skip all the processing indicated above and immediately proceed to the result (Fig. 6.46).

The presence of hyperechogenic inclusions (Fig. 6.47) in cystic cavities are quite similar to the algorithm described above, but to determine the type we should find the size vertically and horizontally. If they coincide (usually in some range) then these points are inclusions. If the dimensions do not exactly match, then the type of inclusion will be linear. At several inclusions each is processed separately.

Echogenicity can be significantly reduced, moderately reduced, unevenly reduced and isoechogenic (Fig. 6.48).

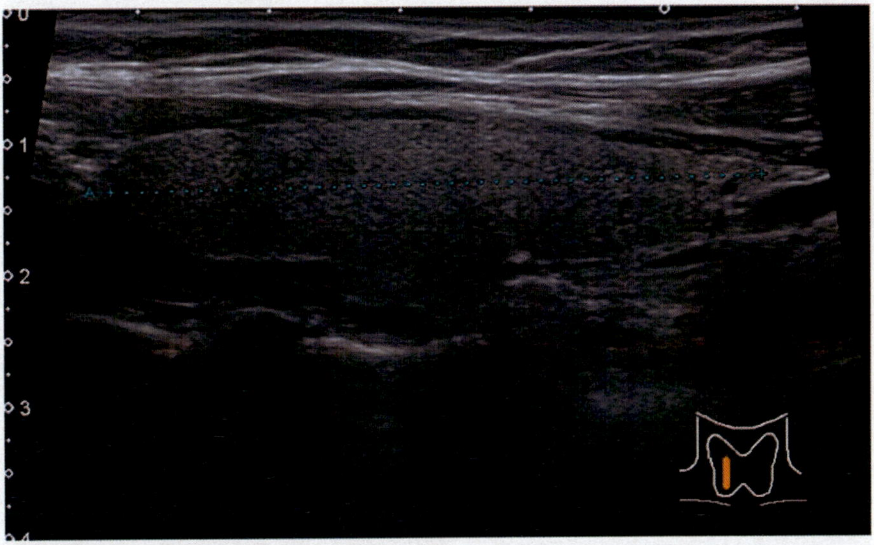

Fig. 6.46 No cystic cavities

Fig. 6.47 Types of hyperechogenic inclusions

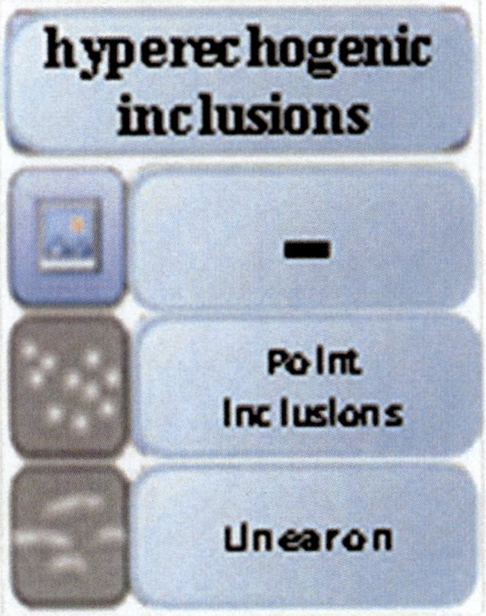

Determination of echogenicity is possible if all possible cystic cavities are excluded from the study area as they distort the result (Fig. 6.49).

Echogenicity can be determined by taking the mean value (Fig. 6.49) in an area that is bounded by a form, but then its irregularity cannot be determined. Echogenicity is determined with help of algorithm for finding the median.

The algorithm for finding the median is as follow:

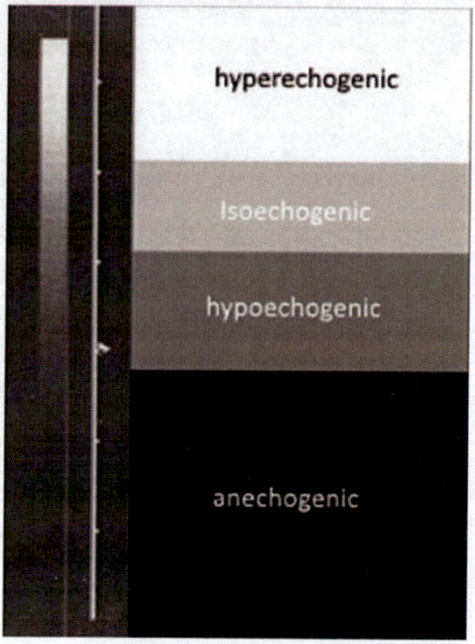

Fig. 6.48 Types of echogenicity

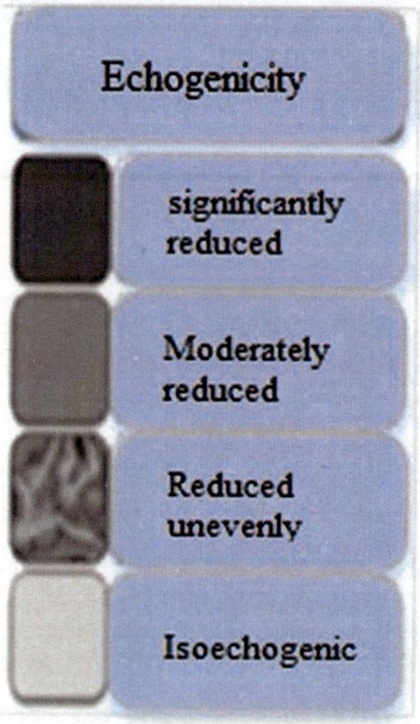

Fig. 6.49 The graduation of echogenicity

1. Determination of the plot size for calculating averages.
2. Conversion an area to an array of plots from step 1.
3. Calculating the average value of the plot from step 1 and comparing this value with the adjacent ones.
4. Depending on the established coefficients for determining the result.

To determine echogenicity, it is necessary to determine the average value in small areas, such as 2 × 2, 4 × 4 or 10 × 10, the smaller the plot, the more accurate the output (in this 4 × 4 system, due to hardware). As the size of the plot decreases, the accuracy of the result increases, but processing time also increases.

The value of echogenicity:

- hyperechoic (255–181);
- hypoechoic (120–51);
- isoechogenic (180–121);
- anechogenic (50–0);
- for an evenly reduced combination of these values.

An algorithm for determining echogenicity.

1. Exclusion of coordinates of cystic cavities from the field of determination of echogenicity.
2. Determining the size of the plot for calculating averages.
3. Imposition of plots from item 2 on the area of computation (the area is limited by the thyroid form with removed areas of cystic cavities).
4. Calculate the average of the plot from item 2 and compare this value with the adjacent ones.
5. If the value is the same (expiration is ± 5%) then the average value for the whole area is determined and the result is determined depending on the boundaries within which it is located.
6. If the values are very different then it is unevenly reduced echogenicity.

An example of determining echogenicity is shown in Fig. 6.50.

The clarity of the boundaries (Fig. 6.51) can be divided into clear and fuzzy and, depending on the type, into subtypes (Fig. 6.52).

The value of echogenicity for networks:

- hyperechoic (255–181);
- hypoechoic/hydrophilic (120–51);
- isoechogenic (180–121);
- anechogenic/hydrophilic with thicknesses (50–0).

Using the algorithm of median determination you can determine the average values for a given curve. In determining the shape of the coordinates obtained boundaries algorithm compares the average values of sections 4 × 4 (Fig. 6.53). If the values are very different (possible exhalation ±5%) the limit is fuzzy (Fig. 6.54).

When defining a shape by getting the boundaries between the background and the object, the bounding system compares the pixel values of the previous one with

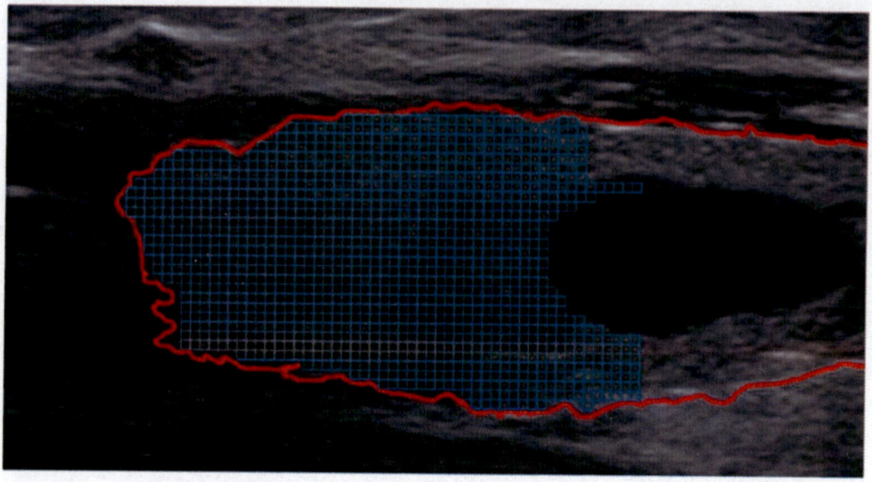

Fig. 6.50 The echogenicity determination

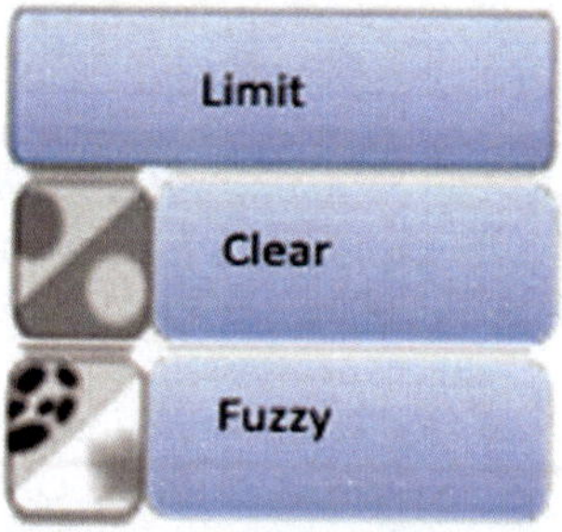

Fig. 6.51 The types of boundaries

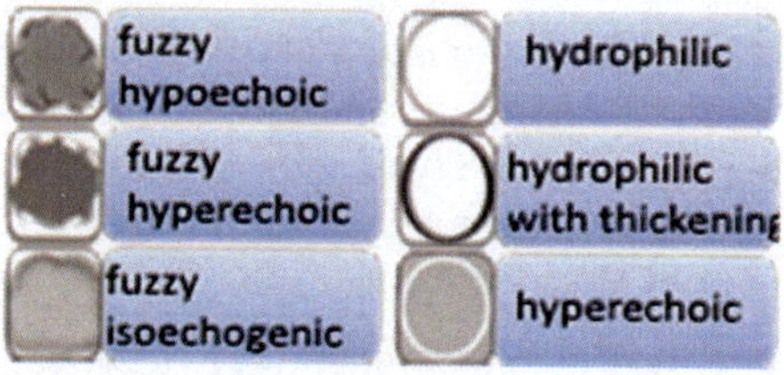

Fig. 6.52 Boundaries subtypes

the next one (Fig. 6.53). If the pixel values of the boundaries are very different, the boundary is considered fuzzy.

If all the pixels are about the same, then the boundary is clear. To determine the subspecies of boundaries, the algorithm similar to the echogenicity calculation algorithm is used for the average value calculation pixels of the boundary pixels. Depending on this average value, the subtypes is determined.

Boundary type determination algorithm is as follows:

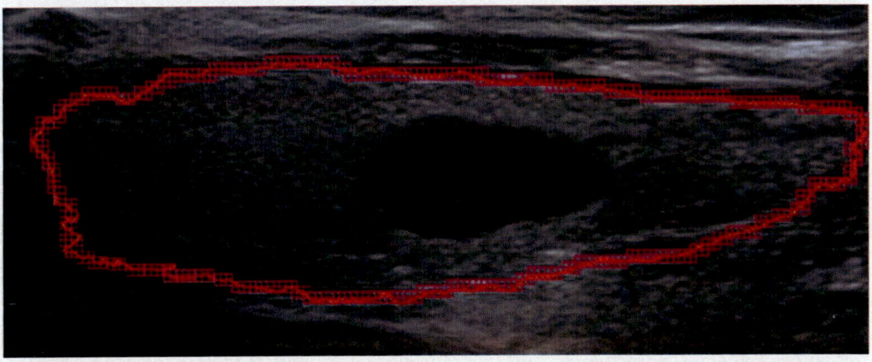

Fig. 6.53 The structure of the video processing system

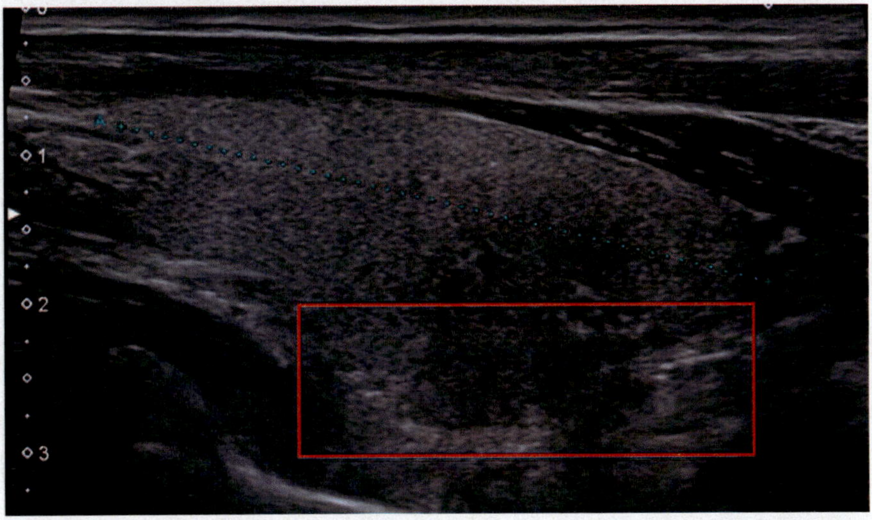

Fig. 6.54 An example of a fuzzy boundary

1. Determination of boundaries coordinates.
2. Overlapping plots (4×4) along the entire boundary of the object.
3. Comparison of plots values from step 2 among themselves.
4. At identical values (expiration $\pm$ 5% possible) the limit is clear.

The algorithm for determining the subtype of boundaries is as follows [14]:

1. Similar to the previous algorithm, steps 1.2.
2. Calculation of the average value for each plot.
3. Determining the average value for the whole boundary.
4. Depending on the average value from step 3 and the range to which it belongs, subtypes of boundaries are determined.

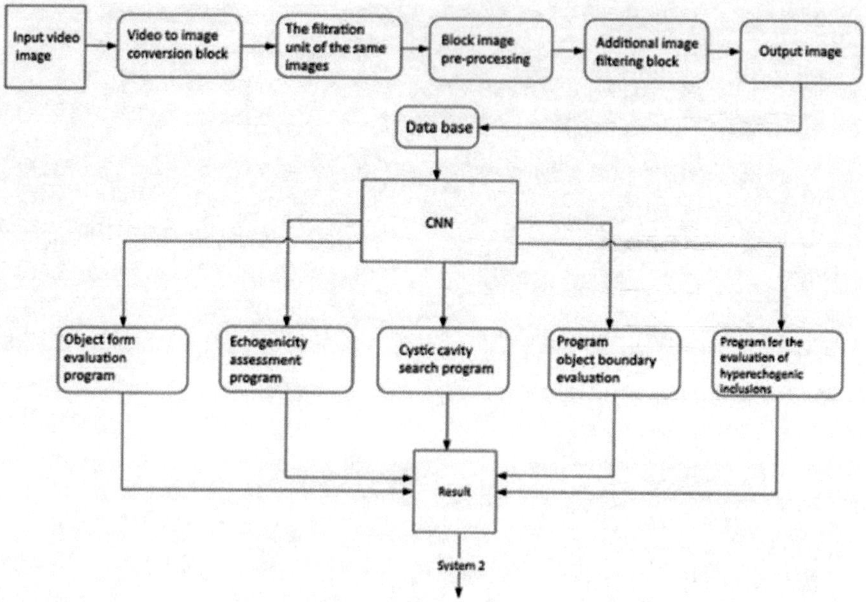

Fig. 6.55 The structure of the video processing system

The average value is compared with that doctor set as well as in the determination of echogenicity. Based on the above image processing subsystem (Fig. 6.11) it is possible to present in full the following way (Fig. 6.45).

All results are transmitted on the basis of the decision support subsystem (Fig. 6.11). For each of the signs of thyroid disease separately, a corresponding value is transmitted to the decision support subsystem. The sign of the shape has 2 values, boundaries 2 and 3 subtypes, echogenicity 4, cystic cavities 5, inclusion 3 (Fig. 6.55).

6.5 Obtaining a Training Sample

Pre-processing the Image

In accordance with Sect. 4.2.4, we consider an example of obtaining a training sample. Due to the fact that the image comes from the ultrasound device in the form of a video sequence and is not sufficiently contrasted, a computer image processing program [25] automatically increases the contrast to the desired level (referring to the echogenicity scale of Fig. 6.49) [26].

An example of image pre-processing (Sect. 6.6.4) is shown in Fig. 6.56.

According to Fig. 4.35 (block diagram of the classifying images algorithm at noisy picture) the next step in building a training sample for deep learning networks processing unstructed images is the operation of contours highlightening.

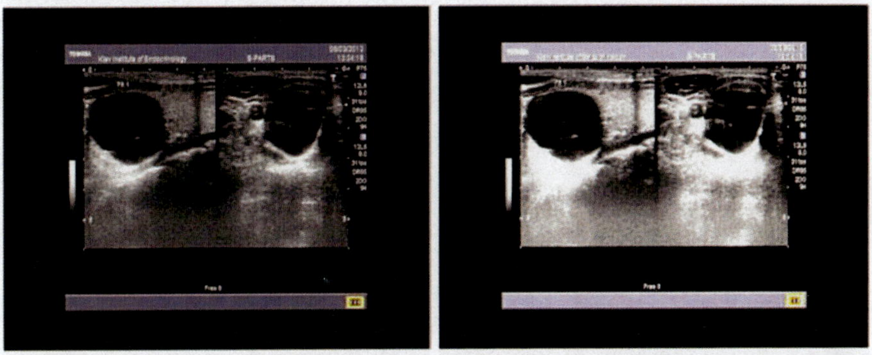

Fig. 6.56 The image pre-processing

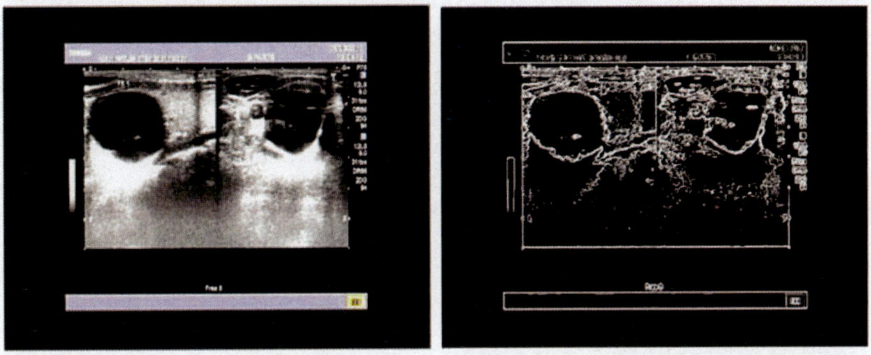

Fig. 6.57 The countour highlightening

Contour Highlightening [12]

An example of contour selection is shown in Fig. 6.57, where the highlighted contours are shown in the right image.

Highlightening of suspicious areas is performed using a special algorithm shown in Fig. 6.58 The suspicious outline is highlighted in red [13]. The problem of discontinuities and irrelevant data inside the circuit is solved.

Obtaining Autocorrelation Functions

Obtaining autocorrelation functions (ACF) [14] corresponding to certain contour shapes is shown in Fig. 6.59.

The results of patient ultrasound scan are shown in Fig. 6.60.

After image processing, four contours were obtained. One of them managed to classify. The neoplasms are isoechogenic, as the middle tone is in the appropriate place on the echogenicity scale. The echostructure is heterogeneous. There are cystic lesions.

Based on the correspondence of a certain pattern of ACF, it was concluded about the “wrong” shape of neoplasm shown in Fig. 6.61.

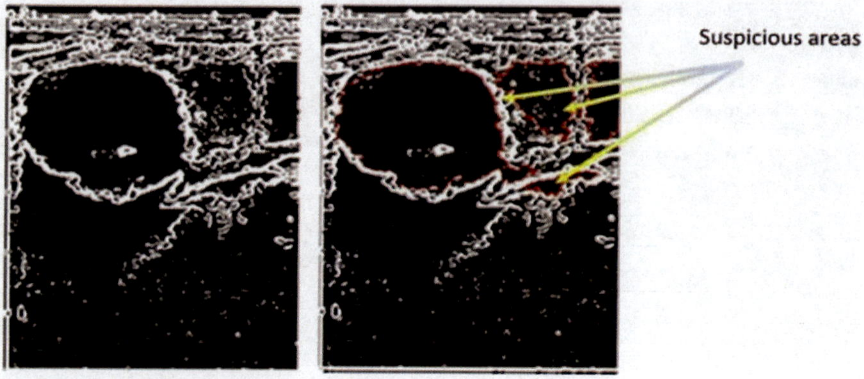

Fig. 6.58 Highlightening of suspicious sites

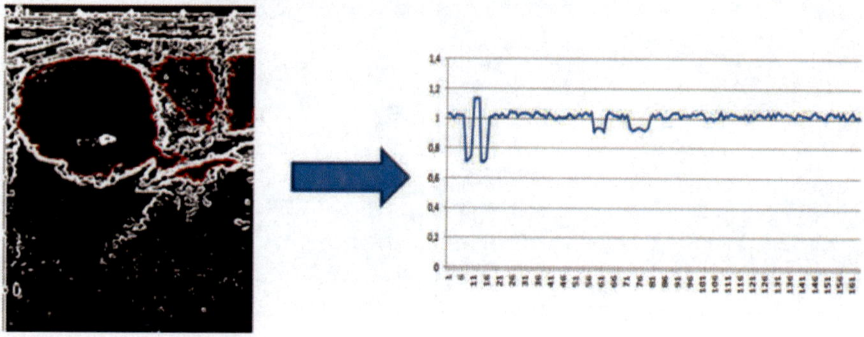

Fig. 6.59 Obtaining autocorrelation functions

Based on the analysis of the contour neighborhood, it was concluded that the contour has clear boundaries. Using the second projection and the Brunn method, the volume of neoplasm was determined: 25.86.

In addition, the following data were entered into the neural network: age: 33 years; gender: female. Diagnosis at the output of the neural network: papillary carcinoma.

The results of ultrasound of the patient BV shown in Fig. 6.62.

After image processing, three contours were obtained. One of them was classified.

The neoplasm is hypoechoic because the middle tone is in the appropriate place on the echogenicity scale. The echostructure is heterogeneous. There are hyperechogenic calcifications.

Based on the correspondence of a certain pattern of ACF, it was concluded that the "wrong" shape of neoplasm shown in Fig. 6.63.

Based on the analysis of the contour neighborhood, it was concluded that the contour is without clear boundaries. Using the second projection and the Brunn method, the volume of the neoplasm was determined: 11.03.

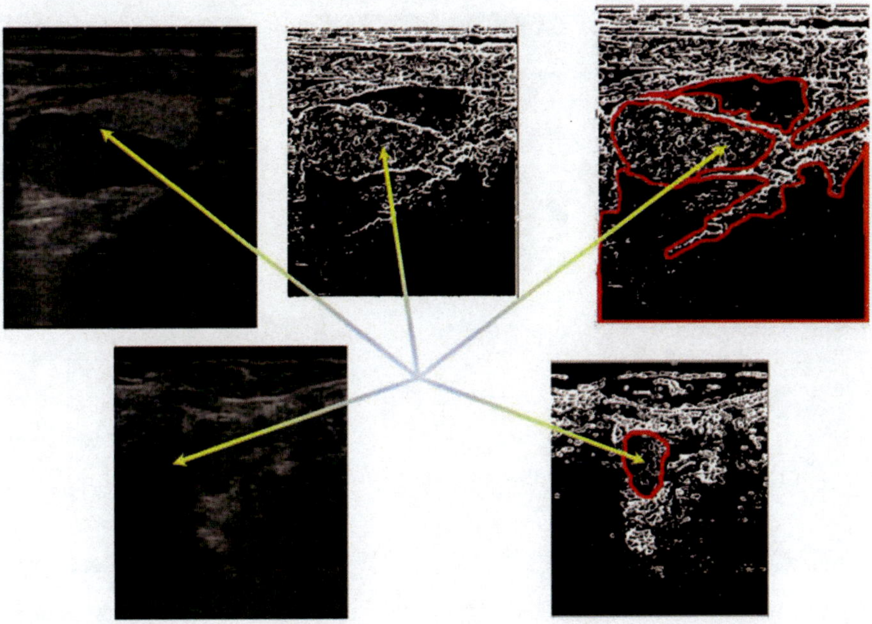

Fig. 6.60 Contours of patient ultrasound scan results

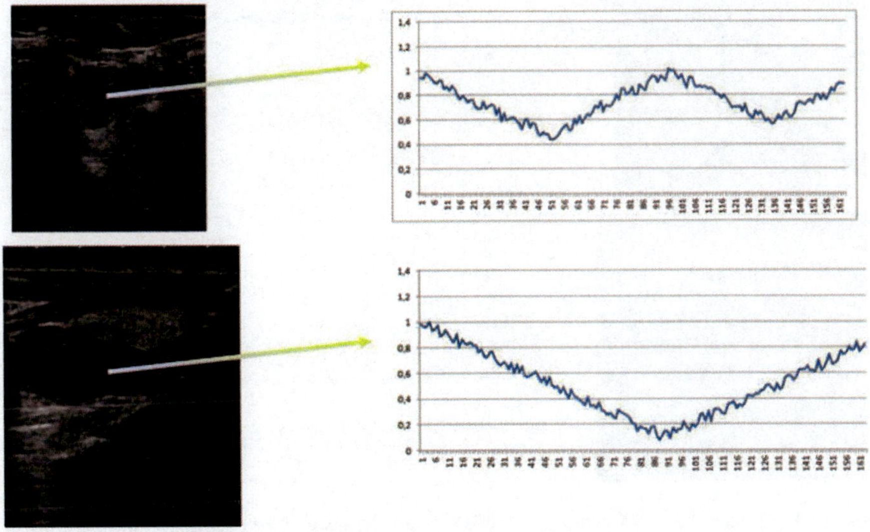

Fig. 6.61 The conclusion about the "wrong" form of neoplasm of the patient's ultrasound scan results

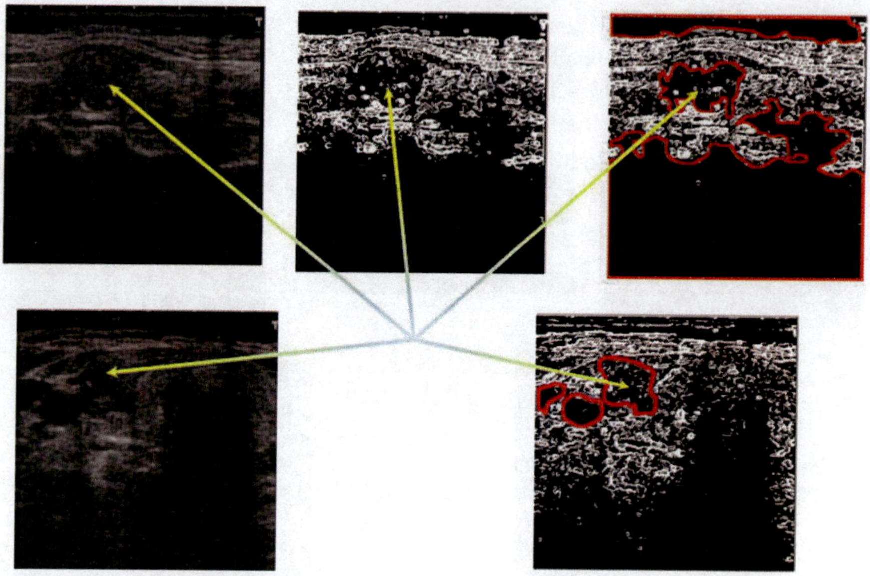

Fig. 6.62 Outlines of patient unltasound results

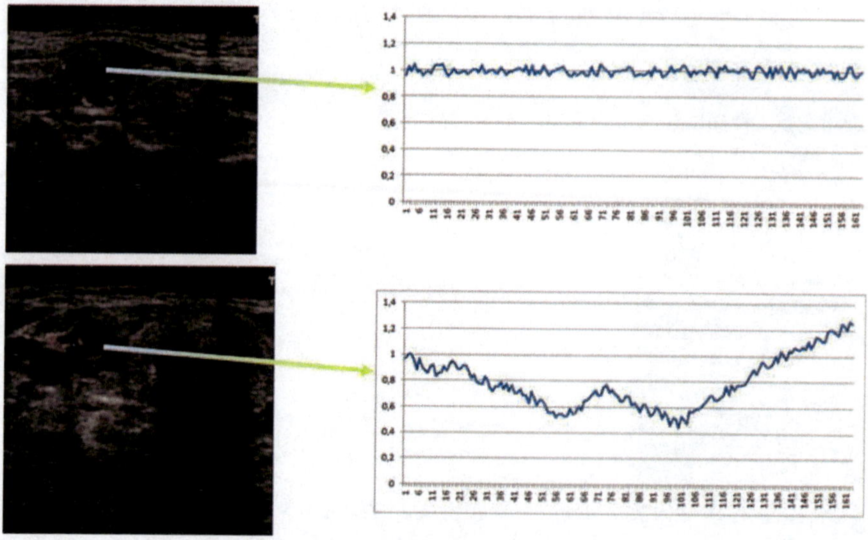

Fig. 6.63 The conclusion about the "wrong" shape of neoplasm of the patient's by ultrasound scan results

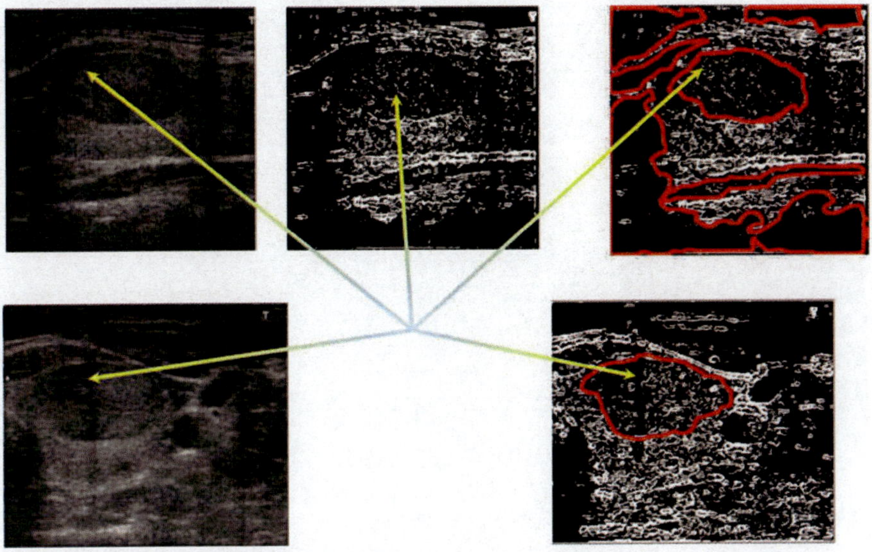

Fig. 6.64 Contours of patient ultrasound scan results

Additionally, the following data were entered into the neural network: age: 58 years; gender: female. Neural Network Output Diagnosis: Papillary Carcinoma.

The results of ultrasound of the patient BV are shown in Fig. 6.64.

After image processing, 6 contours were obtained. One of them was classified.

The neoplasm is hypoechoic because the middle tone is in the appropriate place on the echogenicity scale. The echostructure is homogeneous.

Based on the correspondence of a certain pattern of ACF, it was concluded that the "correct" shape of neoplasm shown in Fig. 6.65.

Based on the analysis of the contour neighborhood, it was concluded that the contour has clear boundaries. Using the second projection and the Brunn method, the volume of neoplasm was determined: 22.18 [6].

Additionally, the following data were entered into the neural network: age: 34 years; gender: female. Neural Network Output Diagnosis: A-Cell Tumor (Follicular Adenocarcinoma).

The results of ultrasound scan of the patient are shown in Fig. 6.66.

After image processing, four contours were obtained. One of them was classified.

The neoplasm is hypoechoic because the middle tone is in the appropriate place on the echogenicity scale. Echostructure is heterogeneous, there are hydrophilic formations.

On the basis of conformity of a certain pattern of ACF it was concluded that the "correct" shape of neoplasm, shown in Figs. 6.66 and 6.67.

Based on the analysis of the contour neighborhood, it was concluded that the contour has clear hydrophilic boundaries. Using the second projection and the Brunn method, the volume of neoplasm was determined: 14.64.

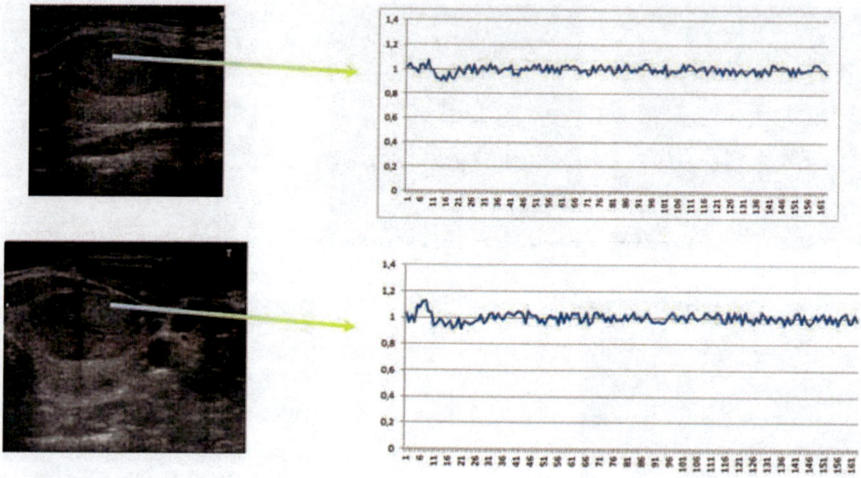

Fig. 6.65 The conclusion about the "wrong" shape of neoplasm results of ultrasound scan of the patient BV

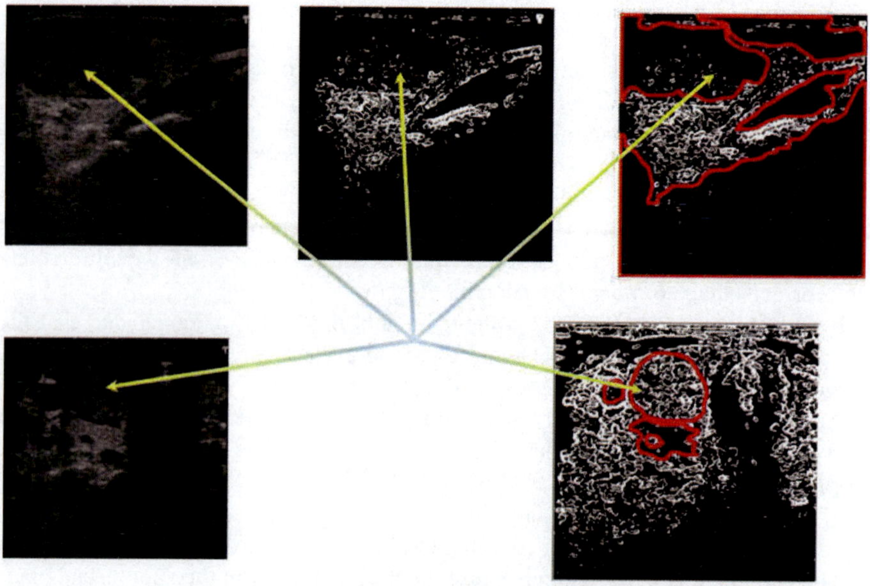

Fig. 6.66 Contours of the results of ultrasound scan of the patient

Additionally, the following data were entered into the neural network: age: 25 years; gender: female. Diagnosis at the output of the neural network: nodal adenomatous goiter.

The ultrasound scan results of the patient are shown in Fig. 6.68.

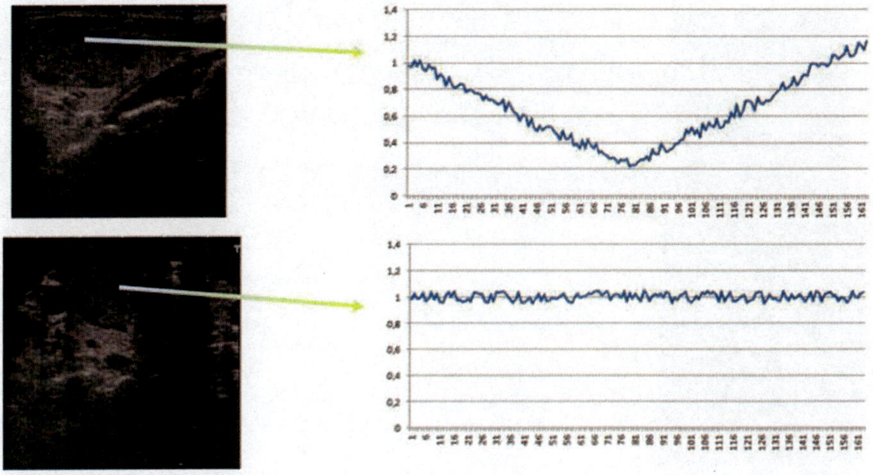

Fig. 6.67 The conclusion about the "correct" shape of neoplasm results of ultrasound scan of the patient

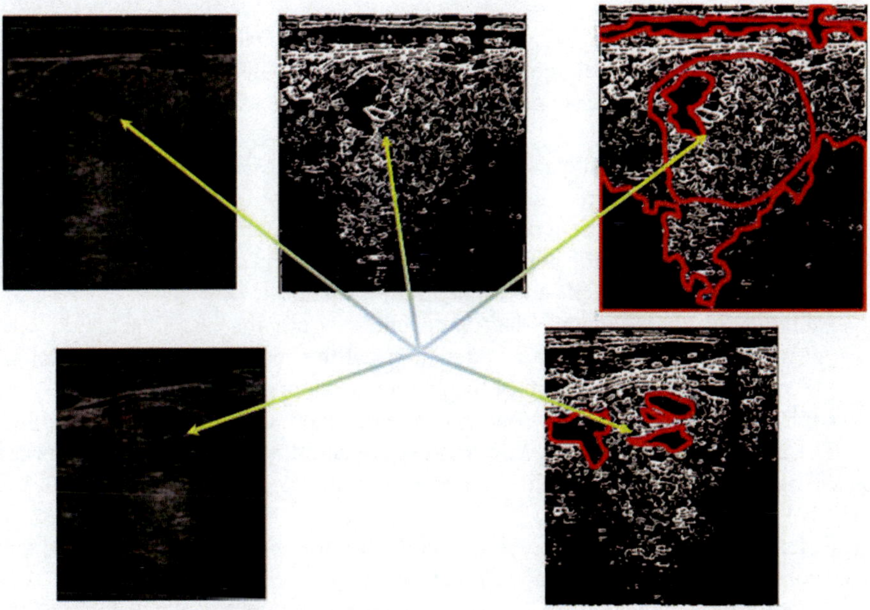

Fig. 6.68 Contours of ultrasound scan results of patient

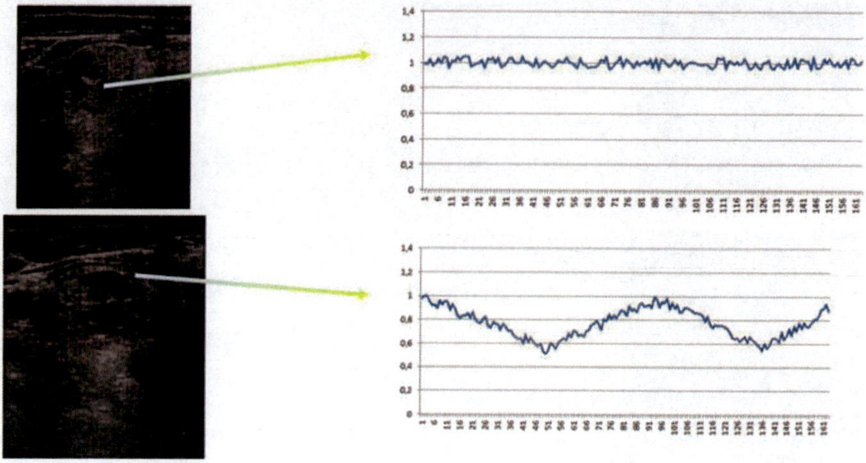

Fig. 6.69 The conclusion about the "correct" shape of neoplasm results of ultrasound scan of the patient

After image processing, four contours were obtained. One of them was classified.

The neoplasms are isoechogenic, as the middle tone is in the appropriate place on the echogenicity scale. The echostructure is heterogeneous. There are massive cystic caves.

Based on the correspondence of a certain pattern of ACF, it was concluded that the "correct" shape of neoplasm shown in Fig. 6.69.

Based on the analysis of the contour neighborhood, it was concluded that the contour has clear hydrophilic boundaries. Using the second projection and the Brunn method, the volume of neoplasm was determined: 43.57.

Additionally, the following data were entered into the neural network: age: 53 years; gender: male. Diagnosis at the output of the neural network: nodal goiter.

The results of ultrasound scan of the patient are shown in Fig. 6.70.

After image processing, four contours were obtained. One of them was classified.

The neoplasms are isoechogenic because the middle tone is in the appropriate place on the echogenicity scale. The echostructure is heterogeneous. There are massive cystic caves.

Based on the correspondence of a certain pattern of ACF, it was concluded that the "correct" shape of neoplasm shown in Fig. 6.71.

Based on the analysis of the contour neighborhood, it was concluded that the contour has clear hydrophilic boundaries. Using the second projection and the Brunn method, the volume of neoplasm was determined: 17.19.

The following data were entered into the neural network: age: 52 years; gender: female. Diagnosis at the output of the neural network: nodal goiter.

The results of ultrasound of the patient are shown in Fig. 6.72.

After processing the image, the 4 contours were removed. One of them was classified.

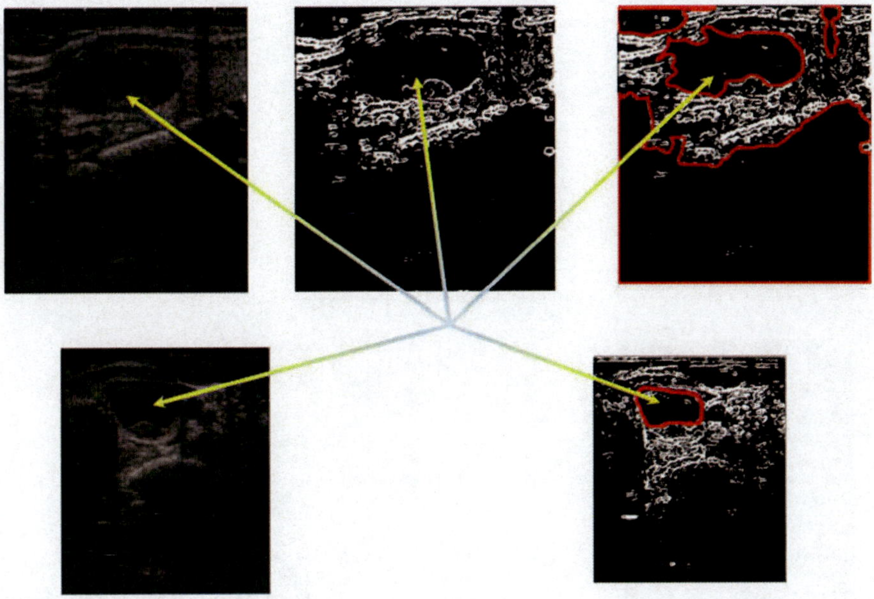

Fig. 6.70 Contours of ultrasound scan results of patient

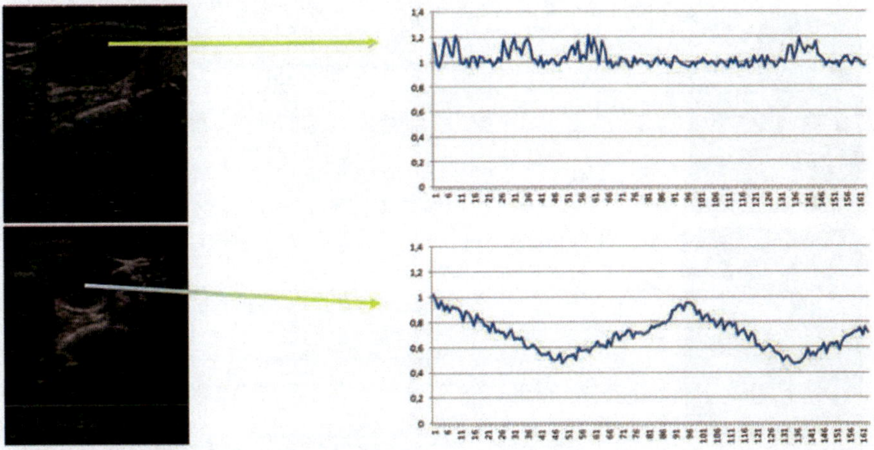

Fig. 6.71 The conclusion about the "correct" shape of neoplasm of the results of ultrasound scan of the patient

The new isoechogenic, so as the middle tone can be found at the mainstream on the exogenous scale. The echogenicity is not uniform. There are massive cystic caves.

Based on the conformity of a certain pattern of ACF, a conclusion was drawn about the "correct" shape of neoplasm, shown in Fig. 6.73.

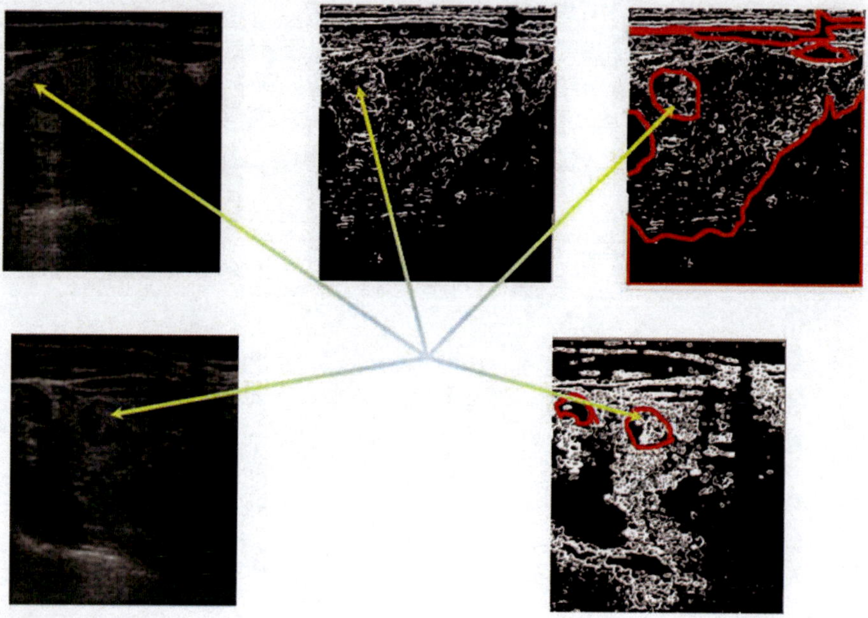

Fig. 6.72 Contours of ultrasound results of the patient

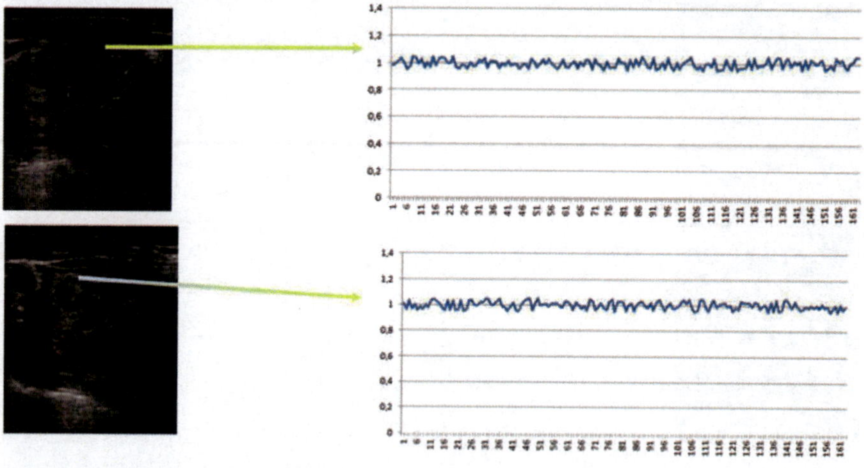

Fig. 6.73 Conclusion about the "right" shape of new creation of the results of the ultrasound examination of the patient

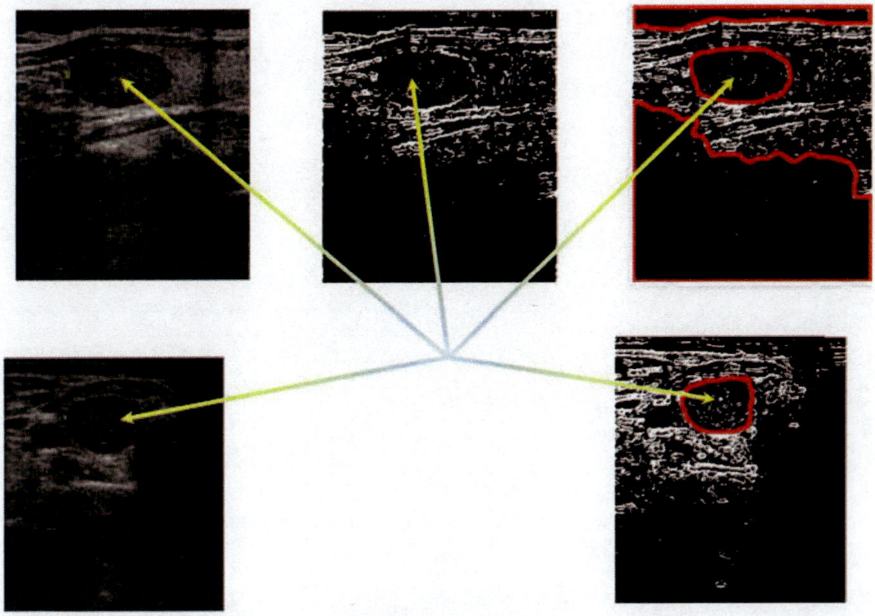

Fig. 6.74 Contours of results of ultrasound of the patient

Based on the analysis of the contour neighborhood, it was concluded that the contour has clear boundaries. Using the second projection and the Brunn method, the volume of neoplasm was determined: 28.87.

Additional in the neural measure I have boules introduced such data: age: 51; gender: male. Diagnosis of the incidence of neuronal measures: nomofollicularadenoma (rightly papillary carcinoma).

The results of the ultrasound scan of the patient PC are depicted on Fig. 6.74.

After processing the image, three contours were shaded. One of them was classified.

The new izoechogenny, so as the middle tone can be found at the mainstream on the exogenous scale. The echo structure is not uniform. There are hydrophilic formations.

Based on the correspondence of a certain pattern of ACF, the conclusion was made about the "correct" shape of neoplasm, shown in Fig. 6.75.

Based on the analysis of the contour neighborhood, it was concluded that the contour has clear hydrophilic boundaries. Using the second projection and the Brunn method, the volume of neoplasm was determined: 13.69.

Additional in the neural measure I have boules introduced so far: age: 44 rock; gender: female. Diagnosis of neuronal outcome: an adenomatous nodular goiter (actually cystadenoma).

Results of ultrasound of the patient AV image on Fig. 6.76.

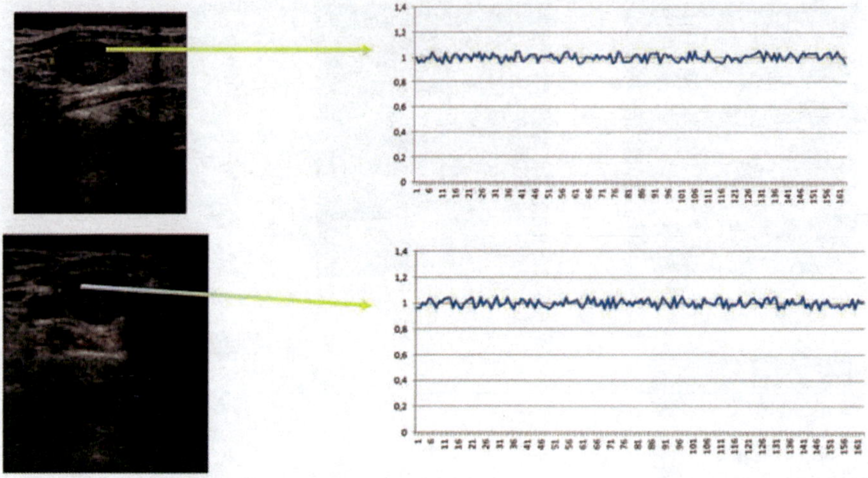

Fig. 6.75 Conclusion about the “right” shape of new creation of results of ultrasound examination of the patient

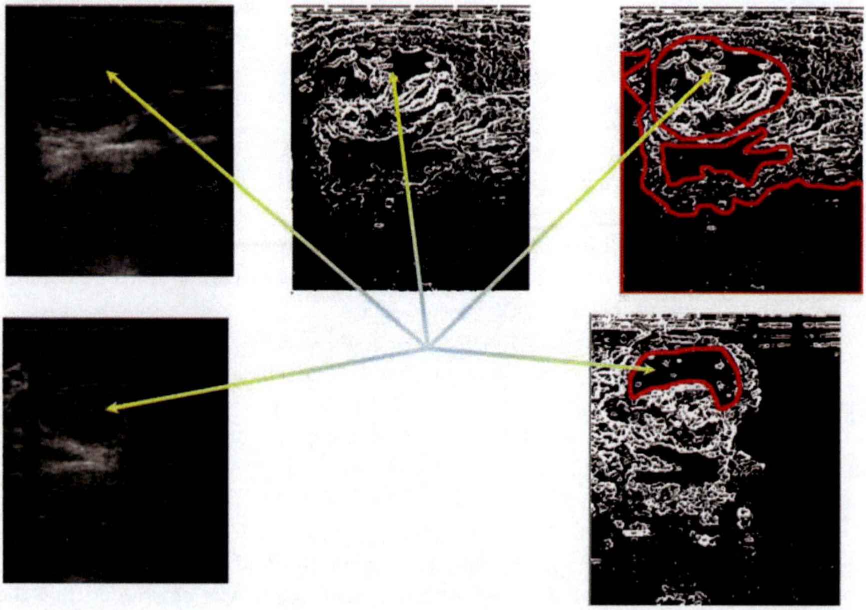

Fig. 6.76 Contours of results in ultrasound

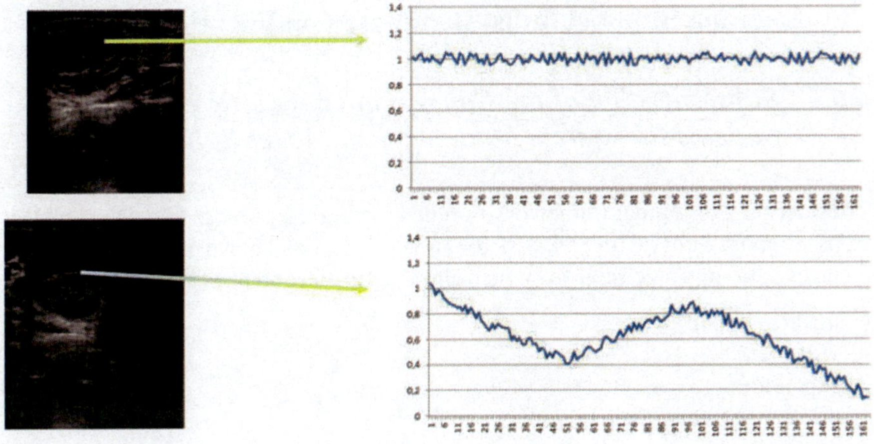

Fig. 6.77 Conclusion about the "right" shape of new creation of the results of the ultrasound examination of the patient

After processing the image, three contours were shaded. One of them was classified.

Neoplasms are isoechogenic, so as the middle tone can be found at the mainstream on the exogenous scale. The echo structure is not uniform. There are massive cystic caves.

Based on the correspondence of a certain pattern of ACF, the conclusion was made about the "correct" shape of neoplasm, shown in Fig. 6.77.

Based on the analysis of the contour neighborhood, it was concluded that the contour has clear hydrophilic boundaries. Using the second projection and the Brunn method, the volume of neoplasm was determined: 15.51.

Additionally, the following data were entered into the neural network: age: 13 years; gender: female. Diagnosis at the output of the neural network: nodal goiter.

In accordance with the structural diagram of the intellectual diagnostic system after isolation of anomalous areas as a result of video processing of ultrasound images, it is necessary to evaluate their parameters, namely: type of rim of tumor, inclusion, structure of tumor, size and echogenicity of tumor. The type of rim, the structure of the neoplasm and the presence of inclusions are determined by constructing autocorrelation functions of fragments of the anomalous region with subsequent comparison with the reference correlation functions corresponding to the possible types of these parameters, for example, for inclusions these are the following options: no, point inclusions, linear inclusions, "tail comets". The echogenicity of the neoplasm is determined by the use of the Piton Image Library (PIL).

6.6 Decision Support Subsystem Based on Fuzzy Inference

6.6.1 *Substantiation of the Concept of Diagnostic Decisions Support Subsystem Mathematical Model Building*

Consider the conceptual framework for constructing the proposed Diagnostic Decisions Support Subsystem (DDSS) for medical. It was shown in Sect. 6.3 that it receives data for processing from the following main sources of information:

- medical history;
- video;
- expert knowledge;
- printed material.

The last two of these are used for parameter system adjustment. The first two sources application is carried out in two stages. Initially, a preliminary diagnostics of the disease based on medical history. Then, depending on the results of this research, it is made a plan for further examination, in which, if it is necessary, includes ultrasound, MRI, CT as one of the sections of this plan. Their results in combination with the accompanying methods of diagnostics have the greatest in formativeness in the detection of pathologies of organs.

Working with medical data involves the use of both quantitative and qualitative values. The need for this is determined by the specifics of the thinking process of the physician-diagnostician, as well as different expert opinions about the same data. In order to reduce the likelihood of misdiagnosis, it is proposed to use a modern soft mathematical apparatus to process such heterogeneous data. It includes fuzzy logic (FL), artificial neural networks (ANN), probabilistic reasoning, and more. Probabilistic considerations relate solely to the future and lose their meaning when a random event occurs (or does not occur) at a certain point in time, so it will not be applied in the developed system. The proposed diagnostic system has the ability to evaluate the patient's current state, taking into account the analysis of his past condition, determined from the medical history. On this basis, the following components of soft calculations are used.

It is suggested to use fuzzy logic apparatus to work with qualitative data. This apparatus is a system of fuzzy inference—a system of logical inference, based on the algorithm of obtaining fuzzy conclusions based on fuzzy preconditions using the basic concepts of fuzzy logic. The fuzzy inference process combines all the basic concepts of fuzzy set theory: membership functions, linguistic variables, fuzzy logic operations, fuzzy implication methods, and fuzzy composition. Fuzzy output systems allow you to solve the problems of automatic control, data classification, pattern recognition, decision making, machine learning, and more.

Fuzzy output systems are designed to convert the values of input process variables into output variables based on the use of fuzzy production rules. For this purpose, fuzzy inference systems should include a fuzzy product policy framework and implement fuzzy inference based on the assumptions or conditions presented in the form of fuzzy linguistic statements.

Based on the above, we can conclude that these systems are widely used in medical diagnostics because they have several advantages:

1. Can process linguistic variables (LV).
2. The output algorithm is transparent for analysis, and it can be controlled simply by changing the fuzzy product rules (FPR) system.
3. It is not necessary to know the model of functioning of the object of research—it is enough only to describe it with the help of the FPR system, which will be its linguistic model.

Fuzzy product rules have different diagnostic significance. It is proposed to determine it in two stages: based on the experts' opinion, to set the initial values of the weights, and then using the apparatus of artificial neural networks to make their adjustments and change if it is necessary. As the opinions of experts are often very different, the use of ANNs in combination with fuzzy logic will allow to adjust the parameters of the diagnostic system in the best way.

The fuzzy inference system (the subsystem of decision support on Fig. 6.11) is presented as two functional blocks: the decision support block and the block of decision rules. This system works both at the stage of preliminary diagnostics (before the ultrasound examination) on the base of data from the medical records and together with obtaining video information.

Currently, most fuzzy inference systems are based on 6 steps:

(1) formation of a base of rules of logical inference;
(2) fuzzification (bringing to fuzziness) of input variables;
(3) aggregation of subconditions;
(4) activation of the inference;
(5) accumulation of inference;
(6) defuzzification (if necessary).

The main stages of fuzzy inference are presented in Fig. 6.78.

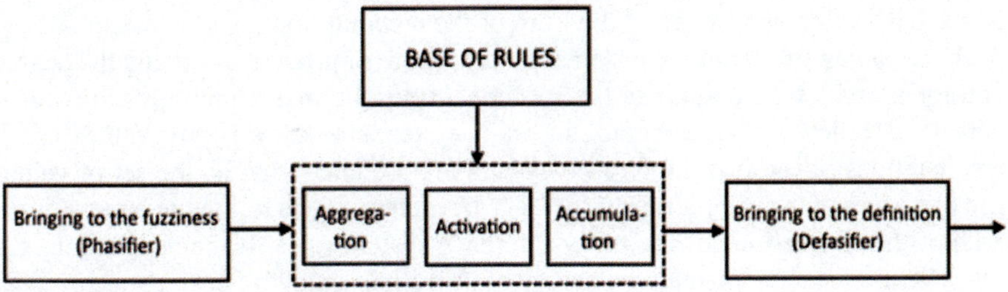

Fig. 6.78 The main stages of fuzzy inference

Formation of knowledge base in the form of production rules

The base of rules of fuzzy inference system is intended to the formal presentation of empirical knowledge or expert knowledge in one or another problem area. In fuzzy inference systems it is used fuzzy rules, in which conditions and inferences are formulated in terms of fuzzy linguistic statements. The set of such rules will hereinafter be called the rule base of fuzzy products. The FPR base is the finite set of FPR agreed upon by the linguistic variables used in them. Most often, the rule base is in the form of structured text:

RULE_1: IF "condition_1, then conclusion_1",
RULE_2: IF "condition_2, then conclusion_2",
...
RULE_n: IF "condition_n, then output_n".

In fuzzy inference systems, the linguistic variables used in fuzzy statements of the used in fuzzy statements the premise of the rules of fuzzy products are often referred to as input linguistic variables, and the variables used in fuzzy statements of fuzzy product rules inferences are often referred to as output linguistic variables.

Fuzzification of input variables. In the context of fuzzy logic, fuzzification means not only a single stage of fuzzy inference execution, but also the procedure of finding values of membership functions of fuzzy sets based on ordinary (NOT fuzzy) input data. Fuzzification is also called the introduction of fuzziness. The purpose of the fuzzification stage is to establish a correspondence between the specific (usually numerical) value of a single input variable of fuzzy inference and the value of the membership function corresponding to its input linguistic variable. Upon completion of this step, it is necessary for all input variables to determine the specific values of the membership functions for each of the linguistic terms used in the preconditions of the fuzzy inference rule base of fuzzy inference system.

Aggregation is the procedure for determining the degree of truth of conditions under each of the rules of the fuzzy inference system. Formally, the aggregation procedure is as follows. By the beginning of this stage, the known truth values of all the preconditions of the fuzzy inference system are assumed. The following describes each of the conditions in the fuzzy inference system rules. If the condition consists of several preconditions, then the degree of truth of a complex expression is determined on the basis of known values of the truth of the preconditions.

Activation in fuzzy output systems is a procedure or process of finding the degree of truth of each of the outputs of fuzzy product rules. Formally, the activation procedure is as follows. By the beginning of this step, the known truth values of all preconditions of the fuzzy inference system are assumed, that is, the set of values and the values of the weighting coefficients for each rule. The following is considered each of the conclusions of the fuzzy inference system rules. If the conclusion of a rule is a fuzzy statement, then the degree of its truth is equal to the algebraic product of the corresponding value by the weighting coefficient.

Accumulation or accumulation in fuzzy inference systems is the procedure or process of finding the membership function for each of the original linguistic variables of the set. The purpose of the accumulation is to combine or accumulate all the degrees of truth of the inferences to obtain the membership function to each of the output variables. The reason for performing this stage is that the inferences related to the same original linguistic variable belong to different rules in the fuzzy inference system.

Defazzification in fuzzy inference systems is a procedure or process of finding a normal (NOT fuzzy) value for each of the original linguistic variables. The purpose of defazzification is to use the results of the accumulation of all output linguistic variables to obtain the usual quantitative value of each of the output variables, which can be used by special devices that do not belong to the fuzzy inference system.

All stages of the algorithm are interconnected and generally allow us to imagine the process of fuzzy inference in the form of a sequence of specific operations, which may in particular be conjunction and disjunction operations. Currently, there are several fuzzy inference algorithms, the most practical of which are the following [15]:

- Mamdani algorithm;
- Tsukamoto algorithm;
- Larsen algorithm;
- Sugeno algorithm;

Mamdani algorithm

At its essence, this algorithm generates the steps discussed above because it most closely matches their parameters.

In the Mamdani model, the mathematical relationship between inputs $x = (x_1, x_2, \ldots, x_n)$ and outputs is determined by the fuzzy rule base of the following format:

if:

$$\{x_1 = a_1, j_1\} AND \{x_2 = a_2, j_1\} AND \{x_3 = a_3, j_1\} AND \ldots AND \{x_n = a_n, j_1\} OR$$
$$\{x_1 = a_1, j_1\} AND \{x_2 = a_2, j_1\} AND \{x_3 = a_3, j_1\} AND \ldots AND \{x_n = a_n, j_1\} OR$$
$$\ldots\ldots\ldots\ldots\ldots\ldots\ldots\ldots\ldots\ldots OR$$

$$\text{THEN } y = d_j, j = \overline{1, m} \tag{6.20}$$

where by means of operations ∪ (OR) and ∩ (AND) the fuzzy rule base can be rewritten in a more compact form:

According to Mamdani algorithm, the logical inference is in the following six stages:

(1) Formation of rules base for fuzzy inference systems.

(2) Fuzziness (fuzzification). The membership functions defined on the input variables are applied to their actual values to determine the membership function for each rule.

(3) Aggregations of precondition in fuzzy production rules. To determine the degree of truth of the conditions of each of the fuzzy product rules, pair fuzzy logic operations are used. Those rules whose degree of truth is non-zero are considered active and are used for further calculations.

(4) Activation of inferences in fuzzy production rules. It is carried out when the membership and output functions are "cut off" by the height of the corresponding computational membership function of the premise of the rule (fuzzy logic "AND"). Moreover, to reduce the inference time, only active rules of fuzzy products are taken into account.

(5) Accumulation of inferences of fuzzy rules of products. All fuzzy subsets obtained for each output variable (in all rules) are combined together to form one fuzzy subset for each output variable.

(6) Bringing to clarity (defuzzification). This procedure is used when it is necessary to convert a fuzzy original set into a clear number. Most often, for the Mamdani model, defuzzification is used by the centroid method, when a clear value of the output variable is defined as the center of gravity for the curve.

The procedure for obtaining a logical conclusion is shown in Fig. 6.79.

Tsukamoto algorithm

Formally, the Tsukamoto algorithm can be defined as follows:

(1) Formation of rules base for fuzzy inference systems. The peculiarities of the rule base formation coincide with the algorithm discussed above.

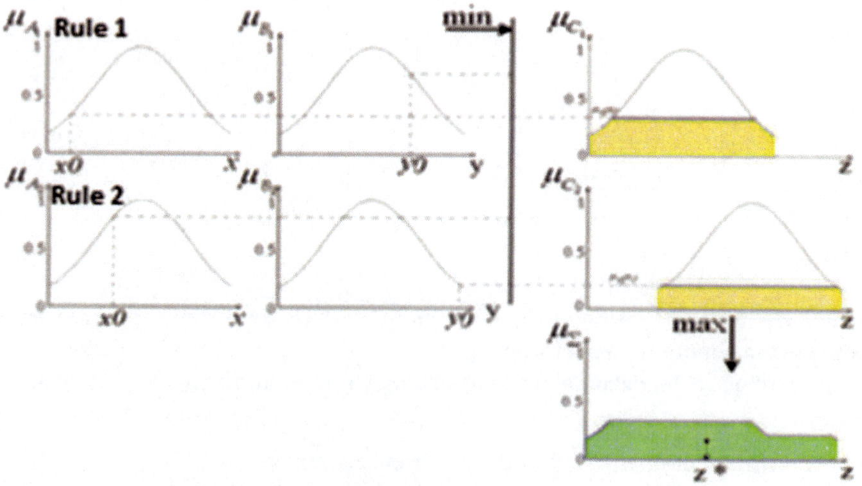

Fig. 6.79 Mamdani logical inference procedure

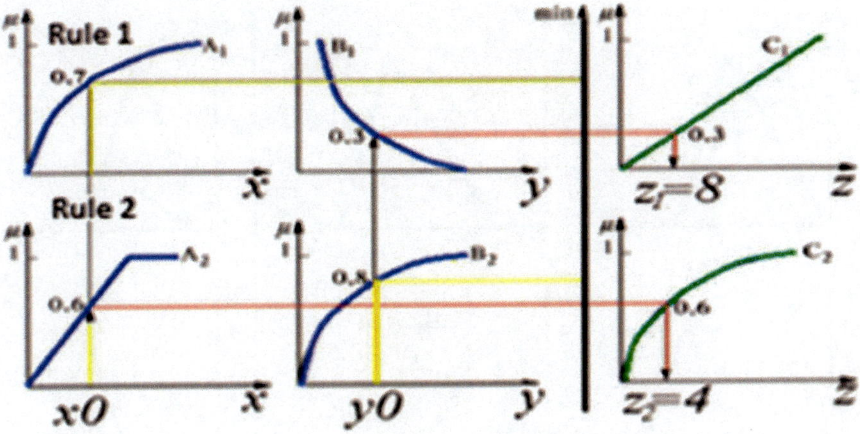

Fig. 6.80 Tsukamoto logical inference procedure

(2) Fuzzification of input variables. Features of fuzzification coincide with those considered above in the description of this stage.
(3) Aggregations of precondition in fuzzy production rules. To determine the degree of truth of the conditions of each of the fuzzy product rules, pair fuzzy logic operations are used. Those rules whose degree of truth is nonzero are considered active and are used for further calculations.
(4) Activation of inferences in fuzzy rules of products is carried out similar to the algorithm of Mamdani.
(5) Accumulation of inferences of fuzzy rules of products is in fact absent, since the calculations are made with ordinary real numbers.
(6) For defuzzification of initial variables it is used a modified version in the form of center of gravity method for one-point sets.

An example of the Tsukamoto algorithm is given in Fig. 6.80.

Larsen Algorithm

Formally, Larsen's algorithm can be defined as follows.

(1) Formation of rules base for fuzzy output systems. The peculiarities of the rule base formation coincide with the algorithm discussed above.
(2) Fuzzification of input variables. Features of fuzzification coincide with those considered above in the description of this stage.
(3) Aggregations of precondition in fuzzy production rules. To determine the degree of truth of the conditions of each of the fuzzy product rules, pair fuzzy logic operations are used. Those rules whose degree of truth is non-zero are considered active and are used for further calculations.
(4) Activation of inferences in fuzzy rules of products.
(5) Accumulation of inferences of fuzzy rules of products. Defuzzification of initial variables Any of defuzzification methods can be used.

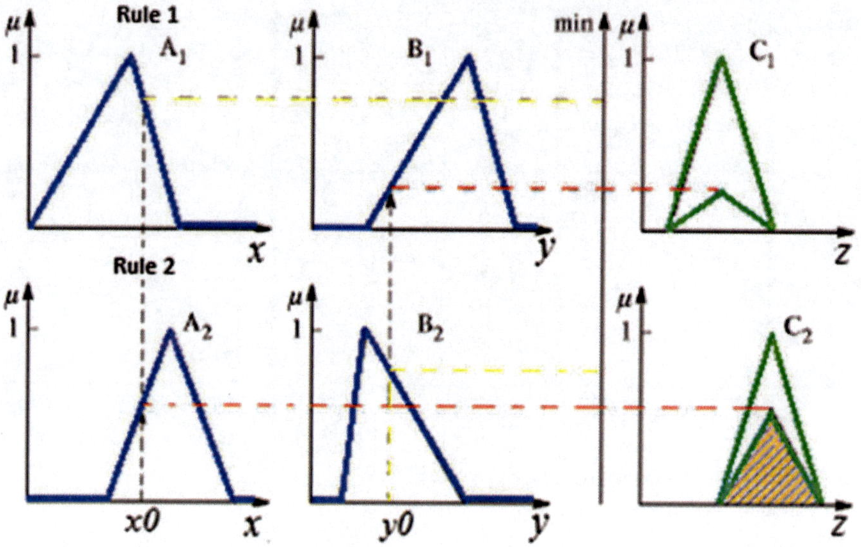

Fig. 6.81 Larsen logical inference procedure

An example of the Larsen algorithm is given in Fig. 6.81.

Sugeno Algorithm

Formally, the Sugeno algorithm can be defined as follows.

(1) Only fuzzy rules are used in the rule base.
(2) Fuzzification of input variables. At this stage, all the known specific values of the input variables of the fuzzy inference system are matched by the fuzzy set.
(3) Aggregations of precondition in fuzzy production rules. To determine the degree of truth of the conditions of all fuzzy product rules, as a rule it is used a logical operation of min-conjunction. Those rules whose degree of truth is non-zero are considered active and are used for further calculations.
(4) Activation of inferences in fuzzy rules of products. For a given (clear) value of an argument, it is determined the degrees of truth for the preconditions of each rule. Then, it is determined the values of the degrees of truth of all inferences of the fuzzy product rules using the min-activation method.
(5) Accumulation of inferences of fuzzy rules of products is in fact absent, since the calculations are made with ordinary real numbers.
(6) For defuzzification of initial variables it is used a modified version in the form of center of gravity method for one-point sets.

In order to select the necessary algorithm, we will consider their advantages and disadvantages with regard to the problem of medical diagnostics.

One of the first to find application in fuzzy inference systems is the Mamdani algorithm. Its distinguishing feature is the use of nonlinear functions min and max for the implementation of conjunction and disjunction operations. Also, this algorithm is characterized by min-activation and defuzzification by the method of center of gravity or center of square. Algorithms Tsukamoto, Sugeno and simplified algorithm of fuzzy inference different from the previous by lack of accumulation stage, since at the stage of inferences it is found clear values of all output LV(OutLV). The difference between the Larsen algorithm and the Mamdani algorithm is the use of *prod*-activation.

Thus, it is suggested to choose its type for the diagnostics of fuzzy logic inference (FLI) system because it satisfies the following requirements.

It is necessary to implement FLI in a clear and fuzzy form at the same time until its last stage, since the output data of the system is presented in both quantitative and qualitative form. For example, as a result of the system work, the output data could be obtained in the following form: "The presence of the disease = Yes" (quality data), "Type of disease = Medullary Thyroid Cancer" (quality data), "quality data = 0.81" (quantitative data). In this regard, the Tsukamoto algorithm and the simplified fuzzy inference algorithm prove to be inapplicable.

Under ultrasound diagnostics of LV in each case can take only one linguistic value (LVal), whose truth coefficient is greater. Based on the principles of the proposed fuzzy system, the number of such membership functions may not be more than two. However, the Sugeno algorithm uses a linear combination of the values of the intersectional membership functions. This is contrary to the logic of decision-making in ultrasound, since there are a number of InpLV, the values of which are initially qualitative and do not have adequate reflection on the ordinal scale. Thus, the Sugeno algorithm in this case also turns out to be inapplicable.

From the point of view of the doctor, to determine the disease with increasing number of detected symptoms the diagnosis is established more accurately, while in the absence of some of them its accuracy does not decrease. This may be the case in the absence of the results of some researches, the inability to conduct them, or the absence of the need to review the high significance of other symptoms. In this regard, it is proposed to use the Mamdani or Larsen algorithm for the most accurate description of the logical inference process. They both satisfy the above conditions. It is proposed to choose the Larsen algorithm to minimize the number of uses of nonlinear min and max functions.

6.6.2 Subsystem of Decision Support

6.6.2.1 Description of Input and Output LV

Currently, the construction of most fuzzy inference systems is based on six steps [15]:

(1) formation of a base of rules of logical conclusion;
(2) fuzzification of input variables;

(3) aggregation of the term;
(4) activation of subshells;
(5) accumulation of conclusions;
(6) defuzzification.

To implement the fuzzy inference algorithm, it is necessary to carry out a number of preliminary actions that are necessary for structuring and concretizing the initial and final data. Consider a fuzzy inference system in the form of a "black box" model. The system performs the function of converting a set of values of symptomatic (SF) and pathogenetic (PF) factors as input variables of the system into recommendations for making the diagnosis as output variables. Consider in more detail the description of InpLV and OutLV made in the following three previous steps.

Preliminary Step 1. Description of the input linguistic variables.

InpLV in this system is a set of factors (SF and PF). The values of the variables are entered by doctors of different specialization in the patient's medical history. The factors can be either qualitative (initially linguistic variables (LVal)) or quantitative (initially numerical variables (NV) that are reduced to linguistic by fuzzification). The set of InpLV for the intellectual diagnostic system in case of thyroid diseases can be presented in the form of Table 6.3.

The factors can be divided into clear and fuzzy, and regardless of symptomatic and pathogenetic. Such classification can be presented in the form of Table 6.4.

The following are considered as thyroid tumors: malignant (papillary (MPT), follicular (MFT), medullary (MMT), anaplastic (MAT)), other diseases (OD) (benign follicular adenoma (BFP), others.

Preliminary step 2. Description of the original linguistic variables.

The OutLV of this system of fuzzy logic inference is the doctor's recommendation at diagnosis. They contain information about the presence of the disease in the patient, the types of pathology and the degree of confidence of the system in the correctness of the given recommendations (their diagnostic reliability). These characteristics are determined for each disease recognized by the system. The degree of confidence is calculated in terms of the numerical confidence factor and the value of the LV "Confidence in Diagnosis" (CD), the calculation of which is described in Step 6.

The system of fuzzy inference allows determine the types of thyroid cancer, namely: benign (BN) (follicular adenoma, etc.) and malignant (MG) (follicular carcinoma, papillary carcinoma, medullary carcinoma, etc.).

The selection of one or more OutLV values is made on the base of a numerical confidence factor, which will be described later. We call the variables presented in Table 6.5, primary OutLV. In addition, to the output variables it is necessary to attribute LV "Type of disease" (type C) and "Presence of the disease" (Presence C) should be assigned. We call them secondary ones. The value of these variables is obtained using clear logical transformations from the values of the previously described OutLV (which can take the values of logical "0" and "1"), which can be schematically shown in Fig. 6.82.

Table 6.3 A set of factors for DSI

No	Name	Abbreviation	Definition	Factor type	Initially, LV or NV
1	Feeling lump or foreign body in throat	FL FBT	+	SF	NV
2	Swallowing disorders	SD	+	SF	LV
3	Neck pain	NP	+	SF	NV
4	Cough and hoarseness of voice	CHV	+	SF	NV
5	Dyspnea	D	+	SF	NV
6	Oncological tests	OT	+	SF	NV
7	Biopsy	B	+	SF	NV
8	Geometric form of pathology	GFP	+	SF	NV
9	Border view	BV	–	SF	NV
10	Echogenicity	E	+	SF	NV
11	The presence of a rim	PR	–	SF	NV
12	Geometric dimensions of the pathology	GDP	+	SF	NV
13	Age	Age	+	PF	NV
14	Gender	Gender	–	PF	LV
15	Burdened heredity	BH	+	PF	LV
16	Smoking	SM	+	PF	NV
17	Consumption of strong alcohol (>10°)	Alc	+	PF	NV
18	Physical and neuro-psychological overload	PNP	–	PF	LV
19	Weakness	WN	+	PF	LV
20	Drowsiness	DN	+	PF	LV
21	Frequent ear pain	FEP	+	PF	LV
22	Autoimmune thyroid	AT	–	SF	NV
23	Iodine level in the body	ILB	–	SF	NV
24	Calciton test	CT	–	SF	LV
25	Thyrotropin hormone	TH	–	SF	NV
26	Triiodothyronine	T3	–	SF	NV
27	Thyroxine	T4	–	SF	NV

Table 6.4 Classification of factors

	Fuzzy	Clear
SF	BV, PR	FL FBT, SD, NP, CHV, D, OT, B, GFP, E
PF	PNP, gender	Age, BH, SM, Alc

Table 6.5 The possible values of the initial LV

OutLV	Possible value
CD MPT	Most likely, there is no disease
	It's more like nothing
	It's faster than not
	Most likely, the disease is
CD MFT	Most likely, there is no disease
	It's more like nothing
	It's faster than not
	Most likely, the disease is
CDMMT	Most likely, there is no disease
	It's more like nothing
	It's faster than not
	Most likely, the disease is
	Most likely, there is no disease
CD MAT	It's more like nothing
	It's faster than not
	Most likely, the disease is
	Most likely, there is no disease
	It's more like nothing
CD OD	Most likely, there is no disease
	Most likely, the disease is

In the figure presented, the blocks "OR" and "AND" indicate the standard operations of disjunction and conjunction of clear logic.

Preliminary Step 3. Determination of InpLV and OutLV Value Areas.

The next preliminary step is to determine the values that each InpLV and OutLV can take. These data can be clearly shown in Tables 6.6, 6.7, 6.8, 6.9 and 6.10.

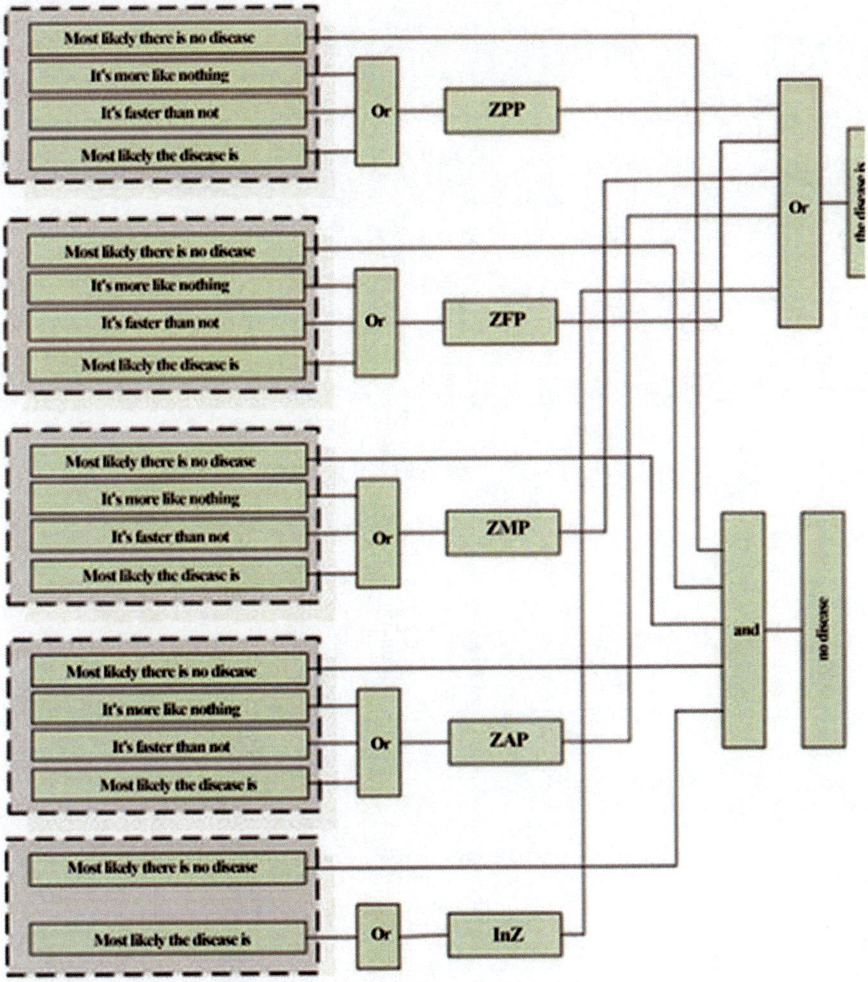

Fig. 6.82 Obtaining secondary OutLV as a result of generalization of primary ones

6.6.2.2 Stages of Fuzzy Logic in Diagnostic Decisions Support Subsystem (DDSS)

Stage 1. Formation of a set of fuzzy production rules

The result of combining input variables to calculate output ones in a system with clear logic can be specified with help of fuzzy production rules (FPR). They are close in structure to linguistic structures and can therefore be easily assembled directly by doctors, as well as technical experts on the base of information about the subject area: data from literary sources, expert opinions. The set of rules may change as the system works.

Table 6.6 Malignant anaplastic tumor, the values of the input LV

Type of LV	Name of LV	Value 1	Value 2	Value 3	Value 4	Value 5	Value 6	Value 7	Value 8	Value 9
InpLV	FL FBT	No	Yes	Yes	Yes	Yes	Yes	Yes	Yes	Yes
	SD	No	Yes	Yes	Yes	Yes	Yes	Yes	Yes	Yes
	NP	No	Yes	Yes	Yes	Yes	Yes	Yes	Yes	Yes
	CHV	No	Yes	Yes	Yes	Yes	Yes	Yes	Yes	Yes
	D	No	Yes	Yes	Yes	Yes	Yes	Yes	Yes	Yes
	OT	N	N	N	N	>N	>N	>N	>N	>N
	B	No	No	No	No	No	No	Так	Так	Так
	GFP	Right	Right	Right	Right	Wrong	Incorrectangular			
	BV	Gibrophilic	Gibrophilic with thickening	Hyperechogenic	Clear	Fiberglass with tears	Fuzzy	Fuzzy hyperechogenic	Fuzzy hypoechoic	Fuzzy isoechogenic
	E	Significantly reduced	Moderately reduced			Isogenicity	Hypochogenicity			
	PR	No				Yes	Yes	Yes	Yes	Yes
	GDP	<N	N	>N		>N	>N	>N	>N	>N
	Age	Baby	Young	Medium	Inclined	Old				
	Gender	Male	Female							
	BH	Unencumbered	Slightly burdened	Heavily burdened						
	SM	No	Few and rare	Many and often						

(continued)

Table 6.6 (continued)

Type of LV	Name of LV	Value 1	Value 2	Value 3	Value 4	Value 5	Value 6	Value 7	Value 8	Value 9
	Alc	No	Few and rare	Many and often						
	PNP	No	Sometimes	Often	More no, than yes	More yes, than no		Most likely the disease is		
OutLV	CD MPT	Most likely the disease is not present			More no, than yes	More yes, than no		Most likely the disease is		
	CD MFT	Most likely the disease is not present			More no, than yes	More yes, than no		Most likely the disease is		
	CD MMT	Most likely the disease is not present			More no, than yes	More yes, than no		Most likely the disease is		
	CDMAT	Most likely the disease is not present			More no, than yes	More yes, than no		Most likely the disease is		
	CDOD	Most likely the disease is not present			More no, than yes					

Table 6.7 Malignant medullary tumor, the values of the input LV

LV type	LV name	Value 1	Value 2	Value 3	Value 4	Value 5	Value 6	Value 7	Value 8	Value 9
Inp LV	FLFBT	No	Yes	Yes	Yes	Yes	Yes	Yes	Yes	Yes
	SD	No	Yes	Yes	Yes	Yes	Yes	Yes	Yes	Yes
	NP	No	Yes	Yes	Yes	Yes	Yes	Yes	Yes	Yes
	CHV	No	Yes	Yes	Yes	Yes	Yes	Yes	Yes	Yes
	WN	No	Yes	Yes	Yes	Yes	Yes	Yes	Yes	Yes
	DN	No	Yes	Yes	Yes	Yes	Yes	Yes	Yes	Yes
	FEP	No	Yes	Yes	Yes	Yes	Yes	Yes	Yes	Yes
	D	No	Yes	Yes	Yes	Yes	Yes	Yes	Yes	Yes
	OT	N	N	N	N	>N	>N	>N	>N	>N
	CT	N	N	N	N	N	>N	>N	>N	>N
	ILB	N	N	N	N	N	>N	>N	>N	>N
	AT	N	N	N	N	N	>N	>N	>N	>N
	B	No	No	No	No	No	No	Yes	Yes	Yes
	GFP	Right	Right	Right	Right	Wrong	Wrong angular			
	BV	Hydrophilic	Hydrophilic with thickening	Hyperechogenic	Clear	Hydrophilic with gap	Not clear	Not clear hyperechogenic	Not clear hyperechogenic	Not clear isoechogenic
	E	Significantly reduced	Moderately reduced			Isogenicity	Hypochogenicity			
	PR	No				Yes	Yes	Yes	Yes	Yes
	GDP	<N	N	>N		>N	>N	>N	>N	>N

(continued)

Table 6.7 (continued)

LV type	LV name	Value 1	Value 2	Value 3	Value 4	Value 5	Value 6	Value 7	Value 8	Value 9
	Age	Children's	Young	Medium	Inclined	Old				
	Gender	Male	Female							
	BH	Unencumbered	Slightly burdened	Heavily burdened						
	SM	No	Few and rare	Many and often						
	Alc	No	Few and rare	Many and often						
	TH	N	N	N	N	N	>N	>N	>N	>N
	T3	N	N	N	N	N	>N	>N	>N	>N
	T4	N	N	N	N	N	>N	>N	>N	>N
	PNP	No	Sometimes	Often	Rather, not what it is	More likely than not		Most likely the disease is		
Out LV	CD MAP	Most likely the disease is gone	Rather, not what it is	More likely than not	Most likely the disease is					

Table 6.8 Malignant papillary tumor, the values of the input LV

LV type	LV name	Value 1	Value 2	Valuue 3	Value 4	Value 5	Value 6	Value 7	Value 8	Value 9
Inp LV	FLFBT	No	Yes	Yes	Yes	Yes	Yes	Yes	Yes	Yes
	SD	No	Yes	Yes	Yes	Yes	Yes	Yes	Yes	Yes
	NP	No	Yes	Yes	Yes	Yes	Yes	Yes	Yes	Yes
	CHV	No	Yes	Yes	Yes	Yes	Yes	Yes	Yes	Yes
	WN	No	Yes	Yes	Yes	Yes	Yes	Yes	Yes	Yes
	DN	No	Yes	Yes	Yes	Yes	Yes	Yes	Yes	Yes
	FEP	No	Yes	Yes	Yes	Yes	Yes	Yes	Yes	Yes
	D	No	Yes	Yes	Yes	Yes	Yes	Yes	Yes	Yes
	OT	N	N	N	N	>N	>N	>N	>N	>N
	CT	N	N	N	N	N	>N	>N	>N	>N
	ILB	N	N	N	N	N	>N	>N	>N	>N
	AT	N	N	N	N	N	>N	>N	>N	>N
	B	No	No	No	No	No	No	Yes	Yes	Yes
	GFP	Right	Right	Right	Right	Wrong	Wrong angular			
	BV	Hydrophilic	Hydrophilic with thickening	Hyperechogenic	Clear	Hydrophilic with discontinuities	Not clear	Not clear hyperechogenic	Not clear hypoechoic	Not clear isoechogenic
	E	Significantly reduced	Moderately reduced			Isogenicity	Hypochogenicity			
	PR	No				Yes	Yes	Yes	Yes	Yes
	GDP	<N	N	>N		>N	>N	>N	>N	>N

(continued)

Table 6.8 (continued)

LV type	LV name	Value 1	Value 2	Valuue 3	Value 4	Value 5	Value 6	Value 7	Value 8	Value 9
	Age	Children's	Young	Medium	Inclined	Old				
	Gender	Male	Female							
	BH	Unencumbered	Poorly fitted	Heavily burdened						
	SM	No	Few and rare	Many and often						
	Alc	No	Few and rare	Many and often						
	TH	N	N	N	N	N	>N	>N	>N	>N
	T3	N	N	N	N	N	>N	>N	>N	>N
	T4	N	N	N	N	N	>N	>N	>N	>N
	PNP	No	Sometimes	Often	Rather, not what it is	More likely than not		Most likely the disease is		
Out LV	CD MAP	Most likely the disease is gone			Rather, not what it is	More likely than not		Most likely the disease is		

Table 6.9 Malignant follicular tumor, the values of the input LV

LV type	LV name	Value 1	Value 2	Value 3	Value 4	Value 5	Value 6	Value 7	Value 8	Value 9
Inp LV	FLFBT	No	Yes	Yes	Yes	Yes	Yes	Yes	Yes	Yes
	SD	No	Yes	Yes	Yes	Yes	Yes	Yes	Yes	Yes
	NP	No	Yes	Yes	Yes	Yes	Yes	Yes	Yes	Yes
	CHV	No	Yes	Yes	Yes	Yes	Yes	Yes	Yes	Yes
	WN	No	Yes	Yes	Yes	Yes	Yes	Yes	Yes	Yes
	DN	No	Yes	Yes	Yes	Yes	Yes	Yes	Yes	Yes
	FEP	No	Yes	Yes	Yes	Yes	Yes	Yes	Yes	Yes
	D	No	Yes	Yes	Yes	Yes	Yes	Yes	Yes	Yes
	OT	N	N	N	N	>N	>N	>N	>N	>N
	CT	N	N	N	N	N	>N	>N	>N	>N
	ILB	N	N	N	N	N	>N	>N	>N	>N
	AT	N	N	N	N	N	>N	>N	>N	>N
	B	No	No	No	No	No	No	Yes	Yes	Yes
	GFP	Right	Right	Right	Right	Wrong	Wrongangular			
	BV	Hydrophilic	Hydrophilic with thickening	Hyperechogenic	Clear	Hidrofilha with discontinuities	Not clear	Not clear hyperechogenic	Not clear hypoechoic	Not Clear isoechogenic
	E	Significantly reduced	Moderately reduced			Isogenicity	Hypochogenicity			
	PR	No				Yes	Yes	Yes	Yes	Yes
	GDP	<N	N	>N		>N	>N	>N	>N	>N

(continued)

Table 6.9 (continued)

LV type	LV name	Value 1	Value 2	Value 3	Value 4	Value 5	Value 6	Value 7	Value 8	Value 9
	Age	Children's	Young	Medium	Inclined	Old				
	Gender	Male	Female							
	BH	Unencumbered	Poorly fitted	Heavily burdened						
	SM	No	Few and rare	Many and often						
	Alc	No	Few and rare	Many and often						
	TH	N	N	N	N	N	>N	>N	>N	>N
	T3	N	N	N	N	N	>N	>N	>N	>N
	T4	N	N	N	N	N	>N	>N	>N	>N
	PNP	No	Sometimes	Often	More likely than not	More likely than not		Most likely the disease is		
Out LV	CD MAP	Most likely the disease is gone			Rather, not what it is	More likely than not		Most likely the disease is		

Table 6.10 Benign follicular adenoma, the values of the input LV

LV type	LV name	Value 1	Value 2	Value 3	Value 4	Value 5	Value 6	Value 7	Value 8	Value 9
Inp LV	FLFBT	No	Yes	Yes	Yes	Yes	Yes	Yes	Yes	Yes
	SD	No	Yes	Yes	Yes	Yes	Yes	Yes	Yes	Yes
	NP	No	Yes	Yes	Yes	Yes	Yes	Yes	Yes	Yes
	CHV	No	Yes	Yes	Yes	Yes	Yes	Yes	Yes	Yes
	WN	No	Yes	Yes	Yes	Yes	Yes	Yes	Yes	Yes
	DN	No	Yes	Yes	Yes	Yes	Yes	Yes	Yes	Yes
	FEP	No	Yes	Yes	Yes	Yes	Yes	Yes	Yes	Yes
	D	No	Yes	Yes	Yes	Yes	Yes	Yes	Yes	Yes
	OT	N	N	N	N	>N	>N	>N	>N	>N
	CT	N	N	N	N	N	>N	>N	>N	>N
	ILB	N	N	N	N	N	>N	>N	>N	>N
	AT	N	N	N	N	N	>N	>N	>N	>N
	B	No	No	No	No	No	No	Yes	Yes	Yes
	GFP	Right	Right	Right	Right	Wrong	Wrong angular			
	BV	Hydrophilic	Hydrophilic with thickening	Hyperechogenic	Clear	Hydrophilic with discontinuities	Not clear	Not clear hyperechogenic	Not clear hypoechoic	Not clear isoechogenic
	E	Significantly reduced	Moderately reduced			Isogenicity	Hypochogenicity			
	PR	No				Yes	Yes	Yes	Yes	Yes
	GDP	<N	N	>N		>N	>N	>N	>N	>N

(continued)

Table 6.10 (continued)

LV type	LV name	Value 1	Value 2	Value 3	Value 4	Value 5	Value 6	Value 7	Value 8	Value 9
	Age	Children's	Young	Medium	Inclined	Old				
	Gender	Male	Female							
	BH	Unencumbered	Poorly fitted	Heavily burdened						
	Alc	No	Few and rare	Many and often						
	TH	N	N	N	N	N	>N	>N	>N	>N
	T3	N	N	N	N	N	>N	>N	>N	>N
	T4	N	N	N	N	N	>N	>N	>N	>N
	PNP	No	Sometimes	Often	Rather, not what it is	More likely than not		Most likely the disease is		
Out LV	CD MAP	Most likely the disease is gone			Rather, not what it is	More likely than not		Most likely the disease is		

The set of FPR can be divided into groups according to InpLV and OutLV, which are included in them.

Let FPR look like:

IF: "Condition$_1$ = True"
And « Condition$_2$ = True»
:
And « Condition$_n$ = True»
Than: "Inference$_1$ = true"
And « Inference$_2$ = true»
:
And « Inference$_m$ = true»

Conditions connected to the FPR through "OR" are not considered. In the presence of this logical operation in the condition of FPR is broken into several smaller rules. If the FPR contains "OR" in the final part, it does not have the specificity necessary for diagnosis.

For brevity and convenience of reading we will record FRP in the form of special tables. In the table column "IF" we will put the expression of the conditional part of each rule: "Condition$_1$", "Condition$_2$", ..., "Condition$_n$". In the column "TO" write "Inference$_1$", "Inference$_2$", ..., "Inference m" for each FPR. Each of these table columns is divided into subcolumns of the names and values of LV that include in the expression through the logical operation "OR". Thus, we obtain a Table 6.11.

FPR table for PF will look like this (Table 6.12).

After defining the system of FPR and VL included in it, first it is necessary to bring clear numerical variables to the linguistic type.

For this it is necessary to match a clear set of values to the set of term-sets of LV in the process of fuzzification.

Stage 2. Fuzzification InpLV

We will determine the membership functions μ clear value to an fuzzy term-set that corresponds to one from values of InpLV, using the methods of expert evaluation. Let the expert E1 consider that the specific value of x^* belongs to a fuzzy term-set for $a_1 \leq x^* \leq b_1$; expert $E_2 - a_2 \leq x^* \leq b_g$; ...; expert $E_g - a_g \leq x^* \leq b_g$. Then the membership function (MF) term can be obtained as follows (Fig. 6.83).

In the presented Fig. 6.83 along the x axis are the values of a clear fuzzy variable $i = \overline{1, g}$ is the expert number, g is a their number. Along the vertical axis μ axis it is plotted the proportion of all experts who consider that this value of x belongs to a given linguistic value of a linguistic variable. In this case, first the sections of the membership function at $x \in \left(\min_i(a_i), \max_i(a_i)\right) \cup \left(\min_i(b_i), \max_i(b_i)\right)$ come out to be curvilinear and are reduced to linear form using the least squares method.

Among the InpLV, the clear variables, and, therefore, requiring the fuzzification for the processing their values with help of fuzzy logic inference system are the follows: Age, BH, OT, SM, Alc. The data obtained for them are presented in Figs. 6.84, 6.85, 6.86, 6.87, 6.88.

Table 6.11 FPR for SF

FPR	IF		THEN	
	LV	Value	LV	Value
FPR_1	FL FBT	+	Possibility of MPT	Much higher
	SD	+		
	NP	+		
	CHV	+		
	D	+		
FPR_2	FL FBT	+	The possibility of MFT	Much higher
	SD	+		
	NP	+		
	CHV	+		
	D	+		
FPR_3	FL FBT	+	Possibility of an MMT	Much higher
	SD	+		
	NP	+		
	CHV	+		
	D	+		
FPR_4	FL FBT	+	Possibility of MAT	Much higher
	SD	+		
	NP	+		
	CHV	+		
	D	+		
FPR_5	FL FBT	+	Possibility of OD	Much higher
	SD	+		
	NP	+		
	D	+		
FPR_6	GFP	Wrong	Possibility of MPT	
		Wrong angular		
	BV	Hydrophilic with tears		
		Fuzzy		
		Fuzzy hyperechogenic		
		Fuzzy hyperechogenic		
		Fuzzy isoechogenic		
	+	Isochogenicity		
		Hypochogenicity		
	PR	+		
FPR_7	GFP	Wrong		

(continued)

Table 6.11 (continued)

FPR	IF		THEN	
	LV	Value	LV	Value
		Wrong angular	The possibility of MFT	
	BV	Hydrophilic with tears		
		Fuzzy		
		Fuzzy hyperechogenic		
		Fuzzy hyperechogenic		
		Fuzzy isoechogenic		
	+	Isochogenicity		
		Hypochogenicity		
	PR	+		
FPR_8	GFP	Wrong angular		
		Hydrophilic with tears	Possibility of an MMT	
	BV	Fuzzy		
		Fuzzy hyperechogenic		
		Fuzzy hyperechogenic		
		Fuzzy isoechogenic		
		Isochogenicity		
	+	Hypochogenicity		
		Wrong angular		
	PR	+		
FPR_9	GFP	Wrong angular	Possibility of MAT	
		Hydrophilic with tears		
	BV	Fuzzy		
		Fuzzy hyperechogenic		
		Fuzzy hyperechogenic		
		Fuzzy isoechogenic		
		Isochogenicity		
	+	Hypochogenicity		
		Wrong angular		
	PV	+		
FPR_{10}	GFP	Correct	Possibility of OD	
	BV	Hydrophilic		
		Hydrophilic with thickening		
		Hyperechogenic		

(continued)

Table 6.11 (continued)

FPR	IF		THEN	
	LV	Value	LV	Value
		Clear		
	+	Significantly reduced		
		Moderately reduced		
	PR	+		
FPR_{11}	OTMPT	Yes	Possibility of MPT	Much higher
FPR_{12}	OTMFT	Yes	The possibility of MFT	Much higher
FPR_{13}	OTMMT	Yes	Possibility of an MMT	Much higher
FPR_{14}	OTMAT	Yes	Possibility of MAT	Much higher
FPR_{15}	BMPT	Yes	Possibility of MPT	Much higher
FPR_{16}	BMFT	Yes	The possibility of MFT	Much higher
FPR_{17}	BMMT	Yes	Possibility of an MMT	Much higher
FPR_{18}	BMAT	Yes	Possibility of MAT	Much higher

The rest of the LV are fuzzy or linguistic from the beginning.

Stage 3. Aggregation

This stage allows to determine the degree of true of BI condition based on the known degrees of true BI subcondition. Based on the results *stage 1*, it is necessary to implement an aggregation stage for 1, 6, 10, 11–18 rules. The rest of the FPR contain the conditions from one subcondition. The result of its aggregation will be the value of membership function InpLV as the degree of truth of the condition. The rules for fuzzy products with numbers 1, 6, 10, 11–18 are as follows:

IF: "$\text{InpLV}_1 = \text{Val}_{11}$" OR "$\text{InpLV}_1 = \text{Val}_{12}$" OR … OR "$\text{InpLV}_1 = \text{Val}_{1m}$"
AND "$\text{InpLV}_n = \text{Val}_{n1}$" OR "$\text{InpLV}_n = \text{Val}_{n2}$" OR … OR "$\text{InpLV}_n = \text{Val}_{nm}$"
:
AND "$\text{InpLV}_n = \text{Val}_{nl}$" OR "$\text{InpLV}_n = \text{Val}_{n2}$" OR … OR "$\text{InpLV}_n = \text{Val}_{nm}$"
THEN: "$\text{InpLV}_1 = \text{Val}_1$"
AND "$\text{OutLV}_2 = \text{Val}_2$"
:
AND "$\text{OutLV}_p = \text{Val}_p$"

Every *n* condition ""$\text{InpLV}_i = \text{Val}_{i_1}$ OR "$\text{InpLV}_i = \text{Val}_{i_2}$" OR … OR "$\text{InpLV}_i = \text{Val}_{ij}$"" consists of m subcondition "$\text{InpLV}_i = \text{Val}_{ij}$", where Val_{ij} is the *j–e* meaning LV in sub-term. Let its number be determined by the number of the input LV value: *ij*. Denote the degree of truth of the sub-term (ST) with the number *i–j*, respectively μ_{ij}. For all the terms of this FPR we obtain a matrix *M*:

Table 6.12 FPR for PF

FPR	IF		THAN	
	LV	Value	LV	Value
FPR_{19}	Age	Average	Possibility of MPT	Higher
FPR_{20}	Age	Inclined	Possibility of MPT	Much higher
		Old		
FPR_{21}	Age	Old	Possibility of MPT	Lower
		Young		
FPR_{22}	Gender	Man	Possibility of MPT	Higher
FPR_{23}	Gender	Women	Possibility of MPT	Lower
FPR_{24}	BH	Nope	Possibility of MPT	Lower
FPR_{25}	BH	Weak	Possibility of MPT	Lower
FPR_{26}	BH	Strongly	Possibility of MPT	Much higher
FPR_{27}	SM	Nope	Possibility of MPT	Lower
FPR_{28}	SM	Few and rare	Possibility of MPT	Higher
FPR_{29}	SM	Many and often	Possibility of MPT	Much higher
FPR_{30}	Alc	Nope	Possibility of MPT	A little lower
FPR_{31}	Alc	Few and rare	Possibility of MPT	A little higher
FPR_{32}	Alc	Many and often	Possibility of MPT	Higher
FPR_{33}	PNP	Nope	Possibility of MPT	A little lower
FPR_{34}	PNP	Sometimes	Possibility of MPT	A little higher
FPR_{35}	PNP	Often	Possibility of MPP	Higher
FPR_{36}	Age	Average	Possibility of MFT	Higher
FPR_{37}	Age	Inclined	Possibility of MFT	Much higher
		Old		
FPR_{38}	Age	Old	Possibility of MFT	Lower
		Young		
FPR_{39}	Gender	Man	Possibility of MFT	Higher
FPR_{40}	Gender	Women	Possibility of MFT	Lower
FPR_{41}	BH	Nope	Possibility of MFT	Lower
FPR_{42}	BH	Weak	Possibility of MFT	Higher
FPR_{43}	BH	Strongly	Possibility of MFT	Much higher
FPR_{44}	SM	Nope	Possibility of MFT	Lower
FPR_{45}	SM	Few and rare	Possibility of MFT	Higher
FPR_{46}	SM	Many and Often	Opportunity of MFT	Extremely high
FPR_{47}	Alc	Absent	Opportunity of MFT	Little lower
FPR_{48}	Alc	Few and rare	Opportunity of MFT	Little higher
$ПНП_{49}$	Alc	Many and often	Possibility MFT	Вищі

(continued)

Table 6.12 (continued)

FPR	IF		THAN	
	LV	Value	LV	Value
FPR_{50}	PNP	Absent	Opportunity of MFT	Little lower
FPR_{51}	PNP	Sometimes	Opportunity of MFT	Little higher
FPR_{52}	PNP	Often	Opportunity of MFT	Higher
FPR_{53}	Age	Average	Opportunity of MMT	Higher
FPR_{54}	Age	Inclined	Opportunity of MMT	Extremely high
		Old		
FPR_{55}	Age	Child	Opportunity of MMT	Lower
		Young		
FPR_{56}	Gender	Man	Opportunity of MMT	Higher
FPR_{57}	Gender	Woman	Opportunity of MMT	Lower
FPR_{58}	BH	Absent	Opportunity of MMT	Lower
FPR_{59}	BH	Weak	Opportunity of MMT	Higher
FPR_{60}	BH	Strongly	Opportunity of MMT	Extremely high
FPR_{61}	SM	Absent	Opportunity of MMT	Lower
FPR_{62}	SM	Few and rare	Opportunity of MMT	Higher
FPR_{63}	SM	Many and often	Opportunity of MMT	Extremely high
FPR_{64}	Alc	Absent	Opportunity of MMT	Little lower
FPR_{65}	Alc	Few and rare	Opportunity of MMT	Little higher
FPR_{66}	Alc	Many and often	Opportunity of MMT	Higher
FPR_{67}	PNP	Absent	Opportunity of MMT	Трохи нижчі
FPR_{68}	PNP	Sometimes	Opportunity of MMT	Little Higher
FPR_{69}	PNP	Often	Opportunity of MMT	Higher
FPR_{70}	Age	Average	Opportunity of MAT	Higher
FPR_{71}	Age	Inclined	Opportunity of MAT	Extremely high
FPR_{72}	Age	Old	Opportunity of MAT	Lower
		Child young		
FPR_{73}	Gender	Male	Opportunity of MAT	Higher
FPR_{74}	Gender	Female	Opportunity of MAT	Lower
FPR_{75}	BI	Absent	Opportunity of MAT	Lower
FPR_{76}	BI	Little	Opportunity of MAT	Higher
FPR_{77}	BI	Big	Opportunity of MAT	Extremely high
FPR_{78}	SM	Absent	Opportunity of MAT	Lower
FPR_{79}	SM	Lightly and rarely	Opportunity of MAT	Higher
FPR_{80}	SM	Heavily and Often	Opportunity of MAT	Extremely high

(continued)

Table 6.12 (continued)

FPR	IF		THAN	
	LV	Value	LV	Value
FPR_{81}	Alc	Absent	Opportunity of MAT	Little lower
FPR_{82}	Alc	Lightly and rarely	Opportunity of MAT	Little higher
FPR_{83}	Alc	Heavily and Often	Opportunity of MAT	Higher
FPR_{84}	PNP	Absent	Opportunity of MAT	Little lower
FPR_{85}	PNP	Sometimes	Opportunity of MAT	Little higher
FPR_{86}	PNP	Often	Opportunity of MAT	Higher

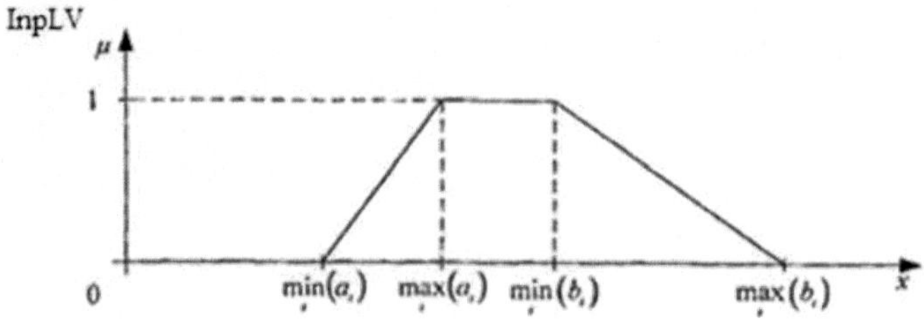

Fig. 6.83 Obtaining MF term

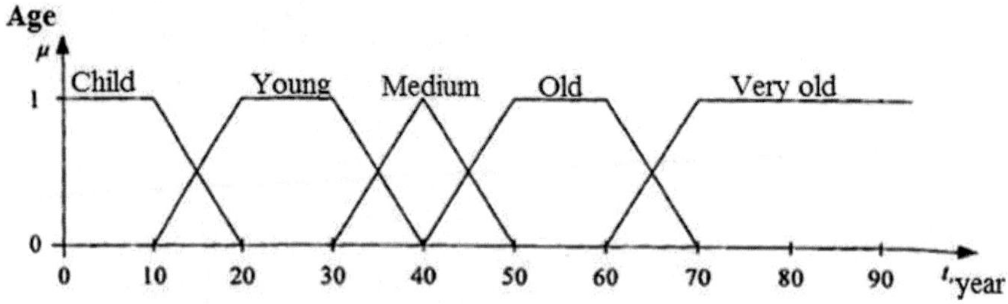

Fig. 6.84 CF for InpLV "Age"

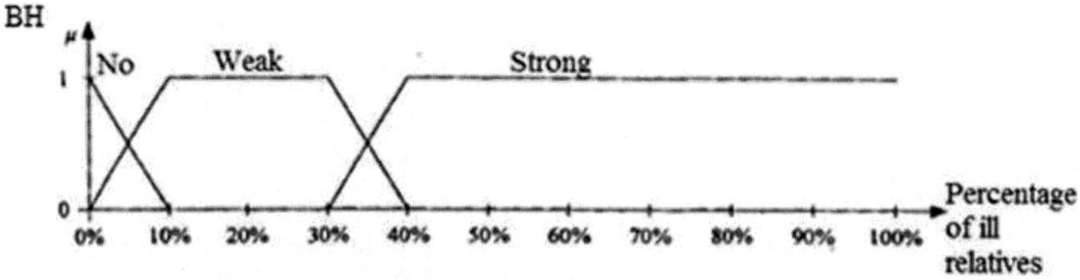

Fig. 6.85 CF for InpLV "BH"

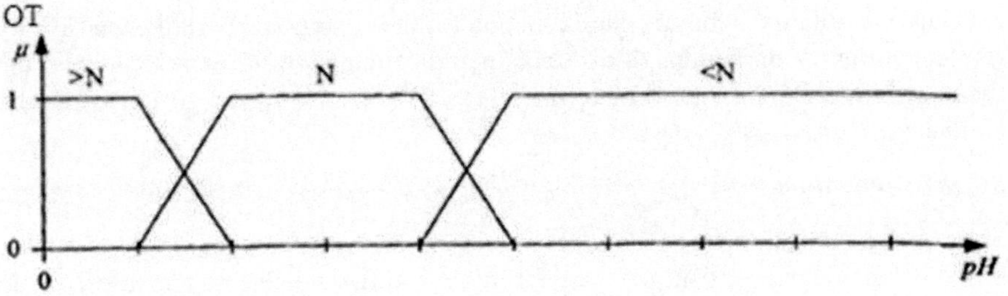

Fig. 6.86 CF for InpLV "OT"

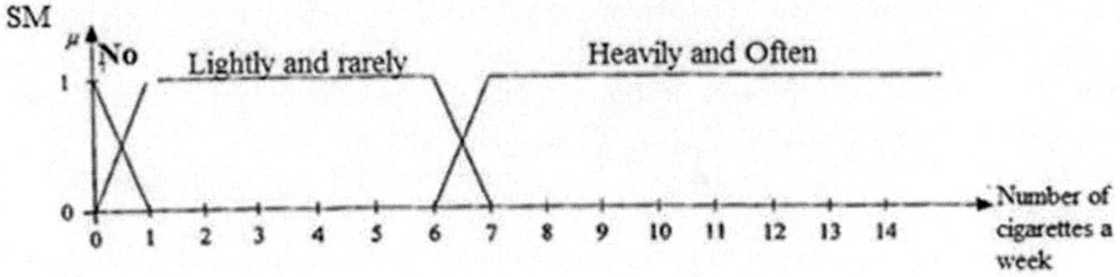

Fig. 6.87 CF for InpLV "SM"

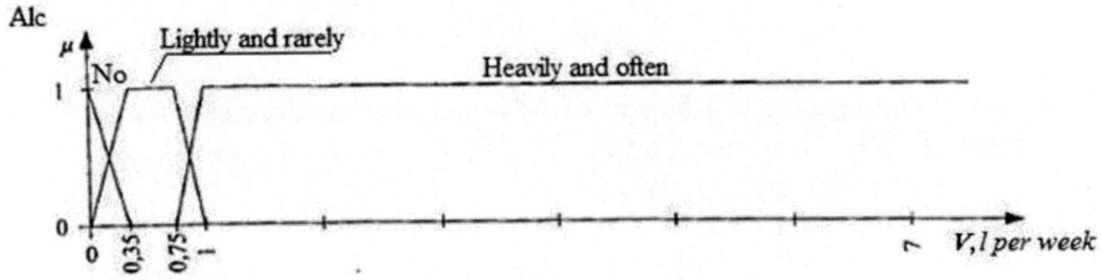

Fig. 6.88 CF for InpLV "Alc"

$$\mathbf{M} = \begin{pmatrix} \mu_{11} & \mu_{12} & \cdots & \mu_{1m} \\ \mu_{21} & \mu_{22} & \cdots & \mu_{2m} \\ \vdots & \vdots & \ddots & \vdots \\ \mu_{n1} & \mu_{n2} & \cdots & \mu_{nm} \end{pmatrix} \tag{6.21}$$

Then for FPR_k can be written for true conditions Y_k:

$$\mu_{\mathrm{U}_K} = \bigwedge_{i=1}^{n}\left(\bigvee_{j=1}^{m} \mathrm{PY}_{ij}\right) = \min_i\left(\max_j(\mu_{ij})\right), \tag{6.22}$$

where $i = \overline{1, n}, j = \overline{1, m}$, $k = \{1, 2, 5, 8, 9\}$.

Thus, we obtain the membership function for the conditional part of each FPR by implementing the operations of disjunction and conjunction of fuzzy logic over the membership functions according to the selected fuzzy inference algorithm, i.e. min and max.

Stage 4. Activation

Next, at the activation stage, the degree of truth of the inferences (I) is calculated based on the degree of truth of the conditions and the weights of the rules. To do this, it is first necessary to determine the diagnostic significance of the rule, which is a fuzzy value. For weights F_k of FPR_k rules MF is based on the same principle as MF in the stage of fuzzification—on the basis of knowledge of experts or literary sources. On the axis z we will postpone the intervals chosen by the experts for the fuzzy weight of the rule. The range of possible weights is limited by the segment $z \in [0; 1]$ ith expert expresses an opinion that the specific weight of the kth rule is approximately equal from F_{ka1} to F_{kb1}. Then, having analyzed the opinions of all the experts, the MF of weight kth FPR is proper can be represented as follows (Fig. 6.89).

The weight F_k of FPR rules in clear form can be obtained by defuzzification of a fuzzy weight factor. Perform this operation by the center of gravity method:

$$F_k = \frac{\int_0^1 z\mu(z)dz}{\int_0^1 \mu(z)dz}. \tag{6.23}$$

Suppose that the known coefficients of the weights of the rules $F_k, k = \overline{1, q}$ where q—number of FPR. Then, membership function for the inferences of rules can be derived from the following expression

$$\mu_{3_k} = F_k\mu_{Y_k} = \min_i\left(\max_j(\mu_{ij})\right), \tag{6.24}$$

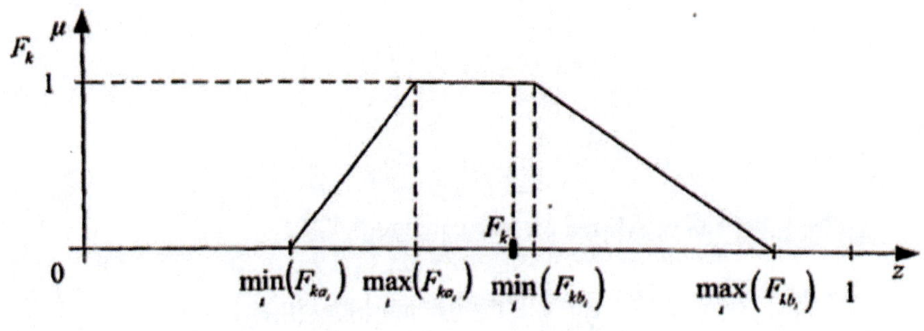

Fig. 6.89 Obtaining MF for weight F_k

where $F_k \in [0; 1]$. However, in fact, it is necessary to calculate the MF of the terms of OutLV for each FPR. For the OutLV term $\mu^*_{uv_k}(y_u)$ it is determined by the rules of fuzzy composition:

$$\mu^*_{uv_k}(y_u) = \tilde{\mu}_{uv}(y_u) F_k \min_i\left(\max_j\left(\mu_{ij}(x_i)\right)\right), \tag{6.25}$$

where $\tilde{\mu}_{uv}(y_u)$ is the MF vth term uth OutLV initially; $\mu^*_{uvk}(y_u)$ is the MF vth term uth OutLV, modified on the basis of kth rule; (y_u) is the variable corresponding to OutLV ω_u, meaning of (y_u) is the rational numbers.

Let q_{uv} is the number of FPR, where at the very end is determined MF term u OutLV $v.F_k \neq 0$ only for one of the terms of one of the OutLV. For other OutLV and other terms of this OutLV $F_k = 0$. Then

$$\mu^*_{uv}(y_u) = \tilde{\mu}_{uv}(y_u) \frac{\sum_{k=1}^{q_{uv}} F_k \min_i\left(\max_j\left(\mu_{ij}(x_i)\right)\right)}{\sum_{k=1}^{q_{uv}} F}. \tag{6.26}$$

Stage 5. Accumulation

The purpose of this stage is to find one of MF for each OutLV. The need for accumulation arises due to the fact that the subs are related to the same OutLV, belong to different FPF.

Let C_{uv} is a fuzzy set that corresponds to vth term uth OutLV. Looking further at each one in succession of OutLV $\omega_u \in W$ and related fuzzy sets $C_{u_1}, C_{u_2}, \ldots, C_{u_v}, \ldots, C_{u_1}, C_{u_2}, \ldots, C_{u_v}, \ldots$, we get MF for output variables. Analytically, the result of accumulation is defined as an association of MF terms $\mu^*_{uv}(y_u)$ for each of OutLV:

$$\omega_u(x) = \max_v\left(\mu^*_{uv}(y_u)\right) = \max_v\left(\tilde{\mu}_{uv}(y_u) F_k \min_i\left(\max_j(\mu_{ij})\right)\right). \tag{6.25}$$

Denote by q_{uv} the number of fuzzy product rules that are contained in the overlays vth term ith OutLV. Then in the case $q_{uv} > 1$ can be calculated MF as follows:

$$\omega_u(y_u) = \max_v\left(\tilde{\mu}_{uv}(y_u) \frac{\sum\limits_{k=1}^{q_{uv}} F_k \min_i\left(\max_j\left(\mu_{ij}(x_i)\right)\right)}{\sum\limits_{k=1}^{q_{uv}} F_k}\right). \tag{6.26}$$

Stage 6. Defuzzification

Next, according to the selected algorithm, the next is stage of defuzzification. In fuzzy output systems, it uses the procedure to find a clear value for each output LV of plural $W = \{\omega_1, \omega_2, \ldots, \omega_s\}$, where s is the total number of OutLV based on the rules of the fuzzy conclusion system.

To carry out defuzzification we will choose the method of the center of gravity:

$$\chi_u = \frac{\int_{\min}^{\max} y_u \mu(y_u) dy_u}{\int_{\min}^{\max} \mu(y_u) dy_u}, \tag{6.27}$$

where y_u is variable with rational values that corresponds OutLV ω_u; $\mu(y_u)$ is the MF, that corresponds OutLV ω_u after the accumulation stage; X_u is the result of defuzzification; min, max is the left and right boundaries of the media spacing considered OutLV ω_u.

Substituting the expressions obtained in the previous stages into this formula, we obtain:

$$\chi_u = \frac{\int_{\min}^{\max} y_u \max\limits_v \left(\tilde{\mu}_{uv}(y_u) \frac{\sum_{k=1}^{q_{uv}} F_k \min\limits_i \left(\max\limits_j (\mu_{ij}(x_i)) \right)}{\sum_{k=1}^{q_{uv}} F_k} \right) dy_u}{\int_{\min}^{\max} \max\limits_v \left(\tilde{\mu}_{uv}(y_u) \frac{\sum_{k=1}^{q_{uv}} F_k \min\limits_i \left(\max\limits_j (\mu_{ij}(x_i)) \right)}{\sum_{k=1}^{q_{uv}} F_k} \right) dy_u}. \tag{6.28}$$

Intermediate variables "The possibility of a MPT" and "The possibility of a MFT", "The possibility of a MMT", "The possibility of a MAT" have the same terms. MF value of the LV data and the corresponding values of OutLV can be represented in the following form (Fig. 6.90).

The value was found to be OutLV are derived from intermediate values LV, taken through one, and the application of logical connection "OR" with adjacent values. The result is a maximum of MF term of OutLV is observed at preference 1, 6, 10, 11–18 terms of the intermediate LV over the neighbors. The rest of the points were completed by calculating what for any y_u sum of MF for all terms is one. This dependency is intermediate of LV and OutLV built on expert judgment.

6.6.3 The Use of Artificial Neural Networks to Determine the Weights of Fuzzy Rules in a System of Neuro-Fuzzy Inference

The above scheme for constructing a system of fuzzy logic inference uses the knowledge of experts to determine the importance of FPR. However, the weights of the rules change depending on the emergence of new facts, as well as inferences, more precisely prove or, conversely, refute previously known data. Along with this, there are such FPR, the weights of which cannot be determined, or they have a very high degree of fuzziness. Experts strongly disagree, and sometimes they are quite

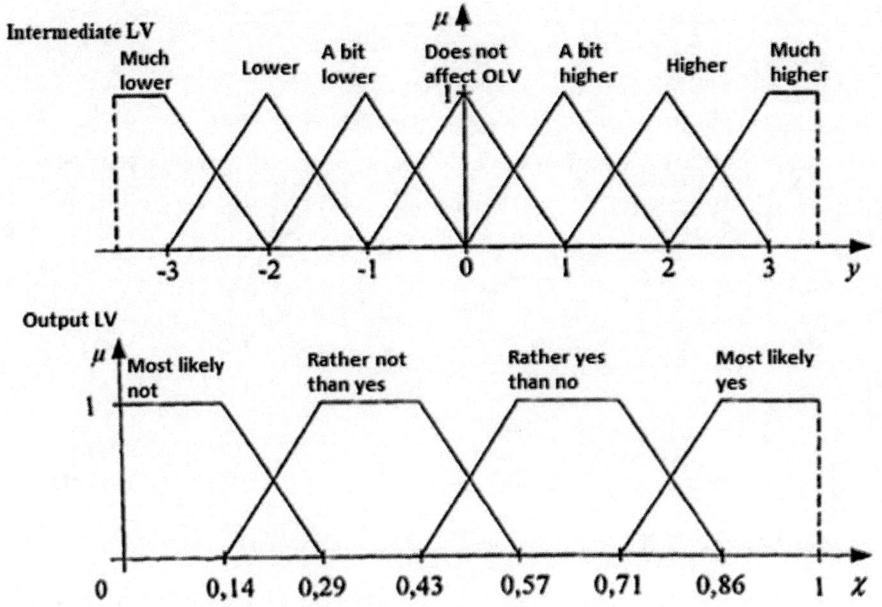

Fig. 6.90 MF of intermediate and output LV

controversial. However, in doubtful cases it is necessary to make diagnoses and their correctness can be evaluated later with periodic examination of the patient.

In order to eliminate this uncertainty, it is suggested that, in order to adjust the FPR weights, to submit a logical inference system in the form of a hybrid ANN.

The structure of a fuzzy neural network is identical to a multilayer one. However, the layers in it correspond to the stages of fuzzy inference [18, 20, 31], which consistently perform the following functions:

- input layer performs the function of phazzification (bringing to fuzziness) based on the specified input membership functions;
- output layer implements the function of defuzzification (bringing to clarity);
- hidden layers reflect the set of FPR and the following output stages: aggregation, activation and accumulation.

Compared to the usual multilayer ANN [1], there is a complicated structure in this network it is seen from Fig. 6.91. However, such a network allows you to clearly trace the functioning logic of the inference system. The lack of such an opportunity in the ordinary ANN is their major disadvantage. In addition, along with the transparency of the structure, obtained from the fuzzy inference system, the main advantage of the network is its adaptability. This advantage of the system is shown in the form of weight adjustment FPR in the learning process of ANN.

Neurons with notation min, max and Σ work as appropriate mathematical functions. The neurons marked with a sign "×", transmit to the output the result of

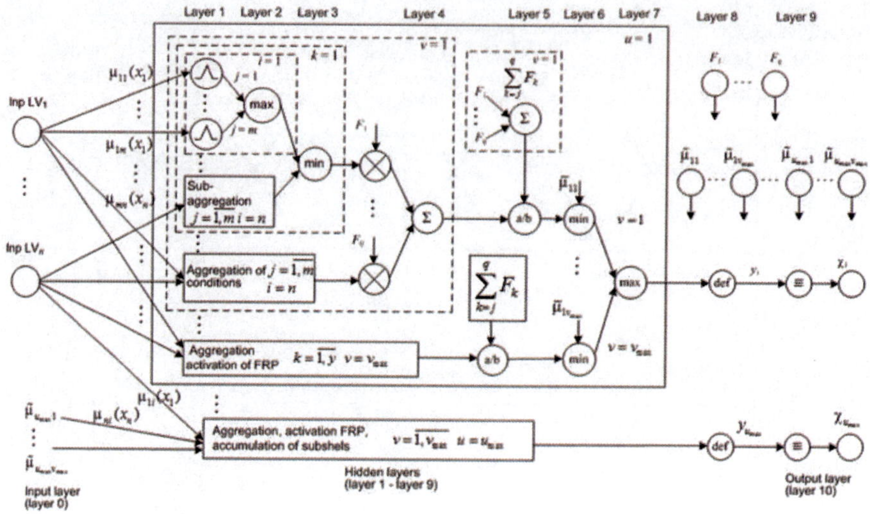

Fig. 6.91 Hybrid neural network structure

the product of the input signals, the sign "≡"—establish correspondence between intermediate of LV and OutLV. " ⋀ "—neurons that carry out the fazzificatuon operation are different for each term of each InpLV. Node "a/b" is intended to be divided by the sum of the weights of the active rules according to the formula at the activation stage. Neuron "def" implements the function of defuzzification previously indicated by the center of gravity method. Each input variable through a fuzzification block is connected to each FPR, as shown in Fig. 6.91. The choice of the entry of InpLV in terms of rules it is carried out in the form of weights of the logical levels "0" and "1", by which the corresponding membership functions are multiplied $\mu_{ij} = (x_i)$ where, $i = \overline{1, n}$ is the number of InpLV, $j = \overline{1, m}$ is the number of term.

Learning this ANN was performed by an error-propagation algorithm modified for use in fuzzy neural networks [19]. The changes were as follows:

- the network was learnt by parts, by each FPR separately;
- only weights were adjusted during training FPR (layer 4), and other parameters were either considered unchanged or modified in accordance with a law known in advance.

Thus, taking into account the fact that layers of neurons with precisely defined parameters were represented by one layer with a complex activation function, then the fuzzy ANN is presented in Fig. 6.91, identical to the three-layer ANN (with one hidden layer). Thus, the training of the neuron-fuzzy network was reduced to the training of a three-layer perceptron. The artificial neural network was trained in the process of tuning the system after executing the algorithm for tuning the DDSS parameters presented in Fig. 6.31. Coefficients F_1–F_q obtained as a result of training are clearly included in the DDSS in the form of degrees of significance for FPR.

A fuzzy neural network is not used during the operation of DDSS and is used only in case of a change in the structure of DDSS (changes in the set of FRP, incoming or outgoing LV) as well as in the case of new facts in the literature or medical practice that prove or disprove previously known data.

References

1. Arbib, M.A.: The Handbook of Brain Theory and Neural Networks, p. 1301. The MIT Press, London (2003)
2. Arsenyev, S.: Extracting knowledge from medical databases. S. Arsenyev (2000). [Online course]. Access mode: http://crystalway.pspu.ru
3. Bryan, R.H, Alexander, E.K., Bible, K.C., Doherty, G.M., Mandel, S.J., Nikiforov, Y.E., Pacini, F., Randolph, G.W., Sawka, A.M., Schlumberger, M., Schuff, K.G., Sherman, S.I., Sosa, J.A., Steward, D.L., Tuttle, R.M., Wartofsky, L.: THYROID vol. 26, Number 1, 2016 American Thyroid Association Management Guidelines for Adult Patients with Thyroid Nodules and Differentiated Thyroid Cancer, The American Thyroid Association Guidelines Task Force on Thyroid Nodules and Differentiated Thyroid Cancer, American Thyroid Association. Mary Ann Liebert, Inc. (2015) https://doi.org/10.1089/thy.2015.0020
4. Chumachenko, H.: Development of an image processing algorithm for diagnostic tasks. In: Chumachenko, H., Levitsky, O. (eds.) Electronics and control systems, pp. 57–65. NAU (2011)
5. Chumachenko, H.: Algorithmic support of distributed databases. Artif. Intell. **3**, 37–42 (2012)
6. Chumachenko, H.: Investigation of the structure of the neural network in the problem of diagnostics. In: Chumachenko, H., Levitsky, O. (eds.) International Scientific and Technical Conference "Avia-2011", pp. 22.40–22.43 (2011)
7. Chumachenko, H.: Construction of distributed databases. Artif. Intell. **2**, 94–98 (2011)
8. Chumachenko, H.: Development of neural network structure in diagnostic problems. Bullet. NAU **2**, 57–65 (2012)
9. Eliseeva, I.I., Yuzbashev, M.M.: General theory of statistics. Finance Stat. 480 (2003)
10. Epstein, E.V., Matyashchuk, S.I.: Ultrasound Examination of the Thyroid Atlas. Kiev (2004)
11. Epstein, E.: Ultrasound examination of the thyroid gland. In: Epstein, E., Matyashchuk, S. (eds.) Atlas-leadership, pp. 43–273. KVITs (2004)
12. Furman, Y.A., Krevetsky, A.V., Peredreev, A.K., et al.: Introduction to contour analysis and its applications to image and signal processing. Fizmatlit 592 (2002)
13. Galushkin, A.: Problem solving in a neural network logical basis. Neurocomput. Dev. Appl. **2**, 49–70 (2006)
14. Gonzalez, R.: Digital image processing. Technosphere 635 (2005)
15. Gonzalez, R., Woods, R.: Digital image processing. Translation from English edited by P. A. Chochia, p. 1072. Technosphere (2005)
16. Gruzman, I.S., Kirichuk, V.S., Kosykh, V.P., et al.: Digital Processing Images of Information Systems: A Training Manual, p. 168. NSTU Publishing House, Novosibirsk (2000)
17. Katkovnik, V.: Spatially adaptive support as a leading model selection tool for image filtering. In: Katkovnik, V., Foi, A., Dabov, K., Egiazarian, K. (eds.) Proceedings of the First Workshop Inf. Th. Methods Sci. Eng., pp. 365–457. WITMSE, Tampere (2008)
18. Kodogiannis, V.S., Chowdrey, H.S.: Multi network classification scheme for computer-aided diagnosis in clinical endoscopy. In: MEDSEP 2004—Advances in Medical Signal and Information Processing International Conference, pp. 262–267. Malta (2004)
19. Kulikowski, C., Ammenwerth, E., Bohne, A.: Medical imaging informatics and medical informatics: opportunities and constraints Findings from the IMIA Yearbook of Medical Informatics. Methods Inf. Med. **41**(2), 183–189 (2002)
20. Michael, A.A.: The Handbook of Brain Theory and Neural Networks, p. 1301. The MIT Press, London (2003)

21. Osovsky, S.: Neural networks for information processing. Translation by I. Rudinsky. Finance Stat. 344 (2002)
22. Propp, R.M.: Clinic and treatment of malignant tumors of the thyroid. Medicine **164**, 100–124 (1966)
23. Schepotina, I.: Algorithms of modern oncology. In: Schepotin, I. (eds.) Academician of the Academy of Medical Sciences of Ukraine GV Bondar, Corresponding Member of the Academy of Medical Sciences of Ukraine V. Ganul, p. 304. Kniga plus (2006)
24. Shmoilova, R.A., Minashkin, V.G., Sadovnikov, N.A., et al.: Theory of statistics: textbook. Finance Stat. **656** (2003)
25. Sonka, M., Hlavac, V., Boyle, R.: Image processing, analysis, and machine vision. PWS 108 (1998)
26. Soyfer, V.A.: Methods of computer image processing. In: Soyfer, V. A. (ed.), p. 698. Fizmatlit, Moscow (2003)
27. Syneglazov, V.M.: Intellectual forecasting methods. Educ Ukraine 236 (2013)
28. Valdina, E.A.: Thyroid Disease: Manual, 3rd edn, p. 368. SPb, St. Peterburg (2006)
29. Valdina, H.: Thyroid Diseases, p. 259. St. Petersburg (2006)
30. Veshotort, A.M., Zuev, Y.A., Krasnoproshin, V.V.: Two-level recognition system with logical corrector. Recognition, classification, forecast. Mathematical methods and their application. Science **2**, 73–98 (1989)
31. Yarushkina, N.: Fuzzy Neural Networks. Artif. Intell. News **2–3**, 47–51 (2001)

Chapter 7
Intelligent Automated Road Management Systems

7.1 Analysis of the Situation at Intersections and Methods of Distribution of Traffic Flows Problem Statement

Currently, there is a sharp increase in the number of road vehicles, with a relatively slow expansion of the road network, which leads to a significant increase in time loss in traffic jams.

Even in those cities where a sharp increase in the number of vehicles was already laid during the construction of the city, significant difficulties occur when driving vehicles, this is due to the fact that the throughput capacity of most carriageways has been exhausted and this contributes to the formation of a large number of traffic jams. In this regard, the problem arises of reducing cars downtime [2].

One way to solve this problem is to expand and modify the existing: road network. Another fundamentally different approach is the creation of new modes of transport. However, both approaches are long-term, requiring significant investment of time and money. Of course, it is necessary to develop both of these approaches, but a method is needed that allows to achieve a quick effect at a relatively low cost.

This way is to increase the efficiency of traffic by creating any distribution systems: traffic flows. This is supposed to allow reducing the loss of time and money due to downtime in traffic jams with minimal costs for the modification of the existing traffic system.

If the intersection has a simple configuration and the main traffic flow is in the direction of travel (transport continues to move in its direction), then the average ratio of traffic flows in different directions is determined and, accordingly, the ratio of the duration of traffic signals is established. When choosing the ratio of the duration of traffic signals, the accepted restrictions are taken into account, for example, the minimum duration of the traffic signal taking into account the switching time of the signals (yellow signal).

If the intersection has a complex configuration, distribution is carried out by using auxiliary traffic signals. At the intersection, the main traffic flows and their

M. Zgurovsky et al., *Artificial Intelligence Systems Based on Hybrid Neural Networks*, Studies in Computational Intelligence 904,
https://doi.org/10.1007/978-3-030-48453-8_7

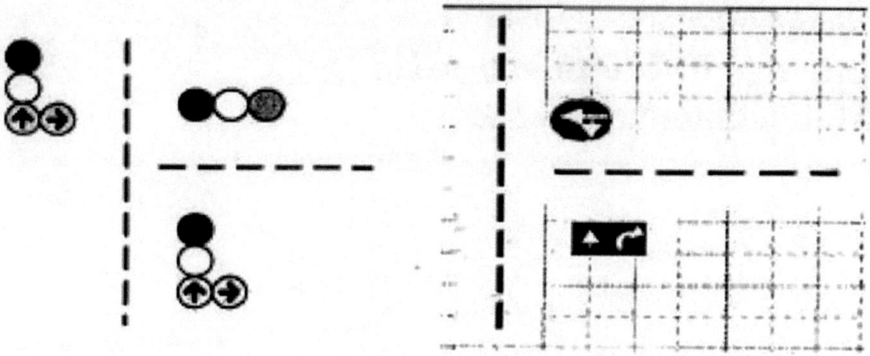

a) b)

Fig. 7.1 An example of an intersection using: **a** is the turn signals; **b** is the road marking

density are highlighted. Depending on the direction and density of traffic, traffic lights are placed. Streams of transport with low intensity are sent using auxiliary traffic signals—arrows. As a rule, the flow of traffic is directed by arrows to resolve the situation of intersection of traffic flows in various directions. An example of a section of a road network using traffic lights is shown in Fig. 7.1a.

In addition to the use of traffic lights, an additional element of traffic control is the use of road markings and traffic signs. An example of a road network section using road marking is shown in Fig. 7.1b.

The advantages of this approach to the distribution of traffic flows include the ease of implementation of the system, high reliability and widespread applicability. The disadvantages include the lack of ability to respond to changing situations on the road. This traffic flow distribution system can work in one of several modes:

- normal mode;
- the standby mode, activated when the traffic is low, signals a flashing yellow traffic signal, usually at night;
- the peak load mode, included during peak hours, differs from the usual short duration of each of the traffic signals.

For example, if in normal mode the duration of the green or red traffic signal is approximately 60 s, then in peak mode it will be approximately 15 s.

The advantages of this approach are ease of implementation and independence of the road network configuration. Automatic traffic lights can be delivered at any intersection.

The disadvantages include a rigidly set operating mode. When the intensity of the traffic flow changes, the traffic lights continue to work in a given mode, even if it becomes ineffective.

In addition, to set up traffic lights with this approach to the distribution of traffic flows, it is necessary to collect sufficiently complete information about traffic at different hours, coordinate traffic flows through a number of intersections, etc.

7.2 Adaptive Management Strategies

Adaptive control strategies use algorithms that are executed in real time to optimize traffic lights based on current traffic conditions, demand and system throughput. An example is the ACS Lite system, which includes software that regulates signal separation, bias, and phase duration.

The main objective of adaptive control is to minimize delays and reduce the number of stops. Systems require large volumes of observations, usually in the form of vehicle detectors and pedestrians that interact with central and/or local controllers. ACS differs from other more traditional traffic control systems due to the possibility of adequate regulation on each cycle based on information received in real time. Theoretically, ACS allows you to create an infinite number of time plans.

Traditional ACS technologies are the Australian SCATS (Sydney Coordinated Adaptive Traffic System) and the UK SCOOT (Split, Cycle, Offset Optimization Technique). In Los Angeles, we developed and implemented a program called ATSC (Automated Traffic Surveillance and Control). New adaptive control algorithms are being developed and tested in the United States under the name ACS. The OPAC (Optimized Policies for Adaptive Control) and RHODES (Real-Time Hierarchical Optimized Distributed Effective System) algorithms are developed and implemented with the support of the Federal Highway Administration (FHWA). Both algorithms are used at arterial crossroads. Another new RTACL (Real-Time Traffic Adaptive Control Logic) adaptive control system was tested in Chicago at the turn of the second millennium. The RTACL algorithm was developed for a network of streets.

Benefits have been demonstrated in several areas where traditional adaptive management technologies have been deployed (e.g., SCOOT, SCATS). However, some researchers argue that systems are no better than good fixed plans for depending on the time of day. These observations may be true, especially in areas where traffic is predictable or there is little traffic growth. Other adaptive control issues include detector maintenance and communication problems. There is currently little information about the benefits of new adaptive strategies like OPAC, RHODES, and RTACL, and conclusions about ACS are drawn mainly from the experience of SCOOT and SCATS. Section 7.1 shows the development of cities and the degree to which systems are used in them (Table 7.1).

Table 7.1 Using ACS

City/District	System	Number of intersections
Los Angeles	ATSC	1170
Auckland	SCATS	350 +
Heneppin	SCATS	71
Arlington	SCOOT	65
Minneapolis	SCOOT	60
Anaheim	SCOOT	20
Darham	SCATS	is unknown

7.3 Analysis of Approaches to the Distribution of Traffic Flows Based on Artificial Intelligence

Traditional methods of distributing traffic flows are of little use for solving problems with a high degree of uncertainty. On the other hand, the accumulated experience of using intelligent (fuzzy logic, neural networks) methods to solve various problems, including poorly formalized ones [4], allows us to apply these methods to the problem of distribution of traffic flows.

7.3.1 Fuzzy Logic Approach

When describing traffic flows, the following assumptions are introduced: transport moves straight through the intersection; traffic flow in oncoming directions is considered as a single one. Two flows are considered, conventionally referred to as "traffic flow on the right" and "traffic flow on top".

To describe the situation at the intersection, the following set of linguistic variables (LP) is proposed: **CarsRight**—traffic flow on the right; **CarsUp**—traffic flow from above; **LightLen**—duration of the green traffic light for the traffic flow on the right; **DeltaLight**—change in the duration of the green traffic light for the traffic flow on the right [1].

The variables **CarsRight** and **CarsUp** characterize the traffic flow coming to the intersection from different directions and are described by the following sets of terms: **CarsRight** = {Zero, Small, Medium, Large}, **CarsUp** = {Zero, Small, Medium, Large}.

Membership functions are introduced to describe linguistic variables. An example of membership functions to describe the linguistic variables **CarsRight**, **CarsUp** is shown in Fig. 7.2.

For LP **LightLen**, **DeltaLight**, the following sets of terms are offered: **LightLen** = {Small, Medium, Large}, **DeltaLight** = {Negative, Zero, Positive}.

Each intersection is described as a combination of entrance and exit roads. In general, their number may be different. A complete set of entrance and exit roads is described respectively by the following sets:

$$I = \{I_1, I_2, \ldots, I_m\}, O = \{O_1, O_2, \ldots, O_n\}.$$

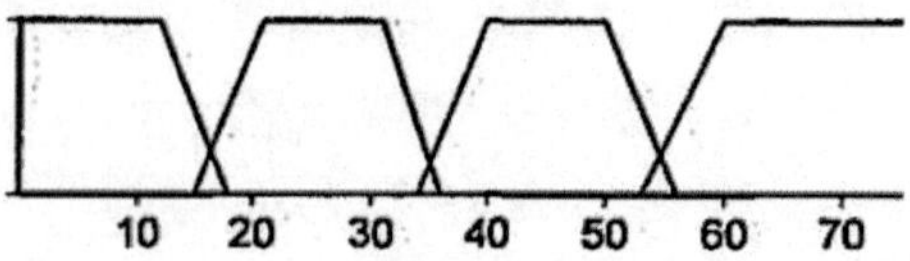

Fig. 7.2 An example of membership functions of the linguistic variables **CarsRight** and **CarsUp**

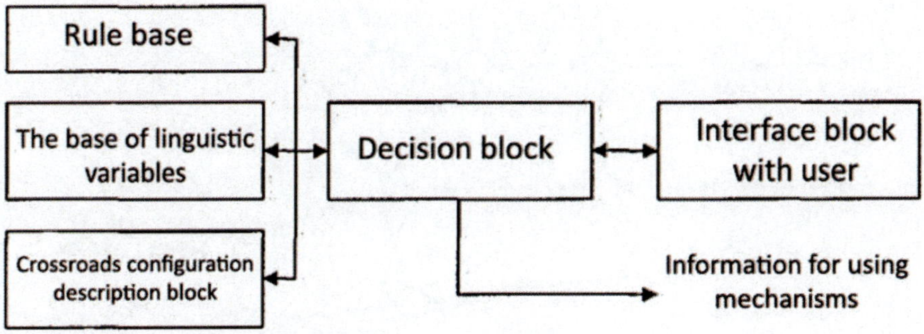

Fig. 7.3 Fuzzy logic tool component for arbitrary crossroads

Analysis of traffic flows through the intersection is carried out by direct enumeration of all possible combinations of traffic flows. Since even for a rather complicated intersection configuration this number of combinations is relatively small, a complete enumeration of all options is performed. For a crossroad of five input and five output arcs, the total number of options will be: $N = 5(C_5^1 + C_5^2 + C_5^3 + C_5^4 + C_5^5) = 5(5 + 10 + 10 + 5 + 1) = 155$.

In [5], the structure of the distribution system for traffic flows based on fuzzy logic (NL) was proposed, shown in Fig. 7.3.

To implement the NL-based approach and verify its effectiveness; In the work, a tool was developed that allows you to set the configuration of the intersection of the characteristics of the transport flows, form a set of linguistic, variables and, a set of rules, perform a model experiment for a given time, collect information about the modeling process.

At an average traffic intensity, a fuzzy algorithm ensured the quality of the distribution of traffic flows similar to automatic traffic lights. With high traffic: a fuzzy algorithm showed; slightly better results compared to automatic traffic lights. In the general case, the number of cars passing through the intersection increases by 7–9%, and the distribution of traffic flows becomes more even (49 and 51% for the fuzzy algorithm and 65 and 35% for the traditional algorithm).

A fuzzy algorithm for the distribution of traffic flows provides a better distribution compared to a traditional automatic controller with fixed intervals, especially at high traffic volumes. The main difficulties in using a fuzzy controller are difficulties in choosing the form and parameters of membership functions, on which the quality of the distribution of traffic flows depends.

7.3.2 *Neural Network Approach*

The road network is described in the form of a directed graph, the arcs of which are roads, and the peaks are intersections [1]. Each arc of the graph is oriented, and

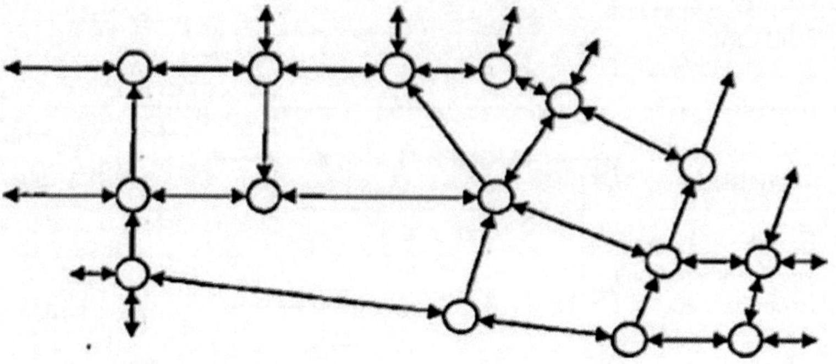

Fig. 7.4 An example of a road network segment, presented as oriented graph

the direction of the arc corresponds to the direction of traffic. Two-way roads are represented as a combination of two counter-directed arcs.

An example of a road network section in this representation is shown in Fig. 7.4.

The following approach to the construction of a system based on neural networks (NN) is proposed.

Each intersection is controlled by its neural network, while it is possible to link different NNs to build a network of controlled intersections. Information on traffic flows at each input arc, as well as information on the intervals of work of traffic signals, is fed to the NN inputs. At the NN outputs, a signal is obtained for changing the duration of traffic signals. In this problem, the NN solves the problem of approximating a complex function. The argument of the function is the normalized values of traffic flows and the intervals of the traffic lights, and the value of the function is the new intervals of the traffic lights.

To build the tool, we used the structure of the NN type "multilayer perceptron". The number of hidden layers and the number of neurons in each hidden layer is determined; based on the quality criterion for the distribution of traffic flows. The more complex a controlled intersection or network of intersections, the more layers and neurons in each layer should have an NN.

Neural Networks training is carried out according to the following scheme.

1. Neural Networks training together with any other system of distribution of traffic flows, for example, with a traditional one based on automatic traffic lights.
2. Retraining of the National Assembly with human participation; if in any situations the quality of management is poor.
3. Retraining of the aircraft at various critical (emergency) situations, which allows for the management of transport even in very difficult situations.

Carrying out a model experiment showed that the neural network algorithm missed, 14–15% more cars compared to the traditional algorithm, and 4–5% compared to the fuzzy algorithm with approximately the same intensity of traffic flows [7]. The distribution of traffic flows was 49 and 51%—similar to the fuzzy algorithm.

7.4 Control Traffic Flow at the Intersection of Arbitrary Configuration

7.4.1 Development of System Requirements

The formulation of the problem of the distribution of traffic flows for an intersection of arbitrary configuration is considered. An approach to the description of an intersection of arbitrary configuration is proposed, which allows us to formulate a control problem; applicable to any intersection and generalize it to implement a system for distributing traffic flows. Approaches to the distribution of traffic flows in a complex intersection configuration are proposed.

In the general case, any intersection can be represented as a set of inputs and outputs. An example of a crossroads is shown in Fig. 7.5.

Each input and output has a certain capacity—the maximum number of machines that can go through this input or output. We denote the input arcs as I, and the output arcs as O. Obviously, the same number of machines must pass along the output arcs as along the input ones. Then for a given intersection we can write the following condition (Fig. 7.6):

$$I_1 + I_2 + I_3 + I_4 + I_n = O_1 + O_2 + O_3 + O_4 + O_n.$$

Although in this case the number of input and output arcs is the same, this is not a prerequisite. Then in general terms for an arbitrary intersection configuration, this can be written as follows:

Fig. 7.5 Custom crossroads example

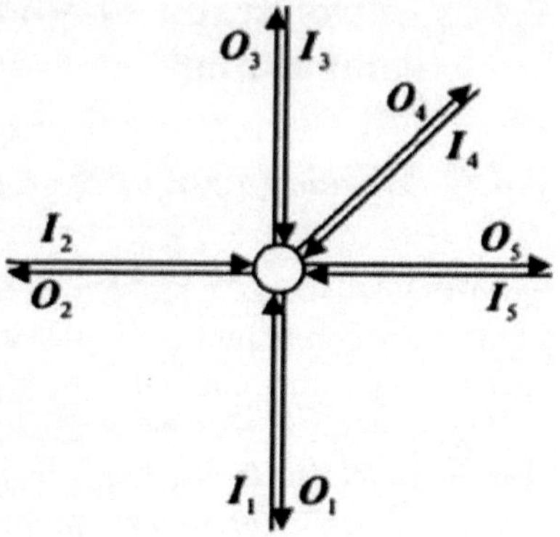

Fig. 7.6 An example of a formal description of an intersection of arbitrary configuration

$$\sum_{i=1}^{m} I_i = \sum_{j=1}^{n} O_j, \tag{7.3}$$

where m, n are the number of input and output arcs, respectively.

Since the maximum flow of cars will not necessarily go along the input arcs, the condition changes as follows:

$$\sum_{i=1}^{m} I_i \leq \sum_{j=1}^{n} O_j, \tag{7.4}$$

The problem of the distribution of traffic flows can be formulated as follows: for each input, it is necessary to find such a set that

$$U^i = \{u_1, u_2, \ldots, u_n\} : \sum_{j=1}^{n} n_j^i = I_i. \tag{7.5}$$

Let $V(I_i)$ be the number of cars at the input I_i. Then the objective function can be described as follows:

$$C = \sum_{i=1}^{m} V(I_i) \rightarrow \min. \tag{7.6}$$

The following method is proposed for distributing traffic flows.

In contrast to the traditional approach, it is proposed to change the ratio of the duration of traffic signals in different directions. Moreover, in the direction where the traffic flow is longer, the duration of the green signal increases, and vice versa, when the flow decreases, the duration of the green signal decreases, and the duration of the red signal increases [7].

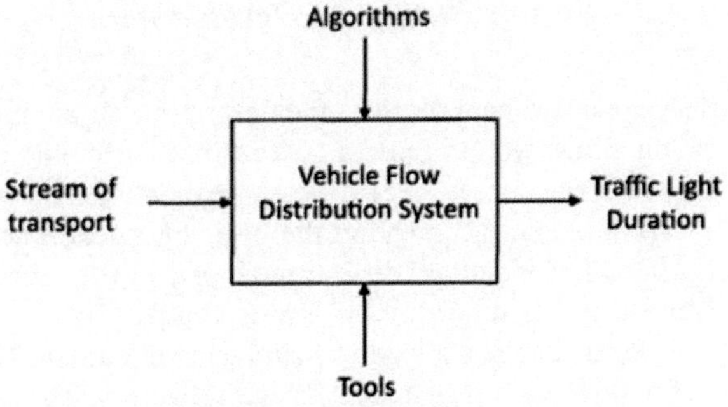

Fig. 7.7 Vehicle flow distribution system

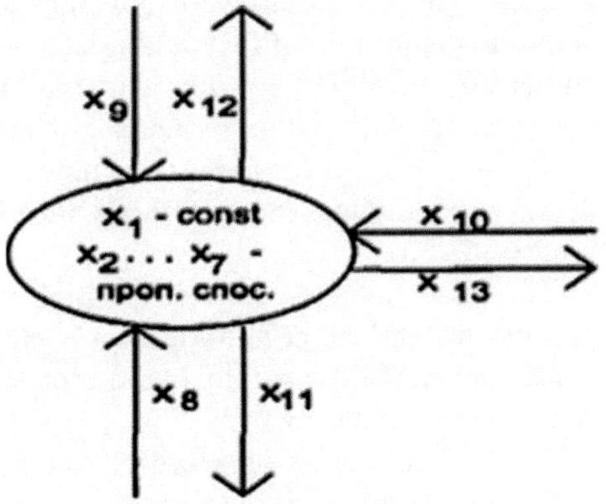

Fig. 7.8 Intersection description

The system of distribution of traffic flows is presented in general form in Fig. 7.7, the input of which receives information about the availability of suitable cars (about their number) to each intersection (in the general case $x_{k+1} \ldots X_1$)—the input stream and $x_{l-1} \ldots x_n$ is the output stream which is input for another intersection in our case $x_8 \ldots x_{10}$ (Fig. 7.8), and for another intersection $x_{11} \ldots x_{13}$ will be the input).

The configuration of the crossroads (x_1 i.e. in our case a T-shaped one) and the streets (in the general case $x_2 \ldots x_k$ and in our case $x_2 \ldots x_7$ is the throughput of each street) connecting them are given. In general, the view is rearranged like this $x = |x_1 x_2 \ldots x_k x_{k+1} \ldots x_l x_{l+1} \ldots x_n|$.

If the exit of the intersection branches, it is assumed that each car approaching it with the same probability will go along each of the branches. The control device needs to decide what traffic signal and what duration (Y) should be at each of the intersections.

$$Y = \frac{1}{1 + e^{-\alpha \sum_{i=l}^{m} x_i w_l}} \tag{7.7}$$

where $w = |w_l \ldots w_n|$ are weighting factors; Y is the duration of the green traffic light.

7.4.2 System Structure Development

The considered approach to the description of an intersection of arbitrary configuration allows you to build a system for distributing traffic flows based on a neural network [6].

To implement control on the basis of a neural network, it is enough to submit the values of input variables to the inputs of the neural network, and then to receive control signals for traffic lights at the output.

The advantages of using a neural network are the relative simplicity of building a traffic flow control system and high reliability. The simplicity of building a control system is explained by the fact that the same type of information conversion device is used—neurons, the structure of which is described by very simple relationships. In addition, the process of functioning of a neural network itself is described by fairly simple relationships. High reliability is ensured by the properties of NN, namely, the preservation of the ability to function in case of failure of some neurons.

The high reliability of the traffic flow distribution system is explained by the fact that in case of failure of any neurons, the overall operability of the control system is preserved.

The disadvantages of the distribution system of traffic flows based on a neural network include the need to train a neural network before practical use. Moreover, depending on the complexity of the controlled intersection system, the training time can be very significant.

Consider the following approach to formalizing the problem.

Let there be a road network consisting of sections of roads and intersections. Such a network can be represented as a graph, the vertices of which are intersections, and arcs are sections of roads between intersections. Since there are one-way roads, the arcs of the graph must be directed. In this case, the road network is described by a directed graph. Two-way roads are proposed to be described as two counter-directed arcs. Given the direction of the roads, an example of a section of the road network will look as shown in Fig. 7.9.

For each intersection and each road, we introduce our own characteristics. For the intersection, these will be the following characteristics:

- the number of intersecting roads D;
- permissible directions of traffic through the intersection W.

The following characteristics are offered for the road:

- the width of the road H;
- the number of lanes n;
- coating quality Q;
- maximum permitted speed V_{max}.

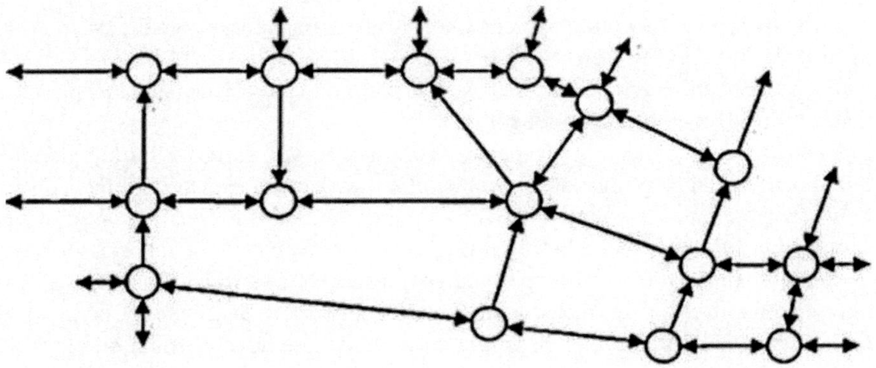

Fig. 7.9 An example of a road network segment, presented as oriented graph

To implement control within an interconnected set of intersections, data such as information on traffic flows moving from neighboring intersections is needed. For a neural network, the input information goes to the inputs directly in the form of data on the number of cars standing in front of the intersection or moving from a neighboring intersection. In this regard, the following control scheme is proposed.

To implement control within an interconnected set of intersections, data such as information on traffic flows moving from neighboring intersections is needed. For a neural network, the input information goes to the inputs directly in the form of data on the number of cars standing in front of the intersection or moving from a neighboring intersection. In this regard, the following control scheme is proposed.

The system of coordination of the work of neural networks (SCR NN) is a computer complex of high power. The SCR NN task is to collect information about the number of cars at the intersections themselves in each direction, information about the traffic flow between intersections, traffic signal duration, information about the status of each neural network and, based on the information collected, adjusts the parameters of each neural network. The neural networks unit is a computing complex that implements a neural network. Based on the information about the traffic flow at the intersection and the data of the SCR NN, the neural network is constantly trained. To collect information about the state of the traffic flow, it is proposed to use sensors that record the number of cars.

Using this interaction scheme allows you to take into account changes in the flow of transport coming to each intersection. If the neural network is not used at every intersection, instead of communicating with other intersections, you can use only information about the traffic flow coming from a neighboring intersection. Hence, various options for including a neural network follow:

- a separate neural network at one intersection;
- interconnection of all neural networks;
- a mixed version, including traditionally controlled intersections, as well as intersections controlled by a neural network.

Possible options for including neural networks, taking into account the assumption made, are shown in Fig. 7.10.

Each connection option can be used in various cases. Consider the possible applications of each of the options.

The option of connecting a separate neural network is used, for example, at the initial one; the stage of introducing a neural network into the distribution of: traffic flows (with gradual introduction) when the surrounding intersections are not yet equipped; a system for distributing traffic flows based on neural networks. Another option is the presence of a sufficiently large transport hub surrounded by low-critical intersections. In this case, introducing a neural network to distribute traffic flows at each of the small intersections is impractical, while at a large distribution of traffic flows is necessary.

The option of connecting all neural networks into a single network is used when distributing traffic flows in a road network consisting of many large transport nodes—intersections with high traffic intensity.

A mixed connection option is used either for boundary intersections—located on the border of a large transport hub, as a result of which part of the neighboring intersections is controlled by a neural network, the other part is not controlled. In practice, this option may also occur in the event of a failure of one of the traffic flow distribution systems at any intersection. In order to maintain the ability of the functioning of the traffic flow distribution system at other intersections, it is proposed to connect the neural network with both the traffic flow distribution system at a neighboring intersection and the source of information about the traffic flow from a neighboring intersection; independent of the distribution system of traffic flows. With this connection, even if the traffic flow distribution system fails, a source of traffic data from a neighboring intersection is saved at a neighboring intersection.

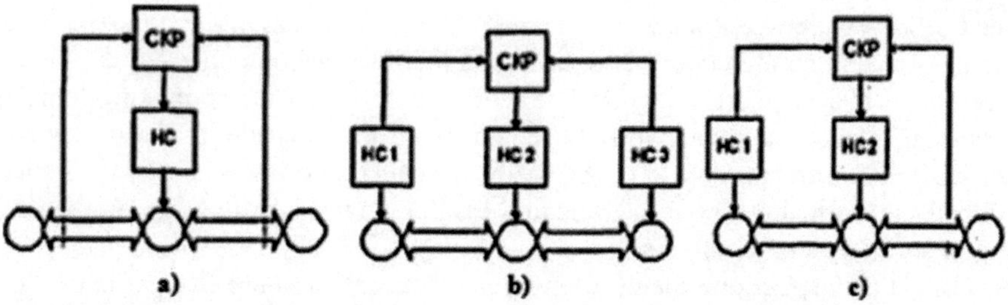

Fig. 7.10 Possible options for including a NN in a road network: **a** neighboring intersections are not controlled by a neural network; **b** neighboring intersections are controlled by a neural network; **c** a mixed version

7.5 Development of Neural Network Models

7.5.1 Analysis of the Capabilities of Neural Network Models and Training Methods

In the case of the problem of the distribution of traffic flows, the NN is used to solve the problem of approximating the function [1]. In this case, the argument of the function is data on the traffic flow, and the function is the control signals for traffic lights. In general, this function can be described as follows:

$$F\{x_1, x_2, \ldots, x_m\} = \{u_1, u_2, \ldots, u_n\}, \tag{7.8}$$

where x_i are arguments of the approximated function; u_n are values of the approximated function.

Crossroads, the scheme of which is shown in Fig. 7.11, a is the simplest type of intersection, and is formed by the intersection of two roads.

We designate each of the roads in accordance with the direction of one of the letters *N*, *S*, *W*, *E*, (north, south, west, east). The following input variables are used for this intersection:

- information about the transport in front of the intersection—only four variables F_N, F_S, F_w, F_E, characterizing the flow of cars in each direction;
- information about the future traffic flow from neighboring intersections—also four variables F_N, F_S, F_w, F_E.

The output variables are control signals for traffic lights, their number depends on the main traffic flows at the intersection.

To implement a "good" system for the distribution of traffic flows, it is necessary to train the neural network on known test data sets. The place of the neural network in the control circuit at the training stage in the framework of the traffic control system is shown in Fig. 7.12.

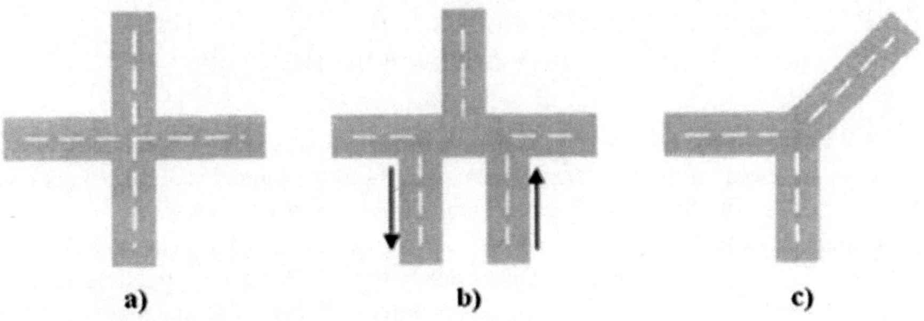

Fig. 7.11 Some common intersection types

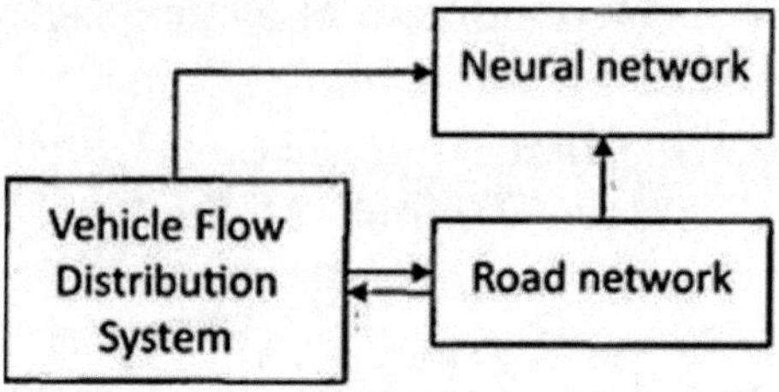

Fig. 7.12 Scheme of including a neural network in the distribution circuit of traffic flows for training

Neural network training is carried out using one of the standard methods (for example, by the back propagation method, the Kohonen algorithm or the Hebb learning rule).

The quality control of the distribution of traffic flows can be carried out by controlling the density of the flow of cars going through the intersection, and the average length of the "traffic jam" in front of the intersection at each input arc. The quality of the distribution of flows; transport can be considered acceptable if, as a result of the implementation of the developed control actions, the situation on the road improves. By improving the situation we mean reducing the length of the "traffic jam" before the intersection, respectively, worsening the situation—increasing the length of the "traffic jam". We can assume that the quality of the distribution of traffic flows is acceptable if the total length of the "traffic jams" on all input arcs has decreased as a result of the distribution of traffic flows. This raises the question of the quality of the distribution of traffic flows in the case when a large traffic flow arriving at once sharply increases the total length of traffic jams at this intersection. Therefore, we will consider the quality of the distribution of traffic flows acceptable if, over a number of traffic light cycles, the distribution system of traffic flows based on a neural network has provided better control than the traditional one based on automatic traffic lights.

To implement a system for distributing traffic flows, we selected the "multilayer perceptron" neural network and the Kohonen network with learning algorithms using the backpropagation method and learning algorithms of Kohonen and Hebb.

7.5.2 *Analysis of Neural Networks Selected for Research and Methods for Their Implementation*

At the entrance of the neural network, traffic information is received, which includes information about incoming transport streams, about the capacity of output arcs, as well as about traffic flows moving from neighboring intersections. At the outputs of the neural network—the intervals of the traffic signals. It is proposed to use a sigmoid function as an activation function. This is due to the fact that the output must receive a signal from "0" to "1", which corresponds to the proportion of the green signal in the full cycle of the traffic light.

The advantages of NN type "multilayer perceptron" for "the problem under consideration are:

- the possibility of further education in the process of functioning;
- provides many input variables and many output variables without qualitatively complicating the structure of the system;
- the system remains operational when; partially incomplete set of input data;
- the system remains operational in case of failure of any elements.

Let's consider these advantages in more detail. Further training in the process of functioning allows you to respond to changes in the road traffic system—for example, when modernizing the road network, the existing one: the system: the distribution of traffic flows remains operational without readjustment and adjustment.

Since a large number of elements of the same type (neurons) are used for the neural network, and all the relationships between them are also of the same type, the increase in the number of inputs and outputs of the system remains unchanged: only the duration of the training changes in the base structure of the neural network.

Since the memory of the neural network is formed by all: the connections between neurons of the neural network: then in case of failure of any neurons, the overall operability of the system of distribution of traffic flows remains, although the failure of some neurons can lead to a decrease in control quality.

Consider the advantages and disadvantages of teaching methods—Hebb and Kohonen learning algorithms.

At the entrance of the neural network, traffic information is received, which includes information about incoming transport streams, about the capacity of output arcs, as well as about traffic flows moving from neighboring intersections. When using the Kohonen network, the range of changes in the duration of traffic signals is divided into several intervals depending on the number of output neurons, while the active neuron at the output will correspond to one of the intervals of the duration of the traffic signal.

The structure of the NN based on the Kohonen network is given in the Sects. 3.8.2 and 3.8.13.

The learning algorithm is given in the Sect. 3.8.2.

7.6 Synthesis of the Intersection Network Coordination System

To implement the task of traffic control, an adaptive coordination system has been developed [1, 2]. The architecture of the system is based on the decomposition of the tasks of evaluation and management into two hierarchical levels, which allows you to divide the system into logical subsystems with different responsibilities (Fig. 7.13):

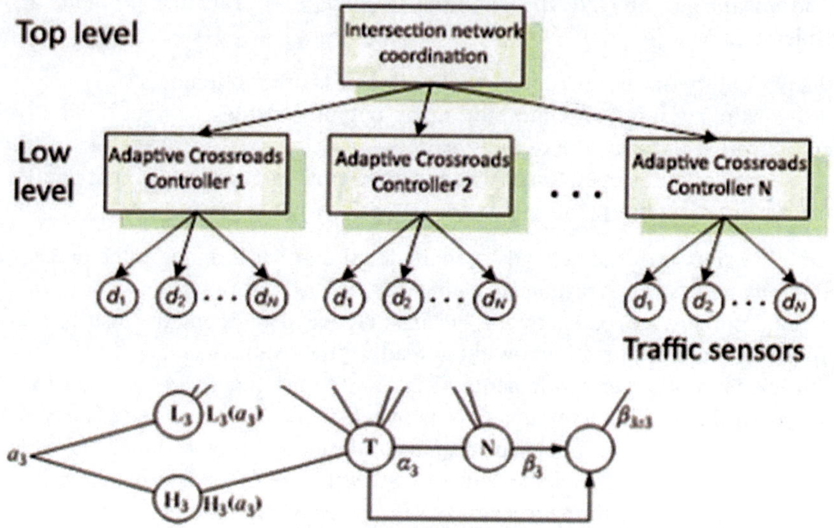

Fig. 7.13 Architecture of a two-level adaptive system

- local intersection management;
- coordination of the intersection network.

The coordination level of the intersection network is a computation center that receives information about traffic congestion, as well as the current values of control parameters from each intersection. At this level, finding the busiest directions and intersections—the base ones. The base length of the traffic light is determined.

Also, for each intersection, knowing the traffic congestion data at the previous intersections, the Q_{IJ} and C_{IJ} control parameters are forecasted corresponding to the forecast of the queue length before the intersection at the moment of the beginning of the corresponding phase (Q_{IJ}) and the intersection traffic intensity during this phase (C_{IJ}).

Based on the value of the Q_{IJ} parameter, a time offset (Offset) is determined for each intersection relative to the base.

The control center coordinates the work of the entire system. From the lower level—the current values of traffic characteristics for each section of B_{IJ} of the road graph, namely: average speed for the section, traffic intensity, queue length, time interval between cars.

Having this data at this level, the most loaded basic direction of movement and the intersection are found. The base length of the traffic light cycle is found.

With the help of a neural network (separate for each intersection), workload forecasting for the next cycle is performed (Fig. 7.14).

The system receives and aggregates detailed data regarding the values of traffic parameters.

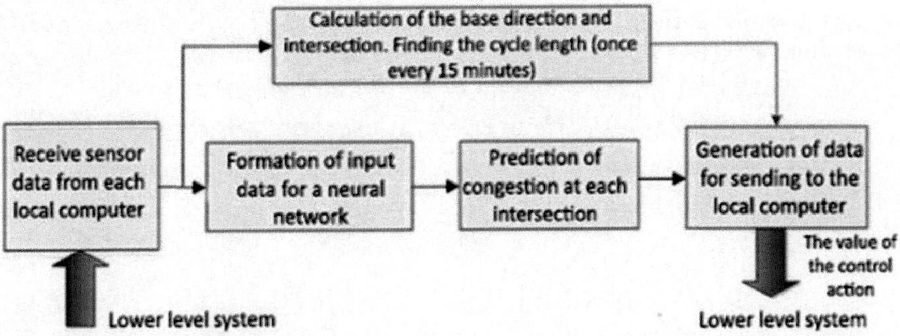

Fig. 7.14 Scheme of the top-level system

7.6.1 Calculation of the Base Congestion Network Intersections

Separately, once every 15 min for the entire system, the busiest directions and the intersection are recalculated. This allows you to respond flexibly to possible complications in some parts of the coordination area and to optimize traffic relative to the busiest section.

We select the total intersection most loaded for all directions. The choice is made using the criterion of the degree of saturation of the intersection. It is defined as

$$X = \left(\sum_{i=1}^{n} x_i\right) / n,$$

where x_i is the degree of saturation of the direction of movement, represents the ratio of the average number of cars arriving in this direction to the intersection during the cycle, to the maximum number of cars that can leave this intersection during the main measure of the corresponding phase.

$$x_i = I_i T_C / (M_{Hi} t_{Oi}),$$

where I_i and M_{Hi} are, respectively, the intensity of motion and the saturation flux.

In this case, the busiest intersections (like a bottleneck) are considered, where there are several directions with a high degree of saturation.

Choose the busiest direction of travel.

To do this, select max (x_i) for the busiest intersection. For the chosen direction, select neighboring intersections (both in the forward and in the opposite direction) until we reach the boundaries of the area.

We form all this data into an array that represents the busiest direction, for example:

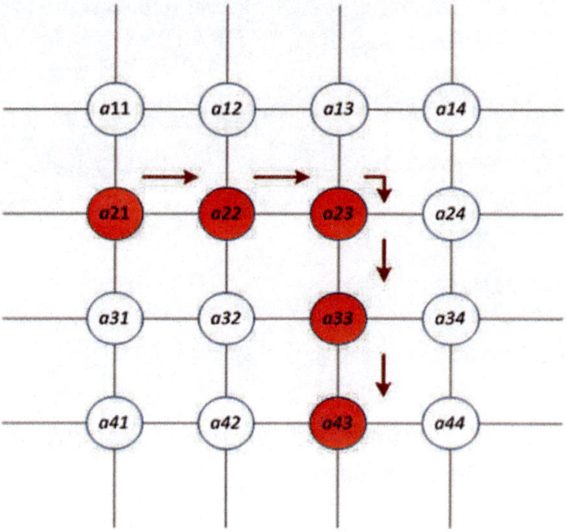

Fig. 7.15 Basic direction of movement in the network

A{a(21), a(22), a(23), and(33), and(43)}

.

A graphical representation is shown in Fig. 7.15.

7.6.2 *Coordination of the Intersection System Using Neural Networks*

For each intersection of the coordination area at the upper level, a control action is found that allows predicting the loading of the intersection for the next cycle [2].

For this, a neural network is used for each direction of movement, which receives data on the intensity of movement at the input, and the result is the value of the control action transmitted to the local level to control the selected intersection (Fig. 7.16).

Streams of cars arrive at the intersection for each direction from previous intersections. Therefore, to predict congestion in the next cycle of the system, we can use

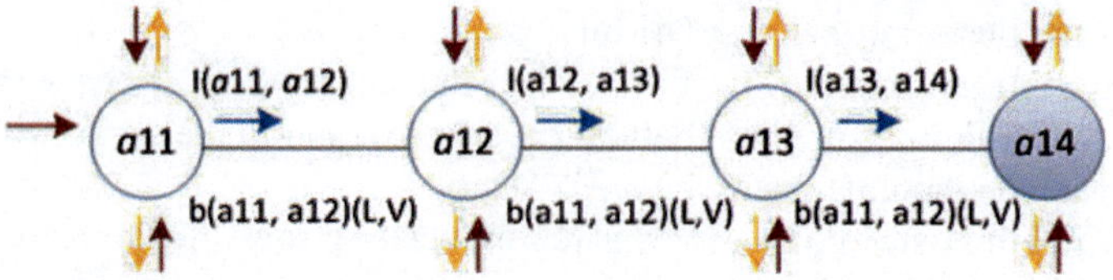

Fig. 7.16 Crossroads network connection

data on the intensity of the flows I and the time between the nodes of the system. The travel time T is determined by information about the length of the haul and the average speed on it. Motion time is a parameter characterizing the influence of flow intensity on predicting the value of control parameters.

7.6.3 Structural and Parametric Synthesis of the Network of the Upper Level of Coordination of Intersections

As the input data for the NN choose:

- traffic intensity I_{ij} at the exit from the three previous intersections;
- the time T_{ij} from each of the three previous intersections to the one for which the calculation is being made (see Fig. 7.16).

Accordingly, this data is normalized.

At the output, we obtain the values of the control parameters relative to the length of the queue TQ at the intersection at the beginning of the phase, as well as the SI value, which represents the control action by the number of cars that will arrive at the intersection with the duration of the corresponding phase. These values are used as control actions for the local neural network, which coordinates the work of a separate intersection. And also the TQ value is used to determine the temporary displacement of the intersection relative to the previous one.

7.6.4 Calculation of the Value of the Time Offset of the Beginning of the Cycle for the Intersection

It makes sense to calculate the time offset of the beginning of the cycle for the intersection relative to the previous intersection at the local level, since it depends on two parameters—the transit time of this section, $T_{Lij} = L_{ij}/V_{ij}$, which depends on the length of the section and average speed, and the predicted value.

The time required for cars to travel in this section is defined as the $T_{Lij} = L_{ij}/V_{ij}$, Optimum time offset between intersections will allow you to skip the queue of cars that are already in front of the intersection, and also take into account that the largest flow of cars from the previous intersection arrives at this intersection when all, or almost all of the cars in the queue have already passed the intersection. This will minimize the loss of the main stream of cars for braking before the intersection and speed gain.

To do this, we define the time offset as

$$T_{Oij} = L_{ij}/V_{ij} - t_{qij},$$

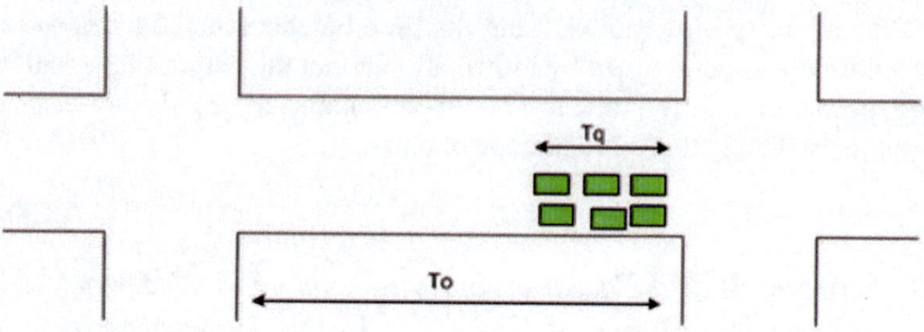

Fig. 7.17 Finding a temporary offset between intersections

where $T_{Oij} = L_{ij}/V_{ij} - t_{qij}$, is the time necessary for the line of cars to pass through the intersection in front of which they stopped (Fig. 7.17).

At the final stage of the calculation, the lengths of each phase, as well as the value of the time offset in seconds, are found for the given cycle length. To do this, data on the length of the plot (l), the average speed on it (V), as well as the number of cars in the queue are analyzed

$$t_O = \frac{l}{v} - t_q,$$

where t_q is the time necessary for the line of cars to pass through the intersection in front of which they stopped.

7.6.5 *Simulation of Adaptive Traffic Management*

To conduct simulation, software was created based on the C # programming language and the .Net Framework technology.

For modeling, it is possible to choose two types of system control—static and neural network. When choosing a static control, flows are modeled according to experimentally obtained data, tuned to the selected traffic intensities, which will be discussed below. If neural network control is selected, then the input data in the form of intensities (converted to the number of machines using software) are fed to the input of the neural network, processed and regulation of the duration of the phases of the traffic light is carried out.

To conduct the simulation, a section of Krasnoarmeyskaya street was chosen.

The total length of the traffic light cycle is set to 120 s. The length of the traffic light phases is set statically for a certain time of day, without taking into account the specific traffic congestion at a given particular moment. Also, about 4 s is spent on phase switching (yellow traffic signal).

Simulation was carried out for two cases.

1. Static regulation.

   The phase lengths and time offset are statically set and do not change.

2. Adaptive control with variable phase lengths:

   (a) the phase lengths are determined depending on the input parameters of the neural network;
   (b) the value of the temporary displacement is determined depending on the speed of the vehicles and the predicted length of the queue.

To carry out the simulation, Poisson flows were created, which were generated for each phase with a certain intensity. For each of the simulated phases, corresponding queues were created. With a green traffic light, one of the streams is serviced. The service intensity is predefined and the same for both flows [3].

As criteria for optimality, we chose to minimize the average time spent by a car in the queue.

These criteria are universal and allow you to optimally control traffic flows to reduce the downtime of the car before the intersection, reduce fuel consumption, as well as reduce environmental damage.

The simulation results are shown in Fig. 7.18.

During the simulation, it was found that the proposed adaptive control system allows to reduce vehicle delays by 15–25% depending on the intensity and characteristics of traffic flows, which allows us to confirm the suitability of this approach to the task of increasing the efficiency of traffic control.

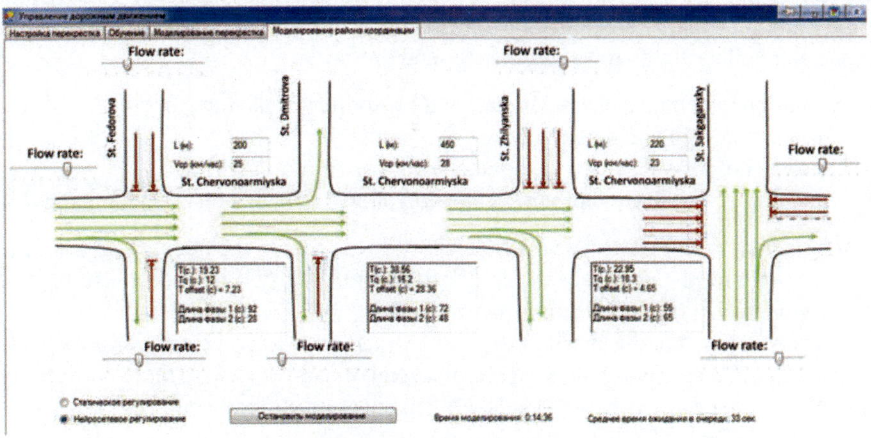

Fig. 7.18 Intersection network simulation

7.7 Implementation of the Proposed Mathematical Software in a Real System

7.7.1 Experimental Quantification of Traffic Flows at the Intersection

The thesis examines the optimal control of traffic flows in a simple intersection. The subject of the study was the intersection of Vladimirskaya and Tolstoy streets. Justifying this choice, we note that at the intersection of these streets a significant number of vehicles accumulate during rush hours.

To train a neural network, it is necessary to obtain the following indicators:

- the number of cars that stand in front of each traffic light with a red signal;
- the number of cars that drove into the green light for each of the phases of movement;
- the number of cars that remained in front of the traffic light after the end of the green signal;
- if all the cars from the queue have passed, but the green signal time has not ended, the number of cars that have traveled without being in the queue before;
- the number of cars that, according to the rules of the road, turned right with a red signal;
- the time of the traffic light cycle and the duration of the green and red signals for each of the phases.

By visual observation of the traffic flow, it can be noted that its only independent characteristic, in the context of one particular intersection, is the speed of arrival of cars. The same characteristics of the traffic flow as the speed of departure of cars on green and yellow light are values that depend on:

- throughput, technical characteristics and equipment of the intersection, laid down during its design and construction;
- the speed of arrival of cars at the intersection;
- the speed of departure, both to green and yellow light, is a reaction to change λ.

Obviously, with growth, the distance between the cars in the queue decreases, the drivers begin to respond faster μ and k grow accordingly. With a decrease λ, after some time, a decrease in k is also observed.

It is also clear that the speed of departure of cars on the green and yellow signals in total should be greater than the speed of arrival of cars, otherwise there is an abrupt increase in queues, which reduces the effectiveness of optimal control to zero.

For convenience, let us mark the starting points of observation on the intersection diagram with marks (Fig. 7.19).

The visual image of the studied intersection is shown in Fig. 7.20.

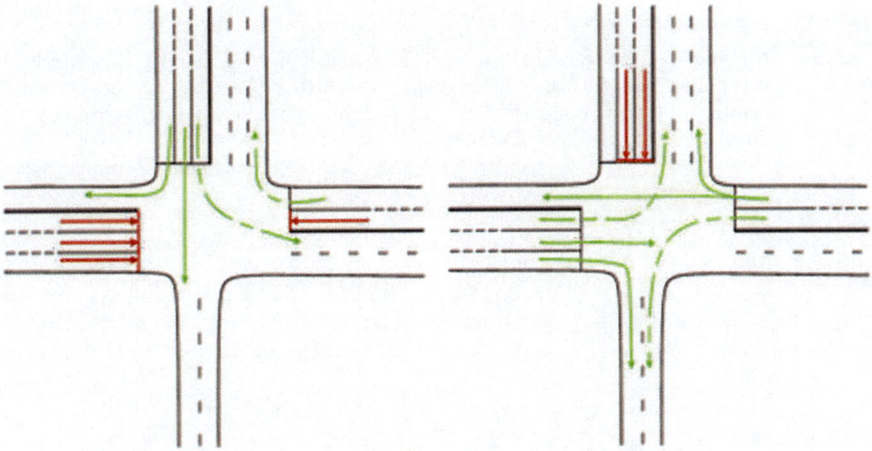

Fig. 7.19 Intersection scheme

Fig. 7.20 Investigated intersection: **1**—direction in a straight street. Tolstoy (east-west); **2**—st. Tolstoy (west-east); **3**—st. Vladimirskaya (Mon-Fri)

A fragment of the studied characteristics is given in Table 7.2, with indicators for each of the points in the same regulatory cycle.

Table 7.2 Test characteristics

Starting point	Number of cars in line before the green light	The number of cars that drove to the right	The number of cars that drove straight	Number of cars that drove left	Number of cars that turned right when the red light	The number of cars that remained after the end of the green signal
1	**2**	**3**	**4**	**5**	**6**	**7**
1	21	8	11	3	7	0
2	20	4	10	7	1	0
1	**2**	**3**	**4**	**5**	**6**	**7**
3	25	8	1	4	0	0
1	23	10	12	2	0	0
2	18	6	7	10	0	0
3	23	5	12	10	0	0
1	28	15	7	3	4	8
2	26	8	12	2	0	11
3	30	5	8	10	1	13
1	33	10	5	2	5	16
2	29	4	8	13	0	12
3	28	2	10	9	1	14

References

1. Bazhin, D.: About one approach to traffic control based on fuzzy logic. Mathematical Modeling in Solving Scientific and Technical Problems, vol. 2, pp. 80–84. Ufa, Technology (2001)
2. Chumachenko, H.: Traffic control system based on neuron networks. Electron. Control Syst. **3**(41), 35–40 (2014). NAU, Kyiv
3. Chumachenko, H., Tishchenko, R.: The choice of the criterion of optimality in traffic control problems. Adapt. Autom. Control Syst. **12**(32), 130–140
4. Sineglazov, V., Chumachenko, V.: Intellectual management of road traffic. Education of Ukraine, p. 192 (2013)
5. Vasiliev, V., Ilyasov, B.: Intelligent control systems using fuzzy logic: a training manual, p. 80. USATU, Ufa (1995)
6. Yakushev, D.: Neural networks in motion control problems. Foreign Electron. **1**, 58–64 (1999)
7. Yusupova, N., Bazhin, D.: Neural network methods of traffic control at intersections. Bull. USATU **3**(1), 126–134 (2002)

Chapter 8
Fire Surveillance Information Systems

8.1 Necessity of Fire Surveillance Information Systems

Today, in Ukraine the number of objects that were displayed on the fire observation panels is 51359, which is 13% of the total number of objects equipped with fire automatics. If we take the analysis for the last year, then in Ukraine 12277 objects were equipped with fire automation systems, and 9440 objects were displayed on the panels. This is despite the fact that it includes not only those objects that are equipped with fire automatics for the current period, but also those objects that were equipped in the previous period.

Today, the situation is such that, if the object is equipped with a modern fire alarm system, the fire signal is, at best, sent to the centralized fire observation panels, and in most cases to autonomous sirens. But even in the first case, it takes tens of seconds or even minutes for the signal to reach the destination. As we can see the process of receiving and processing, the response to fire signals is not locked into a single system, but the framework of the control system that is being implemented today is based on outdated equipment and the same software. As a result of the low technical equipment of the fire department–a late response to the fire signal. Today there are a number of problematic issues in the alarm system that make this system imperfect. First of all, it is the inability to send fire signals to the Operational Dispatch Service (ODS) of the head offices.

Most ODS do not have workplaces for fire observation panels installed in it; alarm notifications are transmitted by the controllers of commercial remote control over the common telephone line, which leads to a delay of signal transmission for 5–9 min, limitation of information about the object from which the signal comes, absence a regulatory framework that would clearly regulate the requirements for such work. Apparently, such an algorithm for transmitting an alarm is completely ineffective and morally outdated. Control centers can only receive verbal fire reports by telephone.

Having analyzed many systems, we came to the conclusion that it is economically and functionally feasible to build a three-levels system consisting of an upper level,

M. Zgurovsky et al., *Artificial Intelligence Systems Based on Hybrid Neural Networks*, Studies in Computational Intelligence 904,
https://doi.org/10.1007/978-3-030-48453-8_8

middle level and a lower level (commercial fire observation panels) [1]. Such a system, in our opinion, will provide a guaranteed and timely transmission of a fire notification that came from a commercial fire observation panel to the operator's automatic workplace. Functions of the global fire observation system [3]:

- registration and archiving of fire reports;
- registration of fire observation panels in accordance with established rules;
- accumulation of information base for operational reporting and further strategic analysis of the occurrence of fires, failures of centralized surveillance systems;
- display information on occurrences of fire-spread objects;
- ability to receive alarm messages both through a single database and through a regional segment;
- possibility of receiving additional information on the object (water supply and other);
- the ability to set the address of the object and route the fire department to it.

Today, the state of fire automation, in the last 5–7 years, has been brought to a state where fire automation systems successfully operate 2 times better. Previously, the figure was 50%, today it is 90–95%.

Centralized fire control system—a complex of technical means, intended for transmission in a given form of messages on the occurrence of fires and the technical condition of installations of fire automatics from the object of fire protection to the point of central fire communication, as well as their acceptance, processing, transmission and registration [9].

Modern centralized fire control systems (CFOS) are a two-tiered separation system built on the principle of open systems.

The structure of the CFOS is illustrated with the help of Fig. 8.1 [4].

The first level is a centralized monitoring panel equipped with computers for various purposes (servers, workstations) and a remote-control unit (RCD).

The second level (facility equipment) provides direct coordination with the fire control panel installed at the facility.

Centralized fire observation panels can be classified according to several characteristics (Fig. 8.2). Observation tactics (automated, nonautomated—outdated, but still in use), communication channels (switched and dedicated wired lines, GSM and radio, there are models that use satellite communication, integrated), capacity (small systems, medium and high capacity). Modern systems are usually built on a modular basis, and characteristics are selected for a specific task. Radio equipment is installed on nontelephoned objects.

On the phone—the leading. For remote sites located in difficult terrain and poor direct radio transmission, GSM is used. Large enterprises may use an internal fiber network.

For objects of particular importance, it is desirable to provide for the possibility of using duplicate channels (for example: radio conductor lines) [6]. This will greatly increase their security.

Given the rapid development and implementation of services on the World Wide Web, it is not difficult to predict that in the near future, channels of this network will

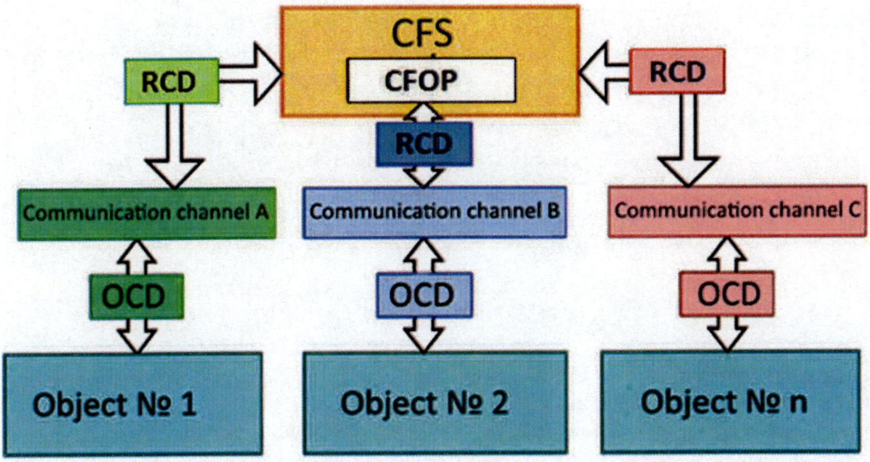

Fig. 8.1 Structure of the Centralized fire observation system: CFS is the Central Fire Station; CFOP is the Centralized Fire Observation Panel; RCD are Remote Connection Devices; OCD are Object Connection Devices

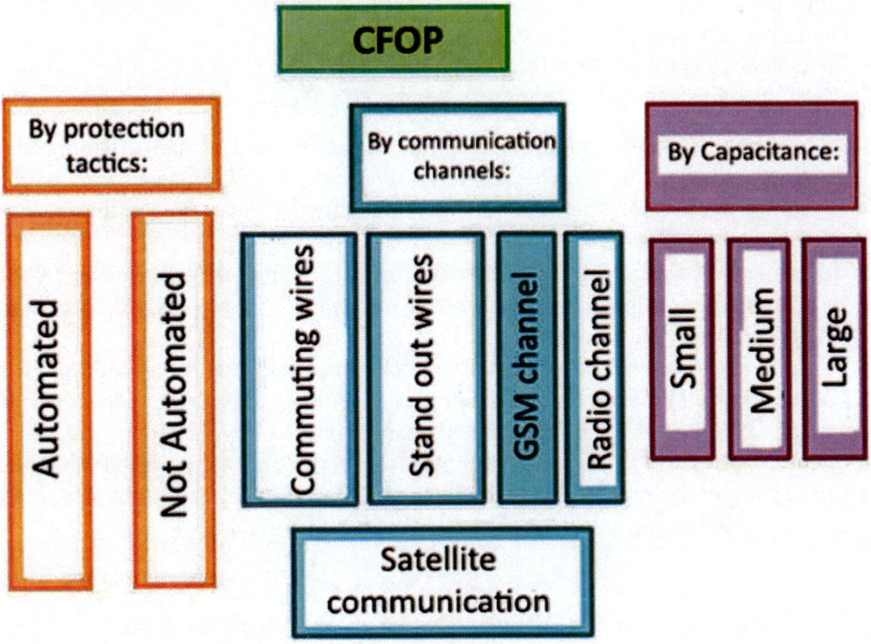

Fig. 8.2 Classification of centralized fire observation panels

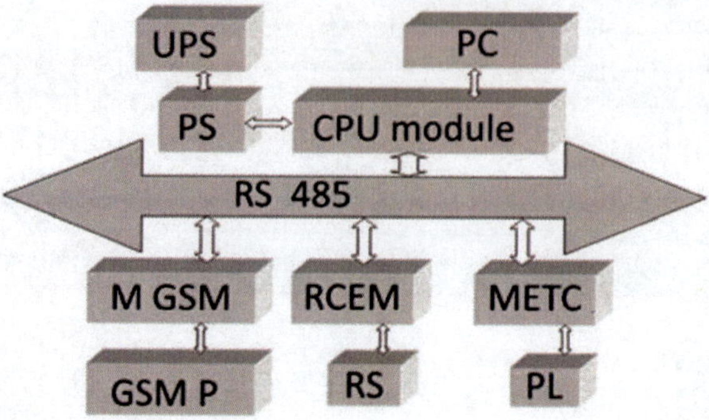

Fig. 8.3 CFOP block diagram: UPS is the Uninterruptible Power Supply; PS is The Power Supply; CPU is the Central Processing Unit; RCEM is the Radio Channel Equipment Module; M GSM is the GSM Channel Equipment Module; METC Is the Module of Equipment With Telephone Channel; RS is the Radio Station; PC is the Personal Computer; PL is the Phone Line; GSM P is the GSM Format Mobile Phone or GSM Modem

be used as means of transmitting information from object devices to the CFCP, for example, there are already so-called IP-cameras, which connect to the Internet and transmit real-time images of the protected object.

The block diagram of the CFCP is shown in Fig. 8.3.

The following requirements are imposed on CFCP equipment:

- automation, is the formation of the signal "Ready to receive" and "Acknowledgment of reception" for the object FACP;
- control of the communication channel, establishment of connection with the FACP in the mode of "Autodial";
- resistance to imitation of alarm messages and crypto protection, achieved by the use of encryption in the transmission of messages by protocols Ademco slow, Franklin, Contact ID, Silent Knight Fast, Radionics;
- high informativeness and selectivity, which ensures separation of fault and fire signals, as well as change of parameters of communication lines;
- the possibility of using different lines (channels) of communication;
- creation, processing and storage of databases on the status of protected objects;
- unification of technical means, that is, the ability to combine different devices into a single centralized surveillance hardware and software.

8.2 Developing an Intelligent Decision Support System

8.2.1 Methodology for Determining Forces and Means

In the event of a fire in a buildings or premises, the fire alarm sensors should be the first to trigger and a signal should appear on the observation panel informing about a fire in a certain room. Based on the received signal, the duty officer must leave (approach) the location of the triggered sensor (s) and check if there is a false alarm. If a fire really occurred, then according to the instructions the duty officer urgently calls the central fire station and reports a fire. On the basis of the received message, one (two) fire brigades leaves for the fire place, on two vehicles, which upon arrival, on the basis of the inspection, determine the fire area, the conditions associated with the spread of the fire and decide on the fire category, which is reported to the central console, where based on the received of the data, the forces (number of fire brigades) and means (number of hoses, volume of water, volume of foam mixture, type of mixture, type of fire escape, length of extendable ladder) required to extinguish heat. After collecting the necessary funds, the fire brigade advances to the fire site. Given the magnitude of the distance to the place of fire, fire extinguishing can begin in a few hours, which is unacceptable.

Another approach is proposed, according to which information on any enterprise, including a floor plan, a list of types of materials that describe both the external and internal sides of the premises, information on the type of production with a list of the presence of combustible liquids involved in the process, the presence of high-voltage electric power transmission lines etc., a diagram indicating the coordinates of the location of the fire alarm sensors is located in the database for a centralized fire monitoring station. Based on the information received and according to the methodology [6]:

1. Definition of the category of premises regarding explosion and fire hazard.
2. Calculation of the fire area according to the location and number of triggered fire alarm sensors.
3. Determination of the category of fire based on the type of premises and area of distribution.
4. Calculation of forces and means necessary to extinguish a fire.
5. The calculation of the route to the point of the epicenter of the fire.

Implementation of the proposed methodology will reduce the time between the moment of the fire and the beginning of its extinguishing by dozens of times.

In accordance with the foregoing, consider the solution to each of the above problems. The determination of the category of fire is carried out in accordance with the provisions set out in the guidelines which are determined by each country individually. They are easily formalized and are in the database; therefore, they are not considered in this paper.

To calculate the route of movement to the point of the epicenter of the fire, it is proposed to use the ant algorithm described in Sect. 2.6.6 and the ant swarm optimization algorithm described in Sect. 2.7.2.3.

8.2.2 Determining the Area of Fire

The object card contains a vector layout of the premises and coordinates of the fire sensors in those premises. Based on this data, you can estimate the area of fire. To begin with, let's determine the area of the entire room according to a known plan.

A polygon (not necessarily convex, since we solve a general problem) is given on a plane with the coordinates of the vertices, in order of circumvention of its sides.

Suppose that an arbitrary polygon ABCDE is given (Fig. 8.4).

The number of vertices $n = 5$.

From each vertex, lower the perpendicular to the *OX* axis.

The area of our polygon can be calculated by integrals:

$$S = \int AB - \int CB + \int CD - \int ED + \int EA.$$

Each integral is the area of the corresponding trapezoid. As follows:

$$S = S_{ABJH} - S_{CBJF} + S_{CDIF} - S_{EDIG} + S_{EAHG}.$$

In other words, we have two vertices: n and $n + 1$. If the vertex $n + 1$ is to the right of the vertex n, then we add this area if subtracted to the left.

$$S_{trap} = \frac{1}{2}(a + b)h,$$

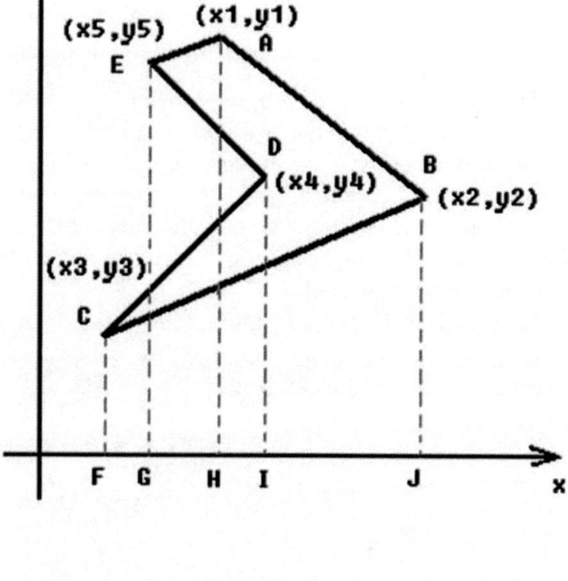

Fig. 8.4 An arbitrary polygon

where *a*, *b* is the based trapezoid; *h* is the height of the trapezoid. To get rid of 1/2, we multiply the expression by 2.

$$2S = (y2 + y1)(x2 - x1) - (y2 + y3)(x2 - x3) + (y4 + y3)(x4 - x3) - (y4 + y5) \\ x(x4 - x5) + (y1 + y5)(x1 - x5) = y2x2 - y2x1 + y1x2 - y1x1 - y2x2 \\ + y2x3 - y3x2 + y3x3 + y4x4 - y4x3 + y3x4 - y3x3 - y4x4 + y4x5 - y5x4 + y5x5 \\ + y1x1 - y1x5 + y5x1 - y5x5 = x1y5 - y2) + x2(y1 - y3) \\ + x4(y3 - y5) + x5(y4 - y1).$$

It is obvious that

$$S = \frac{1}{2}\left|\sum_{i=0}^{n-1}(x_i(y_{i-1} - y_{i+1}))\right|,\ (x_0, y_o) = (x_n, x_n).$$

8.2.3 *Determination of Forces and Means*

As it was mentioned in Chap. 3, neural networks are used to solve the problem of super-situational situations classification. The advantages of using ANN to solve the problem is the following:

- parallelism of calculation. The neural network is not inferior to the speed of the calculations. And for complex models, it even shows better results;
- neural network is resistant to fuzzy or noisy data;
- the neural network itself finds connections between the input and the output. It is not programmed but learns from examples. This means that if the process is very complex, or not explored 100%, it is easier and more reliable to train the neural network than to program the algorithm;
- neural network learns in the course of its functioning. This means that it adapts to new conditions and constantly produces correct data. Her calculations do not become obsolete over time.

The training sample for the decision to catch the optimal choice of forces and means is based on the use of previous experience [4].

The use of the NN to solve the problem implies the need to determine the topology, structure and parameters for the ANN [9, 10]. Thc solution of the set tasks is made as a result of the use of methods of structural-parametric synthesis of hybrid neurons of networks in Chaps. 2, 3, 4.

The simplest solution to this problem is to use a single-layer perceptron with type neurons. In the quality of the inputs:

- coordinates of placement of fire alarm sensors;
- room area;

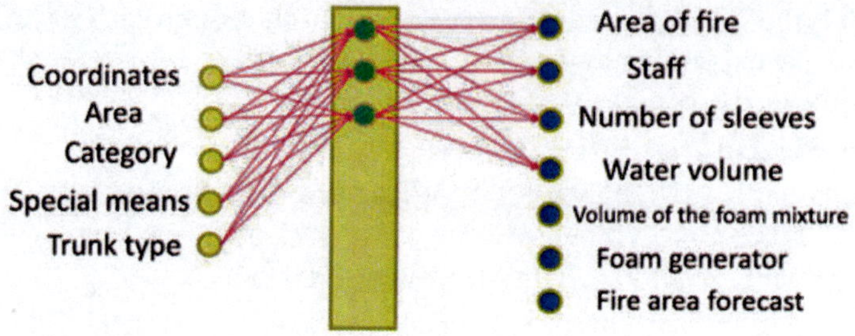

Fig. 8.5 An example of a neural network with one hidden layer

- room category;
- special. Agents (mixtures available);
- trunk type;
- number of sensors triggered;
- the presence of a fire extinguishing system.

As outputs:

- area of fire;
- personnel;
- number of sleeves;
- water volume;
- the volume of the foam mixture;
- foam generator;
- the projected area of fire.

An example is shown in Fig. 8.5.

8.3 Construction of Information System for Organization of Optimal Exit of People from Buildings During a Fire

8.3.1 Problem Statement of Finding the Best Way of Optimal People Evacuation from Buildings During a Fire

Consider a graphical representation of the object for which it is necessary to solve the evacuation problem (Fig. 8.6). The object is a shopping mall located on one of the floors of a multistory shopping center [5].

There are separate shops, blocked by plasterboard structures, which can be located on both sides of the shopping room (numbered from 1 to 18). Inside the shopping room can be outlets without walls, restaurants, cafes (marked with circles).

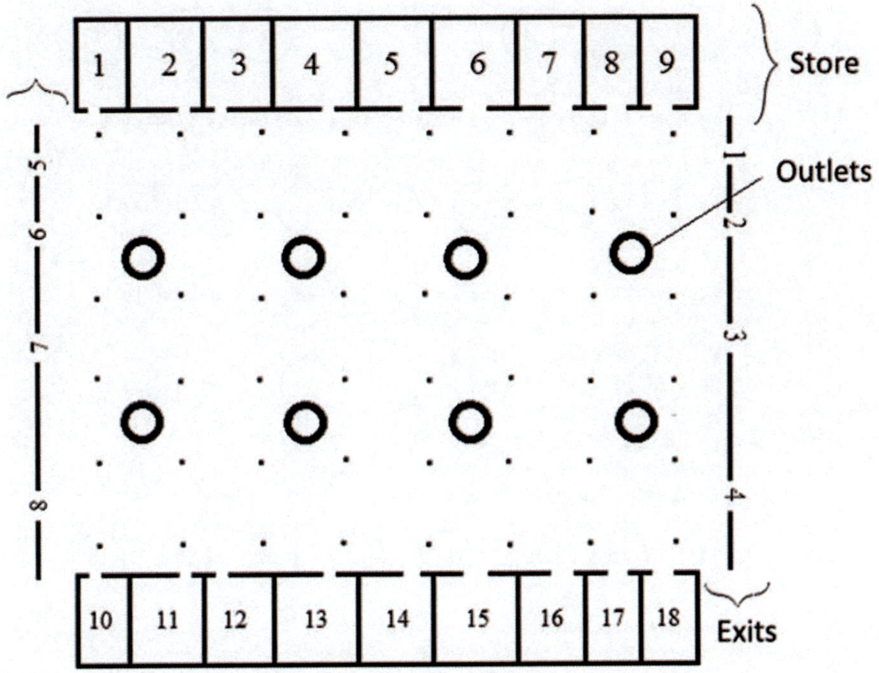

Fig. 8.6 Graphic representation of the evacuation problem

In the floor of the trading floor, with certain discretion, it is built-in LED lamps of different colors (shown as dots on Fig. 8.6), which are included at the time of the fire, making the evacuation route (light one color combines one possible route for each shop). The emergency exits from the trading floor were identified (numbered from 1 to 8), and the door of each store installed sensors input/output, whose data is processed to determine at each time the number of persons in the store and transfer them to the fire alarm observation panel of trading floor. The number of people outside the trading floor can only be estimated.

The trading floor (TF) topology can be represented as a finite, directional and weighted graph G (N, L, W) where N is the set of nodes and L is the set of edges of the graph. The edges of the graph are characterized by their finite nodes, such as the rib $l_{o,e} : o \in N$ is the initial node and the end node (Fig. 8.7), for which the following output data are defined:

(1) $A = \{a_i\},\ i = \overline{1, n}$ many shops and outlets in TF; each element a_i corresponds to a pair $\{x_i, y_i\},\ i = \overline{1, n}$ are coordinates of the store location;
(2) $B = \{b_i\},\ i = \overline{1, n}$ is the number of visitors currently in the shops;
(3) $C = \{c_k\},\ k = \overline{1, m}$ is the multiple points of exit from the TF with coordinates $\{x_k, y_k\},\ k = \overline{1, m}$;
(4) $D = \{d_p\},\ p = \overline{1, l}$ is the many shops where the fire happened with coordinates $\{x_p, y_p\},\ p = \overline{1, l}$;

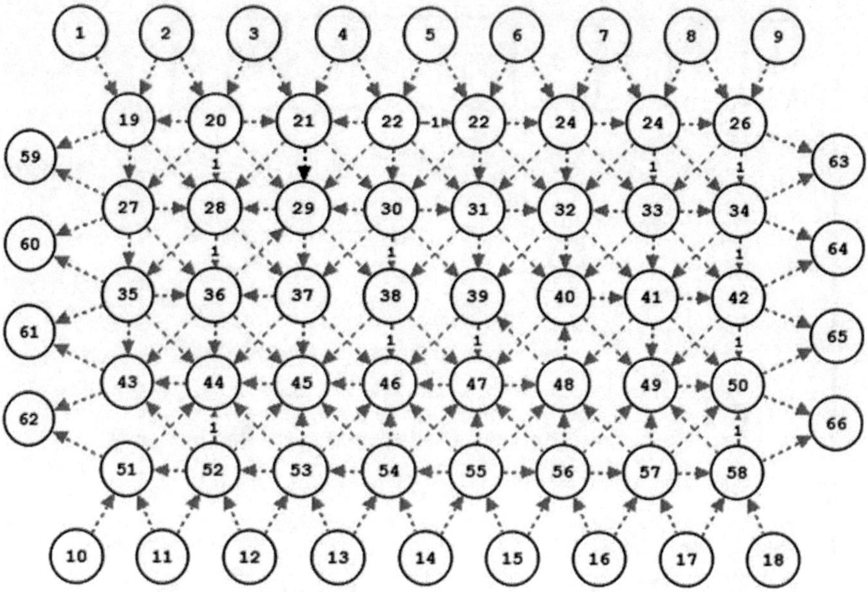

Fig. 8.7 Graphic model connection graph

(5) $E = \{e_j\}$, $j = \overline{1, q}$, is the set of points of occurrence of fire in the TF with coordinates $\{x_j, y_j\}$, $j = \overline{1, q}$;
(6) $F = \{f_h\}$, $h = \overline{1, v}$, is the set of LED fixtures mounted in the floor of the TF with coordinates $\{x_h, y_h\}$, $h = \overline{1, v}$.

The task is to determine the optimal ways of evacuation of visitors from shops and TF in the event of a fire in one or more shops from many D and/or at one or more points of the TF from many E. Each of the paths should be highlighted in a certain color, be of minimum length and provide evacuation of the maximum number of people in the minimum time. Thus, the optimality criterion can be set as: $I = \sum_{i=1}^{n} l_i$ where l_i $\left(i = \overline{1, n}\right)$ is the length of the ith edge of the graph of Fig. 8.7.

8.3.2 *Mathematical Models of Fire Propagation*

The task posed in Sect. 8.3.1 is the task of conditional optimization. As restructions, one can consider the walls and areas of open fire propagation. To determine the areas of open fire, it is necessary to know the coordinates of the source of the open fire, the mathematical model of the fire propagation, the parameters of which take into account the characteristics of the materials of the coatings of the corresponding premises.

Consider a mathematical model of fire spread in closed space in order to determine the areas of open fire in the hall of the TF.

Problem statement of Fire Propagation Prediction

Building structures are classified by their fire resistance and ability to fire propagation. The indicator of fire resistance P_o is the limit that is determined by the time before one of the boundary states occurs:

- loss of carrying capacity;
- loss of integrity;
- loss of insulating ability.

An indicator of the ability of a building structure to fire propagation P_s is the limit of fire propagation, and by this indicator they are divided into three classes:

- the fire propagation limit is zero;
- the limit of the fire propagation $M \leq 25$ cm for horizontal structures and $M \leq$ 40 cm—for vertical;
- the limit of the fire propagation $M > 25$ cm for horizontal structures and $M >$ 40 cm—for vertical.

Thus, for particularly dangerous objects, the fire resistance and the possibility of fire propagation for individual structures are known. Other factors affecting fire dynamics are the average rate of fire propagation. V_0 on different objects (mostly apartmens of a certain type) and the rate of burning of some solid materials V_v.

Existing methods for determining the time of reaching a certain point of fire are based on the experience, intuition of the fire extinguisher and are to summarize the times of its propagation across different rooms and through obstacles. The accuracy of this calculation is quite low due to the uncertainty of the values of many factors, their incompleteness and uncertainty for the decision maker [4].

Let's formalize the problem statement. Let t_0 is the time of ignition, $M(x_0, y_0)$ is the point of fire. Need to be defined t_k is the time of reaching the fire point $K(x_k, y_k)$. We believe that for particularly dangerous objects, the structure of the premises, the location of objects that enhance or slow the fire propagation, and the presence and location of technical openings are known. Note that each point of the object has a coordinate binding on the plane. The point with zero coordinates is in the lower left corner. Each premises, corridors have spatial limitations fixed in the database.

Without limiting the community, suppose that the number of rooms is N, the shape of the fire in the corridors is rectangular, in other rooms looks like a circular sector.

Source information includes:

- average rate of fire V_0^i, $i = \overline{1, N}$ in each of the N premises;
- availability, coordinates and burn rate, V_j^i, $i = \overline{0, k_i}$ jth type of equipment in the Nth room.

Classification of Fire Propagation Process (in Space and Time) Modeling Methods

Band model [9] used to predict fire propagation in partially enclosed spaces (in calculations can be taken into account windows, doorways, ventilation), such as one or more rooms. This model is the first to be widespread, being the least complex.

In this model, the room is divided into homogeneous zones (areas) and equations expressing conservation laws are solved in each of these areas. Typical partitioning consists of two areas—upper and lower. In the upper zone, hot gases (combustion products) are concentrated, and in the lower zone there is cold air that has not yet reacted. The flame in this case transfers the enthalpy from the lower zone to the upper. This approach has its disadvantages and advantages. Yes, the assumption of the division of space into these zones is true only partially, as there is a constant mixing of air masses, and the speed of mixing depends on the specific parameters of the room, such as the shape (especially the ceiling), the features of the ventilation, of the room, etc. But in many cases when you do not need to know the distribution of parameters within the area, this assumption allows you to predict the fire propagation with the required accuracy. This model is also not computationally complex, which extends its scope. The serious disadvantage of the band model is the fact that the adopted simplifications impede the further qualitative development of this model.

The following basic equations are used in the band model:

$$\frac{dm}{dt} = \sum_i m_i, \tag{8.1}$$

$$c_p m\frac{dT}{dt} - A_d Z\frac{dP}{dt} = Q + \sum_i h_i m_i, \tag{8.2}$$

$$p = pRT, \tag{8.3}$$

where m_i is the inflow of mass from the ith region; Q is the total energy inflow into the area due to radiation, convection and thermal conductivity; h_i is the specific enthalpy of the ith zone; $\sum_i h_i m_i$. is the total enthalpy inflow from all areas into the given.

Integral model. The building is considered as a set of rooms connected to each other by openings (doors). The premises may also have openings (windows) outside. The slots provide the flow of hot smoky air from the room with a source of fire to other rooms, as well as from the building into the atmosphere.

The basic equations of the integral model are given below.

Bernoulli's simplified formula was adopted for the flow of air through openings

$$G = F\sqrt{(2p \vartriangle P)}, \tag{8.4}$$

where G is the air flow through the hole, kg/s; F is the intersection area of the air stream, m^2; ρ is the density of the air in the stream, kg/m^3; $\vartriangle P$ is the average pressure drop between rooms, Pa.

The direction of air flow is determined by the pressure ratio. The air moves through the opening from the high-pressure room to the lower pressure room. In some cases, there are two counter airflows in the opening. At the top of the slot, hot smoky air flows from the room with the source of the fire in the lower part of the opening, a compensating stream of cold air flows into the room. The boundary between the outlet and inlet flows lies on the so-called plane of equal pressure.

Differential air mass balance equation is used for each room

$$\frac{dm}{dt} = \psi \sum\nolimits_i G_i, \tag{8.5}$$

where m is the mass of air in the room, kg; ψ is the burn rate of fire load, kg/s;$\sum_i G_i$ is the amount of air flow through the openings, taking into account their sign, kg/s.

The energy balance equation looks like this

$$\frac{dU}{dt} = Q_{\tilde{a}} - Q_{\tilde{e}} + \sum_i c_p T_i G_i, \tag{8.6}$$

where U is the indoor air energy, J; $Q_{\tilde{a}}$ is the rate of heat release during combustion, Jg/s; $Q_{\tilde{e}}$ is the rate of heat absorption by structures, J/s; $\sum_i c_p T_i G_i$ is the sum of the heat fluxes carried by the air through the openings, taking into account their sign.

The band model uses similar equations, but for each zone in each room.

Field model [7]. The development of methods of computational hydrodynamics and increasing the power of computers gave impetus to the development of a new class of fire models—field models based on the Navier–Stokes equations. This model, as well as the zone, is used to predict the development of fire in the premises. The field model also uses space partitioning and solving equations expressing conservation laws. But here the number of areas is much larger (1000–1000000). This improves the accuracy of prediction in more complex rooms (Fig. 8.8). The basis of the field

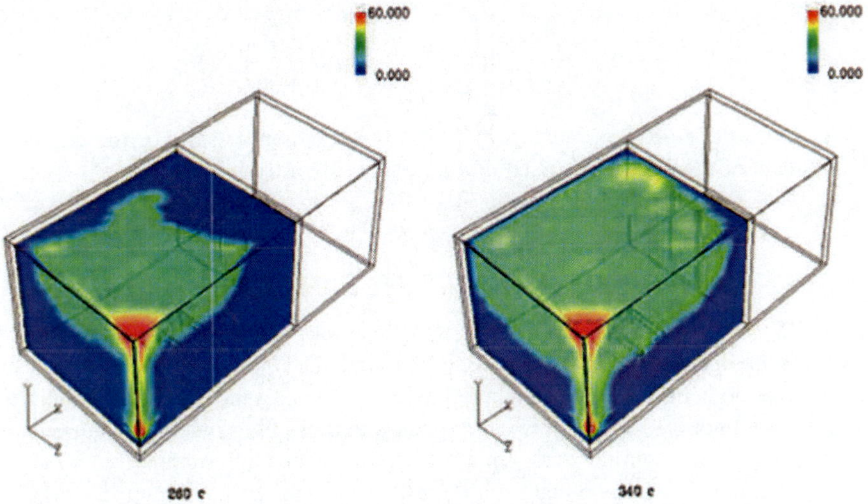

Fig. 8.8 3D-graphical representation of calculation results by field models using SOFIE

model is equations expressing the laws of conservation of mass, momentum, energy, and mass of components in the analyzed small control volume.

Mass conservation equation:

$$\frac{\partial p}{\partial t} + \frac{\partial}{\partial x_j}(pu_j) = 0. \tag{8.7}$$

Impulse conservation equation:

$$\frac{\partial}{\partial t}(pu_j) + \frac{\partial}{\partial x_j}(pu_j u_i) = -\frac{\partial p}{\partial x_j} + \frac{\partial \tau_{ij}}{\partial x_j} + pg_i. \tag{8.8}$$

For Newtonian fluids that obey the Stokes law, the viscous stress tensor is defined by the formula:

$$\tau_{ij} = \mu\left(\frac{\partial u_i}{\partial x_j} + \frac{\partial u_j}{\partial x_i}\right) - \frac{2}{3}\mu\frac{\partial u_k}{\partial x_k}\delta_{ij} \tag{8.9}$$

Energy equation

$$\frac{\partial}{\partial t}(ph) + \frac{\partial}{\partial x_j}(pu_j h) = \frac{\partial p}{\partial t} + \frac{\partial}{\partial x_j}\left(\frac{\lambda}{c_p}\frac{\partial h}{\partial x_j}\right) - \frac{\partial q_j^R}{\partial x_j}, \tag{8.10}$$

where $h = h_0 + \int_{T_0}^{T} c_p dT + \sum_k Y_k H_k$ is the static enthalpy of the mixture; H_k is the heat conversion kth component; q_j^R is the radiation energy flow in the direction x_j.

Equation of conservation of the chemical component k:

$$\frac{\partial}{\partial t}(pY_k) + \frac{\partial}{\partial x_j}(pu_j Y_k) = \frac{\partial}{\partial x_j}(pD\frac{\partial Y_k}{\partial x_j} + S_k) \tag{8.11}$$

To close the system of Eqs. (8.7)–(8.11) the equation of state of the chemical component of the ideal gas is used. For a mixture of gases, it looks like:

$$p = pR_0 T \sum_k \frac{Y_k}{M_k},$$

where R_0 is the universal gas constant; M_k is the molar mass kth component.

Evolutionary modeling used mainly to optimize discrete-value functions. Evolutionary methods have differences, but common is the presence of a target function or a fitness function. There are two approaches to presenting potential solutions. In the first case, such solutions are genotypes, that is, the corresponding binary chromosomes, since it is known that such representation has the maximum information saturation. The second approach is based on a phenotypic representation in which the solutions have a decimal form. It is characterized by obtaining new solutions using normally distributed shifts and without recombination. Neural network models and evolutionary methods have both advantages and disadvantages. In favor of evolutionary modeling, the lack of requirements for the objective functions is evidenced and a mutation operation is provided inside the algorithms, which minimizes the risk of obtaining local optimals. The advantage of neural network technologies is determined by the algorithms of the monotonous pursuit of the objective function to a satisfactory value. The effectiveness of the use of a particular technology depends on the number of sections of the fire, the number of experts and the procedure of using training and control sequences [2].

Modeling Software Tools of Fire Propagation Process in Space and Time

Ensuring people's fire safety requires, with few exceptions, the organization of their safe evacuation. The criteria for safe evacuation of people—timeliness and unobstructed—are now verified on the basis of calculations using various models of human flow (or more widely—models of evacuation implemented in executive algorithms.

Today, there are several dozen models in the world that use different ways to represent the internal environment of the home (fine or rough network), modeling the movement of people (individual, group/streaming), differently take into account the psychological aspects of human behavior (actions when receiving a signal about the fire, the choice of route, the influence of dangerous factors of fire).

Sitis Evatek. This software complex can be used for different types of buildings.

Model type: Partial Behavior Model/Motion Model. By default, only motion is modeled. The user can specify several different profiles, roles, agents, and scenarios for their behavior:

- calculation of the time of evacuation of people taking into account the peculiarities of individual movement of people in the flow based on Russian standards of human velocity from the density of people in a rectangular area around the person;
- input for calculation using the built-in graphical editor, the ability to import geometry from DXF files;
- display the density map, traveled and current paths of all agents;
- ability to reproduce and record calculation results;
- 2D/3D motion visualization modes;
- report generation including output, simulation results, maximum and average density graphs at time points, percentage of outputs used;
- export the prepared report in the DOC file format.

Sitis Flowtek.

Model type: Motion Model.

The main characteristics of the system:

- input for calculation using the built-in graphical editor based on scanned building plans.
- possibility of using parameterization.
- work with a single project file as part of the SITIS software package for fire risk calculation.

- the ability to create multiple evacuation scenarios.
- display of map of settlement areas and evacuation routes.
- 2D/3D animation of human streams with step-by-step view.
- View base parameters for each billing area.
- report generation, including baseline, evacuation time tables from each premises, floor escape times tables, traffic delay sections, summary evacuation time tables for all scenarios, calculation area maps, evacuation route images.

Evacnet 4. This software product can be used for various types of buildings such as offices, stadiums, high-rise buildings, hotels, restaurants and schools.

The main objective of the model is to optimize the evacuation from the building. This means that the evacuation time from the building is minimized.

Model Type: Motion Model.

Model structure: Network model.

Agent behavior: None.

Building Exodus. The purpose of this system is to simulate the evacuation of large numbers of people from different types of buildings. An attempt was made at Building EXODUS to consider "People-People, People-Fire and Human Interaction." The model consists of six sub-models, some of which interact with each other to convey information about the simulation evacuation process, agent data, motion, behavior, toxicity, hazard, and geometry.

Simulex. Evacuation model with the ability to simulate a large number of people from buildings with complex geometric architecture.

Model Type: Partial Behavioral Model. It is based on the distance between the agents, on which their speed depends. In addition, the model allows for overtaking, turns, sideways and backward movements.

Model structure: "regular grid". The floor plan and grids are separated by 0.2×0.2 m cells. The model contains an algorithm that calculates the distance from each block to each exit. The received data is displayed on the map.

Agent Behavior: Hidden behavior. Agent movement: fluctuations in speed, steps to the side and deformation of the body, overtaking etc., based on the results of many video observations and on the analysis of individual movements, and additional results from a number of researchers.

"SIGMA PB"

The SIGMA PB computer software complex is designed to perform calculations of the spread of dangerous factors of fire and evacuation from multi-storey buildings, structures and buildings of different classes of functional fire danger.

The software package is built by the following components:

- builder of a three-dimensional frame of a building, a grid and a geometry of the object;
- evacuation scenario builder;
- module that implements fire development calculation (SigmaFire core);
- module that implements the calculation of evacuation of people (SigmaEva computing kernel);

- module 3D visualization, temporal and spatial analysis of the evacuation and fire propagation.

To calculate the spread of dangerous fire factors (DFF) and evacuation, we use the computational kernels of the domestic programs SigmaFire © and SigmaEva ©, respectively, which implemented a field fire model and an evacuation model individually-stream type.

The SIGMA PB program has the following advantages over Russian and foreign counterparts:

- a single software environment with a single field of information resources and data format for solving the problems of calculating the movement of people and the spread of DFF;
- own builder of objects;
- own calculation modules;
- 3D visualization of evacuation and distribution of DFF in a three-dimensional virtual object environment with the ability to change the position of the observer.

Built-in 3D rendering module allows you to observe the evacuation and propagation of hazardous fire fields in different parts of the building;

(1) heat flow;
(2) ambient temperature;
(3) CO concentration (carbon monoxide);
(4) concentration of CO_2 (carbon dioxide);
(5) HCl concentration (hydrogen chloride);
(6) O_2 concentration (oxygen);
(7) visibility in the smoke, at a height of 1.7 m from the floor, as well as the field of density of the human stream.

At the user's request, statistics are provided according to the script:

- evacuation time from floors and the whole building;
- time to block escape routes;
- dense duration (>6 people/m^2) clusters;
- the number of people affected by DFFs exceeding the limit values.

PyroSim

PyroSim is a graphical user interface for FDS that allows you to quickly, conveniently create, edit, and analyze complex fire development models.

PyroSim allows you to simulate the propagation of hazardous fire factors across a field model, construct hazardous factor fields, and determine the time of blocking of escape routes.

PyroSim lets you import from AutoCAD files in DXF and DWG formats. When imported, 3D faces become obstacles and the rest of all data (lines, curves, etc.) become independent CAD objects.

In addition, PyroSim lets you upload GIF, JPG, or PNG images as substrates, helping you quickly create objects based on them.

PyroSim has tools that help you create and manage multiple grids. Multiple grids in the model allow parallel calculations to be used to speed up calculations, simplify geometry to reduce the number of grid cells in a model (thereby reducing calculation time), and change resolution in different parts of the model.

In PyroSim, you can create and use property libraries for different objects (reactions, surfaces, materials, etc.). These speeds up model creation and reduces the likelihood of errors.

PyroSim allows you to interactively view and edit object properties in a model. Such visual feedback accelerates model creation and reduces the likelihood of errors. The slide shows a surface that uses tangential velocity to model a fan latch.

You can run SmokeView software developed by NIST at any time during model creation or calculation. This program allows you to clearly see the spread of smoke, to build fields of temperature, speed and other dangerous factors. PyroSim also has a built-in tool for plotting two-dimensional graphs of time values.

Pathfinder

Pathfinder is an emergency evacuation simulation program that includes a graphical user interface for creating a model and a module for viewing animated 3D results.

Pathfinder allows you to calculate the evacuation time and cluster dwell time on an individually-flowing traffic pattern.

8.4 Algorithm for Optimal Evacuation of People from the Shopping Center During a Fire Based on the Use of Artificial Intelligence

This task belongs to the class of optimizers. Here are some ways to solve this problem [11]:

1. Ant algorithm (see Sect. 2.6.6).
2. Neural networks [8].

Let's take a closer look at solving the problem based on using the Hopfield neural network. The application of Hopfield networks to solving optimization problems is based on the existence of the Lyapunov function, which decreases while the network spontaneously develops. So, the minima of the energy function must coincide with the stable states of the network [1, 2].

The Hopfield neural network is a recurrent single-layer neural network, where the output of each of the neurons is connected to the inputs of other neurons by feedback (Fig. 8.9).

The dynamic model of each neuron that is used to update weights and outputs is presented as

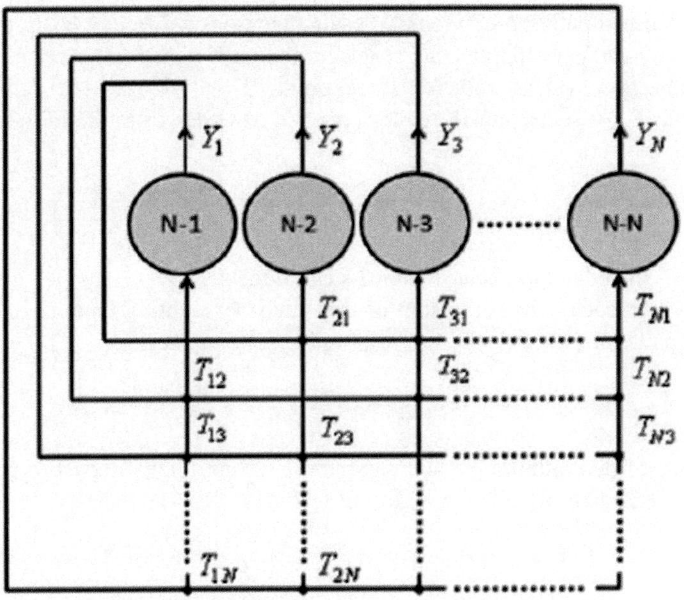

Fig. 8.9 Typical hopfield network with N neurons

$$\frac{dX_j(t)}{dt} = -\frac{X_j(t)}{\tau} + \sum_{i=1}^{N} T_{ij} Y_t(t) + b_j,$$
$$Y_i = f(X_i) = \frac{1}{1-\exp(-a_i X_i)},$$

where X_j (t) is the internal state of the neuron; Y_i (t output of the ith neuron for $i \neq j$. In this model, N is the number of neurons; b_j is the bias of the jth neuron; τ is the time constant; a_i is the amplifier coefficient; T network weight matrix.

Solutions and modeling solutions of path finding problems requires optimization of weight factors, which are subjected to complexes of constraints. Weighted functions can be considered energy functions that are suitable for modeling decisions by neural networks by minimizing the energy function in Hopfield network.

The network topology can be represented as a finite, directed and weighted graph G (N, L, W) where N set of nodes, L is the set of edges of a graph. The edges of a graph are characterized by their finite nodes, such as the rib $l_{o,e}$: $o \in N$ is the initial node and $e \in N$ the end node. The length of each edge $l \in L$ can be represented as the value of the neural network weight coefficient for each edge $W_l \in W$; each edge can be matched a positive number $L \rightarrow R^+$, where R^+ denotes a set of ordered positive integers. The connection of two nodes is denoted as $R_{(s,d)}$, where s, $d \in N$ is the start and end node, and the goal is to find the shortest path from the start node s to the final node d.

In the shortest path search problem, each edge l has a different cost (length) and the shortest path selection algorithm selects the path with the lowest cost of data propagation from the start node to the end point.

The path $R_{s,d}$ is determined by the sequence of nodes from the initial node s to the final node d connected by the edges:

$$P_{s,d} \equiv \{s,\ n_i,\ n_j,\ \ldots,\ n_k\ ,d\} \equiv \{l_{sn_i},\ l_{n_i n_j},\ \ldots,\ l_{n_k d}\},$$

where l_{sn_i} is the edge that connects node s and node n_i.

The cost of each path is the sum of the values (weighting factors) of all edges along the path

$$C_{Ps,d} = C_{L_{sn_i}} + C_{l_{n_i n_j}} + \cdots + C_{L_{n_k d}}.$$

The goal of determining the shortest path is to minimize the cost of connecting paths between a pair of nodes (s, d) for all $s, d \in N$

$$\text{Minimize}\{C_{P_{s,d}}\}\forall s, d \in N.$$

To construct the shortest path using the Hopfield network, the graph (see Fig. 8.13) is represented as an adjacency matrix $\mathbf{A} = [N \times N]$ with zero elements of the main diagonal. If the element of the matrix $A_{ij} = 1$, then there is a connection between the nodes of the graph that corresponds to the paths between the points i, j.

In the structure of the Hopfield neural network, the state of the neuron is matched by the existence of a link between the nodes of the graph (the state of the neuron (output of the neuron) (i, j) corresponds to the inclusion of the edge i, j of the graph as a component of the optimal path).

The dynamic model of a typical neuron at position (i, j) is described as

$$\frac{dX_{ij}(t)}{dt} = -\frac{X_{ij}(t)}{\tau} + \sum_{i=1}^{N}\sum_{j=1}^{N} T_{ijk_i} Y_{ki}(t) + b_{ij},$$

where $X_{ij}(t)$ is the internal state of the neuron at position (i, j); τ is the time constant of the neural network; T_{ijk_i} is the weight of the link between the neuron at position (i, j) and the neuron $(k, 1)$; b_{ij} is the offset value of the neuron (i, j). In addition, a link matrix $\mathbf{K} = [N \times N–1]$ where the element K_{ij} is equal 1, when the link from node to node j does not exist. The cost of connecting a link from node i to link j is denoted as G_{ij} is the final true positive number. The value of nonexistent units is assumed to be zero.

In order to solve the routing problem using the Hopfield model, we first determine the energy function, whose minimization process drives the neural network to its lower energy state. The stable state of the network will meet the solution of the routing problem. In this work we use the energy function that is constructed as follows:

$$E = z_1E_1 + z_2E_2 + z_3E_3 + z_4E_4 + z_5E_5, \tag{8.12}$$

where the component energy functions are described as follows:

$$E_1 = \frac{1}{2}\sum_{i=1}^{N}\sum_{\substack{j=1\\ j\neq i}}^{N} G_{ij}Y_{ij}, \tag{8.13}$$

$$E_2 = \frac{1}{2}\sum_{i=1}^{N}\sum_{\substack{j=1\\ j\neq i}}^{N} K_{ij}Y_{ij}, \tag{8.14}$$

$$E_3 = \frac{1}{2}\sum_{i=1}^{N}\left[\sum_{\substack{j=1\\ j\neq i}}^{N} Y_{ij} - \sum_{\substack{j=1\\ j\neq i}}^{N} Y_{ij}\right]^2, \tag{8.15}$$

$$E_4 = \frac{1}{2}\sum_{i=1}^{N}\sum_{\substack{j=1\\ j\neq i}}^{N} Y_{ij}\left(1 - Y_{ij}\right), \tag{8.16}$$

$$E_5 = \frac{1}{2}(1 - Y_{ds}). \tag{8.17}$$

In Eq. (8.12) the goal of the factor z_1 is to minimize the total cost of the path, taking into account the cost of the existing links, the component z_2 prevents the inclusion of nonexisting links in the path. The component $z_3$states that if a node is entered, it will also be used in the path. The component $z_4$is required to confirm that the state of the neural network converges to a valid path with the lowest energy value. The component z_5 is used to provide a path that must exit the source node s and end at the destination node d.

The network detects internal activation to reduce the total energy and neurons update their state, $\frac{dE}{dY}$ which is calculated directly and is the sum of all energy multipliers. This is a discrete approach to solving a differential equation that describes the relation between the activation function and the energy terms [1, 2].

The equation for the neuron state update is given as:

$$\frac{dX_{ij}}{dt} = -\frac{X}{\tau} - \left(z_1\frac{dE_1}{dY} + z_2\frac{dF_2}{dY} + z_3\frac{dE_3}{dY} + z_4\frac{dE_4}{dY} + z_5\frac{dE_5}{dY}\right), \tag{8.18}$$

where

$$\frac{dE_1}{dY} = \frac{1}{2}G_{ij}\left(1 - \delta_{id}\delta_{js}\right), \tag{8.19}$$

$$\frac{dE_2}{dY} = \frac{1}{2}K_{ij}\left(1 - \delta_{id}\delta_{js}\right), \tag{8.20}$$

$$\frac{dE_3}{dY} = \sum_{\substack{j=1 \\ j\neq i}}^{N}(Y_{ik} - Y_{ki}) - \sum_{\substack{j=1 \\ j\neq i}}^{N}\left(Y_{jk} - Y_{kj}\right), \tag{8.21}$$

$$\frac{dE_4}{dY} = \frac{1}{2}\left(1 - 2Y_{ij}\right), \tag{8.22}$$

$$\frac{dE_5}{dY} = \frac{1}{2}\delta_{id}\delta_{js}, \tag{8.23}$$

where δ_{ij} is the Kronecker symbol that takes the value 1 for $i = j$ and 0, otherwise.

When this system is released, we will change the output voltage of the neuron.

Putting the input values of neurons U_{xi} at time $t = 0$, the evolution of the state of the neural network over time is modeled by numerical solution (8.22). This corresponds to the solution of the system with $n^*(n - 1)$ nonlinear differential equations, where the variables are the output voltages of the Y_{xi} neurons. For this purpose, we choose the fourth-order Runge–Kutta method, because it is quite accurate and easy to implement δt. Accordingly, the simulation consisted of observing and updating the output voltages on neurons in incremental steps over time τ. In addition, the time constant for each neuron is set to 1 and for simplicity we assume that yes $\lambda_{xi} = \lambda$ and $g_{xi} = g$, all independent of the index (x, i). Simulation showed that the best value for δt is 10^5. Further decreasing this value increases the simulation time without improving the results. Another important parameter in the simulation is the initial input voltage of the U_{xi} neurons. If the neural network does not have a preference for a particular path, all U_{xi} must be set to zero. However, some initial random noise of $-0.0002 \leq \delta t \leq U_{xi} \leq +0.0002$ will help to break the symmetry caused by symmetrical network topologies, as well as being able to have more than one shortest path. Simulation stops when the system reaches a stable end state. This is assumed to occur when all output voltages of the neurons do not change more than the threshold $\Delta Y_{th} = 10^5$ from one update to the next. In this stable state, each neuron either enters the path for $(Y_{xi} \geq 0.5)$ or not $(Y_{xi} < 0.5)$.

For example, we use the network shown in Fig. 8.13. The proposed network consists of 4 nodes and 5 units. The shortest path between a pair of nodes (1, 4) is proposed where $S = 1$ and $D = 4$. The model of the proposed network includes 12 neurons. The initial state of all neurons is randomly considered in the range [0, 1], the constant $\tau = 1$ for all neurons, and the other network parameters are considered as follows:

$$z_1 = 950;\ z_2 = 2500;\ z_3 = 1500;\ z_4 = 475;\ z_5 = 2500;\ \Delta t = 10^{-5}.$$

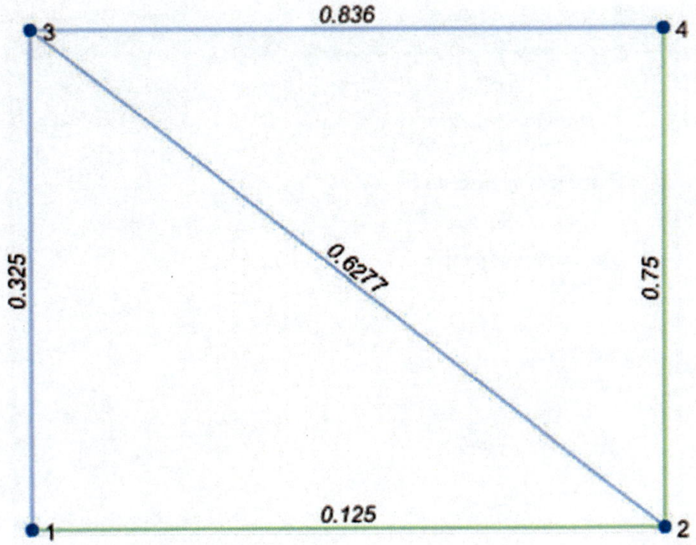

Fig. 8.10 Example network topology

For the network in Fig. 8.10 the path $P_{1,4} = \{1;\ 2;\ 4\}$ was found to be the shortest, and the cost (length) of the shortest path $C_{P_{1,4}} = 0.875$, was the minimum between the cost of all possible paths. The shortest path was found after 2558 iterations.

8.4.1 Software Structure for Organizing the Optimum Exit of People from Buildings During a Fire

The input panel looks like the one shown in Fig. 8.11.

The panel for displaying the scheme of the building is shown in Fig. 8.12.

8.5 A Software Example of How to Optimize People's Escape from Buildings During a Fire

Suppose there was a fire in store # 7. The number of people in the store was 700 people, in the trading room (SR)—500 people.

The first is the removal of people in the immediate vicinity of the place of ignition.

According to the proposed algorithm and based on the operation of the neural network evacuation route is shown, shown in Fig. 8.13.

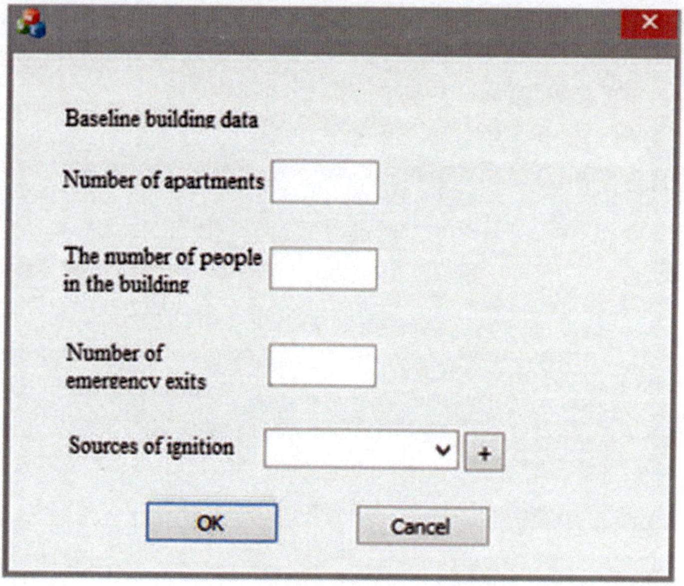

Fig. 8.11 Input panel

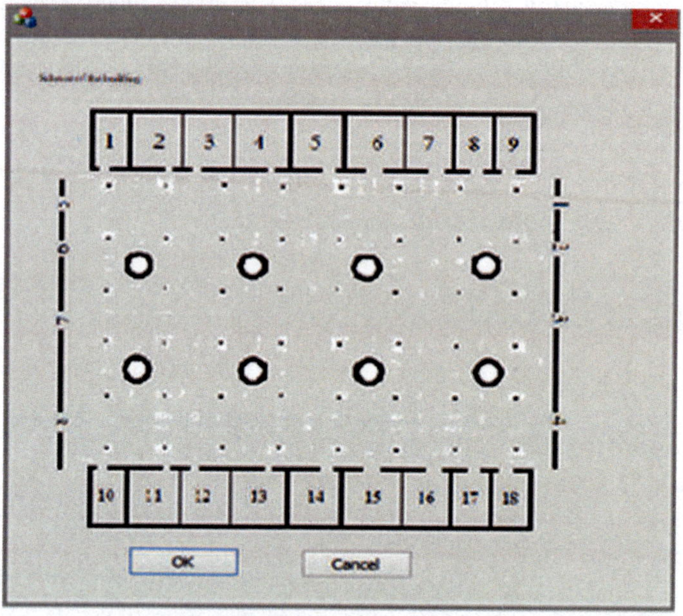

Fig. 8.12 Panel for displaying the scheme of the building

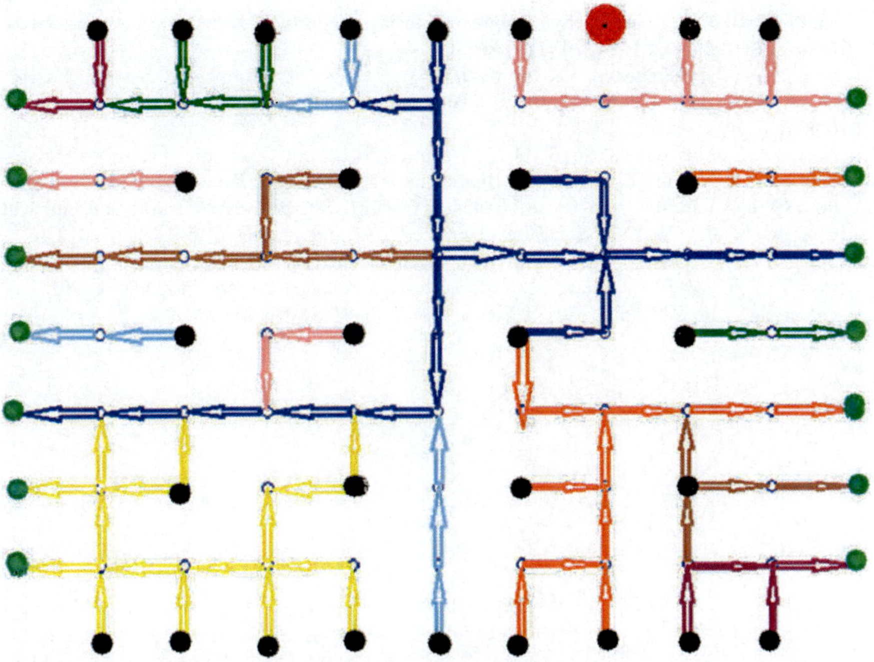

Fig. 8.13 Scheme of evacuation in case of fire in one of the shops

Here, the red dot indicates the outbreak, the black dots indicate shops and outlets, the green dots indicate escape routes. Arrows of different colors indicate evacuation routes for different shops and outlets.

References

1. Bratishchenko, V.: Design of information systems, p. 84. BSUEP, Irkutsk (2004)
2. Chumachenko, H., Kryvenko, I.: Neural networks module learning. Electron. Control Syst. (48), 76–80 (2016). NAU, Kyiv
3. Chumachenko, H., Ledovsky, A.: Fire forecasting based on the use of neural networks. Electron. Control Syst. **2**(28), 142–148 (2011). NAU, Kyiv
4. Chumachenko, H., Tsilitsky, V., Biliy, M.: Analysis of the distributed information system of fire surveillance as a queuing system. Electron. Control Syst. **3**(29), 116–119 (2011). NAU, Kyiv
5. Chumachenko, H., Tsilitsky, V., Biliy, M.: Information system of fire surveillance. In: The X International Scientific and Technical Conference "Avia-2011". pp. 22.56–22.59 (2011). Last accessed 15–21 April 2011
6. Chumachenko, H., Tsilitsky, V., Biliy, M.: Construction of information system of high reliability of one class. Electron. Control Syst. **4**(30), 127–134 (2011). NAU, Kyiv
7. Chumachenko, H., Luzhetskyi, A.: Building a system of simulation modeling for spatially-distributed processes. Electron. Control Syst. **1**(39), 108–113 (2014), NAU, Kyiv

8. Chumachenko, H., Kupriyanchyk V.: Fire monitoring intellectual information system. Electron. Control Syst. **2**(44), 81–84 (2015). NAU, Kyiv
9. Grekul, V., Denishchenko, G., Korovkina, N.: Design of information systems. Internet University of Information Technologies, [Electronic resource] Acces mode: intuit.ru, p. 1245 (2005)
10. Hoffman, E., Oliynyk, A., Subbotin, S.: Method of structural-parametric synthesis of Neuro-Fuzzy networks. In: Dovbysh, A., Borysenko, O., Baranova, I. (eds.) Modern Information Systems and Technologies: Materials of the First International Scientific and Practical Conference, pp. 175–176. SSU, Sumy, Ukraine (2012). Last accessed 15–18 May 2012
11. Sineglazov, V., Chumachenko, H., Krivenko, I.: Intelectual system of optimal evacuation route searching. In: XXIV International Conference on Automated Control "Automation 2017" of Proceedings. Kyiv, Ukraine (2017). Last accessed 13–15 Sept 2017

Index

M. Zgurovsky et al., *Artificial Intelligence Systems Based on Hybrid Neural Networks*, Studies in Computational Intelligence 904,
https://doi.org/10.1007/978-3-030-48453-8